International Workshop on

DYNAMICAL FEATURES OF NUCLEI AND FINITE FERMI SYSTEMS

International Workshop on

DYNAMICAL FEATURES OF NUCLEI AND FINITE FERMI SYSTEMS

September 13 – 17, 1993
Sitges, Barcelona (Spain)

Editors

Xavier Viñas, Martí Pi & Angels Ramos

Universitat de Barcelona

World Scientific
Singapore • New Jersey • London • Hong Kong

Published by

World Scientific Publishing Co Pte Ltd
5 Toh Tuck Link, Singapore 596224
USA office: 27 Warren Street, Suite 401-402, Hackensack, NJ 07601
UK office: 57 Shelton Street, Covent Garden, London WC2H 9HE

British Library Cataloguing-in-Publication Data
A catalogue record for this book is available from the British Library.

DYNAMICAL FEATURES OF NUCLEI AND FINITE FERMI SYSTEMS

ISBN-13 9789810215569
ISBN-10 9810215568

Preface

The International Workshop on Dynamical features of Nuclei and Finite Fermi Systems was held in Sitges (Barcelona), in September 1993, in continuation of the Workshops at Grenoble (1979,1981,1984), Trieste (1982,1985), Aussois (1987,1992), Bad Honnef (1988) and Elba Island (1990).

The subjects discussed in the workshop covered several aspects of nuclear dynamics such as fluctuations, nuclear fragmentation, transport theories and particle production in heavy-ion reactions. A full session was devoted to the physics of metallic and atomic clusters which has many points in common with the physics of the nucleus.

We are grateful to all the participants for their presentations which contributed to maintain the scientific quality of the workshop. The numerous enthusiastic and stimulating discussions, helped by the beauty of the Maricel Palace and the surrounding town of Sitges, created a very nice and warm atmosphere throughout the workshop. We would also like to thank the International Advisory Committee: J.A. Alonso, R. Arvieu, J. Bondorf, X. Campi, U. Mosel, J. Németh, W. Nörenberg, J. da Providência, B. Rémaud and S. Stringari, for their valuable suggestions and comments. The organization of the workshop has been possible with the financial support of the DGICyT (Ministry of Education and Science, Spain), the Comissionat per a Universitats i Recerca (Generalitat de Catalunya), the University of Barcelona and the Universidad Internacional Menéndez y Pelayo, the Director and the Administrative officer of which, Ms. M. Duran and Ms. M.A. Marnet, are deeply thanked for their technical help. Finally, the kind hospitality of the Ajuntament de Sitges that allowed us to use the Maricel Palace and the staff of the Hotel Subur are warmly acknowledged.

The Organizing Comittee:

M. Barranco, S. Hernández, M. Lozano,
S. Marcos, E. Moya de Guerra, M. Pi,
A. Polls, A. Ramos and X. Viñas

CONTENTS

Dynamical Fluctuations and Nuclear Fragmentation

FLUCTUATIONS IN FRAGMENT SIZE DISTRIBUTIONS

X. CAMPI and H. KRIVINE
Division de Physique Théorique[1]
Institut de Physique Nucléaire
91406- Orsay Cedex, France

ABSTRACT

We present a method to analyse the fluctuations in fragment size distributions. The size dispersion of the mean fragment distribution is compared with the statistical average of the size dispersions of single events. Large fluctuations are found in a finite size percolation lattice. Our suggestion is that these fluctuations are mainly due to boundary constraints on the percolation cluster.

1 Introduction

Fragmentation is a process in which an object made of S constituents breaks in m_0 pieces $s_1, s_2, ...s_{m_0}$. The number of fragmentation phenomena that can be defined is only limited by fantasy. It may concern objects as diverse as galaxies, elementary particles, polymers, colloids or atomic nuclei. [1–4] One can imagine to class these phenomena according to length scales, to time scales, to the nature of the links between the elementary constituents, etc... However what makes the interdisciplinary study of fragmentation most exciting is the possibility to discover "universal" classes of fragmentation mechanisms that would apply to real systems, regardless of the space or time scales or of the details of the linking.

Fragmentation is caused by a sudden input of energy which drives the system from an equilibrium state with a single piece of size S to an equilibrium state with several pieces s_i. Because in general this transition takes place out of equilibrium it is very difficult to predict the evolution of the system using the first principles of mechanics. In addition, fragmentation is a *statistical* process in the sense that the initial conditions are never completely determined. At a microscopic level, the position of the constituents,

[1]Unité de Recherche des Universités de Paris 11 et Paris 6 associée au CNRS

the distribution of the breaking forces are, for example, not reproducible. In any case it is mandatory to begin the phenomenological study of fragmentation by a determination of the relevant observables. We have studied recently [5] some observables that are connected with the mean fragment size distribution $< n_s >$. As a step further, we suggest a natural extension of the method to study the fluctuations of n_s.

This contribution is organized as follows. In Section 2 we discuss the criteria we use for performing the statistical analysis. In Section 3 we recall the results concerning the mean values. In Section 4 we present the method to study fluctuations. Some applications to simple models are in Section 5.

2 Statistical Averages and Fluctuations

From a theoretical point of view, the basic quantity in a fragmentation process is the probability $P\{n\}$ to observe a partition $\{n_1, n_2, ...n_S\}$. The experiments (or the numerical simulations) give only the frequency $f\{n\}$ of the partitions. Because of statistical fluctuations, $f\{n\}$ is only a rough representation of $P\{n\}$. However, in a first step we do not need the full information carried by $P\{n\}$. Performing partial integrations on $f\{n\}$ (statistical averages) it is almost possible to eliminate the statistical fluctuations while still preserving the basic information. For example, the mean fragment size distributions $< n_s, \alpha >$ of partitions of class α are a natural compression of the information. The choice of α is very important. In general it will be a single parameter. We will call it the driving parameter. Let's discuss this point in some detail.

Consider a random mechanism that generates partitions of an integer number and some choice for the criteria of the statistical averages. Let's take for instance, partitions with the same multiplicity of fragments. For a given multiplicity m_0 the mean fragment size distribution $< n_s, m_0 >$ would look like the thick line of fig 1. The shape of the curve depends of course on m_0. One could perform similar averages taking partitions according to other criteria, for example the same number of intermediate mass fragments (IMF), the mass of the largest fragment s_{max}, the sum of all observed fragments mass but $s = 1$ (S_{bound}), etc. On fig 1 we have also shown the mean fragment size $\bar{s}$ and the width $\sigma^2_{[s]}$ of the mean distribution.

The doted curve represents the fragment size distribution $n_s^{(j)}$ of an event (partition) j with same multiplicity m_0. Because of the finitness of the system

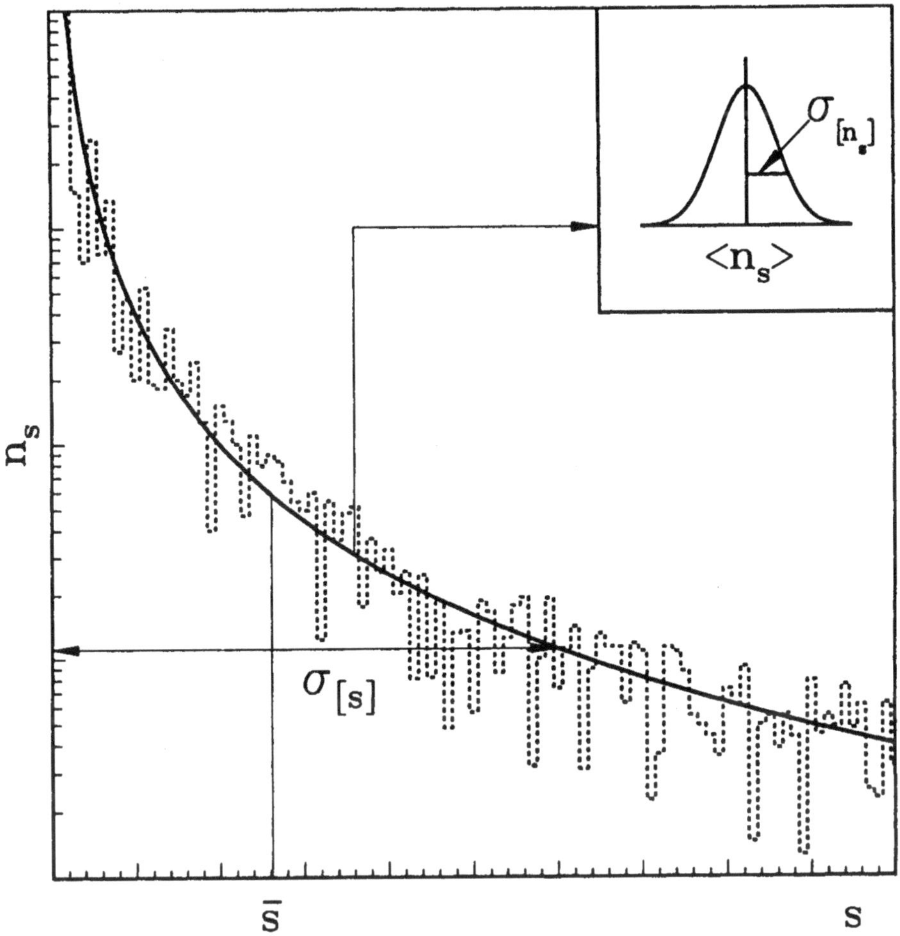

Figure 1: Schematic representation of the various quantities discussed in the text. The mean fragment size distribution (thick line) with mean fragment size $\bar{s}$ and width $\sigma_{[s]}$. The dotted line is the distribution $n_s^{(j)}$ of a particular event j. For a fixed size s, the distribution of n_s is also represented, with mean value $< n_s >$ and width $\sigma_{[n_s]}$ (window).

every realisation is rather far from $< n_s >$. For a fixed size s it is interesting to look at the frequency distribution of $n_s^{(j)}$. One expects a curve centered around $< n_s >$ with some width $\sigma_{[n_s]}$ (see window of fig 1). We know that for a Bernouilli process the ratio $\frac{\sigma_{[n_s]}}{<n_s>}$ goes to zero as $1/\sqrt{m_0}$. Any deviation from this behaviour indicates the existance of non-statistical fluctuations. In contrast with statistical fluctuations that are just a "noise", these fluctuations may contain interesting information.

Let's be more specific by taking a simple example. Consider an ensemble of m_0 perfect coins, i.e. the probability distribution of sorting "heads" is $P(p) = \delta(p - 1/2)$, the mean value of p, $< p >$ is $1/2$ and its variance $\sigma^2_{[p]} = 0$.

Calling n the number of times that we get "heads" ($0 \leq n \leq m_0$), $< n >= m_0/2$ and

$$\frac{\sigma^2_{[n]}}{< n >^2} = \frac{< n^2 > - < n >^2}{< n >^2} = \frac{1}{m_0} \tag{1}$$

The width of the distribution is only due to the finite size m_0 of the system. These are the purely statistical fluctuations. Now supose that the coins are "biased", for example that $P(p)$ is such that again $< p >= 1/2$ but $\sigma^2_{[p]} \neq 0$.

Then

$$\frac{\sigma^2_{[n]}}{< n >^2} = \frac{1}{m_0} + \sigma^2_{[p]}(1 - \frac{1}{m_0}) \tag{2}$$

The new term in r.h.s is the signature of fluctuations that are due to the fluctuation of the random mechanism itself. In complex systems these fluctuations may reveal the presence of hidden variables, correlations, random constraints, etc. Some times these are called dynamical fluctuations, to make the distinction with the purely statistical fluctuations. Finally, we consider the case in which both the multiplicity m_0 and the probability p fluctuate. One gets [2]

$$\frac{\sigma^2_{[n]}}{< n >^2} = \frac{1}{< m_0 p >} - 1 + \frac{< m_0(m_0 - 1)p^2 >}{< m_0 p >^2} \tag{3}$$

---

[2]Notice that the quantity $\frac{<m_0(m_0-1)p^2>}{<m_0 p>^2}$ appears when looking for intermittency fluctuations [14,16].

We see that now it is much more difficult to disclose the non-statistical fluctuations and that the distribution of m_0 will affect the determination of $\sigma^2_{[p]}$.

Obviousely the best choice of the driving parameter α is the one that minimizes the fluctuations. That is the reason why we choose $\alpha = m_0$. In what follows we will make this choice.

The choice of the statistical average dictated by statistical considerations may very well not coincide with other choices based on physical arguments, prejudices or experimental constraints. However it is important to keep in mind the above considerations before raising definite conclusions about the origin of eventual non-statistical fluctuations.

3 Mean Values

The data of a fragmentation experiment can be represented as a list of partitions $\{n_s^{(j)}\}$ (where $j = 1, 2, \cdots N$ is the event number) forming a matrix. This matrix can be examined by rows (horizontal analysis) or by columns (vertical analysis).

A natural way to examine the $\{n_s^{(j)}\}$ horizontally is by the moments technique. The moments of n_s are defined as usual by

$$m_k = \sum_{s=1}^{S} s^k n_s \tag{4}$$

Some of these moments are related to usual quantities. For example, m_0 is the multiplicity of fragments and

$$\bar{s} = m_1/m_0 \tag{5}$$

is the mean fragment size (see fig 1). The variance of the distributions is

$$\sigma^2_{[s]} = \frac{m_2}{m_0} - \left(\frac{m_1}{m_0}\right)^2 \tag{6}$$

We have also introduced [6] a "reduced variance"

$$\gamma_2(m_0) = \frac{m_2 m_0}{m_1^2} = \frac{\sigma^2_{[s]}}{\bar{s}^2} + 1 \tag{7}$$

The moments technique is usefull if a few of them suffice to characterize n_s. For example m_0, m_1 and m_2. However if one assumes that the data concern completely determined fragmentations, (the sum of the sizes of all fragments equals the total size S of the system) then $m_1 = S$ does not bring new any information. It is then more convenient to substract from n_s, event by event, the contribution of the largest fragment s_{max}

$$n_s \Rightarrow n_s - \delta(s - s_{max}) \tag{8}$$

Then, for instance $m_1 = S - s_{max}$. This procedure [7] has other advantages. If n_s is such that a heavy fragment is present (U-shape distributions for example), without this substraction, the moments with $k \geq 1$ are rather insensitive to the distribution of the other fragments. This substraction is particularly important to study the signals of critical phenomena [8,9] related to the emergence of a power law behaviour. This emergence manifests itself by a rise (divergence in an infinite system) of moments with $k \geq 2$. This behaviour would be completely hidden without the substraction. The moments analysis can be applied to average fragment size distributions (see below) or to single partitions. We will denote the late by $m_k^{(j)}$, $\gamma_2^{(j)}$...

The vertical analysis will consist in the statistical average of events of same class α (same multiplicity m_0, same s_{max}, etc) as discussed in the previous Section. We will denote this vertical average by the symbol $<>$. The mean fragment distribution is

$$< n_s > = \frac{1}{N} \sum_j^N (n_s^{(j)} - \delta(s - s_{max}^{(j)})) \tag{9}$$

that can be characterized by the moments

$$\bar{m}_k = \sum_s s^k < n_s > \tag{10}$$

from which we calculate

$$\bar{\gamma}_2(m_0) = \frac{\bar{m}_2 m_0}{\bar{m}_1^2} \tag{11}$$

One can alternatively perform statistical averages of the moments of single events

$$< m_k > = \frac{1}{N} \sum_j^N m_k^{(j)} \tag{12}$$

event						
1	n_1^1,	$n_2^1 \cdots$	n_S^1	$\Rightarrow$	m_k^1 ,	γ_2^1
2	n_1^2,	$n_2^2 \cdots$	n_S^2	$\Rightarrow$	m_k^2 ,	γ_2^2
		$\cdots$	$\cdots$	$\cdots$		
		$\cdots$	$\cdots$	$\cdots$		
j	n_1^j,	$n_2^j \cdots$	n_S^j	$\Rightarrow$	m_k^j ,	γ_2^j
		$\cdots$	$\cdots$	$\cdots$	$\Downarrow$	$\Downarrow$
					$< m_k >$,	$< \gamma_2 >$
	$< n_1 >$,	$< n_2 > \cdots$	$< n_S >$	$\Rightarrow$	$\bar{m}_k$	$\bar{\gamma}_2$

Table 1: Matrix of the data

and

$$< \gamma_2 >= \frac{1}{N} \sum_j^N \gamma_2^{(j)} \tag{13}$$

Obviously, for quantities depending linearly on n_s it is indifferent which analysis (horizontal or vertical) is performed first. This is true in particular for the moments

$$\bar{m}_k =< m_k > \tag{14}$$

However, because of the m_1^2 dependence in the denominator, quantites like $\bar{\gamma}_2$ and $< \gamma_2 >$ do not necessarily coincide, as it was erroneously assumed by one of the authors [6]. We will see in Section 4 that the difference $\bar{\gamma}_2 - < \gamma_2 >$ may be used to study the presence of non-statistical fluctuations.

In short, the above procedures are listed in Table 1. Starting with the horizontal analysis, we calculate event by event the moments $m_k^{(j)}$, $\gamma_2^{(j)}$ etc. Then we make the statistical average of these quantities to get $< m_k >$, $< \gamma_2 >$. But we can also start with the vertical analysis, calculating $< n_s >$ and then $\bar{m}_k$, $\bar{\gamma}_2$.

Fig 2 shows $< \gamma_2 >$ and $< S_{max} >$ as a function of the reduced multiplicity $m = m_0/S$ for some simple models of fragmentation (for more details see [5]). We see that the shape of these curves is characteristic of the models (specially the height and the position of the peak of $< \gamma_2 >$). This method is much more sensitive than the standard χ^2 fits (for example determining

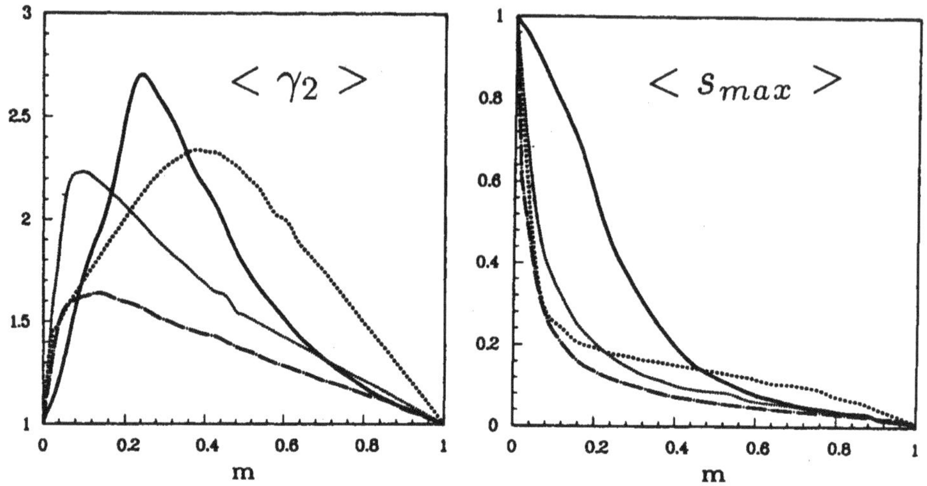

Figure 2: The quantities $< \gamma_2 >$ and $< s_{max} >$ as a function of the reduced multiplicity $m = m_0/S$ for various simple models : Percolation (3d)(cont. line) ; equiprobable partitions (big dots) ; Mekjian model (small dots) ; cascade (dashed-dot). The number of particles of the systems is $S = 125$.

"effective" τ parameters assuming power-law forms). In fact, this method has been widely used to compare experimental data with theoretical calculations [11,12,13] and has proven to be useful to improve the parametrisations of some nuclear fragmentation models [10].

4 Fluctuations

Fluctuations in the fragmentation of a classical system of finite size arise from the indetermination of the initial conditions. The question we want to understand is how $< n_s >$ is realized from single events $n_s^{(j)}$. One can look directly at the variance of n_s

$$\sigma^2_{[n_s]} = < (n_s - < n_s >)^2 > \tag{15}$$

In general we do not need all the information contained in these S quantities. It is more convenient to compress the information performing horizontal averages.

Following the work of Bialas and Peschanski [16,17] on intermittency, Ploszajczak and collaborators [14,15] have investigated in fragmentation problems quantities of the type

$$F_2(\delta) = \frac{\sum_s < n_s(\delta)(n_s(\delta) - 1) >}{\sum_s < n_s(\delta) >^2} \tag{16}$$

where $\delta = 1, 2, \cdots$ is a window of fragment sizes. A power law dependence of F_2 on δ is a necessary condition for the existence of intermittency in the fragment size distribution. If these fluctuations, which are of very specific nature, were to be observed in the experimental study of fragmentation, then very intriguing theoretical questions would be raised.

We suggest a different method. The first moment m_1 (or s_{max}) is the simplest horizontal compression of the data. Its variance can be written

$$\sigma^2_{[m_1]} = \sum_s s^2 \sigma^2_{[n_s]} + \sum_{ss'} ss'[< n_s n_{s'} > - < n_s >< n_{s'} >] \tag{17}$$

The firts term of the r.s.h. is sensitive to both statistical and non-statistical fluctuations (see eq 2). Because of the finite size of the system it is always finite. The second term is generated by two body correlations. The variance of m_1 (which obvioulsy coincides with $\sigma_{[s_{max}]}$) has been studied for nuclear fragmentation data and for percolation theory in ref [6].

Another possibility to quantify the fluctuations is to use the quantity γ_2. We propose to look at how the mean $< \gamma_2 >$ of the the reduced variances of single events converges towards the reduced variance of the mean, $\bar{\gamma}_2$. These two quantities differ because of the non linear dependance on n_s, as already mentioned. Without the substraction of s_{max}, both γ_2 would be identical. This suggests that

$$\delta\gamma_2 =< \gamma_2 > -\bar{\gamma}_2 \tag{18}$$

varies in phase with $\sigma^2_{[m_1]}$. This hypothesis [18] has been verified numerically for a large variety of fragmentation mechanisms (cascade models, percolation, maximum entropy, classical molecular dynamics...) [19,20]. In other words, one can indifferently use $\sigma^2_{[s_{max}]}$ or $\delta\gamma_2$ as measure of fluctuations.

5 Fluctuations in simple models

Once the ability of $\delta\gamma_2$ to analize the fluctuation properties of the data has been established, the question to adress is : do simple fragmentation models exhibit non-statistical fluctuations ? We have studied this question using the models of section 3. Fig 3 shows the quantity $\delta\gamma_2$ as a function of the reduced multiplicity m_0/S for the equiprobable partition model (dots), binary cascade (small dots) and bond percolation in a cubic lattice (solid line).

The size of the systems is $S = 125$. One may observe that the first and the second models exhibit a weak signal and for the lowest multiplicities only. These fluctuations are generated simply by correlations due to particle number conservation, rapidly vanishing when increassing the multiplicity. The signal is in contrast much stronger for the percolation model.

The origin of the fluctuations in the percolation remains to be understood. Our present suggestion is that they originated from the random surface constraints that limit the size of the clusters located at the periphery of the system. This effect may be particularly important if a large fragment s_{max} is present in the system. Because of mass conservation, large fluctuations of s_{max} means large fluctuations of m_1 and *a fortiori* large $\delta\gamma_2$, as discussed at the end of the preceding section. The arguments that support qualitatively this hypothesis are the following. $\delta\gamma_2$ vanishes rapidly beyond the critical multiplicity ($m_0/S \simeq 0.25$), where the constraints on the size of s_{max} also vanish. The maximum amplitude of $\delta\gamma_2$ goes to zero when the size of the system $S \to \infty$, i.e. when the ratio of the number of surface to volume sites goes to zero. The influence of this ratio can also be investigated for a fixed size S. Fig 4 shows $\delta\gamma_2$ for cubic lattice systems with $S = 216$ particles, one with a 6x6x6 lattice (dotted line) and the other with a 3x6x12 lattice (continous line). We see that for the former, that has a larger surface, $\delta\gamma_2$ is much larger.

We believe that the problem of the origin of non-statistical fluctuations in finite size geometrical models of fragmentation is very interesting by itself. It is almost unexplored. It would also be fascinating to find that these fluctuations do show up in the fragmentation of real microscopic systems. We are working in these directions.

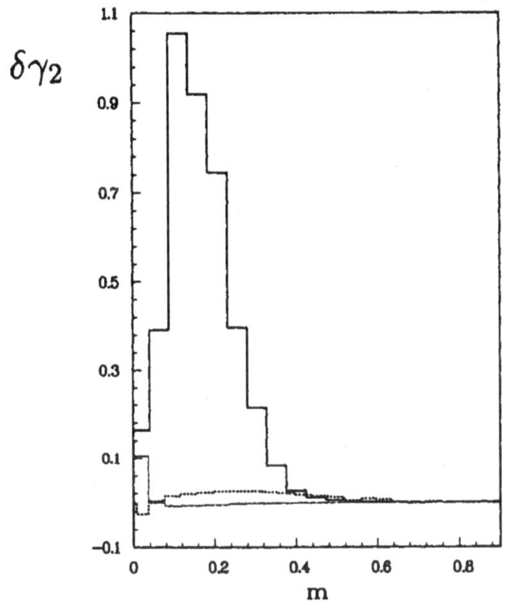

Figure 3: The quantity $\delta\gamma_2$ (eq 18) as a function of the reduced multiplicity m : Equiprobable partitions (dots), cascade (small dots) and cubic lattice percolation (cont. line). The number of particles is $S = 125$.

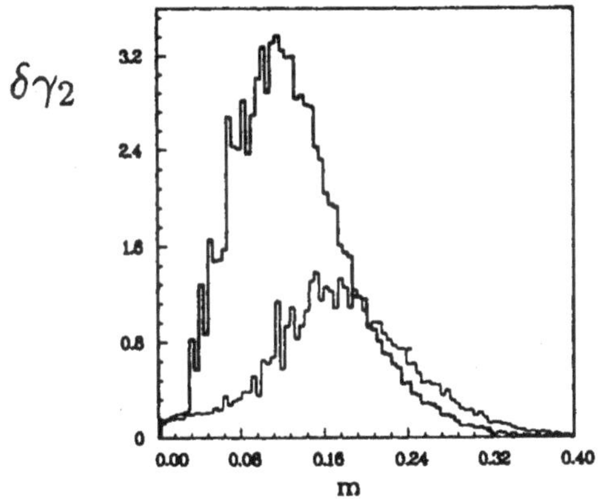

Figure 4: The quantity $\delta\gamma_2$ (eq 18) as a function of the reduced multiplicity m for a percolation lattice 6x6x6 (dashed) and for a lattice 3x6x12 (continous). ($S = 216$ in both cases).

6 Summary

The main points we wanted to raise in this contribution are the following.

- Fragmentation is a statistical process because the initial conditions are not reproducible.

- In finite size systems, this uncertainty in the initial conditions may create fluctuations in the fragment size distributions. These fluctuations depend on the way one performs the statistical averages. From a purely statistical point of view, the best procedure is to perform the statistical average of events with same multiplicity of fragments.

- The fluctuations can be studied comparing the width of the mean fragment distribution with the mean of the widths of the distributions of single events, or by looking at the variance of the largest fragment.

- Simple cascade and maximum entropy fragmentation models show only very small fluctuations due to particle conservation. Percolation fragmentation exhibits large fluctuations. These are mainly due to the fluctuation of the size of the percolation cluster. Our suggestion is that this size fluctuation arises from boundary constraints.

7 Acknowledgements

We are indebted to A. Bialas, P. Desesquelles, R. Peschanski, E. Plagnol and M. Ploszajczak for stimulating discussions.

8 References

[1] D.E. Grady and M.E. Kipp, J. Appl. Phys. 58 (1985) 1210

[2] S. Redner, in Statistical Models for the Fracture of Disordered Media, ed. by H. J. Herrmann and S. Roux (Elsevier, New York, 1990), Chap. 10 and references therein.

[3] P. Kreutz et al. Nucl. Phys. A556 (1993) 672

[4] L. Oddershede,P. Dimon and J. Bohr, Riso National Lab. preprint (1993).

[5] X. Campi and H. Krivine, Z. Phys. A344 (1992) 81.

[6] X. Campi, Phys. Let. B208 (1988) 81.

[7] X. Campi, J. Phys A19 (1986) L917.

[8] D. Stauffer, in Introduction to Percolation Theory, ed. Taylor and Francis (London and Philadelphia, 1985).

[9] D. Stauffer, A. Coniglio and M. Adam, Adv. in Pol. Sci. 44 (1982) 103.

[10] Bao-An Li, A. R. DeAngelis and D. H. E. Gross, Phys. Let. B303 (1993)225.

[11] P. Desesquelles et al. Phys. Rev. C48 (1993)

[12] K. Hagel et al. Phys. Rev. Let. 68 (1992) 2141.

[13] B. Jacobsson et al. Nucl. Phys. A509 (1990) 195.

[14] M. Ploszajczak and A. Tucholski, Phys. Rev. Let. 65 (1990) 1539.

[15] P. Bozek, M. Ploszajczak and A. Tucholski, Nucl. Phys. A539 (1992) 693.

[16] A. Bialas and R. Peschanski, Nucl. Phys. B273 (1986) 703.

[17] A. Bialas and R. Peschanski, Nucl. Phys. B308 (1988) 857.

[18] X Campi and H. Krivine to be published.

[19] B. Elattari private communication

[20] E. Plagnol private communication

FINGERPRINTS OF DYNAMICAL INSTABILITIES

Ph. Chomaz[1], *M. Colonna*[1,2], *A. Guarnera*[1,2]

[1]) *GANIL, B.P. 5027, F-14021 Caen Cedex, France*

[2]) *LNS, Viale Andrea Doria, Catania, Italy*

ABSTRACT

We first explain why any reduced descriptions, such as mean field approximation, are stochastic in nature. We show that the introduction of this stochastic dynamics leads to a predictive theory in a statistical sens whatever the individual trajectories are characterized by the occurrence of bifurcations,instabilities or phase transitions. Concerning nuclear matter, we discuss the spinodal instability. In such a critical situation, we explore the possibility to replace the stochastic part of the collision integral in the Boltzmann-Langevin model by the numerical noise associated with the finite number of test particles in ordinary *BUU* treatment. We finally show that the fingerprints of these instability are kept during the evolution because of the relatively long recombination time compared with the typical time scales imposed by the Coulomb repulsion and the possible collective expansion.

1. Introduction

The complete treatment of a many-body dynamics requires the consideration of the correlated evolution of a huge number of degrees of freedom which is in general impractical. A tractable approximation is to retain only a small part of the dynamical information (namely the most pertinent variables such as the one-body observables or collective modes) and to project the evolution onto the associated manyfold[1]. In particular, conserving only the one-body density leads to the well known mean field approximation. In general, the projected equation of evolution for the retained variables will be non-linear and therefore the dynamics may exhibit features such as instabilities, bifurcations and chaos. A typical example of such a complex evolution is the well known fusion-fission dynamics which possesses at least two valleys with two instability points. In the fission process the system will follow the fission valley until its trajectory becomes unstable and then the system will fall in the other valley (the fusion valley). If we now reverse the process, the system will follow a different path because it will stay in the second valley until the other instability point is reached and will then come back to the first valley. This is a typical hysteresis circle, the system following a different path when an external constraint is increased or decreased.

However, the lack of full information about the state of the system, both initially and in the course of time, acts as a stochastic term in the effective equation of motion for the retained variables, because they characterize an entire ensemble of microscopically different states of the system. In effect, this ensemble acts as the heat

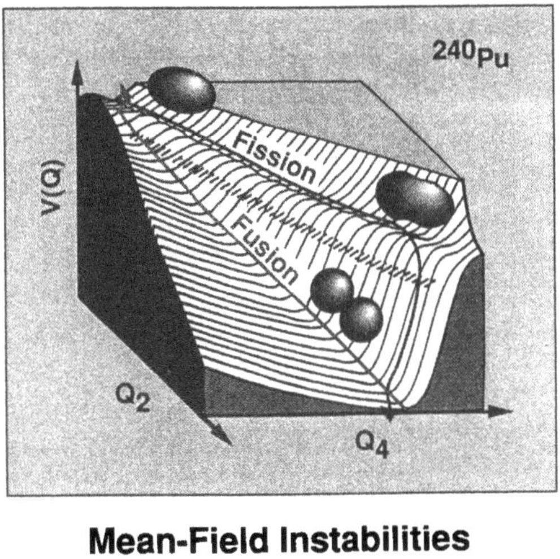

Fig. 1: The potential energy of a constrained Pu nucleus as a function of the quadrupole and hexadecapole deformation calculated in a constrained Mean-Field approach from [2]. The fusion and the fission paths are represented exhibiting a typical hysteresis circle due to the presence of two instability points.

bath in Brownian motion and the equation of motion will be of the Langevin type. In this case one is forced to consider not only one single trajectory, in the projected manyfold, but an ensemble of trajectories, which differ because small fluctuations in the evolution may be propagated very differently by the deterministic but non-linear part of the effective equation of motion. In such a case, when chaos and bifurcations are encountered during the evolution, the mean average trajectory (mean field trajectory in the case of the projection on one-body observables) is no more defined. However, it must be noticed that the stochastic ensemble of trajectories is perfectly defined.

This is in fact a general property of stochastic approaches: even if the underlying (without stochasticity) dynamics is unpredictable (such as in deterministic chaos), the introduction of a stochastic term will suppress its deterministic character but will restore its predictive power. Let us explain this apparent paradox in the next section.

2. Stochastic Treatment of Chaos:

2.2 On the Treatment of Chaos in Stochastic Theories

Let us consider any deterministic evolution of a system described by some dynamical variables $\mathcal{F}$

$$\mathcal{F}(t) = \mathcal{G}[\mathcal{F}](t) \tag{1}$$

or any mapping of a discreet quantity $\mathcal{F}_i$

$$\mathcal{F}_i = \mathcal{G}[\mathcal{F}]_i \tag{2}$$

where $\mathcal{G}$ is the result of a general evolution operator or mapping operator. In general $\mathcal{G}$ is an integro-differential equation depending on the past history of $\mathcal{F}$. These evolutions are in general non-linear and therefore can be very complex. In particular, they may exhibit instabilities, bifurcations and deterministic chaos which may reduce the predictive power of these approaches.

However, the actual physical processes described by these evolutions often present a certain degree of stochasticity. In particular, these non-linear evolutions are often obtained as a reduction of more complex theories which may be perfectly linear. In these cases the non-linearities are arising from the projection of these "master equations" on a reduced set of degrees of freedom. This suppression of the non-interesting degrees of freedom generally introduces some entropy because of the lost of information on these disregarded correlations. In return, this creation of entropy introduces a stochastic term in the evolution.

Therefore, the actual problem is to akin to a stochastic evolution (such as a generalized Langevin process)

$$\mathcal{F}^{(n)}(t) = \mathcal{G}[\mathcal{F}^{(n)}](t) + \delta\mathcal{G}^{(n)}[\mathcal{F}^{(n)}](t) \tag{3}$$

or for a discrete mapping

$$\mathcal{F}_i^{(n)} = \mathcal{G}[\mathcal{F}^{(n)}]_i + \delta\mathcal{G}[\mathcal{F}^{(n)}]_i \tag{4}$$

where the upper script, n, refers to a particular realization of the stochastic dynamics (called in the following an event or a system). The ensemble of N events can be conveniently described by the distribution in the space of the dynamical variables $\mathcal{F}$:

$$\phi[\mathcal{F}] = \frac{1}{N}\sum_{n=1}^{N} \delta[\mathcal{F}^{(n)} - \mathcal{F}] \ , \tag{5}$$

which gives the probability that a randomly selected event n has the specified variables $\mathcal{F}$. The transport of the distribution $\phi[\mathcal{F}]$ is given by a master equation which expresses both the deterministic evolution and the transition rates between two $\mathcal{F}$'s due to the random evolution. Formally, the propagation of $\phi[\mathcal{F}]$ is described by an equation of evolution

$$\phi[\mathcal{F}](t) = \Gamma\left\{\phi[\mathcal{F}]\right\}(t) \tag{6}$$

or for discrete mapping

$$\phi[\mathcal{F}]_i = \Gamma\left\{\phi[\mathcal{F}]\right\}_i \tag{7}$$

These new equations have the simple property to be linear. Indeed, they describe the evolution of an ensemble of independent systems, the $\mathcal{F}^{(n)}$'s. These systems are completely independent since their equations of evolution depend only on $\mathcal{F}^{(n)}$. Therefore, if ϕ_1 and ϕ_2 are two solutions of the equations (6) or (7), i.e. two ensembles of trajectories, $\phi_1 + \phi_2$ is also an ensemble of trajectories and so is also a solution of the equations (6) or (7):

$$\Gamma\left\{\phi_1 + \phi_2\right\} = \Gamma\left\{\phi_1\right\} + \Gamma\left\{\phi_2\right\} \tag{8}$$

Therefore this evolution of the distribution $\phi[\mathcal{F}]$ do not present neither instabilities, nor bifurcation nor chaos.

2.2 Application to the Stochastic Mean Field Approach.

Let us discuss a particular application of the previous general considerations: the stochastic mean field approaches.

Mean field theories such as time dependent Hartree Fock (TDHF) or its classical analog Vlasov are associated with strongly non linear equations which may exhibit solitons, instabilities, bifurcation and chaos. However, since they are the result of a projection of the N-body problem on to the one-body degrees of freedom, they must be considered as stochastic in nature, the unknown correlations generating fluctuations. An example of such stochastic evolution is the well known Boltzmann-Langevin equation which treats the fluctuations induced by the stochastic nature of the two-body collisions.

This approach considers an ensemble of N identical systems, labelled by $n = 1, ..., N$. Each one is described by its reduced one-body phase-space density $f^{(n)}(\mathbf{r}, \mathbf{p}, t)$. The time evolution of each phase-space density is given by an equation of the form

$$\frac{\partial f^{(n)}}{\partial t} + \left\{H^{(n)}, f^{(n)}\right\} = I[f^{(n)}]. \tag{9}$$

Here H is the effective one-body Hamiltonian and $\{\cdot, \cdot\}$ the Poisson bracket, so that the left-hand side describes the Vlasov evolution for $f^{(n)}$ as it evolves in its own self-consistent mean field $H^{(n)} = H[f^{(n)}]$. The term on the right-hand side represents the effect of the residual two-body collisions among the nucleons in the system. This collision term is generated by a stochastic sequence of individual transitions in the system. Because of this feature, the evolution of the phase-space distribution is akin to Brownian motion, with the collision integral playing the role of the fluctuating Langevin force.

The mean number of transitions Δn in which nucleons are scattered from two phase-space elements into two other phase-space elements is defined in accordance with the usual Boltzmann approximation for fermions. Since the collisions are regarded as random processes analogous to a random walk, the number of transitions actually occurring fluctuates according to the corresponding Poisson distribution and the variance of the quantity Δn is therefore given by $\sigma(\Delta n)^2 = \Delta n$.

Since this stochastic nature of the two-body interaction causes each individual many-body system to evolve differently, the equation (9) will generate an entire bundle of dynamical trajectories $f^{(n)}(\mathbf{r}, \mathbf{p}, t)$. This ensemble of trajectories can be described by the following distribution of one-body phase-space densities [3],

$$\phi[f] = \frac{1}{N} \sum_{n=1}^{N} \delta[f^{(n)} - f] , \tag{10}$$

which gives the probability that a randomly selected system n has the specified occupancy function $f(\mathbf{r}, \mathbf{p})$.

Since the two-body collision rate is assumed to depend only on $f^{(n)}$ at the actual time t, the evolution of each system of the ensemble is a typical Markovian process, *i.e.* it does not depend on the particular history experienced by the system (nor does it depend on the state of the other systems in the ensemble). In transport theory, the Markovian evolution of the distribution $\phi[f]$ is given by a master equation which expresses the rates of transition between one density f' and another f''. Under suitable smoothness conditions, the master equation can be written as a differential equation of the Fokker-Planck type,

$$\frac{d\phi[f]}{dt} = \sum_{n>0} \frac{(-)^n}{n!} \int ds_1 \cdots ds_n \frac{\partial}{\partial f(s_1)} \cdots \frac{\partial}{\partial f(s_n)} (\alpha_l[f](s_1, \cdots, s_n)\phi[f]) \ , \tag{11}$$

where $s = (\mathbf{r}, \mathbf{p})$. Since the basic stochastic process is of two-body form, only four phase-space locations are involved at any one time and consequently the sum over n in eq. (11) terminates at the fourth order. As shown in ref. [3], the transport coefficients $\alpha_n[f]$ can be derived from the early growth rate of the n-body correlations.

This Fokker-Planck equation is linear as discussed in the previous section therefore all the particularities (or pathologies) of the deterministic non linear treatment have disappeared and the stochastic description is both more physical and predictive.

3. Test in idealized systems

The previous approach have been carefully tested in many physical scenarii. In particular it was shown that the ensemble of trajectories does converge to the right statistical equilibrium. In presence of instability the Langevin term provides a mechanism to break the possible symmetries and to test the mean field instabilities. Let us first illustrate this property.

3.1 Symmetry breaking

To illustrate these ideas we have considered an idealized case of a two-dimensional torus at half normal density and at a temperature T=3 MeV. For the considered force, (see ref. [4] for more details) this system lies in the spinodal instability region, i.e. the region where the system is unstable against variation of its density. Fig 2a presents a lattice simulation [5] of the Boltzmann-Langevin evolution in which the one-body distribution function f experiences a stochastic evolution and therefore fluctuates, in response to the random effect of the Langevin force. Since the mean field is unstable, these fluctuations are amplified and eventually lead to the formation of fragments. It is important to remark that at the initial time the density distribution is flat; as the time increases, fluctuations coming from the stochastic part of the collision integral are breaking this initial symmetry, leading to clusterization.

3.2 Simulating the Langevin Force by a Simple Noise

In Fig.2b, we show the same time evolution in the case of a *BUU* calculation using 90 test particles. Now fluctuations are present since the beginning due to the finite sampling of the phase space distribution but, starting from a given time (

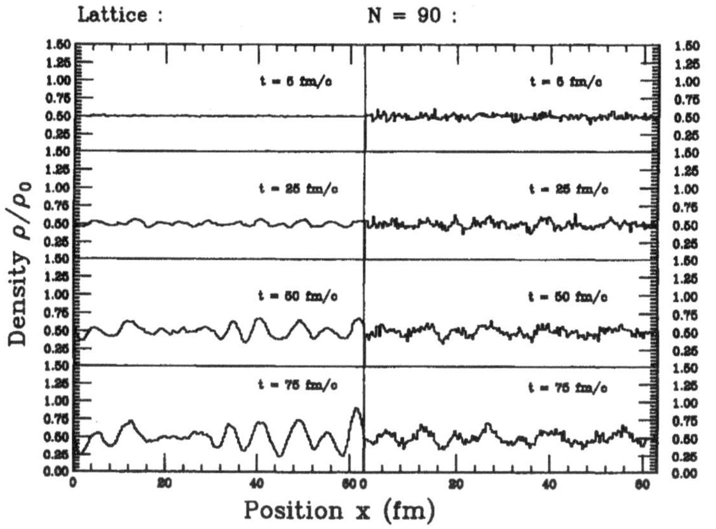

Fig. 2: The density profile $\rho(x,t)$ associated with one particular trajectory at four different times t, for the *BL* lattice calculation (part *a*) and for the *BUU* test-particle simulation (part *b*).

$t \approx 50\ fm/c$), the evolution becomes comparable to that of the lattice calculation (part a), reaching finally a similar structure. The comparison between lattice and *BUU* calculations using a different number of test particles is represented in fig.3. One can see on this figure that if the used number of test particles is suitable, it is possible to reproduce the asymptotic results of the lattice calculations. In our case it happens using 90 test particles per nucleon. Anyway, there is a range of test particle numbers which could be used obtaining not very much different results. Indeed, as we can see from Fig. 3, the advance that we observe in the clusterization time is only 20 fm/c using a smaller number of test particles ($N_{test} = 20$). This is due to the fact that since the instabilities are exponentially increasing the time to reach a given amplitude of the fluctuation is a logarithmic function of the number of test particles per nucleon.

4. Linear Response Theory

In order to deeply investigate the growth of fluctuations, we start from the Linear Response Theory, which is valid for small fluctuations and therefore it is suitable to describe the first evolution of the system.

4.1 Dispersion Relation

Let us first consider the Vlasov equation,

$$\frac{\partial f}{\partial t} + \frac{\partial f}{\partial \mathbf{r}} \cdot \frac{\partial H}{\partial \mathbf{p}} - \frac{\partial f}{\partial \mathbf{p}} \cdot \frac{\partial H}{\partial \mathbf{r}} = 0 \ , \tag{12}$$

which describes the collisionless evolution of f in the self-consistent effective field. In the following, the effective one-body Hamiltonian is assumed to have the form

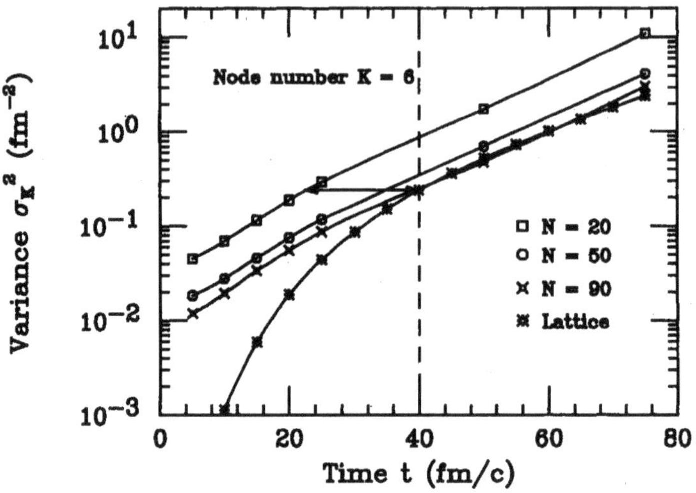

Fig. 3: The time evolution of the Fourier transform of the correlation function, σ_k^2, for the most unstable mode (which has node number $K = kL_x/2\pi = 6$), for the *BL* lattice calculation (part *a*) and the *BUU* test-particle simulation with three different values of N, the number of test-particles per nucleon (part *b*). This quantity is the variance of fluctuations having wave number equal to k. In this Figure the arrow represents the typical delay when the number of test particle is changed by a factor 5.

$H(\mathbf{r},\mathbf{p}) = p^2/2m + U[\rho]$, with $\rho(\mathbf{r}) = h^{-D}\int d\mathbf{p} f(\mathbf{r},\mathbf{p})$ being the spatial density of the nucleons. Expanding the disturbance δf around the static uniform solution f_0, we obtain the RPA equations for the perturbation $\delta f(\mathbf{r},\mathbf{p},t)$,

$$\frac{\partial \delta f}{\partial t} + \frac{\partial \delta f}{\partial \mathbf{r}} \cdot \frac{\mathbf{p}}{m} - \frac{\partial f_0}{\partial \mathbf{p}} \cdot \frac{\partial \delta U}{\partial \mathbf{r}} = 0 \; . \tag{13}$$

To solve the above equation, it is convenient to replace the $\mathbf{r}$ representation by a wave number representation $\mathbf{k}$, i.e. to perform a Fourier transform with respect to the position $\mathbf{r}$, and to introduce the frequency ω. We find

$$(\omega - \frac{\mathbf{k}\cdot\mathbf{p}}{m})\delta f(\mathbf{k},\mathbf{p}) + \frac{\mathbf{k}\cdot\mathbf{p}}{m}\frac{\partial f_0}{\partial \varepsilon}\frac{\partial U_k}{\partial \rho}\delta\rho(\mathbf{k}) = 0 \; . \tag{14}$$

In the region where $\mathbf{k}\cdot\mathbf{p} \neq m\omega$ (as is the case, in particular, whenever when the energy is not real, such as for unstable modes) we can solve this relation formally:

$$f_{\mathbf{k},\omega}(\mathbf{p}) = \frac{\partial U_k}{\partial \rho}\frac{\mathbf{k}\cdot\mathbf{p}}{\mathbf{k}\cdot\mathbf{p} - m\omega}\frac{\partial f_0}{\partial \varepsilon}\,\rho_{\mathbf{k},\omega} \; . \tag{15}$$

The dispersion relation is obtained by expressing the self-consistency condition $\rho_{\mathbf{k}} = h^{-D}\int d\mathbf{p} f_{\mathbf{k}}(\mathbf{p})$. This integration over the momentum $\mathbf{p}$ yields the following condition on ω,

$$1 = \frac{\partial U_k}{\partial \rho}\int \frac{d\mathbf{p}}{h^D}\frac{\mathbf{k}\cdot\mathbf{p}}{\mathbf{k}\cdot\mathbf{p} - m\omega}\frac{\partial f_0}{\partial \varepsilon} = \frac{\partial U_k}{\partial \rho}\int \frac{d\mathbf{p}}{h^D}\frac{(\mathbf{k}\cdot\mathbf{p})^2}{(\mathbf{k}\cdot\mathbf{p})^2 - m^2\omega^2}\frac{\partial f_0}{\partial \varepsilon} \; , \tag{16}$$

where the last relation has been obtained by using the fact that only terms that are symmetric under the transformation $\mathbf{p} \to -\mathbf{p}$ contribute to the integral. We note that ω depends only on the magnitude of $\mathbf{k}$, since all directional information is eliminated by the integration over $\mathbf{p}$. Moreover, for those values of k for which the dispersion relation can be solved, there are two possible frequencies $\omega_{\mathbf{k}}$ which are related by complex conjugation.

To gain more familiarity with the properties of the unstable modes, we consider the zero-temperature limit. In this case one can solve the dispersion relation (16) analytically, due to the fact that the distribution f_0 describes a sharp Fermi sphere so that $\partial f_0/\partial\varepsilon = -\delta(\varepsilon - \varepsilon_F)$, where ε_F is the Fermi energy associated with the specified uniform density ρ. One obtains imaginary eigenenergies when the following relation is satisfied,

$$\frac{2}{D}\varepsilon_F + \rho\frac{\partial U_{\mathbf{k}}}{\partial\rho} < 0 \ , \tag{17}$$

where D is the dimension of the physical space. When condition (17) is satisfied, the frequency of the eigenmode is imaginary and we obtain the solutions $\omega_{\mathbf{k}}^{\pm} = \pm i/t_{\mathbf{k}}$, where $t_{\mathbf{k}}$ is the corresponding characteristic time.

For a two-dimensional system we then find:

$$-\omega_{\mathbf{k}}^2 = \frac{1}{t_{\mathbf{k}}^2} = \frac{k^2}{m}\rho\left(\frac{\varepsilon_F}{\rho} + \frac{\partial U_{\mathbf{k}}}{\partial\rho}\right)^2 \Big/ \left(\frac{\varepsilon_F}{2\rho} + \frac{\partial U_{\mathbf{k}}}{\partial\rho}\right) \ . \tag{18}$$

4.2 Finite Range, Surface Energy and Dynamical instabilities

In the limit of long wave lengths, $k \to 0$, or for zero-range interaction, the relation (17) reduces to the condition for mechanical instability,

$$\left(\frac{\partial P}{\partial\rho}\right)_{T=0} = \frac{\partial}{\partial\rho}\left(\rho^2\frac{\partial}{\partial\rho}\bar{\varepsilon}\right) = \frac{2}{D}\varepsilon_F + \rho\frac{\partial U}{\partial\rho} < 0 \ . \tag{19}$$

and the dispersion relation (18) predicts an instability time proportional to the wave length λ. This is evidently an unphysical result since the smaller the fluctuation the faster their growing time. This result is coming from the fact that a zero-range interaction does not possess any scale.

A physical situation can be recovered if we introduce a finite range for the mean field. This finite range will generate a surface energy which will prevent from getting too small clusters and will introduce a natural cut-off in the unstable length. As we will see, it will also make the spinodal k-dependent.

Let us illustrate this point pictorially by graphically solving the instability condition in the case of a k-dependent mean-field.

The simplest way to introduce a finite range mean-field is to fold the zero-range one with function g (such as a gaussian),

$$U \to U \otimes g \tag{20}$$

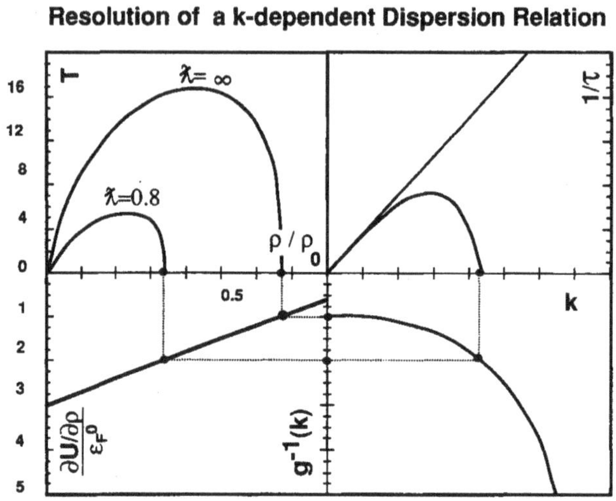

Fig. 4: Graphic resolution of the instability condition (lower-part), representation of the resulting dispersion relation (upper-right) and of the k-dependent spinodal in the temperature/density plane (lower part).

the range of this function being selected under the requirement of getting a good surface energy. In such a case the instability condition simply reads

$$\rho\left(\varepsilon_0 + \rho_0 \frac{\partial U}{\partial \rho} g_k\right) < 0 \;, \tag{21}$$

This equation can be solved by comparing g_k^{-1} and $(\partial U/\partial\rho)/(\varepsilon_0/\rho_0)$. This is graphically done in Fig. 4: for a given value of ρ/ρ_0 we can compute $(\partial U/\partial\rho)/(\varepsilon_0/\rho_0)$ (lower-left figure). If this value is bigger than 1, it corresponds to a value of g^{-1} for a given k (lower-left figure). Therefore the system is unstable at this density and the value of k that we found corresponds to the largest unstable wave number. Therefore the dispersion relation of instabilities (upper-right figure) ends up at this value of k.

If we now let the temperature take on a finite value, the dispersion relation (18) will change somewhat while retaining its overall character. Therefore, we will be able to draw a spinodal contour which will be k-dependent (upper-left figure).

The introduction of the collision term mainly adds a damping (*i.e.* a positive imaginary part) to the frequency. This can be most clearly seen in the relaxation-time approximation where $\bar{I}[f]$ is replaced by $-\delta f/t_{\rm relax}$, so that the average effect of the collisions is to change the frequency $\omega_{\rm k}^{\pm}$ into $\omega_{\rm k}^{\pm}$-$i/t_{\rm relax}$.

4.3 Linearization of the Boltzmann-Langevin Equation

If δf is the fluctuating part of the one-body distribution function and $\{\delta f_\nu\}$ represents the set of eigenmodes of the linearized problem we can write $\delta f = \sum_\nu A_\nu \delta f_\nu$. It is possible to demonstrate that the time evolution of the fluctuation $\sigma_\nu^2 = \prec A_\nu A_\nu^* \succ$

follows the Lalime's equation:

$$\frac{d}{dt}\sigma_\nu^2 = -2i\omega_\nu\sigma_\nu^2 + 2D_\nu \tag{22}$$

where ω_ν is the eigenvalue associated to the considered mode ν and D_ν represents the correlation of the langevin force projected on the normal modes. If this mode is unstable and we define $\tau_\nu = i/\omega$, the solution of the previous equation reads:

$$\sigma_\nu^2(t) = D_\nu\tau_\nu(e^{\frac{2t}{\tau_\nu}} - 1) + \sigma_0^2 e^{\frac{2t}{\tau_\nu}} \tag{23}$$

where σ_0^2 is the initial value of σ_ν^2. If the time is long enough, the relation (23) can be approximately written as follows:

$$\sigma_\nu^2(t) \approx (D_\nu\tau_\nu + \sigma_0^2)e^{\frac{2t}{\tau_\nu}} \tag{24}$$

The terms $D_\nu\tau_\nu$ and σ_ν^2 play a similar role in the evolution of fluctuations: therefore it is possible to replace the fluctuating source term by any simple noise of the same amplitude. In particular, using a finite number of test particles in ordinary *BUU* calculations, we can introduce a noise associated to the related phase space discretization. This noise scales like the inverse of the number of test particles per nucleon, N_{test}, and therefore N_{test} may be tuned in order to reproduce the physical amount of fluctuations. This is exactly what we discussed in section 3.

4.4 Dynamical Spinodal and Results for a 3D Space

Let us now present some results obtained in a real 3D calculation. First Fig. 5 displays the spinodal contour lines in a temperature versus density plane. One can see that the spinodal associated with a finite wave length is reduced in comparison with the static instability condition. Moreover, isocontour of instability time are also presented on this figure. One can see that when the mode starts to be unstable its instability time is infinite. Therefore it is dynamically stable. In fact a given mode can be considered as dynamically unstable only when its instability time becomes shorter than a characteristic time for the global evolution of the system. If we consider that the good time scales are around 50 fm/c, we realize that the dynamically unstable region is much smaller than the actual static spinodal one.

From this dispersion relation it should be notice that the most unstable modes happen to be of a rather short wave length. Therefore the resulting partitions will favor small nuclei around the mass 20.

From the Boltzmann-Langevin equation we get the diffusion coefficient D_ν for a given unstable mode, for different values of temperature and density, as it is shown in Fig.6. Finally Fig.7 shows the number of test particles that one has to use in order to match the Boltzmann-Langevin results. This number decreases both with density and temperature, since the collision number, and then the fluctuation source term, increase going to higher density and temperature. In this figure we can see that 20 to 40 test-particles is a good choice in order to mimic the Boltzmann-Langevin dynamics and indeed full calculations of heavy ion reactions exhibit spinodal instabilities and clusterization.

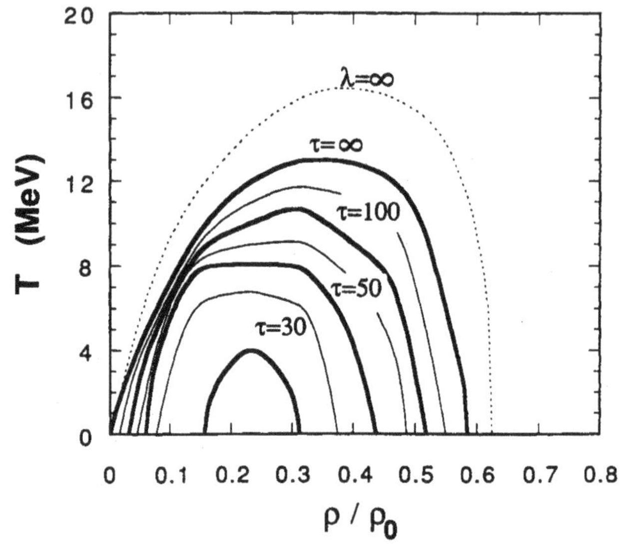

Fig. 5: Static spinodal region and contour of equal instability time for the wave length $\lambda/2\pi = 1.6fm$.

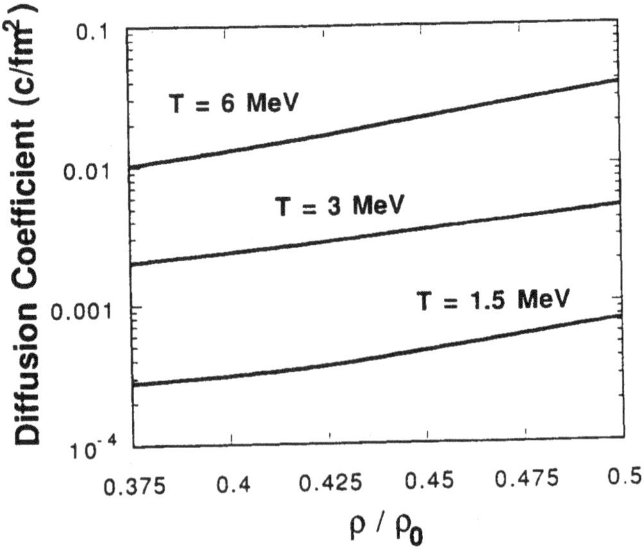

Fig. 6: The diffusion coefficient D_ν for the most unstable mode as a function of the average density in an initially uniform three-dimensional torus. Three different values of the temperature have been considered.

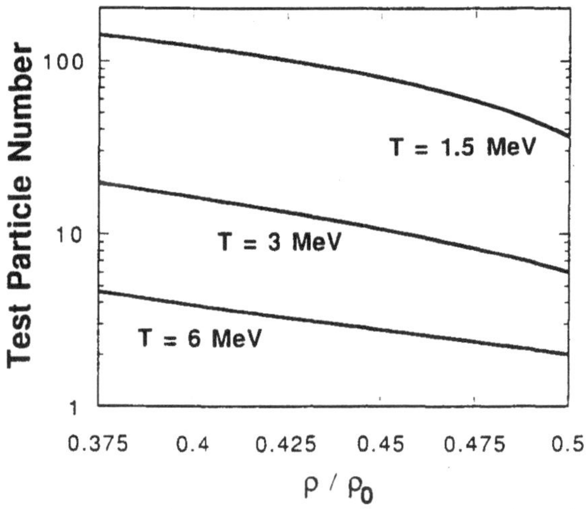

Fig. 7: The number of test particles reproducing the Langevin dynamics in the spinodal region as a function of density and temperature (see Fig. 6).

From this study we have seen that the partition of the system is dominated by few unstable modes therefore we may keep some fingerprints of the spinodal instabilities in the fragment size distribution. However, we are dealing with a strongly interacting system and one may worry about the possible final state interaction. Therefore we must ask ourself about the freeze-out time.

5. Finale State Interaction, Statistical Equilibrium and Freeze-out: Test in an idealized system

In order to test how the memory of the spinodal instabilities survives the clusterization process we have studied an idealized 2D system. Figure 8 presents the evolution of piece of infinite nuclear matter initially in the spinodal region. One can see that the fragments are formed soon, after the beginning of the process and that the resulting partition is characteristic of the most unstable mode. After this fast clusterization the system relaxes very slowly toward statistical equilibrium and, indeed, on the numerical simulation, one observes that the system does not evolve much during the last 100 fm/c.

This can be understood by simply computing the time the system needs to sample the different accessible partition. This time is related to the time a fragment needs to undergo a nuclear reaction. This time is simply given by:

$$\tau_{sample} = 1/\sigma v \rho \approx 200 fm/c \tag{25}$$

where σ is the typical reaction cross-section and ρ the cluster density. The velocity v can be computed from the system temperature. This velocity of the cluster happens to be small because of the energy equi-partition. Therefore the sampling time is found

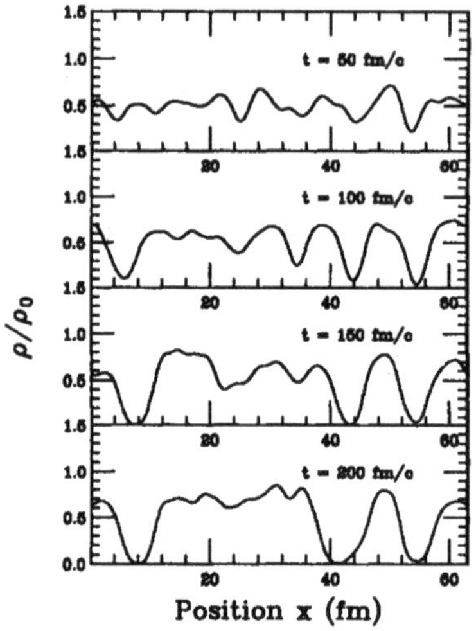

Fig. 8: Clusterization process in a periodic box

to be very long of the order of 200 fm/c in agreement with the numerical simulation.

In real nuclei several effects are modifying this picture. First the finite size effects which are introducing surfaces which tend to "catalyze" the equilibration process. This is illustrated in Fig 9 where we show the clusterization dynamics in a finite nucleus. We can see that the clusterization is still dominated by the instabilities but the memory of this mechanism is lost because of the presence of strong interactions.

However in the previous simulation the Coulomb interaction was not included. As we can see in fig 10 this repulsive interaction strongly reduces the final recombination. Indeed, it is easy to realize that the time a cluster needs to leave the system (*i.e.* to move of $d = 1$ fm) is given by:

$$\tau_{Coul} = \sqrt{MRd/V_c} \approx 10 - 30 fm/c \tag{26}$$

where M is the cluster mass, R the distance to the charge center and V_c the Coulomb potential. Therefore the Coulomb interaction introduces a fast freeze-out time and the memory of the dynamical instabilities can be kept.

Another way to introduce a freeze-out time is to consider an expanding box. Indeed the collective velocity will push the fragments one apart from the other so that they will soon stop interacting. This is illustrated in Fig. 11 where we can see that a small expansion velocity is very efficient as far as the freeze-out is concerned. The system is keeping a complete memory of the instability. It must be noticed that the most unstable modes have always the same wave length. However, since the system is expanded during the time of the clusterization, this unstable mode generates more fragments.

The various scenarii are summarized in Fig. 12 which presents the mean number of fragment as a function of the time. One can see that many physical effects

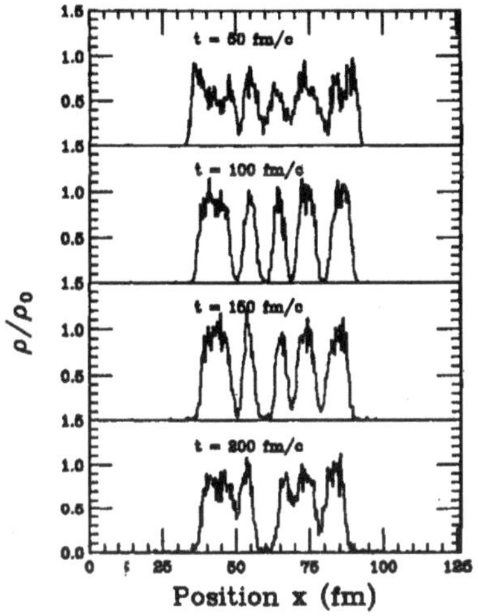

Fig. 9: Clusterization process of a finite nucleus.

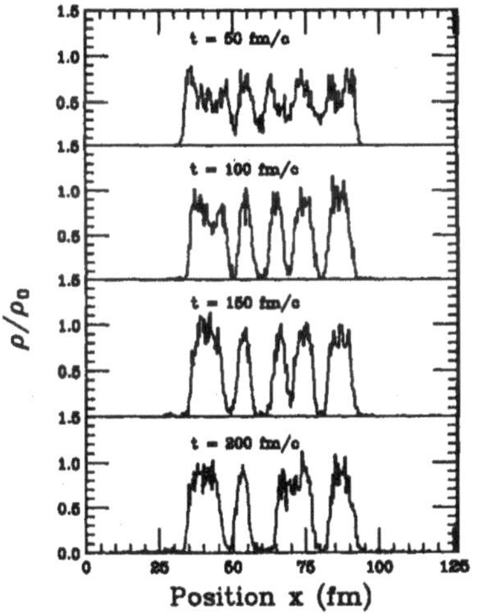

Fig. 10: Clusterization process of a charged finite nucleus.

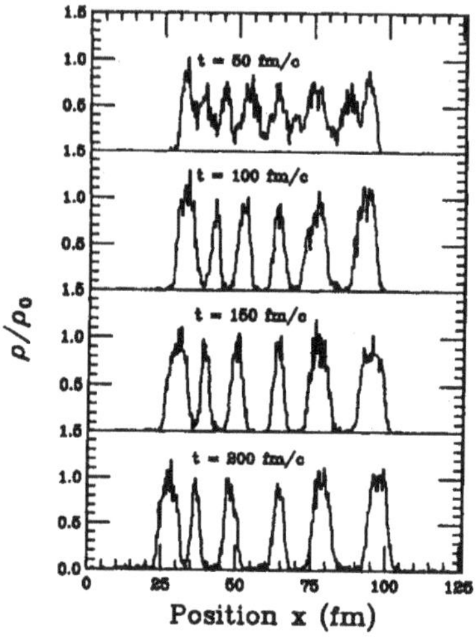

Fig. 11: Clusterization process of an expanding nucleus.

are introducing a fast freeze-out time so that the clusterization pattern due to the instability is kept.

6. Conclusion

In this paper we have justified the use of stochastic mean-field approach and we have clarified the possible role of chaos. We have discussed the dynamical instabilities and their signature. We have introduced the concept of region of dynamical instabilities by looking at the values of the smallest instability time. We have shown that the most unstable modes do not depend strongly on the expansion velocity.

In particular we have shown that the Coulomb interaction is freezing the partition of the system at the time of the clusterization. Therefore the fragmentation pattern can be considered as a fingerprint of the dynamical instabilities. We have also discussed the fact that a typical pattern would be the decay in several light ions (of mass around 20).

However one question remains: are these partitions real signatures of a dynamical process or can they be obtained in a statistical description of the multifragmentation?

Part of this work is a fruit of collaboration with F. Burgio, M. Di Toro and J. Randrup. We warmly thank all of them. Many thanks are due to P. Fox for his important role in the elaboration of this Work.

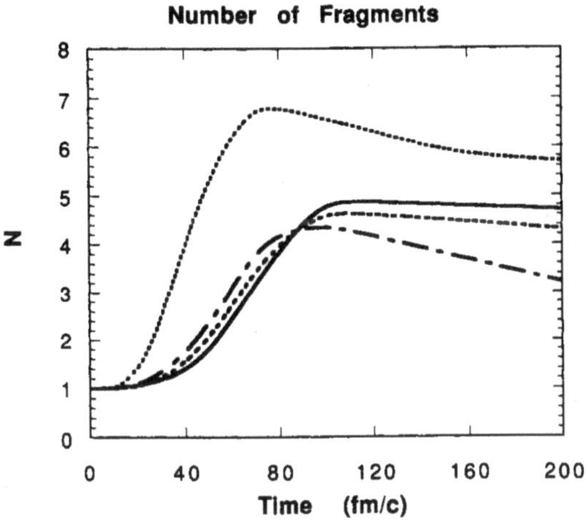

Fig. 12: The mean number of fragment as a function of the time for the various scenarii presented before: the periodic box (solid line), the finite system (dot-dashed line), the charged finite system (dashed line) and the expanding system (dotted line).

References

[1] R. Balian, Y. Alhassid, and H. Reinhart, Phys. Rep. **131**(1986)1.

[2] P. Ring and P. Schuck, *The Nuclear Many Body Problem* (Springer, Berlin, 1981).

[3] J. Randrup and B. Remaud, Nucl. Phys. A514 (1990) 339

[4] M. Colonna, G.F. Burgio, Ph. Chomaz, M. Di Toro, and J. Randrup, Phys. Rev. C47 (1993) 1395.

[5] Ph. Chomaz, G.F. Burgio, and J. Randrup, Phys. Lett. B254 (1991) 340

Fermionic Molecular Dynamics for Colliding and Decaying Nuclei

HANS FELDMEIER and JÜRGEN SCHNACK
Gesellschaft für Schwerionenforschung, D-64220 Darmstadt

ABSTRACT

Fermionic Molecular Dynamics models a system of fermions by means of a trial many–body state composed of an antisymmetrized product of single–particle states which are localized gaussians in coordinate and momentum space. The parameters specifying them are the analogue to the variables in classical molecular dynamics. The time–dependent variational principle yields the equations of motion which are solved for collisions of $^{12}C+^{12}C$ and deexcitations of ^{12}C. The collisions show a great variety of phenomena including explosion, sequential fragmentation and multifragmentation. The deexcitation for nuclei with $E^*/A \approx 5MeV$ is dominated by particle evaporation on time scales of the order of 10^{-20} s or longer.

1. The Fermionic Molecular Dynamics (FMD) Model

The equations of motion in FMD[1] are derived from the time–dependent variational principle of quantum mechanics

$$\delta \int_{t_1}^{t_2} dt \, \langle \, Q(t) \, | \; i\frac{d}{dt} - \underset{\sim}{H} \; | \, Q(t) \, \rangle \; = \; 0 \; , \tag{1}$$

where the trial many–body state $|\, Q(t)\,\rangle = \;|\, q_1(t), \, q_2(t), \cdots \rangle$ is an antisymmetrized product of single–particle states $|\, q_k(t)\,\rangle$ with a gaussian spatial distribution, spin and isospin.

$$\langle\, \vec{x} \mid q_k(t) \rangle = \exp\left\{ -\frac{(\vec{x} - \vec{r}_k(t))^2}{2a_k(t)} + i\vec{p}_k(t)\cdot\vec{x} \right\} \otimes \;|\, \chi_k(t) \, \phi_k(t) \,\rangle \otimes \;|\, \xi_k \,\rangle \tag{2}$$

$q_k(t) = \{ \vec{r}_k(t), \, \vec{p}_k(t), \, a_k(t), \, \chi_k(t), \, \phi_k(t), \, \xi_k \}, \; k = 1, \cdots, A$ denotes a set of time dependent parameters containing the mean position, mean momentum, complex width, spin and isospin of the single–particle state with number k. The equations of motion which result from Eq. (1) for these parameters can be summarized as

$$\sum_{\nu} \mathcal{A}_{\mu\nu} \, \dot{q}_\nu = -\frac{\partial \mathcal{H}}{\partial q_\mu} \quad , \tag{3}$$

where $\mathcal{H} = \langle\, Q(t) \mid \underset{\sim}{H} \mid Q(t) \,\rangle$ is the energy. The hamiltonian $\underset{\sim}{H}$ contains the kinetic energy $\underset{\sim}{T}$ for which the expectation value is given by[2]

$$\langle\, Q(t) \mid \underset{\sim}{T} \mid Q(t) \,\rangle = \sum_{k,l}^{A} \langle\, q_k(t) \mid \underset{\sim}{t} \mid q_l(t) \,\rangle \, O_{lk} \tag{4}$$

and the two-body interaction $\underset{\sim}{V}$ with the expectation value

$$\langle\, Q(t) \,|\, \underset{\sim}{V} \,|\, Q(t) \,\rangle = \frac{1}{2} \sum_{k,l,m,n}^{A} \langle\, q_k(t)\; q_l(t) \,|\, \underset{\sim}{v} \,|\, q_m(t)\; q_n(t) \,\rangle (O_{mk} O_{nl} - O_{nk} O_{ml}) \quad . \tag{5}$$

O_{kl} is the inverse of the overlap matrix

$$\left(O^{-1}\right)_{kl} := \langle\, q_k(t) \,|\, q_l(t) \,\rangle \tag{6}$$

and depends on time via the parameters $q_k(t)$. $\mathcal{A}_{\mu\nu}(Q(t))$ is a nontrivial metric given by

$$\mathcal{A}_{\mu\nu}(Q(t)) = 2\,\mathrm{Im}\,\langle\, \frac{\partial}{\partial q_\mu}\, Q(t) \,|\, \frac{\partial}{\partial q_\nu}\, Q(t) \,\rangle \quad . \tag{7}$$

The greek indices μ, ν identify both, the number of the wave packet and the particular parameter of this packet.

For the results presented here we used the Volkov 2 interaction[3] which has a soft repulsive core and a long range attraction multiplied by a Wigner term and a Majorana exchange. In order to save computer time we used the following product approximation for the expectation value of the two-body interaction[4]

$$\begin{aligned} \langle\, Q \,|\, \underset{\sim}{V} \,|\, Q \,\rangle &\approx \sum_{k<l}^{A} \frac{\langle\, q_k q_l \,|\, \underset{\sim}{v} \,|\, q_k q_l \,\rangle - \langle\, q_k q_l \,|\, \underset{\sim}{v} \,|\, q_l q_k \,\rangle}{\langle\, q_k q_l | q_k q_l \,\rangle} f(c_{kl}) \\ c_{kl} &= \sum_{m\neq k,l}^{A} \left(\frac{\langle q_k | q_m \rangle \langle q_m | q_k \rangle}{\langle q_k | q_k \rangle \langle q_m | q_m \rangle} + \frac{\langle q_l | q_m \rangle \langle q_m | q_l \rangle}{\langle q_l | q_l \rangle \langle q_m | q_m \rangle} \right) \\ f(c_{kl}) &= \exp\{-\alpha\; c_{kl}\} \end{aligned} \tag{8}$$

Instead of the four-fold summation in Eq. (5) we deal in Eq. (8) with a product of a two-fold sum times a function which contains a single summation. The ansatz is motivated by the identity

$$det(\{\langle q_k | q_l \rangle\}) = \exp\left\{Tr(\ln(\{\langle q_k | q_l \rangle\}))\right\} \tag{9}$$

and contains a parameter α which was adjusted to $\alpha = \frac{1}{4}$ for least deviations between the exact expression Eq. (5) and the approximate in Eq. (8). This approximation reduces the computational time by about a factor A (number of wave packets).

In FMD the effects of the Pauli principle enter in the exchange term of the two-body interaction. If there is no interaction there is also no influence of the Pauli principle on the motion of the wave packets, as it should be. On the other hand the skew symmetric matrix $\mathcal{A}_{\mu\nu}(Q)$ is strongly dominated by the antisymmetrization. It couples the velocities $\dot{q}_\nu$ of all particles. Only in the classical limit, when the wave packets have no overlap, $\mathcal{A}_{\mu\nu}(Q)$ becomes independent on Q and assumes the classical canonical form where for example $\vec{p}_k$ and $\vec{r}_k$ are pairs of canonical variables. For overlapping packets $\mathcal{A}_{\mu\nu}(Q)$, which is a purely geometrical object not depending on the hamiltonian, determines locally pairs of canonical variables which are complicated nonlinear functions of all q_μ. We are not using these canonical variables but they can be calculated in principle. The interaction energy $\langle Q \mid \underset{\sim}{V} \mid Q \rangle$ expressed in these canonical variables is the correct Pauli potential. This is at variance with molecular dynamics models which try to incorporate the Pauli exclusion principle by adding a so-called "Pauli potential" to the classical Hamilton function[7,8,9]. There the "Pauli potential" is also active for free particles and the canonical variables are simply the classical $\vec{p}_k$ and $\vec{r}_k$.

2. Shell Model Properties and Ground States in FMD

In FMD the many-body system is represented by an antisymmetrized product of single-particle gaussian wave-packets which are localized in coordinate and momentum space. Therefore it is not immediately obvious that FMD contains shell-model features like the nodal structure of the single-particle orbits. One should however keep in mind that any unitary transformation among the single-particle states which occupy an antisymmetrized product leaves the many-body state invariant. Therefore, after antisymmetrization, any representation which is complete in the occupied area of phase space is as good as any other. To illustrate this we take four one-dimensional real gaussians with the same real width parameter a and displace them by $d = 0.75\sqrt{a}$. The mean momenta are all zero (see l.h.s. of Fig. 1). We use these four gaussians as the occupied states in a Slater determinant and calculate the eigenstates of the one-body density operator which is given by

$$\rho = \sum_{k,l=1}^{4} |q_k\rangle O_{kl} \langle q_l| \quad . \tag{10}$$

The orthonormal eigenstates of this operator are shown in coordinate representation on the r.h.s. of Fig. 1 and compared to the eigenstates of a harmonic oscillator potential (dashed lines). The difference between both can be made arbitrarily small by letting $d/\sqrt{a}$ approach zero. It can be proven analytically that the Slater determinant spanned by N gaussians located very close together is just the harmonic oscillator shell-model state. This illustrates nicely that the FMD trial Slater-determinant contains the harmonic oscillator shell-model states. The reason is, as mentioned already, that a Slater determinant is invariant under unitary transformations among its occupied states.

The ground state $\mid Q_{GS} \rangle$ for a given particle number is in FMD the trial state which gives the lowest expectation value of $\underset{\sim}{H}$, therefore

$$\frac{\partial}{\partial q_\mu} \langle Q_{GS} \mid \underset{\sim}{H} \mid Q_{GS} \rangle = 0 \qquad \forall\, q_\mu \quad . \tag{11}$$

Since all the derivatives vanish it follows from the equation of motion Eq. (3) that in the ground state the time derivatives of all parameters $\dot{q}_\nu$ are zero, which means that the ground state is absolutely stationary. This might seem trivial and it is but

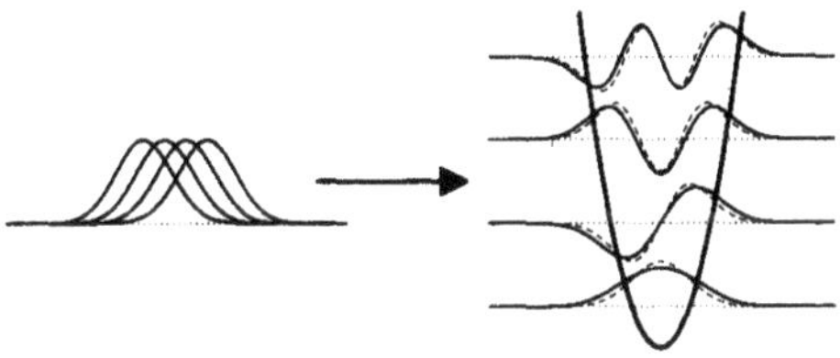

Fig. 1: Comparison between lowest four eigenstates of a harmonic oscillator (dashed lines, right hand side) with orthonormal eigenfunctions of the one–body density operator (full lines, right hand side) which result from occupying four non–orthogonal gaussian wave packets (left hand side) with distance $d = 0.75\sqrt{a}$.

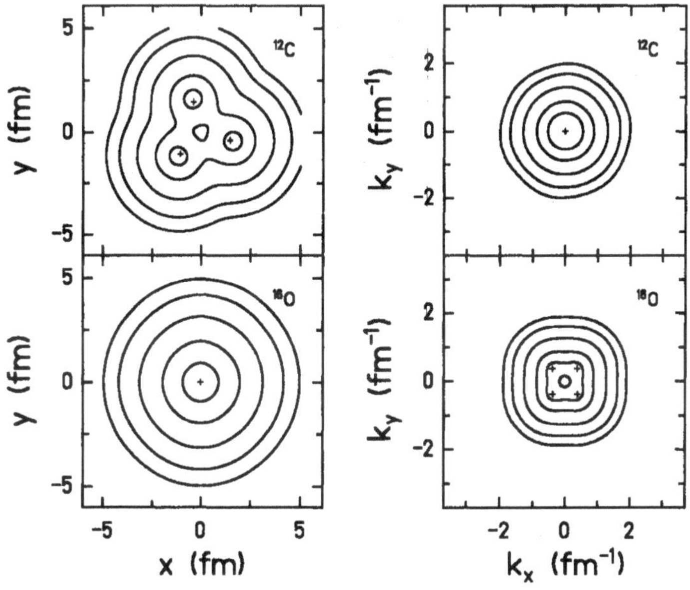

Fig. 2: Densities in coordinate and momentum space of FMD ground states. Crosses indicate centroids of wave packets.

most molecular dynamics models used for heavy–ion collisions do not possess this property.

In Fig. 2 the density plots are displayed for the ground states of ^{12}C and ^{16}O. On the left hand side the coordinate space densities are shown while the right hand part represents the momentum distributions. Since the Volkov interaction, we are using in this work, has besides the direct term only an exchange in the coordiante space α–clusters are strongly favoured. This is obvious in the ^{12}C case where one clearly sees three α–particles. The crosses indicate the mean positions $\vec{r}_k$ and mean momenta $\vec{p}_k$ of the wave packets, respectively. Please note that the crosses are not exactly at the maxima of the density which is due to the antisymmetrization. In momentum space all $\vec{p}_k = 0$ for ^{12}C. This does of course not mean that there is no Fermi motion. The ground state has an extended momentum distribution as can be seen from the r.h.s. of Fig. 2. The Fermi motion in FMD is essentially the zero–point motion in the wave packets and not a random motion of classical particles[5].

In ^{16}O the interaction likes to put all $\vec{r}_k = 0$ and place the momenta of the four α–clusters at the corners of a tetrahedron. The crosses which indicate the $\vec{p}_k$ are not in the x–y plane that is why the tetrahedron appears in the projection as a square. As discussed above this configuration is equivalent to a harmonic oscillator shell model.

One has to regard the FMD ground states as intrinsic states like a deformed Hartree Fock solution. The true ground state is a superposition of the deformed state averaged over all angles. However, in the simulation of the collision we pick one orientation as a realization of the many possibilities present in the many–body ground state.

3. Collisions of ^{12}C + ^{12}C

In Fig. 3 we display the time evolution of a central collision of ^{12}C + ^{12}C at E/A= 28.7MeV in coordinate (upper line) and momentum space (lower line). In this and all other density contour–plots the densities are integrated over the z–direction so that fragments or particles which are escaping out of the reaction plane can also be seen. The initial orientations of the two ^{12}C are chosen at random. If one starts with the ground state the momentum distribution is time independent up to the moment when the nuclei come close and attract each other. Here we excited each initial ^{12}C by a total energy of 1 MeV misplacing slightly the ground state positions of the wave packets up to 0.1 fm. The deviation from the exact ground state is so small that it cannot be seen in the contour plots. The advantage of such an initial state is that the strong α–cluster symmetry which is an artifact of our simple Volkov interaction is initially broken. Around time 70 fm/c the system reaches the highest compression which is followed by an expansion leading to densities below the saturation density. The corresponding momentum space density has shrunk which means less Fermi motion than in the ground state. As can be seen from the crosses which indicate the centroids of the wave packets the initial collective motion has been stopped around 150 fm/c and all wave packets are now close to $\vec{p} = 0$.

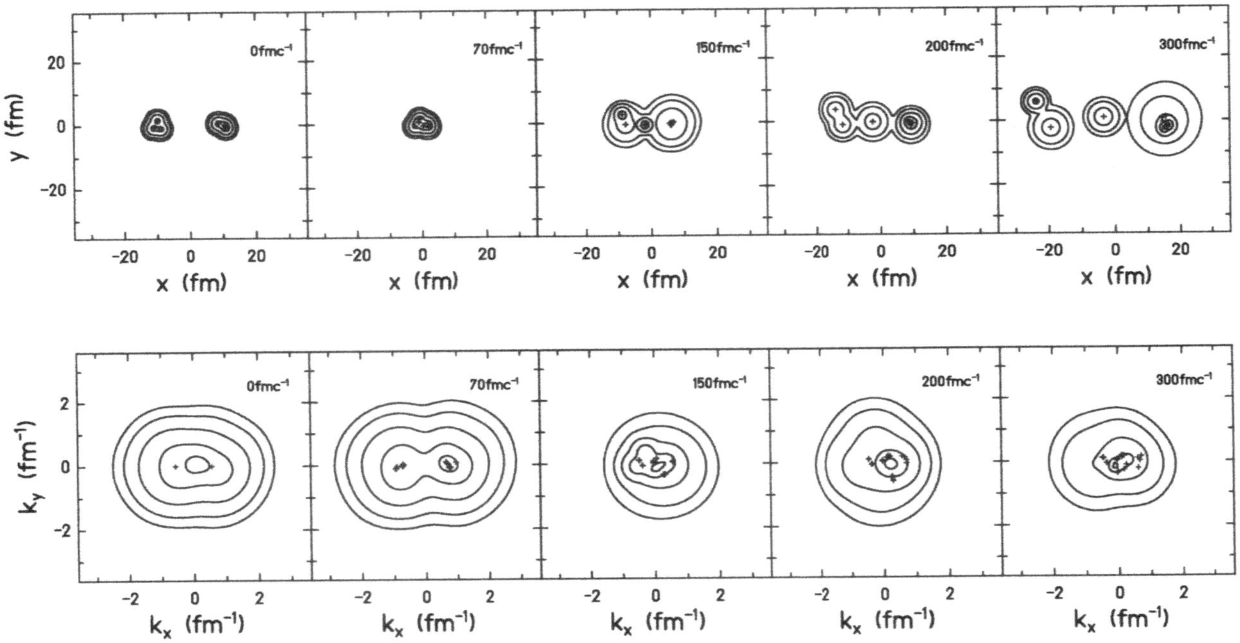

Fig. 3: Density in coordinate (upper line) and momentum space (lower), integrated over z–direction for $^{12}C + ^{12}C$ central collision at E/A = 28.7 MeV. The collision leads to multifragmentation of one ^{12}C.

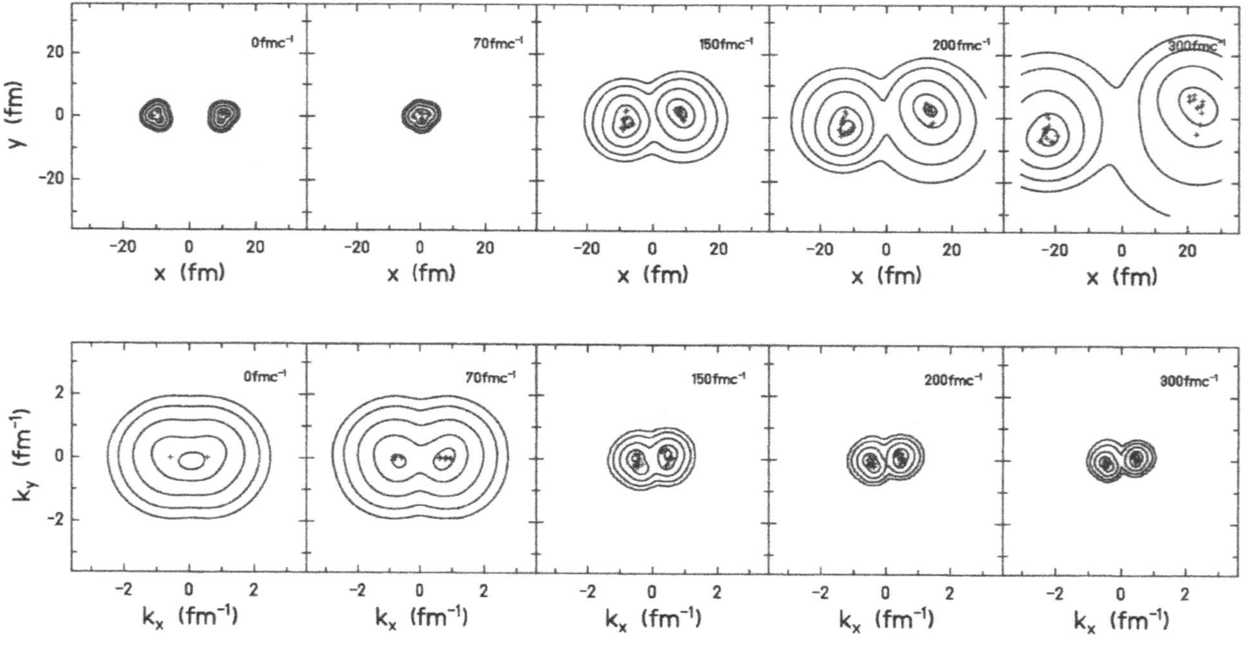

Fig. 4: Same as Fig. 3 but for different orientations of the initial intrinsic states of the ^{12}C nuclei. The collision leads to a complete desintegration.

After 150 fm/c the system starts forming clusters out of the stretched configuration which have again normal densities and hence more Fermi motion. At 300 fm/c we encounter a ^{8}Be on the left which will decay after some time, an excited α–particle in the centre and a highly excited ^{12}C on the right hand side. This ^{12}C will most probably emit particles. In order to verify this we should follow the time evolution further. In section 4 we investigate the decay of excited ^{12}C nuclei for times up to 2000 fm/c.

Figure 4 shows the same reaction but for another choice of the initial orientations. Up to 70 fm/c it looks very similar to the collision shown in Fig. 3. But at time 150 fm/c it becomes clear that the expansion is not reversed. The system keeps on expanding in coordinate space and shrinking in momentum space. It seems that in this case the interaction between the packets has destroyed all coherence which is necessary to form clusters. In the end all particles are unbound and a complete explosion has taken place. If one pounders on the fact that the total energy of the system is 30 MeV it becomes clear, why the momentum distribution has to be very narrow after the explosion. Since the total energy is conserved, these 30 MeV represent the total kinetic energy of all 24 particles so that on the average only 1.25 MeV is available for each particle. The initial Fermi motion is almost completely used up to pay for the loss of binding energy.

The completely different outcome of these two examples, although both collisions are central and at the very same energy, shows that FMD is able to produce large fluctuations qualitatively in accord with the observations. In order to see how much the final outcome depends on small deviations in the initial state we study the following two cases.

Figure 5 displays the coordinate space density of two collisions where for both ^{12}C the α–clusters are in the reaction plane. The difference is that in the lower case the left hand ^{12}C is rotated by $\pi/6$ compared to the upper case. The final results at 300 fm/c are quite different. The upper case ends up with two excited ^{12}C while the lower case results in multifragmentation of one ^{12}C and an excited second ^{12}C.

In the second example the two initial states differ by their excitation energy. One is the ground state and the other one contains two ^{12}C nuclei where each of them is excited by 1 MeV. The overlap between the ground state and excited one is $|\langle\, ^{12}\mathrm{C}\,|\,^{12}\mathrm{C}^*\,\rangle| = 0.87$. The upper part of Fig. 6 depicts the many–body overlap $|\langle\, Q_1(t)\,|\,Q_2(t)\,\rangle|^2$ of the two states as a function of time. Initially it is $0.75 = (0.87)^2$ which is quite large for a 24 particle state. When the nuclei touch and the interaction sets in around 75 fm/c the overlap drops rapidly to zero and remains zero for all later times. The densities still look rather similar at 150 fm/c. But the two channels are already completely orthogonal. The reason is that the many-body overlap for 24 particles is essentially the product of all 24 single particle overlaps. Therefore it is enough if one or a few single particle states deviate from their partners in the other many–body state to make the two many–body states orthogonal. For all practical purposes the final states are thus orthogonal and one does not have to take interference effects into account. The final result is just an incoherent sum over all final states which were evolved according to the FMD equations from the initial ensemble

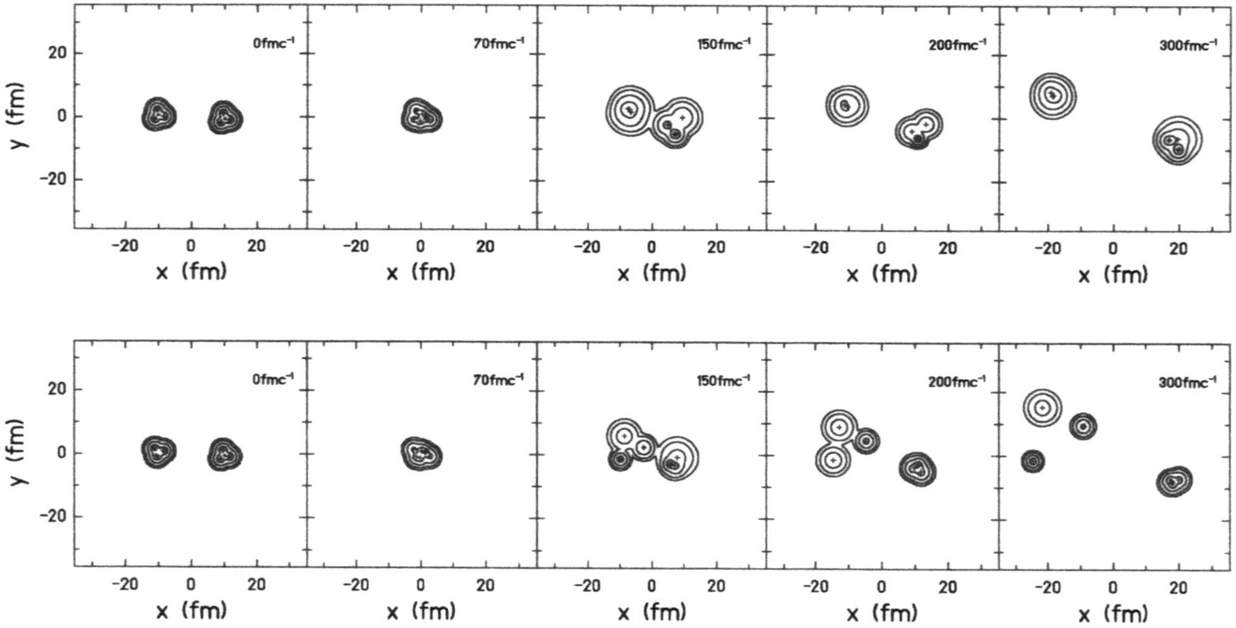

Fig. 5: Densities in coordinate space, integrated over z-direction, for two collisions which differ only by rotating the r.h.s. nucleus by 30^0 in the reaction plane.

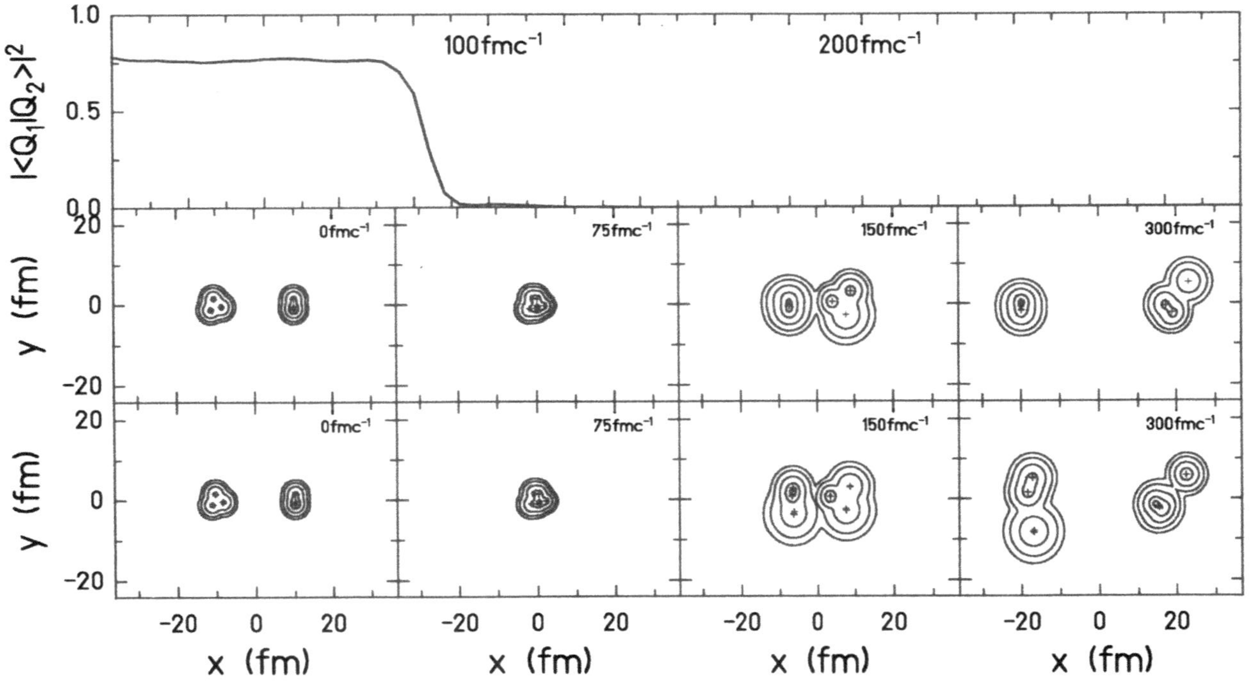

Fig. 6: Upper part: Many–body overlap as function of time for two initial states $|\,Q_1\,\rangle$, $|\,Q_2\,\rangle$ for central $^{12}C + ^{12}C$ collisions at E/A = 28.7 MeV. Lower part: Density in coordinate space, integrated over z–direction, for $|\,Q_1\,\rangle$ = ground states and $|\,Q_2\,\rangle$ = excited states with $\langle\, Q_2\,|\, \underset{\sim}{H}\, |\,Q_2\,\rangle - \langle\, Q_1\,|\, \underset{\sim}{H}\, |\,Q_1\,\rangle =$ 2 MeV excitation.

composed of the coherent superposition of all orientations of the intrinsic states.

4. Deexcitation of excited ^{12}C

Within the FMD model we investigate the time evolution of excited ^{12}C nuclei. These excited states are prepared by displacing the mean positions $\vec{r}_k$ of each wave packet and squeezing the whole nucleus, i.e. shrinking the width and the distance between the centre of mass and the mean positions. This procedure keeps the expectation values of the total momentum and of the total angular momentum at zero. The way of preparing an excited state does not play an important role for the deexcitation process if the nucleus equilibrates before particle evaporation. At excitation energies of 5 MeV per particle in FMD the many-body state evolves long enough in order to loose all memory on the initial conditions before the first evaporation occurs. In about 80 runs evaporation of single nucleons dominates the deexcitations of ^{12}C. A typical deexcitation is displayed in Fig. 7 where the emission of the first nucleon takes place around 200 fm/c. After the excitation at 0 fm/c the nucleus first expands (upper line of contour plots) and equilibrates at a lower density. After this short initial phase the wave packets perform a random movement. This cannot be seen in Fig. 7 as we can show only a few snap shots. But in a movie, which is available upon request as a PC-program running on DOS, one clearly sees that each packet bounces around several times before one wave packet by chance has received enough energy from the others to leave the nucleus. In the case shown here a packet identified by its centroid (cross) and the lowest contour line is ejected towards north-north-east at 200 fm/c. Its spatial width is rather large and the momentum width has shrunk so that the zero point energy is only of the order of 1–2 MeV. This situation is very simular to the explosion seen in Fig. 4 except that here only one wave packet becomes unbound and travels away. The upper part of the Fig. 7 demonstrates that for times later than 200 fm/c the real part of the complex width parameter a (thick solid line) belonging to the emitted wave packet is stationary in time while the imaginary part (thick dashed line) grows linearly in time with a rate $1/m$ (m = nucleon mass). Since this is the exact solution of the Schrödinger equation for a freely moving gaussian wave-packet we regard this particle as free. In the frame at 200 fm/c one sees a second particle which is spulled along with the evaporated one but later it falls back into the nucleus. Careful inspection between 100 and 200 fm/c shows that the real part of a is not yet completely stationary which is due to the final state interacting with the dragged particle.

The lower part of Fig. 7 demonstrates that the width parameters a_k must be dynamical. Here we show the time evolution of the same initial state as above but we fix artificially all a_k at their respective values at $t = 0$ (see upper part, thin full and thin dashed lines). The physical mechanism for particle evaporation in FMD is a random concentration of energy on one packet due to the interaction with all other particle which is then followed by an expansion in coordinate space and a shrinking in momentum space such that the loss of binding is paid with the loss of Fermi or zero point motion. Finally the particles leave with a mean kinetic energy of about

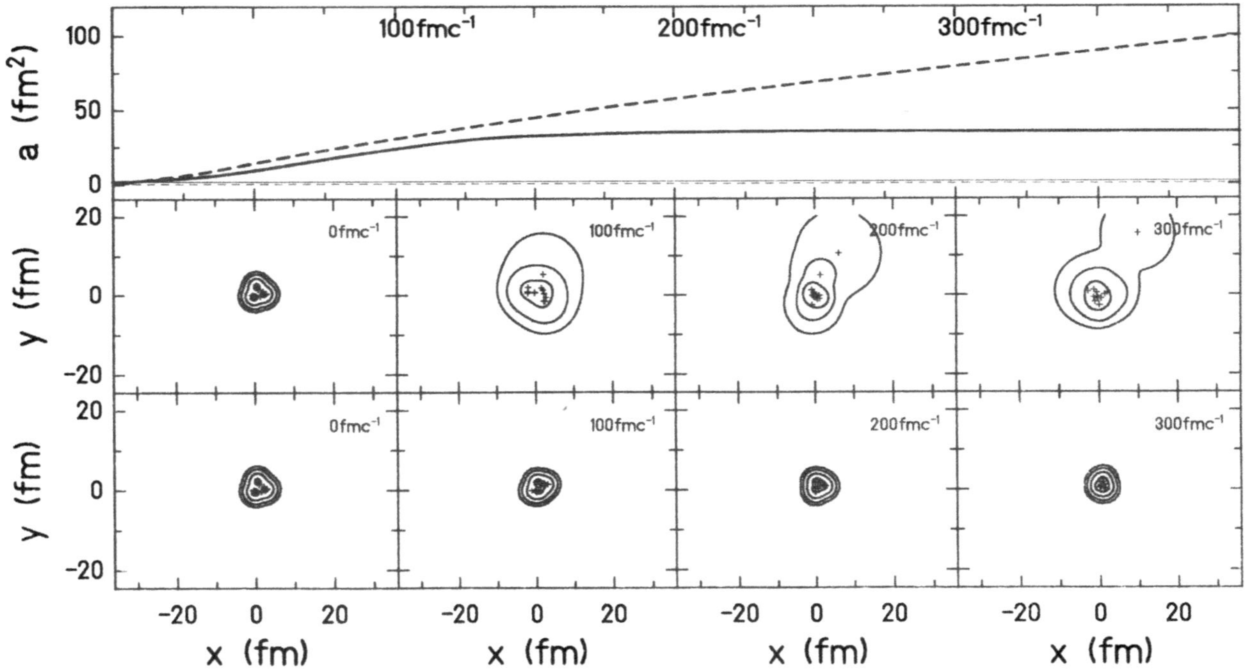

Fig. 7: Influence of the dynamical width parameters on the evaporation process. Upper part: real (full line) and imaginary (dashed line) part of the width a belonging to the evaporated wave packet. Middle part: density in coordinate space, integrated over z-direction, for time evolution with dynamical widths. Lower part: time evolution of density with time–independent width parameters which shows no evaporation.

2 MeV which together with the separation energy of about 8 MeV cools the residual nucleus by about 10 MeV, quite in accordance with experimental findings.

The simulations were performed up to 2000 fm/c. Within this time the rate of evaporated particles does not drop considerably. Longer simulations are necessary to see how the FMD nuclei behave close to the particle threshold. Furthermore, we shall include the Coulomb interaction in forthcoming studies to see the difference between neutrons and protons.

Summary

Fermionic Molecular Dynamics sets an example how to treat the Pauli principle in "classical" trajectory calculations. The model is quantal in the sense that it cannot violate the uncertainty principle as it works with wave packets. The model by construction must also obey the Pauli principle because the many-body trial state is always antisymmetric with respect to particle exchange. These two fundamental properties guarantee that FMD has the correct treatment of Fermi motion and shell model properties. On the other hand, in the limit of narrow non-overlapping wave packets, the FMD equations reduce to Newton's equations of motion for classical particles. For that and other reasons FMD is not only an approximation to time-dependent Hartree Fock.

Although FMD has deterministic equations of motion it exhibits large fluctuations in the final channels when averaging over the orientations of the intrinsic states in the initial ground state. One sees multifragmentation, sequential fragmentation, explosion and sequential decay, all in the same reaction at the same beam energy and at the same central impact parameter.

The width degree of freedom which allows each wave packet to determine its shape in phase space — narrow in coordinate, wide in momentum space, or vice versa — is essential for the evaporation of particles and the formation of fragments. A fixed width implies that each particle (better wave packet) carries always about 10 MeV zero-point kinetic energy along, which cannot be given away to the other particles[6].

First results on long-time simulations of excited nuclei show that FMD is capable of describing particle evaporation with qualitatively correct deexcitation energies and evaporation times. The present results are so encouraging that we shall include Coulomb and improved two-body interactions in future investigations.

References

1. H. Feldmeier, *Fermionic Molecular Dynamics*, Nucl.Phys. **A515** (1990) 147
   H. Feldmeier, K. Bieler, Proc. of the 1. Europ. Biennial Workshop on Nucl. Phys., Megeve/France, World Scientific (1991) p.125

2. K. Bieler, *Fermionische Molekulardynamik*, diploma thesis, TH Darmstadt (1991)

3. A.B. Volkov, Nucl.Phys. **74** (1965) 33

4. J. Schnack, *Untersuchungen zur Fermionischen Molekulardynamik*, diploma thesis, GSI report **GSI-93-21** (1993)

5. J. Aichelin, Phys. Rep. **202** (1991) 233

6. A. Ono, H. Horiuchi, T.Maruyama, A. Ohnishi, Prog. of Theor. Phys. **87** (1992) 1185

7. D.H. Boal, J.N. Glosli, Phys. Rev. **C38** (1988) 1870, 2621

8. C. Dorso, J. Randrup, Phys. Lett. **B215** (1988) 611

9. L. Wilets, E.M. Henley, M. Kraft, A.D. MacKellar, Nucl. Phys. **A282** (1977) 341

Correlations and Fluctuations in Nuclear Multifragmentation

J. Pochodzalla,(1) J.C. Adloff,(3) M. Begemann-Blaich,(1) P. Bouissou,(4) T. Hofmann,(1) J. Hubele,(1) G. Imme,(5) I. Iori,(6) J. Kempter,(1) P. Kreutz,(2) G.J. Kunde,(1) W.D. Kunze,(1) S. Leray,(4) V. Lindenstruth,(1) Z. Liu,(1) U. Lynen,(1) M. Mang,(1) R.J. Meijer,(1) U. Milkau,(1) A. Moroni,(6) W.F.J. Müller,(1) C. Ngô,(4) B. Ocker,(1) C.A. Ogilvie,(1) G. Raciti,(5) F. Rosenberger,(1) G. Rudolf,(3) H. Sann,(1) R. Scardoni,(6) A. Schüttauf,(2) W. Seidel,(7) V. Serfling,(1) L. Stuttge,(3) W. Trautmann,(1) A. Tucholski,(1) A. Wörner,(1) and B. Zwieglinski(8)

(1) *Gesellschaft für Schwerionenforschung, D-64220 Darmstadt, Germany*
(2) *Institut für Kernphysik, Universität Frankfurt, D-60486 Frankfurt, Germany*
(3) *Centre de Recherches Nucléaires, F-67037 Strasbourg, France*
(4) *Laboratoire National Saturne, CEN Saclay, F-91191 Gif-sur-Yvette, France*
(5) *Dipartimento di Fisica dell' Universitá and I.N.F.N., I-95129 Catania, Italy*
(6) *Instituto di Scienze Fisiche, Universita Milano and I.N.F.N., I-20133 Milano, Italy*
(7) *Forschungszentrum Rossendorf, Postfach 510119, D-01314 Dresden, Germany*
(8) *Soltan Institute for Nuclear Studies, 00-681 Warzawa, Poland*

ABSTRACT

Latest results from the first ALADIN experiment studying peripheral collisions in Au + C, Al, Cu and Pb collisions at E/A = 600 MeV are reviewed. The observed charge distribution favour statistical decay models which include an expansion of the excited projectile spectator. Whereas several statistical multifragmentation models are able to reproduce the element distributions, with high accuracy the element distributions, the predicted isotopic yield distributions differ significantly. Kinematic correlations between three heavy fragments are compared to three-body trajectory calculations. Although these correlations do not allow to unequivocally determine the breakup configuration, they significantly limit the possible parameter space of the breakup scenario. Within the assumed scenarios the data suggest a fast breakup. Source parameters derived from the statistical decay models are confronted with the source parameters needed to describe kinematic observables.

1. The Quest of Nuclear Matter Properties

The determination of fundamental properties of nuclear matter far from the ground state is one of the principal motivations for studying energetic collisions between two complex nuclei. At densities slightly above the normal nuclear density transverse flow phenomena are regarded as fingerprints of the equation of state [1, 2]. At higher densities of typically 2-3 times normal density particle production - in particular particle production via multistep processes - is believed to be a sensitive probe of nuclear matter [3]. At subsaturation densities on the other hand, the fragmentation pattern of the nuclear system is expected to be influenced by the bulk properties of nuclear matter.

The special interest in this low density regime , however, arouse from the conjecture that the nuclear many-body system might undergo a phase-transition during the disassembly. This idea [4, 5] was motivated by the Van der Waals - like behaviour of the nucleon - nucleon interaction. Of course, such a phenomenon is not only fascinating by itself but it is also of more general interest: despite many quantitative differences - also in ultra relativistic heavy ion studies we face the problem to identify a phase transition in a finite, transient system.

2. The Rise and Fall of Multifragmentation

For a nucleus at low excitation energies, $E_x/A \leq 1$ MeV, the dominant decay channels are the evaporation of light particles and fission. If it is possible to excite a nucleus far beyond its total binding energy, a complete vaporization into individual nucleons is anticipated. Between these two limits, we expect a regime where the production of several intermediate mass fragments (IMFs) is the dominant exit channel. It is obvious that events with multiple fragment production may provide the expected glimpse of a nuclear liquid-gas phase transition since they are the favorite candidates for a true multifragmentation.

2.1. Departure from the Evaporation Regime

In the first experiment at the ALADIN we, therefore, investigated which collisions lead to the emission of several IMFs. We have measured the reactions of Au on targets of C, Al, Cu and Pb at 600 MeV/nucleon and have studied the impact parameter dependence of multi-fragment emission from the excited projectile spectator. In order to characterize the events a quantity called Z_{bound}, measuring the number of charges contained in the projectile fragments of $Z \geq 2$, has been introduced. Z_{bound} gives the atomic number of the excited projectile spectator reduced by the number of hydrogen nuclei emitted from it. It thus depends on the primary abrasion *and* the secondary deexcitation stages and reflects the violence of the collision in a two-fold way. The most striking result of this first analysis was that by ordering the collisions according to the summed charge Z_{bound} a remarkable target independence of the charge

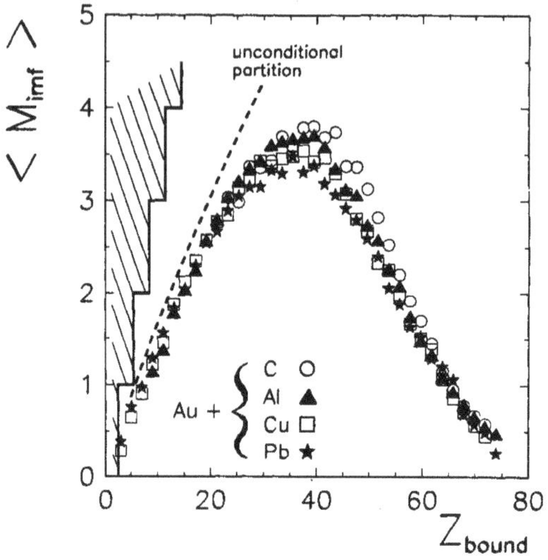

Figure 1: $\langle M_{IMF} \rangle$ - Z_{bound} correlation observed for Au + C, Al Cu and Pb reactions at E/A = 600 MeV [6].

distributions was found [7].

From a nearly complete measurement of the resulting projectile fragments we have observed the different modes of nuclear disassembly, ranging from spallation and evaporation to multi-fragment emission and finally to complete vaporization of the projectile spectator (Fig. 1). It was found for the C target that the largest number of IMFs is observed for central collisions. For heavier targets the maximum of the IMF multiplicity occurs for more peripheral reactions. On all three targets this maximum of 3 to 4 IMFs is observed at approximately the same size of the spectator of 150 nucleons.

2.2. The Fall towards Vaporization

For a more quantitative interpretation of the data displayed Fig. 1, one has to keep in mind that the mass of the pre-fragment decreases whereas the excitation energy per nucleon increases when going from peripheral to the most violent reactions. For illustration, the solid line in Figure 2 shows the relation between the size and the specific excitation energy of the projectile spectator as predicted by microscopic BUU simulations.

To disentangle size and excitation energy effects and explore the evolution from multifragmentation towards a complete vaporization, central collisions of symmetric

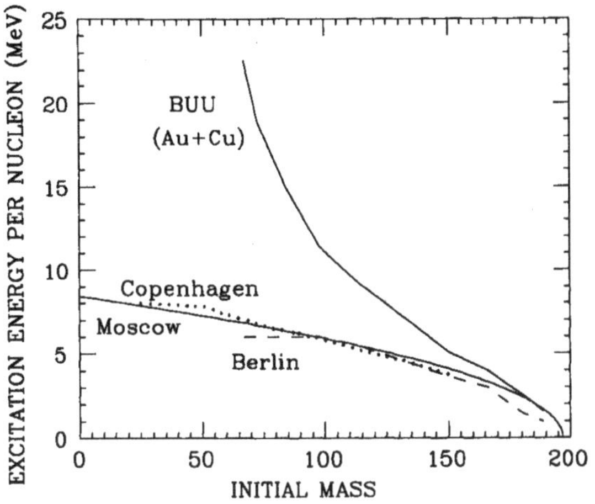

Figure 2: Excitation energy per nucleon of the decaying projectile spectator as a function of its mass. The upper solid line represents the deposited energies predicted by BUU simulations of Au+Cu collisions at E/A = 600 MeV. The solid, dashed and dotted lines are the theoretical inputs required by the MOSCOW [20], BERLIN [8] and COPENHAGEN [9] multifragmentation codes in order to reproduce the observed charge distributions.

target-projectile combinations are ideally suited. In such collisions the (not necessarily thermal) excitation energy can be controlled by the incident beam energy thereby keeping the size of the excited system rather constant.

In a joined experiment of the ALADIN/LAND and the MINIBALL/WALL collaborations we, therefore, investigated the system Au+Au at E/A = 100, 250 and 400 MeV. Fragments with charge $Z \geq 2$ and rapidity close to the projectile rapidity were detected in the ALADIN forward spectrometer. Fragments originating from the target spectator and the participant region were detected in the MSU MINIBALL/WALL array which was complemented by the Si-CsI hodoscope of the ALADIN facility.

Figure 3 shows the observed mean multiplicity of intermediate mass fragments with $3 \leq Z \leq 30$ as a function of the total charged particle multiplicity N_c detected in the MINIBALL/-WALL for the three beam energies [10]. Provided that N_c is a monotonous function of the impact parameter, the data show that at E/A = 100 MeV the most central collisions are associated with the highest average multiplicity of IMFs. With increasing energy a larger number of free nucleons and/or light particles and a corresponding smaller fragment multiplicity is observed in central collisions, thus signaling the onset of vaporization of a system that is too highly excited to produce significant numbers of fragments. Confirming our qualitative interpretation

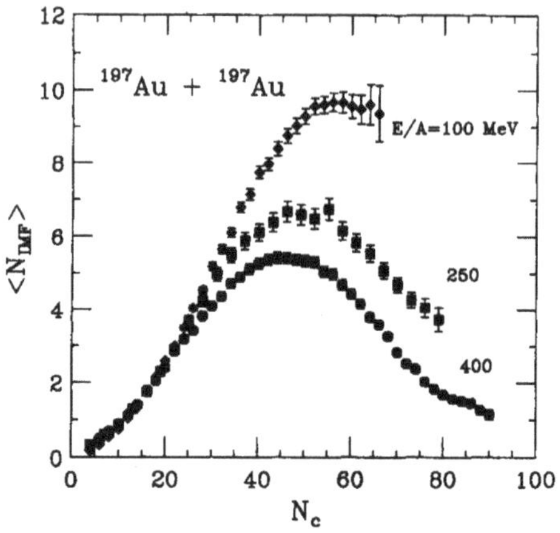

Figure 3: Mean multiplicity of intermediate mass fragments $\langle N_{IMF} \rangle$ as a function of the charged particle multiplicity N_c observed in the MINIBALL/-WALL for Au+Au collisions at E/A = 100, 250 and 400 MeV [10].

of the data presented in figure 1, the maximal IMF production is shifted towards larger impact parameters.

Despite the fact that the charge distributions and the maximum observed multiplicity of IMF's normalized to the total size of the decaying system are rather similar in central ($\approx$ 10/400) and peripheral ($\approx$ 3.5/150) fragmentation processes, it is important to realize a potential difference between these two types of reactions. The large compression reached in central collisions between two heavy nuclei and the resulting large radial collective flow [39, 12] may lead to fast explosion of the system. Because of the short time scale only short range fluctuations will therefore determine the fragmentation pattern. On the other hand, the projectile spectator produced in central collisions are presumably not significantly compressed (see also below) and the expansion of the system is predominantly driven by the thermal pressure. As a consequence a relatively slow expansion might take place and the system might have enough time [13] to develop instabilities in a low density phase or even to undergo a liquid-gas phase transition.

3. Charge and Mass Partition in Multifragmentation

3.1. Charge Distributions

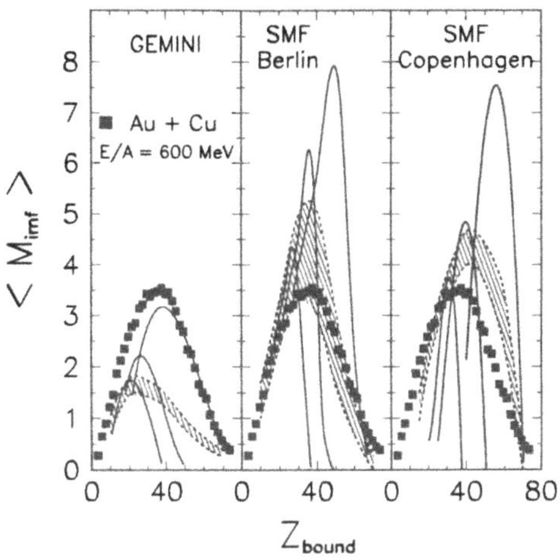

Figure 4: Comparison of the $\langle M_{IMF} \rangle$ - Z_{bound} correlation observed for Au + Cu reactions (squares) with predictions of the sequential decay code GEMINI (left part) and of the SMF models of Gross and co-workers (center part) and of Bondorf and co-workers (right part). The lines represent excitation functions for a given initial systems (A_1,Z_1) = (100,40), (131,54) and (190,75). The shaded bands are predictions based on initial conditions provided by BUU simulations.

As mentioned already above, a given value of Z_{bound} corresponds to different impact parameters [14] if different target nuclei are involved. Therefore, the observed universal behavior seen in Fig. 1 provided a necessary – although not sufficient – condition for a chemical equilibrium being established during the fragmentation process. It seemed, therefore, worthwhile to confront the observed fragment distributions with predictions of statistical models. In particular, we performed calculations with the sequential evaporation code GEMINI [15], which is based on a transition state theory [16], and the COPENHAGEN [17, 18] and BERLIN [19] statistical multifragmentation (SMF) models which both describe the simultaneous decay of an expanded nuclear system.

The shaded areas in Fig. 4 mark the results of calculations using the primary projectile spectators predicted by the BUU simulations (see solid line in Fig. 2). Qualitatively, all calculations exhibit a rise and fall of the IMF multiplicity. Whereas the BERLIN [19] and COPENHAGEN [18] multifragmentation codes predict rather similar relations between Z_{bound} and $\langle M_{IMF} \rangle$ which are in reasonable agreement with the data, the IMF multiplicities deduced from the sequential decay program GEMINI are lower by a factor of 2 – 3. Clearly, the observed charge distributions favor statistical decay calculations which assume an expanded decaying system [6].

In the mean time, these results have triggered a series of theoretical activities, where the complete ALADIN data (i.e. not only $\langle M_{IMF}\rangle$ but also higher moments of the charge distributions) where used to fix the input parameters of SMF models [20, 8, 9]. Despite several minor differences in the calculations, all three groups agree that a significantly lower excitation energy of the decaying projectile spectator than expected from the microscopic BUU simulations is required in order to reproduce the measured charge distributions (see Fig. 2).

Qualitatively, such a decrease of the temperature by a factor $T_f/T_{BUU} \approx 0.8$ is consistent with the assumed expansion of the system prior to the breakup. During this expansion the entropy is predicted [21, 5] to increase by about $\Delta S/A \approx 0.3$-0.6. Passing through the region of the mechanical instability, the entropy of the system might increase by an additional amount of $\Delta S/A \approx 0.5$-1.0 [22, 23]. Using the relation between temperature, density, and entropy for a Maxwell-Boltzmann gas [24], $\rho_f/\rho_i = (T_f/T_i)^{3/2} \cdot e^{-\Delta S/A}$, and with $\Delta S/A = 1.2$ and $T_f/T_{BUU} = 0.8$ one expects a break up density of about 1/5 of the initial density. Such a value is comparable to the breakup densities assumed in the statistical multifragmentation calculations.

3.2. Isotope Distribution

The isotopic composition of complex fragments reflects the neutron-to-proton ratio of the decaying system [25]. In addition, the isotope distributions depend strongly on the sequential decay processes after freeze-out. Therefore, isotope distributions represent a stringent quantitative test ground for statistical decay models.

The masses and momenta of the fragments are determined by measuring their trajectories with the TP-MUSIC detector located behind the ALADIN magnet. Combining the TOF and the TP-MUSIC information, the mass of fragments with $Z \geq 8$ was evaluated with a relative resolution of $\Delta A/A \approx 3\%$ (FWHM). Figure 5 shows the measured mean neutron-to-proton ratio of the fragments as a function of the fragment charge measured for the reaction Au+Al at E/A = 600 MeV. Generally the fragments are on the more neutron rich side of the valley of stability. Note, however, that for fragments with charges greater than about 65, this ratio may be distorted by our experimental trigger condition.

Predictions of the MOSCOW, BERLIN and COPENHAGEN statistical multifragmentation models are given by the solid, dashed and dotted lines, respectively. Despite the fact that these three models give similar charge distributions [20, 8, 9] they show clear differences in the isotopic distributions. Whereas the MOSCOW code agrees

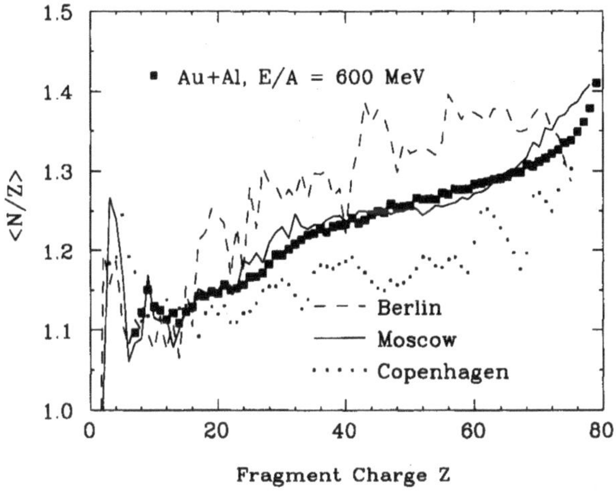

Figure 5: Mean N/Z ratio as a function of the fragment charge [26]. The solid, dashed, dotted lines represent the predictions of the MOSCOW, BERLIN and COPENHAGEN SMF codes, respectively. The large fluctuations in the latter two calculations are due to the discrete steps for the input mass and excitation energy corresponding impact parameter steps of 0.5 fm.

rather well with the experimental observations, the COPENHAGEN and the BERLIN codes underestimate resp. overestimate the average N/Z ratios significantly. It was checked, that these deviations cannot be compensated by any reasonable variation of the N/Z ratio of the decaying system nor by an increase or fluctuation of the excitation energy. Most likely, the different theoretical N/Z ratios reflect the different treatment of the final sequential decay stage after freeze out [27]. In turn, the observed discrepancies with the data call for a more detailed treatment of this secondary stage.

4. Dynamical Observables

4.1. Breakup Configurations from Proximity Correlations

In the previous section indirect evidence for a simultaneous breakup of excited nuclear systems has been accumulated by comparing measured fragment yields with predictions based on statistical decay models [28, 29, 6]. More characteristic fingerprints of the disintegration dynamics may show up in kinematic correlations which are governed by the long-range Coulomb force [31-40]. In the following we will consider only those events where exactly three fragments were tracked in the TP-MUSIC. For

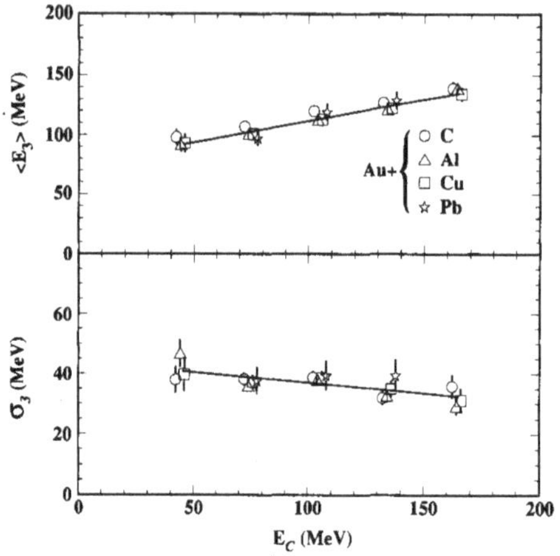

Figure 6: Mean center-of-mass total kinetic energy $\langle E_3 \rangle$ (top) and the standard deviation σ_3 (bottom) as a function of the nominal Coulomb energy for the three largest projectile fragments emitted in Au + C, Al, Cu and Pb collisions at E/A = 600 MeV. The points were obtained by sorting the data in 30 MeV wide bins.

this event class Z_{bound} varies between 45 and 70 and peaks at $Z_{bound} \approx 55$. Typically 80 % of Z_{bound} is contained in the charges of the three heaviest fragments.

The momenta of the fragments are expected to reflect the breakup dynamics. In the present analysis the momenta $\vec{p}_{c.m.,i}$ of the three fragments in their center-of-momentum (c.m.) frame are used to determine the summed c.m. - kinetic energy

$$E_3 = \sum_{i=1}^{3} p^2_{c.m.,i}/(2 \cdot m_0 \cdot A_i) \tag{1}$$

where m_0 is the atomic mass unit. In Figure 6 the average value $\langle E_3 \rangle$ (top part) and the standard deviation σ_3 of E_3 (bottom part) are displayed as a function of the sum of a nominal Coulomb repulsion

$$E_c = e^2 \cdot \sum_{i<j} Z_i Z_j / (r_0(A_i^{1/3} + A_j^{1/3})). \tag{2}$$

For the radius parameter r_0 a value of 1.4 fm was used in all calculations. Within the statistical uncertainties no target dependence is apparent and the data depend linearly on E_c.

In order to test the sensitivity of our observations to the breakup geometry we performed classical three-body trajectory calculations. We examined two schematic disintegration configurations which mimic the main categories of breakup scenarios discussed in the literature [28, 29, 6]:

(i) The first class simulates two sequential, binary splittings of the initial nucleus. In both steps, the surfaces of the two produced fragments were initially separated by a fixed distance 2·D. The time delay t between the two decays was chosen from an exponential distribution $P(t) \propto exp(-ln2 \cdot t/\tau)$. The initial relative energy of the two fragments was selected from a distribution

$$P(E) \propto E^{\alpha} \cdot exp(-E/T) \tag{3}$$

where for the exponent α values of 0.5 or 1 were used. The direction of the corresponding initial relative velocity was isotropically distributed.

(ii) With the second type of simulations a simultaneous emission out of a given volume is modelled. The centers of the three non-overlapping fragments are distributed randomly within a sphere of radius R. To each fragment an isotropically distributed initial velocity was assigned. Constrained by total momentum conservation, these velocities were selected according to the energy distribution described by Eq. 3. In addition to this random motion, an initial radial flow velocity $\vec{v}_{f,i} = (2\epsilon_f/m_0)^{1/2} \cdot \vec{d_i}/R$ was taken into account. Here, $\vec{d_i}$ is the position of the center of the fragment i with respect to the center-of-momentum, and the parameter ϵ_f denotes the flow energy per nucleon for fragments located at d_i = R.

The charges and masses of the fragments were obtained by a Monte Carlo sampling of the measured events, thus reducing the uncertainties associated with the fragment distribution. For each event the temperature parameter T was chosen according to the experimental value of Z_{bound} from the relation $T = f_T \cdot (2 \cdot (79 - Z_{bound}))^{1/2}$, where f_T is a free parameter. For $f_T = 1$ this relation describes within the relevant range of Z_{bound} the impact parameter dependence of the temperature of the initial projectile spectators as predicted by microscopic transport calculations [6] reasonably well. The paths of the fragments were calculated under the influence of their mutual Coulomb field and two-fragment proximity forces according to Ref. [40]. Since for the further analysis those trajectories were rejected for which the fragments did overlap during the propagation, the influence of the proximity force turns out to be rather small (see also below). In order to account for the recoil from light particles emitted sequentially from the initial fragments, the measured charges and masses were increased prior to the trajectory calculations in accordance with the initial temperature. After the fragment-fragment interaction has ceased, the sequential emission of light particles leading to the observed masses and charges was assumed to take place.

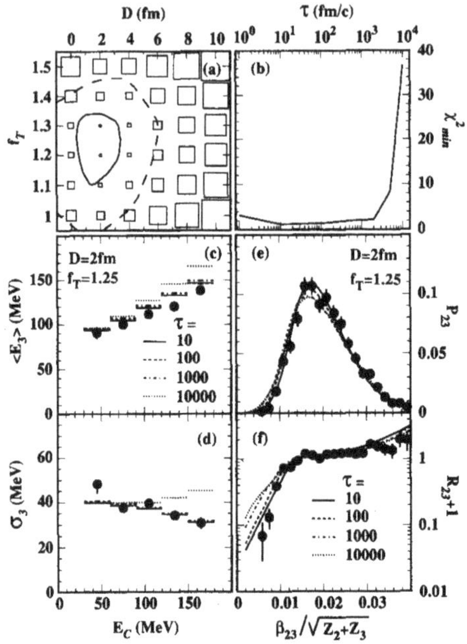

Figure 7: Summary of the sequential decay calculations. Part a: Two-dimensional display of χ^2 as a function of the initial fragment-fragment separation D and the scaling factor f_T of the temperature for a life time τ = 100 fm/c. The area of the largest box corresponds to $\chi^2 = 50$. Part b: Minimum χ^2 as a function of the lifetime τ. Parts c and d: Optimal predictions for $\langle E_3 \rangle$ and σ_3 for different lifetimes τ. Parts e and f: Coincidence yield and two-fragment correlation function of fragment 2 and 3 as a function of their scaled relative velocity for the optimum parameter set with D=2 fm and f_T = 1.25 and the indicated values of τ.

In order to quantify the agreement between the simulations and the experimental observations we define a reduced χ^2

$$\chi^2 = \frac{1}{5}\sum_{i=1}^{5}(x_i - y_i)^2/\delta_i^2 \ . \tag{4}$$

Here, x_i are the four coefficients characterizing the linear fits to the data in Fig. 1 and, in addition, the mean scaled [35] relative velocity between the two lighter fragments 2 and 3. δ_i and y_i denote the experimental uncertainties of these quantities and the corresponding model predictions, respectively.

The results of the sequential type of calculations are summarized in Fig. 7. The box-plot in part (a) displays for a given lifetime of τ = 100 fm/c the variation of χ^2

as a function of the two other model parameters f_T and D. For orientation, contour lines for $\chi^2 = 4$ and 10 are shown. A distinct minimum at D = 2 ± 0.5 fm and f_T = 1.25 ± 0.05 can be discerned. Within the quoted uncertainties, similar minima were found at the *same* values of D and f_T for all considered values of $\tau \leq 10^4 fm/c$. However, the height of this minimum, χ^2_{min}, increases strongly for large values of τ (Fig. 7b) and clearly excludes emission times $\tau > 2000$ fm/c. The good agreement of the simulations with the data for smaller values of τ is illustrated in parts c and d of Fig. 7. Whereas the E_c dependences of $\langle E_3 \rangle$ and σ_3 provide no further constraint on τ, the probability distribution P_{23} of β_{23} is best reproduced by calculations using small $\tau \sim 10$ fm/c (solid line in Fig. 7e). This preference for very small relative emission times is even more evident in Fig. 7f where we compare the measured and predicted correlation functions [32] of fragment 2 and 3, $R_{23}(\beta_{23})+1$.

Results of calculations modelling the simultaneous volume emission of the three fragments are compiled in Fig. 8. Similar to the case of the sequential simulations a clear minimum of χ^2 can be determined for each given flow parameter ϵ_f by varying independently the other two model parameters R and f_T. Part a of Fig. 8 shows in a R - f_T plane the contour lines with $\chi^2 = 2$ for $\epsilon_f = 0$ (at R ≈ 15 fm), 0.5 (R ≈ 22 fm) and 1 MeV (R ≈ 26 fm) and for two different values for the exponent α. The corresponding minimal χ^2 are displayed in part b as a function of ϵ_f. Values of ϵ_f larger than 1 MeV are ruled out whereas smaller values of ϵ_f show no significantly different χ^2_{min}. Nonetheless, for each given value of α there exists a rather unique relation between ϵ_f, R and f_T. For large emission radii R the data can only be described if a collective radial velocity is taken into account. This radial motion can be provided by the Coulomb repulsion between the fragments if the breakup occurs out of a volume with R = 15 fm. The corresponding average initial distance between fragment 2 and 3 is about $\langle d_{23} \rangle \approx 18$ fm and is comparable to a value of $\langle d_{23} \rangle \approx 16$ fm obtained for the sequential scenario using $\tau = 10$ fm/c.

The interplay between the Coulomb repulsion on one hand and the initial velocity - composed of the random thermal motion and the radial flow - on the other hand, is exemplified in Fig. 8c and d. Adopting the optimal parameter set for the case ϵ_f = 0.5 MeV and $\alpha = 1$, i.e. R=22.5 fm and f_T = 1.2, the sequential evaporation (E), flow (F) and the Coulomb repulsion (C) have been added successively to the initial thermal energy (T). The decrease of the thermal contribution by about 5 MeV when going from $\epsilon_f = 0$ (f_T= 1.35) to ϵ_f = 0.5 MeV (f_T=1.2) is small compared to the additional contribution of 20 MeV which is due to the radial flow. Finally, we show

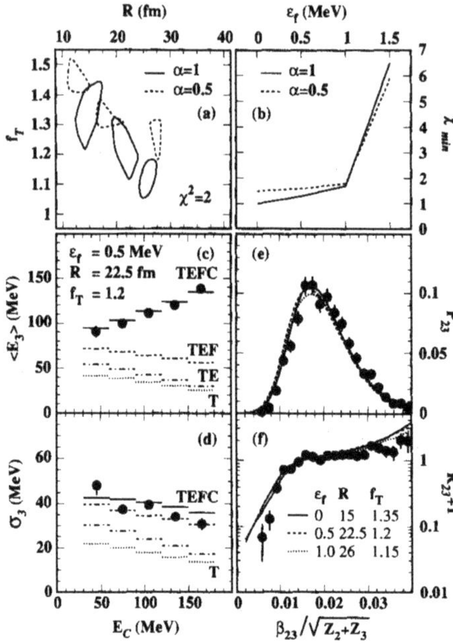

Figure 8: Summary of calculations assuming a volume emission. Part a: Contour lines for χ^2 in a plane defined by the volume radius R and the scaling factor f_T of the temperature for flow parameters ϵ_f = 0, 0.5 and 1 MeV (increasing from left to right). Part b: Minimum χ^2 as a function of ϵ_f. Parts c and d: Predictions for $\langle E_3 \rangle$ and σ_3 for the case $\epsilon_f = 0.5$ MeV. Parts e and f: Coincidence yields and two-fragment correlation function of fragment 2 and 3 as a function of their scaled relative velocity adopting the optimum parameter set for the indicated flow energy.

in panels e and f of Fig. 8 that also for this emission geometry, the relative velocity distribution of fragment 2 and 3 and the correlation function $R_{23} + 1$ are reasonably well reproduced for all values of $\epsilon_f \leq 1$ MeV.

4.2. Kinematic Observables from Multifragmentation Models

The trajectory calculations presented in the previous section - taken at its face values - suggest a fast disintegration process of an highly excited (f_T = 1.35) and rather extended (R $\approx$ 15 fm) nuclear system. Intuitively, for such an extended system a temperature below the initial temperature, i.e. a value for $f_T < 1$, would have been expected. If, nonetheless, these large values for f_T and R are adopted in the

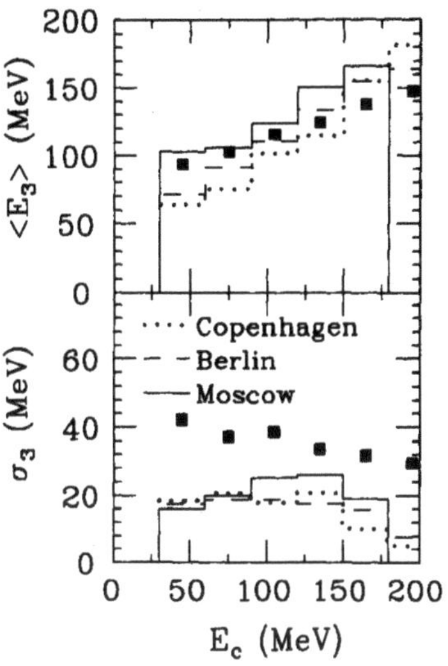

Figure 9: Mean center-of-mass total kinetic energy $\langle E_3 \rangle$ (top) and the standard deviation σ_3 (bottom) as a function of the nominal Coulomb energy averaged over the 4 reactions Au + C, Al, Cu and Pb at E/A = 600 MeV. The dotted, dashed and solid histograms present predictions of the COPENHAGEN [18], BERLIN [19] and MOSCOW [20] multifragmentation models, respectively.

statistical multifragmentation codes, charge distributions are predicted which are clearly inconsistent with the experimental observations.

On the other hand, the statistical multifragmentation models discussed in section 3 require a temperature parameter of $f_T \approx 0.8$ and a break-up volume corresponding to a radius parameter of typically 7 - 8 fm. Despite these different source parameters, the statistical models are able to reproduce $\langle E_3 \rangle$ reasonable well (see top part in figure 9): in these calculations the reduced thermal contribution to E_3 ($\sim T$) due to the smaller value of f_T is compensated by an increased Coulomb repulsion ($\sim 1/R$). The width σ_3, however, reflects in addition to thermal fluctuations, also fluctuations due to the position sampling within the breakup volume. Since the latter contribution decreases for a decreasing volume, both, the lower temperature *and* the smaller radius lead to a significant underprediction of σ_3 by the statistical decay calculations (lower part in Fig. 9). It has been verified, that this underprediction of σ_3 can not be compensated by allowing fluctuations of the initial excitation energy. This deviation also persists if the excitation energy is increased to values corresponding to $f_T = 1$.

(Note, that the MOSCOW [20] multifragmentation code takes also fluctuations of the initial excitation energy into account.) It, therefore, seems that a thermal (random) motion and the Coulomb repulsion alone are not able to describe the dynamics in the final disassembly stage.

The further analysis of these data is still in progress. At the present stage, we can therefore only speculate under what assumptions it will be possible to reconcile the kinetic observables (E_3 and σ_3) and the statistical decay models. It might be important to realize that both kind of calculations ignore the nuclear interaction during the breakup dynamics: in the statistical model calculations of section 3 interaction between the fragments is limited to the Coulomb repulsion. Although the three-body trajectory calculations include a nuclear proximity potential, its influence is strongly suppressed by the requirement that the fragments do not overlap during the propagation. If one would, however, allow a stronger overlap of the produced fragments in an earlier stage of the breakup, the nuclear attractive force between the fragments may partially compensate the repulsive Coulomb interaction. Smaller radii would therefore not necessarily lead to an overestimation of the kinetic energies. At the same time large fluctuations - perhaps akin to dissipative phenomena or shape fluctuations known to be important in binary fission processes [42, 43, 41]- may arise. First steps to implement the nuclear interaction between the fragments in a consistent manner in the rather successful statistical decay models have indeed been already undertaken [44, 45]. If in future studies the sensitivity of kinematic observables to the nuclear interaction - and the associated fluctuations - will be confirmed, we might be able to catch the first rather undistorted glimpse of the growth of fluctuations in an expanding nuclear system.

5. Acknowledgements

Enlightening discussions are gratefully acknowledges with D.H.E. Gross and A.S. Botvina. This work was partly supported by the Bundesministerium für Forschung und Technologie and the Deutsche Forschungsgemeinschaft.

6. References

[1] For a recent reviews see, e.g., *The Nuclear Equation of State*, NATO ASI Series B: Physics Vol. 216A, ed. W. Greiner and H. Stöcker (Plenum Press, New York 1989); K.-H. Kampert, J. Phys. **G15**, 691 (1989); H.H. Gutbrod, A. Poskanzer, and H.G. Ritter, Rep. Prog. Phys. **52**, 1267 (1989).

[2] W. Schmidt *et al.*, Phys. Rev. C **47**, 2782 (1993)

[3] For a recent review see Gy. Wolf, GSI-preprint, GSI-93-55 (1993)

[4] P.J. Siemens, Nature (London) **305**, 410 (1983).

[5] G.F. Bertsch and P.J. Siemens, Phys. Lett. B **126B**, 9 (1983).

[6] J. Hubele *et al.*, Phys. Rev. C **46**, R1577 (1992).

[7] P. Kreutz, PhD thesis, Univ. Frankfurt (1992); P. Kreutz *et al.*, Nucl. Phys. **A556**, 672 (1993).

[8] Bao-An Li, A.R. DeAngelis, D.H.E. Gross, Phys. Lett. B **303**, 225 (1993).

[9] H.W. Barz *et al.*, Nucl. Phys. **A561**, 466 (1993).

[10] M.B. Tsang *et al.*, Phys. Rev. Lett. **71**, 1502 (1993).

[11] W. Bauer *et al.*, Phys. Rev. C **47**, R1838 (1993).

[12] N. Hermann, Nucl. Phys. **A553**, 739c (1993).

[13] D. Idier *et al.*, Phys. Rev. C **48**, R498 (1993).

[14] J. Hubele *et al.*, Z. Phys. A **340**, 263 (1991).

[15] R.J. Charity *et al.*, Nucl. Phys. **A483**, 371 (1988).

[16] L.G. Moretto, Nucl. Phys. **A247**, 211 (1975).

[17] A.S. Botvina, A.S. Il'inov, and I.N. Mishustin, Yad. Fiz. **42**, 1127 (1985) [Sov. J. Nucl. Phys. **42**, 712 (1985)].

[18] J. Bondorf *et al.*, Nucl. Phys. **A443**, 321 (1985).

[19] For a recent review see D.H.E. Gross, Rep. Prog. Phys. **53**, 605 (1990).

[20] A.S. Botvina and I.N. Mishustin, Phys. Lett. B **294**, 23 (1992) and private communication.

[21] J. Kapusta, Phys. Rev. C **24**, 2545 (1981).

[22] J.A. Lopez and P.J. Siemens, Nucl. Phys. **A431**, 728 (1984).

[23] D.H. Boal and A.L. Goodman, Phys. Rev. C **33**, 1690 (1986).

[24] L.D. Landau and E.M. Lifshitz, *Course of Theoretical Physics: Statistical Mechanics* (Pergamon, New York, 1980), Vol. 5, Pt. 1, pp. 125ff.

[25] For an overview see O.V. Lozhkin and W. Trautmann, Phys. Rev. C **46**, 1996 (1992) and references therein.

[26] V. Lindenstruth *et al.*, to be published.

[27] D.H.E. Gross and K. Sneppen, HMI-Preprint, HIM 1993/P1-Gros 4 (1993).

[28] D.R. Bowman *et al.*, Phys. Rev. Lett. **67**, 1527 (1991).

[29] K. Hagel *et al.*, Phys. Rev. Lett. **68**, 2141 (1992).

[30] P. Glässel *et al.*, Z. Phys. A - Atoms and Nuclei **310**, 189 (1983).

[31] D. Pelte *et al.*, Phys. Rev. C **34**, 1673 (1986).

[32] R. Trockel *et al.*, Phys. Rev. Lett. **59**, 2844 (1987).

[33] D.H.E. Gross *et al.*, Phys. Lett. B **224**, 29 (1989).

[34] R. Bougault *et al.*, Phys. Lett. B **232**, 291 (1989).

[35] Y.D. Kim *et al.*, Phys. Rev. Lett. **67**, 14 (1991).

[36] Y.D. Kim *et al.*, Phys. Rev. C **45**, 338 (1992)

[37] G. Bizard *et al.*, Phys. Lett. B **276**, 413 (1992).

[38] D.R. Bowman *et al.*, Phys. Rev. Lett. **70**, 3534 (1993).

[39] E. Bauge *et al.*, Phys. Rev. Lett. **70**, 3705 (1993).

[40] J. Lopez and J. Randrup, Nucl. Phys. **A503**, 183 (1989).

[41] For a recent review see F. Gönnenwein, *The Nuclear Fission Process*, Edt. C. Wagemans, CRC Press, Boca Raton (1991), pg. 287 ff.

[42] O.K. Serdyuk *et al.*, Sov. J. Nucl. Phys. **46**, 399 (1987).

[43] S. Grossmann *et al.*, Nucl. Phys. **A481**, 340 (1988).

[44] L. Satpathy *et al.*, Phys. Lett. B **237**, 181 (1990).

[45] A. Das *et al.*, J. Phys. G **19**, 319 (1993).

FLUCTUATIONS IN MICROSCOPIC NUCLEAR SIMULATIONS*

JØRGEN RANDRUP

Nuclear Science Division, Lawrence Berkeley Laboratory
University of California, Berkeley, California 94720

and

SAKIR AYIK

Department of Physics, Tennessee Technological University
Cookeville, Tennessee 38505

ABSTRACT

We briefly recall the Boltzmann-Langevin theory of nuclear dynamics and its treatment within the Fokker-Planck approximation; the structure of the correlated fluctuations in thermal equilibrium is especially considered. We then summarize recent progress in deriving approximate analytical expressions for the associated transport coefficients and describe a numerical method for simulating the stochastic dynamical evolution of the one-body phase-space density on the basis of these transport coefficients.

1. Introduction

The need for including dynamical fluctuations in numerical simulations of nuclear reactions at intermediate energies has become increasingly apparent in the past few years, as drastically improved instrumentation has allowed the detailed experimental exploration of multifragmentation processes. For central collisions between massive nuclei, it is expected that the bulk matter is initially compressed significantly above the saturation value. The resulting decompression may then lead the system to expanded configurations that are mechanically unstable and so a given system may develop into a wide spectrum a final manifestations [1]. Indeed, a variety of numerical studies have indicated that a nuclear collision may, under suitable conditions, lead to configurations in which bulk or surface instabilities may be agitated [2, 3, 4].

Over the past several years, significant progress has been made on the dynamical description of macroscopic nuclear dynamics within the framework of semi-classical theory. These models have their origin in effective one-body theories that have proven very useful for understanding a wealth of nuclear properties. The pure one-body description is best justified at low energies where the exclusion principle is most effective. However, for many processes of current interest the effective temperatures amount to several or many MeV, and so it is necessary to take account of the residual interactions between the nucleons. This is most conveniently done in the semi-classical description along the lines pioneered by Boltzmann for the description of dilute gases of classical particles, namely by including a two-body collision term in the equation of motion for the one-body phase-space density $f(\mathbf{r},\mathbf{p})$. For fermionic systems, this approach was first adapted by Nordheim [5] for the description of transport of electrons in solids, by simply inserting the Fermi-Dirac blocking factors in the collision integral; related work was done subsequently by Ühling and Uhlenbeck [6].

In recent years, this approach has been employed in the area of nuclear dynamics. Early on, it was used to study the equilibration of the momentum distribution in a uniform gas of nucleons, Δ resonances, and pions [7]. The combination of this treatment with the Vlasov evolution in the self-consistent effective one-body field $H[f]$ leads to the so-called BUU (or VUU or Landau-Vlasov) model [8]. This combined model has proved to be a very good basis for understanding a variety of features associated with nuclear collision dynamics, such as energy dissipation, density compression, the occurrence of collective flow, and particle production (photons, pions, kaons, anti-protons, *etc.*) [9, 10].

In the BUU model, the effect of the residual interactions is taken into account only in an average manner, and the corresponding dynamics is then deterministic, though dissipative. The description is therefore best suited for processes that do not exhibit macroscopic bifurcations, such as the evolution from the saddle point towards fission or the early stages of a central nuclear collision. The incorporation of stochasticity into the semi-classical one-body description can be accomplished rather naturally by retaining the fluctuating effect of the two-body collisions. When this is done, the evolution continually displays dynamical branching and catastrophic processes can thus be addressed. The situation is conceptually similar to brownian motion in which the average effect of the random force produces friction while the fluctuating part generates diffusion.

The formal groundwork for incorporating stochasticity into the Boltzmann dynamics was made by Ayik and Gregoire [11] who coined the term *Boltzmann-Langevin* theory for the extended description. Randrup and Remaud [12] then recast the transport problem into a Fokker-Planck form in which the entire distribution of phase-space densities, $\phi[f]$, is addressed. Subsequently, Burgio, Chomaz, and Randrup [13, 14] developed a specific method for solving the Boltzmann-Langevin equation by a direct simulation of the elementary two-body collision processes, employing a representation of the density f on a lattice in phase space. Here we briefly review this treatment and then summarize very recent progress in deriving simple approximate expressions for the associated transport coefficient functionals, on the basis of which the numerical simulation of the stochastic dynamics is brought significantly closer to practicality.

2. Nuclear One-Body Dynamics

The basic object of study is the reduced one-body phase space density $f(\mathbf{r},\mathbf{p})$ which can be obtained from the full A-body density by projection. The matter density can then be calculated by integration over the momentum, $\rho(\mathbf{r}) = h^{-D}\int d\mathbf{p} f(\mathbf{r},\mathbf{p})$; the local flow velocity and the kinetic energy density can be obtained similarly. It is often convenient to use the dimensionless quantity s to denote $(\mathbf{r},\mathbf{p})$, with $ds = h^{-3}d\mathbf{r}d\mathbf{p}$.

The one-body density f evolves in its self-consistent field described by the effective single-particle Hamiltonian $h[f](\mathbf{r},\mathbf{p}) = p^2/2B(\mathbf{r}) + U(\mathbf{r})$, where $B(\mathbf{r})$ is the effective mass and $U(\mathbf{r})$ the effective potential, both being functionals of $f(\mathbf{r},\mathbf{p})$. The effect of the residual interactions are treated as resulting from direct two-body collisions between the constituent nucleons. Depending on the level of refinement, the equation of motion for $f(\mathbf{r},\mathbf{p})$ is then given by

$$\dot{f}(\mathbf{r},\mathbf{p}) \equiv \frac{\partial}{\partial t}f - \{h[f],f\} = \begin{cases} 0 & \text{(Vlasov)} \\ \bar{I}[f] & \text{(Boltzmann)} \\ I[f] & \text{(Boltzmann-Langevin)} \end{cases} . \tag{1}$$

In the simplest description the residual interactions are neglected, leading to the Vlasov equation, which is the semi-classical analog to the time-dependent Hartree-Fock model. The effect of the two-body collisions is described by the collision integral $I[f] = \bar{I}[f] + \delta I[f]$ which can be decomposed into an average part $\bar{I}$ and its fluctuating remainder δI.

The basic two-body collision process responsible for the collision term $I[f]$ can be regarded as an agency for moving probability around in phase space. During a short time interval Δt the expected number of nucleons that scatter out of two phase space cells located at s_1 and s_2 and into cells located at s_1' and s_2' is given by

$$d\nu_{1,2;1',2'} = f_1 f_2 \bar{f}_{1'} \bar{f}_{2'}\ \delta(\mathbf{r}_{12})\delta(\mathbf{r}_{1'2'})\delta(\mathbf{r}_{11'})\ w_{12;1'2'}\ ds_1 ds_2 ds_{1'} ds_{2'} \Delta t\ , \tag{2}$$

where $f_k \equiv f(s_k)$ and the vacancy factors $\bar{f}_k \equiv 1 - f_k$ take account of the Pauli blocking of the final nucleon states. Furthermore, $w_{12;1'2'} = (4h^6/m^2)\delta(\Delta\mathbf{p})\delta(\Delta\epsilon)d\sigma/d\Omega$ is the elementary transition rate. The variance on the number of such elementary transitions is of the same size, $\sigma_{d\nu}^2 = d\nu$, due to the statistical nature of the individual scattering process [12, 14].

Inclusion of the average collision integral leads to the Boltzmann model, which was first adapted to fermionic systems by Nordheim [5] (and subsequently by Ühling and Uhlenbeck [6]). In the nuclear context it is variously referred to as the Boltzmann-Ühling-Uhlenbeck, Vlasov-Ühling-Uhlenbeck, or Landau-Vlasov model [8]. The average effect of the two-body collisions on the phase-space occupancy is then

$$\bar{I}[f]: \qquad \dot{f}_1 = \int \frac{d\mathbf{p}_2}{h^D} v_{12} \int d\Omega'_{12} (\underbrace{\bar{f}_1 \bar{f}_2 f_{1'} f_{2'}}_{\text{gain}} - \underbrace{f_1 f_2 \bar{f}_{1'} \bar{f}_{2'}}_{\text{loss}}) \frac{d\sigma_{12\leftrightarrow 1'2'}}{d\Omega'_{12}} . \tag{3}$$

The last factor is the differential cross section for (elastic) nucleon-nucleon scattering in the medium, at the appropriate density and excitation. Thus the in-medium properties of the nucleons (and any other hadrons present) must be known to allow quantitative results to be calculated. Conversely, in so far as the calculated observables depend on this physical ingredient, confrontations with data may help to extract this microscopic information from experiment.

The average collision integral $\bar{I}[f]$ vanishes when the local momentum distribution is of Fermi-Dirac form. Consequently, $f^0(\mathbf{r},\mathbf{p}) \equiv 1/(1+\exp[(h(\mathbf{r},\mathbf{p})-\mu(\mathbf{r}))/\tau])$ is a stationary solution to the BUU equation and corresponds to the finite-temperature Thomas-Fermi description. Hence the BUU model (1b) reproduces all the associated macroscopic nuclear features, including the equation of state of nuclear matter. In the limit of zero temperature, $\tau = 0$, the general nuclear properties emerge, including nuclear bindings, matter densities, deformation energies, and fission barriers.

A natural route towards introducing fluctuations into nuclear dynamics is to retain the *fluctuating* part of the collisions integral, $\delta I[f]$, in the equation of motion. This quantity acts to produce a stochastic evolution of $f(\mathbf{r},\mathbf{p},t)$ and is akin to the Langevin force acting on a brownian particle. Therefore, the ensuing description has been denoted the Boltzmann-Langevin equation [11].

Since the elementary scattering processes are stochastic in nature, a given initial density $f(t=0)$ evolves into an entire bundle of different histories $\{f(t)\}$. The resulting ensemble of trajectories must then be described by the associated *distribution* of phase-space densities, $\phi[f]$. In dynamical simulations, this distribution is represented by a sample of $\mathcal{N}$ representative densities $\{f^{(n)}\}$, with $f^{(n)}$ being the phase-space density associated with the particular manifestation n, *i.e.* $\phi[f] \approx \mathcal{N}^{-1}\sum_{n=1}^{\mathcal{N}} \delta[f^{(n)} - f]$.

Once the distribution $\phi[f]$ is known, the expectation value of a one-body observable $\mathcal{A}(s)$ can be calculated [12],

$$\prec \mathcal{A} \succ = \int \mathcal{D}[f]\; \mathcal{A}[f]\; \phi[f] = \int ds\; \mathcal{A}(s)\; \prec f(s) \succ \approx \mathcal{N}^{-1} \sum_n \mathcal{A}^{(n)} \,. \tag{4}$$

The first relation is a functional integral over all possible phase-space densities $f(s)$ and $\phi[f]$ then gives the associated (normalized) weight. The second relation expresses the fact that the ensemble average expectation value of the one-body observable $\mathcal{A}(s)$ is equal to the expectation value calculated on the basis of the ensemble average density $\prec f \succ = \mathcal{N}^{-1}\sum_n f^{(n)}$. Finally, the third relation shows how the quantity can be calculated in terms of the expectation values $\mathcal{A}^{(n)}$ associated with each of the $\mathcal{N}$ representative densities $f^{(n)}(s)$.

Moreover, the correlation between the values of two given one-body observables $\mathcal{A}$ and $\mathcal{A}'$ can be obtained as [12]

$$\sigma_{\mathcal{A}\mathcal{A}'} \equiv \prec \mathcal{A}\mathcal{A}' \succ - \prec \mathcal{A} \succ \prec \mathcal{A}' \succ = \int ds \int ds' \mathcal{A}(s)\; \sigma(s,s')\; \mathcal{A}'(s') \,, \tag{5}$$

where the correlation function is $\sigma(s,s') \equiv \prec \delta f(s)\delta f(s') \succ$, with δf denoting the deviation of a given density f from the ensemble average density $\prec f \succ$. The correlation function is singular but can be decomposed into regular diagonal and off-diagonal parts [13, 14],

$$\sigma(s,s') = \sigma^2(s)\delta(s-s') + \sigma^{\text{cov}}(s,s') \,, \tag{6}$$

where $\sigma^2(s)$ expresses the variance in the occupancy of a single cell, while $\sigma^{\text{cov}}(s,s')$ represents the correlation between the fluctuations in occupancy of two different cells.

3. Fokker-Planck Treatment

The dynamical evolution of the ensemble of phase-space densities can be approximately described by suitable extensions of standard transport-theoretical methods, and the evolution of the distribution $\phi[f]$ is then governed by a generalized Fokker-Planck equation [12],

$$\dot{\phi} = -\int ds \frac{\partial}{\partial f(s)} V[f](s)\phi[f] + \int ds \int ds' \frac{\partial^2}{\partial f(s)\partial f(s')} D[f](s,s')\phi[f] \ . \quad (7)$$

The dynamics is thus determined by the transport coefficients $V[f](s) = \bar{I}[f](s)$ and $D[f](s,s') = \frac{1}{2} \prec \delta I[f](s)\, \delta I[f](s') \succ$, which have simple physical meanings: Starting from a given phase-space density $f(s,t)$, the drift coefficient expresses the mean rate of the collision-induced change in f, so

$$\prec f(s,t') \succ \approx f(s,t) + V[f](s)\Delta t \ , \quad (8)$$

while the diffusion coefficient expresses the rate at which the correlations between these changes grow,

$$\sigma(s,s',t') \ \equiv \prec \delta f(s,t')\delta f(s',t') \succ \ \approx \ 2D[f](s,s')\Delta t \ , \quad (9)$$

where $\prec \cdot \succ$ is the average over the ensemble of possible densities, $\{f(t')\}$, into which $f(t)$ may have developed at the slightly later time $t' = t + \Delta t$.

In analogy with the decomposition of the correlation function (6), the diffusion coefficient can be naturally split into two parts [14],

$$D[f](s_1,s_2) = [D_1^{\rm var}\delta_{12} + D_{12}^{\rm cov}]\,\delta(\mathbf{r}_{12}) \ . \quad (10)$$

Here the subscripts provide a convenient means of indicating the momentum values, and δ_{12} means $h^D\delta(\mathbf{p}_{12})$. The transport coefficients are local in space and so we need only be concerned with one spatial location at a time. The coupling between different points in space and time is produced by the Vlasov propagation in the effective one-body field. The main effort in solving the BL equation is associated with the stochastic but correlated rearrangement of the population in momentum space

For this discussion it suffices to consider a uniform gas of free Fermi-Dirac particles, and it is then convenient to introduce *contracted* quantities obtained by integration over the spatial separation $\mathbf{r}_{12}$ [12, 15]. For example, the contracted diffusion coefficient D_{12} is given by the expression within the brackets in eq. (10).

The covariance part of the contracted diffusion coefficient is given by [14]

$$\begin{aligned} 2D_{12}^{\rm cov} &= f_1\hat{W}_{12}^{--}f_2 + \bar{f}_1\hat{W}_{12}^{++}\bar{f}_2 - f_1\hat{W}_{12}^{-+}\bar{f}_2 - \bar{f}_1\hat{W}_{12}^{+-}f_2 = 2D_{12}^{+} + 2D_{12}^{-} \quad (11)\\ &= \sum_{34}\left[\tfrac{1}{2}(f_1f_2\bar{f}_3\bar{f}_4 + \bar{f}_1\bar{f}_2f_3f_4)w_{12;34} - (f_1\bar{f}_2f_3\bar{f}_4 + \bar{f}_1f_2\bar{f}_3f_4)w_{13;24}\right] \ . \end{aligned}$$

We have indicated the explicit formulas for the four basic rate functions $W_{12}^{\sigma_1\sigma_2}$ in terms of the elementary transition rates $w_{12;34}$ and, moreover, we have grouped the four distinct contributions to $D_{12}^{\rm cov}$ into $D_{12}^{\pm}$ according to their sign.

The functionals $W^{\sigma\sigma'}(s,s')$ measure the rate of transitions that involve the two specified phase-space locations s and s', either as initial states ($\sigma=\sigma'=-$), as final states ($\sigma=\sigma'=+$), or with one being the initial state and the other the final state,

$$f(s)\,W^{--}(s,s')\,f(s_b) \;=\; \frac{1}{2}\int d\nu_{1,2;1',2'}\,\delta(s_1-s)\,\delta(s_2-s')\,, \tag{12}$$

$$\bar{f}(s)\,W^{++}(s,s')\,\bar{f}(s') \;=\; \frac{1}{2}\int d\nu_{1,2;1',2'}\,\delta(s_{1'}-s)\,\delta(s_{2'}-s')\,, \tag{13}$$

$$f(s)\,W^{-+}(s,s')\,\bar{f}(s') \;=\; \int d\nu_{1,2;1',2'}\,\delta(s_1-s)\,\delta(s_{2'}-s')\,, \tag{14}$$

where the factor $\frac{1}{2}$ serves to avoid double counting of the final or initial two-nucleon states, respectively, and we note that $W^{+-}(s,s')=W^{-+}(s',s)$.

The diagonal part of the diffusion coefficient, $D_1^{\rm var}$, and the drift coefficient, V_1, are given then given by

$$2D_1^{\rm var} = \bar{f}_1 W_1^+ + f_1 W_1^-\,, \tag{15}$$

$$V_1 = \bar{f}_1 W_1^+ - f_1 W_1^-\,, \tag{16}$$

where the rate coefficients W_1^σ can be obtained from $W_{12}^{\sigma_1\sigma_2}$ by integration [14],

$$W_1^- = \sum_2 \hat{W}_{12}^{--} f_2 = \frac{1}{2}\sum_2 \hat{W}_{12}^{-+}\bar{f}_2\,, \tag{17}$$

$$W_1^+ = \sum_2 \hat{W}_{12}^{++}\bar{f}_2 = \frac{1}{2}\sum_2 \hat{W}_{12}^{+-} f_2\,. \tag{18}$$

We note that $\sum_2 D_{12}=0$, due to particle-number conservation. It is also true that $\sum_2 D_{12}\epsilon_2=0$ and $\sum_2 D_{12}\mathbf{p}_2=\mathbf{o}$, since the elementary two-body collision processes conserve energy and momentum as well.

By expanding around the average trajectory $\prec f(\mathbf{r},\mathbf{p},t)\succ$, it is possible to derive closed equations of motion for the moments of the distribution $\phi[f]$.[12, 14] The equation of motion for the ensemble average is

$$\frac{d}{dt}\prec f(s,t)\succ = \prec V[f]\succ \approx V[\prec f\succ] = \bar{f}(s)W^+(s) - f(s)W^-(s)\,, \tag{19}$$

where the gain and loss rates $W^\pm$ are evaluated on the basis of the average distribution $\prec f\succ$. This equation yields the solution to the usual *BUU* equation, in which the fluctuating part of the collision term is ignored. With a similar approximation, the equation for the variance in occupancy becomes [12, 14],

$$\frac{d}{dt}\sigma^2(s) = \bar{f}(s)W^+(s) + f(s)W^-(s) - 2W(s)\sigma^2(s)\,, \tag{20}$$

where $W\equiv W^-+W^+$. It is easy to show that the deviation $\Delta(s)\equiv\sigma^2-f\bar{f}$ is governed by the equation $\dot{\Delta}=-2W\Delta$, which brings out the fact that the variance σ^2 relaxes towards $f\bar{f}$ at the rate $2W$, even when f has not yet reached its equilibrium form. By similar means, it is possible to derive an equation of motion for the covariance $\sigma_{12}^{\rm cov}$, but it is more complicated [14, 15]. The moment expansion is useful when the distribution $\phi[f]$ is narrowly peaked, but breaks down for multimodal distributions, such as occur when instabilities are present.

4. Equilibrium Fluctuations in a Fermi-Dirac Gas

Under suitable conditions of confinement and stability, the Boltzmann-Langevin equation of motion, as well as the corresponding Fokker-Planck transport equation, yield dynamical solutions that approach stationarity and which display properties characteristic of the associated quantum-statistical equilibrium. It is therefore instructive to consider the equilibrium features of the model.

For a spatially uniform Fermi-Dirac gas of nucleons that interact only via their stochastic two-body collisions, the correlated equilibrium fluctuations are given by a simple analytical expression. By adapting a method known for classical gases [16], the Nantes group derived the expression for a nucleon gas [17]; a more detailed derivation was subsequently given [15]. In the latter work is was shown explicitly how the equation of motion for the correlation function leads to the appropriate equilibrium form. That work also introduced a convenient formalism which is employed in the following.

The basic two-body collision process conserves particle number A, energy E, and momentum $\mathbf{P}$. These one-body observables are briefly denoted by F^α, with $\alpha = A, E, \mathbf{P}$, and at a given phase-space point s_1 they take on the values $F_1^A = 1$, $F_1^E = \epsilon_1$, and $F_1^{\mathbf{P}} = \mathbf{p}_1$, respectively. It is convenient to introduce the 5×5 matrix Φ whose elements are given by

$$\Phi^{\alpha\beta} = \sum_1 F_1^\alpha f_1^0 \bar{f}_1^0 F_1^\beta \ , \quad \alpha, \beta = A, E, \mathbf{P} \ , \tag{21}$$

where $f^0(\epsilon)$ denotes the equilibrium phase-space density. The matrix Φ decouples into a 2×2 matrix involving A and E and a 3×3 matrix proportional to the unit matrix $\overleftrightarrow{I}$,

$$\Phi = \rho \begin{pmatrix} \phi_0 & \phi_1 \\ \phi_1 & \phi_2 \end{pmatrix} \otimes \frac{2m}{3} \rho \phi_1 \overleftrightarrow{I} \ . \tag{22}$$

The elements of the matrix Φ have been expressed in terms of energy moments over the Fermi surface [15],

$$\phi_n = \frac{1}{\rho} \sum_1 \epsilon_1^n f_1^0 \bar{f}_1^0 \approx \frac{3}{2} \frac{\tau}{\epsilon_F} \epsilon_F^n \left[1 + \frac{\pi^2}{6}(n-1)(n-1+\frac{3}{2}) \frac{\tau^2}{\epsilon_F^2} \right] \ , \tag{23}$$

where the last relation has been obtained by performing a Sommerfeld expansion [19]; we note that $\phi_1 = \frac{3}{2}\tau$.

The Fermi energy ϵ_F in (23) is the usual one determined from the density ρ, $\epsilon_F \sim \rho^{2/3}$. It provides a convenient reference energy, since it is independent of the temperature τ. However, for finite values of the temperature, the Fermi surface is diffuse and the energies must be defined with care, somewhat in analogy to the discussion of the geometrical properties of the diffuse nuclear density distribution [18]. The Fermi surface profile function is given by $-\tau \partial f^0 / \partial \epsilon = f^0 \bar{f}^0$ and has the variance

$$\sigma_\tau^2 = \frac{\phi_2}{\phi_0} - (\frac{\phi_1}{\phi_0})^2 \approx \frac{\pi^2}{3} \tau^2 \ . \tag{24}$$

Because of this finite width, the effective Fermi energy $\epsilon_0 = \phi_1/\phi_0$ is increased and the chemical potential μ is correspondingly reduced,

$$\epsilon_0 \approx \epsilon_F[1 + (\frac{\sigma_\tau}{2\epsilon_F})^2] \approx \epsilon_F[1 + \frac{\pi^2}{12}\frac{\tau^2}{\epsilon_F^2}] , \tag{25}$$

$$\mu \approx \epsilon_F[1 - (\frac{\sigma_\tau}{2\epsilon_F})^2] \approx \epsilon_F[1 - \frac{\pi^2}{12}\frac{\tau^2}{\epsilon_F^2}] . \tag{26}$$

Moreover, the inverse of the matrix $\mathbf{\Phi}$ is denoted by Θ and is thus given by

$$\Theta = \frac{1}{\rho\phi}\begin{pmatrix} \phi_2 & -\phi_1 \\ -\phi_1 & \phi_0 \end{pmatrix} \otimes \frac{3}{2m\rho\phi_1}\overleftrightarrow{I} , \tag{27}$$

where we have introduced the determinant $\phi = \phi_0\phi_2 - \phi_1^2 = \phi_0^2\sigma_\tau^2 \sim \tau^4$.

After this preparation, we are now ready to give the expression for the contracted correlation function associated with a fermion gas in statistical equilibrium [15, 17],

$$\tilde{\sigma}_{12} = \tilde{\sigma}_1^2\delta_{12} + \tilde{\sigma}_{12}^{\rm cov} = f_1^0\bar{f}_1^0\delta_{12} - f_1^0\bar{f}_1^0 S_{12} f_2^0\bar{f}_2^0 . \tag{28}$$

Here we have used the tilde over σ to indicate that it pertains to equilibrium. The coefficient S_{12} expresses how the fluctuations at two different phase-space points are correlated and it can be expressed as a bilinear form in the observables F^α that are left invariant by the two-body collision process [15],

$$S_{12} = \sum_{\alpha\beta} F_1^\alpha \Theta^{\alpha\beta} F_2^\beta = \frac{1}{\rho\phi_0}\left[1 + \frac{\epsilon_1 - \epsilon_0}{\sigma_\tau}\frac{\epsilon_2 - \epsilon_0}{\sigma_\tau} + 3\frac{\mathbf{p}_1\cdot\mathbf{p}_2}{2m\epsilon_0}\right] , \tag{29}$$

where we recall that the 5×5 matrix Θ is the inverse of $\mathbf{\Phi}$ defined in (21).

We also note the following sum rules,

$$\sum_2 S_{12} f_2^0 \bar{f}_2^0 = 1 , \tag{30}$$

$$\sum_2 S_{12} f_2^0 \bar{f}_2^0 \epsilon_2 = \epsilon_1 , \tag{31}$$

$$\sum_2 S_{12} f_2^0 \bar{f}_2^0 \mathbf{p}_2 = \mathbf{p}_1 . \tag{32}$$

With the help of these relations, it is easy to verify that the above expression (28) for $\tilde{\sigma}_{12}$ does not lead to fluctuations in the conserved quantities A, E, and $\mathbf{P}$, since $\sum_2 \tilde{\sigma}_{12} F_2^\alpha = 0$ for all values of s_1.

Furthermore, the following sum rule holds,

$$\begin{aligned}\sum_2 S_{12} f_2^0 \bar{f}_2^0 S_{23} &= \sum_{\alpha\beta\gamma\delta} F_1^\alpha \Theta^{\alpha\beta}\left(\sum_2 F_2^\beta f_2^0 \bar{f}_2^0 F_2^\gamma\right)\Theta^{\gamma\delta} F_3^\delta = \sum_{\alpha\beta\gamma\delta} F_1^\alpha \Theta^{\alpha\beta}\Phi^{\beta\gamma}\Theta^{\gamma\delta} F_3^\delta \\ &= \sum_{\alpha\beta\delta} F_1^\alpha \Theta^{\alpha\beta}\delta_{\beta\delta} F_3^\delta = \sum_{\alpha\beta} F_1^\alpha \Theta^{\alpha\beta} F_3^\beta = S_{13} .\end{aligned} \tag{33}$$

We finally note that the angular correlation $C(\theta_{12})$ has only monopole and dipole components in equilibrium, $\tilde{C}(\theta_{12}) = -P_0 - 3P_1(\cos\theta_{12})$, with the monopole term resulting from the conservation of particle number and energy and the dipole term arising from momentum conservation.

5. Approximate Expressions for the Transport Coefficients

It is possible to derive simple approximate expressions for the rate functions $W_{12}^{\sigma_1\sigma_2}$ at equilibrium, where the average phase-space density is of Fermi-Dirac form, $f_1^0 = 1/[1+\exp((\epsilon_1-\mu)/\tau))]$. For this, one may employ a method developed in connection with the study of Fermi liquids [20, 21] and which is useful when the temperature is low in comparison with the Fermi kinetic energy, $\tau \ll \epsilon_F$, as is the case in the applications we have in mind. We summarize below this treatment and refer to [22] for more details.

It is convenient to split the mixed diffusion coefficient $D_{12}^{\rm cov}$ into its positive and negative parts, as already anticipated in eq. (11), each consisting of two terms, which are identical in equilibrium. For both parts, the integration over $\mathbf{p}_4$ can be done trivially, because the transition rate w contains a momentum-conserving δ-function. The remaining integral over $\mathbf{p}_3$ can be written as an integral over the energy ϵ_3 and the direction $\hat{\mathbf{p}}_3 = (\theta_3, \phi_3)$. We shall assume that the differential cross section is isotropic, $d\sigma/d\Omega = \sigma_0/4\pi$, and the ϕ_3-integration is then trivial. The specification of θ_3 implies a definite value of ϵ_4, due to the momentum conservation mentioned above, and since only momenta lying in the Fermi surface contribute. We then obtain the following approximate expressions,

$$D_{12}^+ : \ \int d\mathbf{p}_3\,(\cdot) \ = mP_F \int d\epsilon_3 d\phi_3 d\cos\theta_3\,(\cdot) \approx 4\pi\frac{m^2}{P_F}\,\frac{1}{2\cos\frac{1}{2}\theta_{12}} \int d\epsilon_3 d\epsilon_4\,(\cdot)\ , \qquad (34)$$

$$D_{12}^- : \ \int d\mathbf{p}_3\,(\cdot) \ = mP_F \int d\epsilon_3 d\phi_3 d\cos\theta_3\,(\cdot) \approx 4\pi\frac{m^2}{P_F}\,\frac{1}{\sin\frac{1}{2}\theta_{12}} \int d\epsilon_3 d\epsilon_4\,(\cdot)\ , \qquad (35)$$

where the first relation applies to D_{12}^+ and the second to D_{12}^-.

It then follows that the dependence of $D_{12}^{\rm cov} = D_{12}^+ + D_{12}^-$ on the angle θ_{12} between $\mathbf{p}_1$ and $\mathbf{p}_2$ is given approximately by

$$G(\theta) = G^+(\theta) + G^-(\theta) \approx \frac{1}{2\cos\frac{1}{2}\theta} - \frac{1}{\sin\frac{1}{2}\theta} = \sum_{\ell\geq 0}\left[(-1)^\ell - 2\right] P_\ell(\cos\theta)\ . \qquad (36)$$

The normalization has been chosen such that $G(\theta)$ is equal to minus one on the average. The Legendre expansion is very simple, with the coefficients alternating between minus one and minus three, and we note that the first two terms are identical to those associated with the equilibrium correlations (28) discussed above.

The energy integrals are constrained by the energy-conserving δ-function occurring in the transition rate w and can be evaluated approximately as shown in ref. [21]. For example, for the positive term we have the following relations,

$$\int dx_3 dx_4\ f_1^0 f_2^0 \bar{f}_3^0 \bar{f}_4^0\ \delta(x_1+x_2-x_3-x_4) \approx f_1^0 f_2^0\ \frac{x_1+x_2}{1-e^{-x_1-x_2}}\ , \qquad (37)$$

$$\int dx_2 dx_3 dx_4\ f_1^0 f_2^0 \bar{f}_3^0 \bar{f}_4^0\ \delta(x_1+x_2-x_3-x_4) \approx f_1^0 \bar{f}_1^0\ \frac{x_1^2+\pi^2}{2}\ , \qquad (38)$$

where $x_i \equiv (\epsilon_i - \mu)/\tau$. The expression (37), considered as a function of ϵ_1 and ϵ_2, is peaked in the Fermi surface: when either ϵ_1 or ϵ_2 exceeds the Fermi energy, the corresponding Fermi-Dirac occupancy factor f_1^0 or f_2^0 tends to zero, and when either one of these energies is below the Fermi level, the last factor falls off equally rapidly. In consequence, the result (37) is approximately proportional to the product $f_1^0\bar{f}_1^0 f_2^0\bar{f}_2^0$, as can be verified by direct numerical comparisons. This feature is very convenient.

Although the approximate result (37) exhibits the right overall behavior, the relatively gentle energy modulation on top of the product $f_1^0\bar{f}_1^0 f_2^0\bar{f}_2^0$ is not accurately accounted for, since small deviations from the Fermi surface have been ignored in the derivation. As a consequence, direct use of that expression leads to certain inaccuracies, including a diffusive violation of energy and momentum conservation. Though relatively small, the occurrence of such errors is inconvenient and we therefore modify the expression slightly, in order to restore the conservation laws and, at the same time, gain further simplification.

Our resulting approximate expressions for the transport coefficients are then

$$\begin{aligned} \text{Drift}: \quad & V_1 \approx W_0\,[f_1^0 - f_1 - f_1^0\bar{f}_1^0 \textstyle\sum_2 C_{12}(f_2^0 - f_2)]\,, \\ \text{Diffusion}: \quad & D_{12} \approx W_0\,[f_1^0\bar{f}_1^0\delta_{12} - f_1^0\bar{f}_1^0 C_{12} f_2^0\bar{f}_2^0] = D_1^{\text{var}}\delta_{12} + D_{12}^{\text{cov}}\,. \end{aligned} \tag{39}$$

The rate of the transport process is governed by the relaxation time t_0,

$$t_0^{-1} = W_0 \approx \pi^2\frac{T^2}{\epsilon_F^2}[1 - \pi^2\frac{T^2}{\epsilon_F^2}]\rho\sigma_0 V_F\,, \tag{40}$$

in agreement with expressions derived for Fermi liquids [21]. Studies of nuclear dynamics carried out within the relaxation-time approximation have employed similar expressions, see for example [23, 24].

The overall diffusion rate, as obtained by integrating D_1^{var} in eq. (39) over $\mathbf{p}_1$, is then given by

$$D_0 \equiv \sum_1 D_1^{\text{var}} \approx \frac{3}{2}\pi^2\frac{T^3}{\epsilon_F^3}[1 - \pi^2\frac{T^2}{\epsilon_F^2}]\rho^2\sigma_0 V_F\,. \tag{41}$$

It should be noted that the temperature-dependent modulation in the bracket in eqs. (40-41) has been obtained on the basis of direct numerical integration of the exact expression (11). It is accurate to within one per cent for temperatures up to 5 MeV, at which point the correction term amounts to nearly 20 per cent. For applications at higher temperatures, the expression should be improved, which can be done relatively easily, but it should be kept in mind that our analytical approximations are not expected to remain accurate much beyond such a temperature.

The general form of the diffusion coefficient D_{12} follows from the decomposition (10) and the above results which show that the angular dependence is given by the Legendre expansion (36), to leading order in the temperature, and that $D_{12}^{\text{cov}} \sim f_1^0\bar{f}_1^0 f_2^0\bar{f}_2^0$ for the energy dependence. The coefficient C_{12} expresses the correlation induced between the occupancies at the two momenta $\mathbf{p}_1$ and $\mathbf{p}_2$ as a result of the two-body collisions and is given by

$$\rho\phi_0 C_{12} = \frac{\epsilon_1 - \epsilon_0}{\sigma_T}\frac{\epsilon_2 - \epsilon_0}{\sigma_T} + \sum_{\ell\geq 0}\frac{\phi_0}{\phi_\ell}(\frac{p_1 p_2}{2m})^\ell[2 - (-1)^\ell]P_\ell(\cos\theta_{12}) = \rho\phi_0\sum_{\ell\geq 0} C_{12}^\ell\,. \tag{42}$$

The form of the monopole and dipole terms are dictated by the demand that particle number, energy, and momentum be strictly conserved throughout each separate history. In fact, $C_{12}^0 + C_{12}^1 = S_{12}$, where S_{12} is the coefficient expressing the equilibrium correlation $\tilde{\sigma}_{12}^{\rm cov}$ in eq. (28). Consequently, $\sum_1 V_1 F_1^\alpha = 0$ and $\sum_2 D_{12} F_2^\alpha = 0$, and so the approximate transport coefficients (39) indeed conserve particle number, energy, and momentum exactly.

As for the higher multipoles, the energy dependence implied by (36) has been modified slightly to achieve a more convenient form. We note in this context that $\phi_\ell \approx \phi_0 \epsilon_0^\ell$, so the associated arbitrariness is of higher order in the temperature. It also follows that the angular dependence of $D_{12}^{\rm cov}$ is still given by the function $G(\theta)$ in (36), up to small finite-temperature corrections of the order of $(\tau/\epsilon_F)^2$.

Once the diffusion coefficient D_{12} has been specified, the form of the drift coefficient V_1 can be derived by demanding that the transport dynamics leads to the correct quantum-statistical equilibrium. For this analysis is it convenient to perform the the standard moment expansion around the mean trajectory [12, 14],

$$\dot{\sigma}_{12} \approx 2D_{12} + \sum_3 \left(\frac{\partial V_1}{\partial f_3} \sigma_{32} + \sigma_{13} \frac{\partial V_2}{\partial f_3} \right) . \tag{43}$$

By inserting the expressions for the transport coefficients (39) and the form of the equilibrium correlations $\tilde{\sigma}_{12}$ (28), it is straightforward to verify that σ_{12} indeed relaxes towards $\tilde{\sigma}_{12}$ (when the system is stable). In particular, when equilibrium has been reached, $\sigma_{12} = \tilde{\sigma}_{12}$, there is no further change in the correlations,

$$\begin{aligned} \dot{\sigma}_{12} &= 2W_0 (f_1^0 \bar{f}_1^0 \delta_{12} - f_1^0 \bar{f}_1^0 C_{12} f_2^0 \bar{f}_2^0) - 2W_0 \sigma_{12} + W_0 \sum_3 \left(f_1^0 \bar{f}_1^0 C_{13} \sigma_{13} + \sigma_{13} C_{32} f_2^0 \bar{f}_2^0 \right) \\ &= 2W_0 \Big[f_1^0 \bar{f}_1^0 \delta_{12} - f_1^0 \bar{f}_1^0 C_{12} f_2^0 \bar{f}_2^0 - f_1^0 \bar{f}_1^0 \delta_{12} + f_1^0 \bar{f}_1^0 S_{12} f_2^0 \bar{f}_2^0 + f_1^0 \bar{f}_1^0 C_{12} f_2^0 \bar{f}_2^0 \\ &\quad - \frac{1}{2} f_1^0 \bar{f}_1^0 \sum_3 C_{13} f_3^0 \bar{f}_3^0 S_{32} f_2^0 \bar{f}_2^0 - \frac{1}{2} f_1^0 \bar{f}_1^0 \sum_3 S_{13} f_3^0 \bar{f}_3^0 C_{32} f_2^0 \bar{f}_2^0 \Big] = 0 , \end{aligned} \tag{44}$$

using the sum rule $\sum_2 C_{12} f_2^0 \bar{f}_2^0 S_{23} = S_{13}$. It is also evident that the mean density relaxes towards the equilibrium density f^0.

The energy dependence of $D_{12}^{\rm cov}$ is given by

$$D(\epsilon, \epsilon') \equiv \frac{1}{\rho \phi_0 W_0} \sum_{12} \delta(\epsilon_1 - \epsilon) \delta(\epsilon_2 - \epsilon') D_{12}^{\rm cov} \approx - \frac{f_0 \bar{f}_0}{\tau} \frac{f_0' \bar{f}_0'}{\tau} \left(1 + \frac{\epsilon - \epsilon_0}{\sigma_\tau} \frac{\epsilon' - \epsilon_0}{\sigma_\tau} \right) \frac{\sqrt{\epsilon \epsilon'}}{\epsilon_F} . \tag{45}$$

The first expression can be calculated by direct numerical (Monte Carlo) evaluation of the integrals in eq. (11) for $D_{12}^{\rm cov}$, while the approximate expression follows from our analytical approximation (39). The function $D(\epsilon, \epsilon')$ is normalized such that its total integral is equal to minus one. The corresponding projection $D(\epsilon)$ is obtained by integrating the above expression for $D(\epsilon, \epsilon')$ over ϵ', $D(\epsilon) \approx -\sqrt{\epsilon/\epsilon_F} f_0 \bar{f}_0 / \tau$. This simple expression is an excellent approximation to the result of integrating the approximate formula numerically. Both are slightly narrower than the result of the exact calculation but the overall normalizations are the same within one per cent.

The peak value of our approximation for $D(\epsilon)$ exceeds the exact result by about a factor of $\frac{4}{3}$, while it vanishes more rapidly as one moves away from the Fermi surface. We also note that the conditional mean value of $D(\epsilon, \epsilon')$ is given by ϵ to a very good approximation, while its conditional variance is given approximately by σ_τ^2.

It is also instructive to consider the angular dependence of $D_{12}^{\rm cov}$,

$$D(\theta) \equiv \frac{2}{\rho\phi_0 W_0} \sum_{12} \delta(\cos\theta_{12} - \cos\theta) D_{12}^{\rm cov} \approx G(\theta) \; . \tag{46}$$

The approximate expression is implied by the approximation (39). The trigonometric expression for $G^+(\theta)$ is an excellent approximation, and the one for $G^-(\theta)$ is also quite good. These expressions diverge at $\theta = \pi, 0$, respectively, while the exact result remains finite. However, there is relatively little angular weight near these two extreme points and, furthermore, the divergencies are integrable, so this deficiency is relatively unimportant. It can also be demonstrated that the Legendre expansion converges quickly to a reasonable global approximation to $G(\theta)$.

The approximate transport coefficients (39) are expressed in terms of the appropriate local Fermi-Dirac equilibrium density $f^0(\mathbf{r}, \mathbf{p})$, which in turn is determined by the local density $\rho(\mathbf{r}) = \sum_1 f(\mathbf{r}, \mathbf{p}_1)$, and the local effective temperature $\tau(\mathbf{r})$ which follows from the local kinetic density $\sum_1 \epsilon_1 f(\mathbf{r}, \mathbf{p}_1)$ (expressed, of course, in the local flow frame in which $\sum_1 \mathbf{p}_1 f(\mathbf{r}, \mathbf{p}_1)$ vanishes). The anticipated use of the derived expressions is then as follows: For a given phase-space density $f(s)$, determine at each position $\mathbf{r}$ the density $\rho(\mathbf{r})$ and effective temperature $\tau(\mathbf{r})$, leading to the local Fermi-Dirac equilibrium density $f^0(\mathbf{r}, \mathbf{p})$, and then use the expressions (39) to obtain the corresponding transport coefficients. In this manner, the expressions can be employed for arbitrary non-equilibrium densities $f(s)$, although their accuracy presumably diminishes away from equilibrium.

It is instructive to note that the derived approximation (39) can be viewed as the commonly employed *harmonic* approximation to the Fokker-Planck description of the Boltzmann-Langevin transport problem, in which the diffusion coefficient is a constant, *i.e.* at each position $\mathbf{r}$ it is independent of the actual state f (provided it has the same density, flow velocity, and effective temperature) and the restoring force is proportional to the deviation from equilibrium, $f - f^0$ (often denoted the linear-response approximation). Of course, when the Vlasov propagation is included the local equilibrium density f^0 becomes a functional of the dynamical density f via the local density $\rho[f]$ and the local effective temperature $\tau[f]$, as explained above, so the approximation is less restrictive than it might at first appear. The harmonic approximation is expected to be reasonable in situations where the deviations from local equilibrium are relatively small. Although the drift coefficient is linear in the deviation from local equilibrium, the induced average change at a given point $(\mathbf{r}, \mathbf{p}_1)$ in phase-space depends not only on the deviation at that momentum $\mathbf{p}_1$ but on the deviations at all the other momenta $\mathbf{p}_2$ as well (at the same position $\mathbf{r}$), so the expression for V_1 in (39) can be regarded as a generalized relaxation-time approximation in which the restoring coefficient is a matrix rather than a constant.

6. Numerical Simulation of the Fokker-Planck Equation

The BL equation can be simulated directly on a lattice in phase space, as discussed in detail in refs. [13, 14, 25]. However, the simulation of all the distinct two-particle-two-hole processes is extremely time demanding. As a consequence, actual simulations have so far been limited to idealized two-dimensional systems and the extension to more realistic scenarios appears beyond reach with the present computer power. There is therefore a need for developing more approximate simulation methods, and the Fokker-Planck formulation of the problem offers such a possibility, since it describes the correlated changes only at the two-body level (as opposed to the four-body correlations generated by each actual collision process). We describe below a straightforward method for this task.

Assume that we know the phase-space density $f(s,t)$ at a given time t. The calculational task is then to calculate the density $f(s,t+\Delta t)$ after a suitably short time interval Δt has elapsed. According to the BL equation of motion (1), this change has two parts, the evolution associated with the collisionless propagation of the nucleons in the selfconsistent effective field described by the Hamiltonian $h[f]$, and a change $\Delta f(s)$ resulting from the occasional two-body collisions. The former part can be treated by a variety of well-developed methods and need not concern us here; this part is anyway trivial in the case of a free nucleon gas.

The possible collision-induced changes constitute an ensemble $\{\Delta f\}$ which is constrained by the transport coefficients, in accordance with the relations (8) and (9). So it must be ensured that

$$\prec \Delta f(s_1) \succ = V[f](s_1)\Delta t\ , \quad \prec \Delta f(s_1)\Delta f(s_2) \succ = 2D[f](s_1,s_2)\Delta t\ . \tag{47}$$

We may therefore proceed as follows. From the given density $f(s,t)$ we calculate the corresponding transport coefficients $V[f](s)$ and $D[f](s,s')$, using the expressions (39). At each point in space $\mathbf{r}$, we then pick random increments $\Delta f(\mathbf{r},\mathbf{p})$ from the ensemble of possible increments, making sure that the above constraints on mean values and correlations are in accordance with the local values of the transport coefficients, as expressed by (47). In this manner one of the possible new densities $f(s,t+\Delta t) = f(s,t) + \Delta f(s,t)$ is obtained. The procedure is then repeated on the basis of new this density, and in this manner one entire history $f(t)$ is generated. An ensemble of $\mathcal{N}$ independent histories $\{f^{(n)}(t)\}$ can be obtained by performing $\mathcal{N}$ such "runs", each time starting with the same specified initial density $f(s,t=0)$.

The computational effort depends directly on the efficiency with which the correlated sampling problem can be solved. (If we wish to generate a sample of $\mathcal{N}$ histories, using M time steps for each, then we need to calculate the transport coefficients a total of $\mathcal{N}M$ times and each time pick random changes $\Delta f(s)$ under the appropriate moment constraints (47).) The simple approximate forms derived for the transport coefficients makes it possible to calculate these extremely fast. Consequently, the numerical simulation of realistic scenarios has been brought within the realm of practicality, provided that a suitably fast method can be devised for picking the random changes $\Delta f(s)$ according to (47).

7. Concluding Remarks

Nuclear collisions at intermediate energies create systems that may be reasonably described on the basis of semi-classical one-body theory, augmented by the effects of the residual stochastic nucleon-nucleon collisions. Considered as an interacting Fermi-Dirac gas of nucleons, these systems are fairly unique in that their excitation is comparable to the Fermi energy, so that the systems are too degenerate for classical statistics to apply, and yet too agitated to render the zero-temperature limit sufficiently accurate. Moreover, during the most interesting part of the dynamics the system is still far from global equilibrium and so statistical methods are of limited utility. The development of transport descriptions for these complicated processes may therefore bring novel theoretical advances.

The nuclear Boltzmann-Langevin theory provides a well-based and consistent description of nuclear dynamics and the reported recent advances in developing approximate treatments give grounds for hope that numerical simulations can be carried out for realistic scenarios, such as intermediate-energy nuclear collisions.

It should be kept in mind that the Boltzmann-Langevin theory, even though it provides a physically sound basis for understanding nuclear dynamics at intermediate energies, should only be expected to describe the macroscopic features, in analogy with the liquid drop model of nuclear bindings, and effects arising from specific nuclear structure, such as premordial many-body correlations, are beyond its scope. The importance of such complications may, in principle, be assessed by confronting the model with suitable experimental data.

This work was supported by the Director, Office of Energy Research, Office of High Energy and Nuclear Physics, Nuclear Physics Division of the U.S. Department of Energy under Contracts No. DE-AC03-76SF00098 and DE-FG05-89ER40530.

1. G. Batko and J. Randrup, Nucl. Phys. A563 (1993) 97
2. W. Bauer, G.F. Bertsch, and H. Schultz, Phys. Rev. Lett. 69 (1992) 1888
3. L.G. Moretto, K. Tsuo, N. Colonna, and G.J. Wozniak, Phys. Rev. Lett. 69 (1992) 1884
4. D.H.E. Gross, Bao-An Li, and A.R. DeAngelis, Ann. Physik 1 (1992) 467; Bao-An Li and D.H.E Gross, Nucl. Phys. A554 (1993) 257
5. L.W. Nordheim, Proc. Roy. Soc. A119 (1928) 689
6. E.A. Ühling and G.E. Uhlenbeck, Phys. Rev 43 (1933) 552
7. J. Randrup, Nucl. Phys. A314 (1979) 429
8. G.F. Bertsch and S. Das Gupta, Phys. Rep. 160 (1988) 190
9. P. Schuck, R.W. Hasse, J. Jaenicke, C. Gregoire, B. Remaud, F. Sebille, and E. Suraud, Prog. Part. Nucl. Phys. 22 (1989) 181

10. W. Cassing, V. Metag, U. Mosel, and K. Niita, Phys. Rep. 188 (1990) 363

11. S. Ayik and C. Gregoire, Phys. Lett. B212 (1988) 269; Nucl. Phys. A513 (1990) 187

12. J. Randrup and B. Remaud, Nucl. Phys. A514 (1990) 339

13. Ph. Chomaz, G.F. Burgio, and J. Randrup, Phys. Lett. B254 (1991) 340

14. G.F. Burgio, Ph. Chomaz, and J. Randrup, Nucl. Phys. A529 (1991) 157

15. J. Randrup, *Interdisciplinary Workshop on Statistical Description of Transport in Plasma, Astro-, and Nuclear Physics*, Les Houches, France (1992), Nova Science

16. N.G. van Kampen, *Stochastic Processes in Physics and Chemistry*, North Holland, New York, 1981

17. B. Benhassine, M. Farine, E.S. Hernandez, D. Idier, B. Remaud, and F. Sebille, Nucl. Phys. A545 (1992) 81c

18. R.W. Hasse and W.D. Myers, *Geometrical Relationships of Macroscopic Nuclear Physics*, (Springer, Berlin, 1988), p. 17

19. S. Chandrasekhar, *Stellar Structure* (Dover, New York, 1957) p. 389

20. P. Morel and P. Nozières, Phys. Rev. 126 (1962) 1909

21. G. Baym and C. Pethick, *The Physics of Liquid and Solid Helium, Part II*, Eds: K.H. Bennemann and J.B. Ketterson, John Wiley & Sons, New York (1976) 1

22. J. Randrup and S. Ayik, Nucl. Phys. A (1994) in press (preprint LBL-34484)

23. H.S. Köhler, Nucl. Phys. A378 (1982) 159; *ibid.* 181; Phys. Scripta 26 (1982) 51

24. H.S. Köhler and B.S. Nilsson, Nucl. Phys. A417 (1984) 541

25. F. Chapelle, G.F. Burgio, Ph. Chomaz, and J. Randrup, Nucl. Phys. A540 (1992) 227

STRONG DISSIPATION IN FISSION

YASUHISA ABE
Yukawa Institute for Theoretical Physics
Kyoto University, Kyoto 606, Japan

and

TAKAHIRO WADA
Department of Mathematical Sciences, Osaka Dental University
Makino-Honmachi, Hirakata, Osaka 573, Japan

and

NICOLAE CARJAN
Centre d'Etudes Nucléaires Bordeaux
33175 Gradignan Cedex, France

ABSTRACT

Fission dynamics of excited nuclei is analysed from the viewpoint of Brownian motion. A time evolution of the fissioning motion is solved by Langevin equation, taking into account of cooling due to particles evaporation. Anomalous multiplicities of pre-scission neutrons and kinetic energies of fission fragments require a strong dissipation of one-body friction type.

1. Introduction

Heavy ion reactions produce highly excited compound nuclei, which decay through evaporation and fission. Fission is a typical example of collective motions. In high excitation, collective degrees of freedom could be considered as Brownian particles interacting frequently with thermalized nucleons which could be considered as a heat bath. So fission of highly excited compound nuclei provides us a good physical object for a study of transport properties of excited nuclear matter.

A frictional force exerting on the fissioning degree of freedom makes the motion to slow down as an average and hence makes life time of fission longer. Actually, fission of excited nuclei is not so simple, because all the energy brought in is immediately transferred to intrinsic excitation of nuclei to form a heat bath. In such a situation, fission process can be considered to be an escape over the barrier under the influence of the heat bath, as Kramers first discussed in 1940.[1]

Anyhow, an interesting physical quantity is the life time. In fusion-fission processes, it is usually difficult to measure fission life directly. Almost unique way, though indirect, is to measure neutron multiplicities prior to scission, because the life time of neutron emission is shorter than that of fission and can be used as a clock to measure the latter.[2] As a rough estimation, the multiplicity of pre-scission neutrons $< \nu_{pre} > = \tau_f/\tau_n = \Gamma_n/\Gamma_f$, where $\Gamma_n(\tau_n)$ is the width (life) of neutron emission and

$\Gamma_f(\tau_f)$ the width (life) of fission decay. Then $\tau_f =< \nu_{pre} > \cdot \tau_n =< \nu_{pre} > \cdot \hbar/\Gamma_n$. The width Γ_n can be reliably calculated with Weisskopf formula. The width Γ_f can be also calculated with Bohr-Wheeler theory, but was turned out not to be consistent with the measured $< \nu_{pre} >$ as excitation energy goes up. In higher excitation, multi-step treatments like the master equation or the cascade model is necessary to calculate $< \nu_{pre} >$, because at each step of evaporation an excitation energy etc change. Even after taking into account the effect, discrepancies remain of about factor 2. Therefore, the fission life is inferred to be longer than that given by Bohr-Wheeler theory. In order to reproduce the observed neutron multiplicities, Hinde *et al.*[3] had to introduce phenomenologically a long "delay" time ($\sim 5 \times 10^{-20}s$) during which fission cannot occur. Possible effects which make the fission life longer are dynamical effects which are not taken into account in Bohr-Wheeler theory, i.e., the transient time during which the collective degree of freedom (Brownian particle) attains "equilibrium", the saddle-to-scission time, and so-called Kramers' factor, i.e., the pre-exponential factor to Bohr-Wheeler formula which takes into account an effect of frictional force. All of the above threes make the life longer as a friction or a dissipation is stronger. In order to take them into account, we have to solve a dynamical equation for fissioning motion under the influence of the heat bath, such as Fokker-Planck equation. As is well-known, fission is a multi-dimensional process for which Fokker-Planck equation is not practical. This is the reason why the present author proposed to use Langevin equation and showed its applicability to fission.[4] Many works,[5] some of them using Langevin equation, have been made to deduce dissipation constants for nuclear collective motions from the anomalous multiplicity data, γ-ray yields etc, but so far a quantitative conclusion commonly accepted has not yet been obtained. In addition to the multiplicities, there is another relevant physical quantity, i.e., kinetic energy distribution of fission fragments, which mainly reflects dynamical evolution from the saddle to the scission. Thus the purpose of the present talk is to present important results obtained by analysing both the neutron multiplicities and the kinetic energies at the same time with the multi-dimensional Langevin equation coupled to particles evaporation.

2. Langevin Equation and the Macroscopic Collective Models

The multi-dimensional Langevin equation which describes nuclear collective motions is given as follows,

$$\frac{dp_i}{dt} = -\frac{\partial V}{\partial q_i} - \frac{1}{2}\frac{\partial}{\partial q_i}(m^{-1})_{jk}p_j p_k - \gamma_{ij}(m^{-1})_{jk}p_k + g_{ij}R_j(t)$$
$$\frac{dq_i}{dt} = (m^{-1})_{ij}p_j,$$

where repeated indices are assumed to sum up. $V(q)$ is the potential energy, which is calculated as the sum of a generalized surface energy, a Coulomb energy for diffused surface and a centrifugal potential. The temperature dependence of the nuclear surface energy is included in the form; $E_s(q,T) = E_s(q,T=0)\cdot(1-\xi T^2)$. The parameter ξ is calculated with the extended Thomas-Fermi model.[6] but is yet rather ambiguous.

We, thus, take two values within the ambiguity. The temperature T is calculated from the excitation energy with $E^* = a \cdot T^2$, where the parameter a is taken to be that of Töke and Swiatecki[7] in the spherical shape, 23.2, for ^{200}Pb. $m_{ij}(q)$ is the inertia tensor, which is calculated with Werner-Wheeler approximation for the velocity field.[8] $\gamma_{ij}(q)$ is the shape-dependent dissipation tensor, for which we take the Wall-and-Window one-body dissipation[9] and the hydrodynamical two-body dissipation.[8] The strength of the random force g_{ij} is related by $\Sigma_k g_{ik} g_{jk} = T \cdot \gamma_{ij}$, which guarantees systems to reach the equilibration after certain time in the case of the closed system. In the present open case the system approaches the quasi-stationary regime, which is first discussed by Kramers.[1] In the schematic case with an oscillator potential and parabolic barrier, Kramers' stationary limit (for moderate or strong friction) is given as follows,

$$\Gamma^K = \frac{\hbar\omega_0}{2\pi\omega_b} \cdot \left[\sqrt{\omega_b^2 + (\beta/2)^2} - \beta/2\right] \cdot e^{-B_f/T} = \left[\sqrt{1+x^2} - x\right] \cdot \Gamma_{BW},$$

$$\Gamma_{BW} = \frac{\hbar\omega_0}{2\pi} \cdot e^{-B_f/T}, x = \beta/2\omega_b,$$

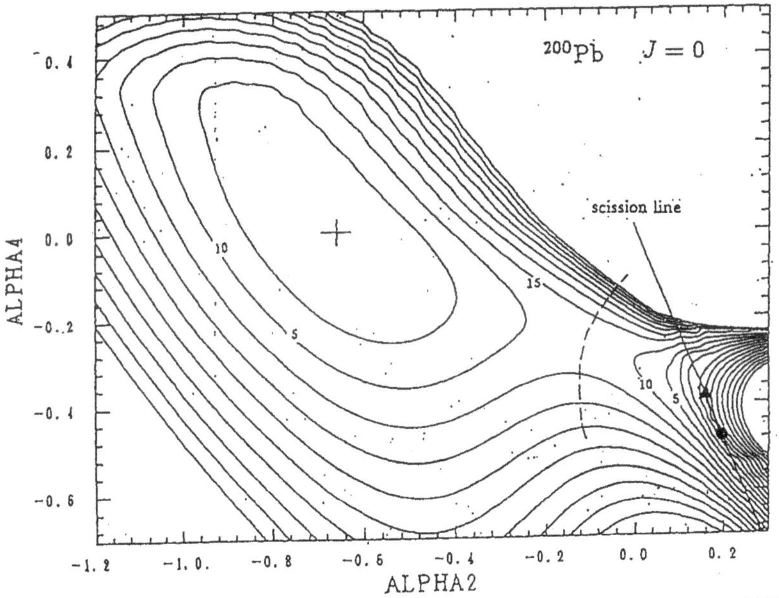

Fig.1. Potential energy surface for ^{200}Pb with $J = 0$ and $T = 0$. The dot on the scission line denotes the averaged scission point over the trajectories with the one-body friction, while the triangle denotes that with the two-body viscosity ($\mu = 0.2TP$). The cross denotes the potential minimum and the dashed line the ridge line including the saddle point.

where ω_0 and ω_b are the frequencies at the potential minimum and at the barrier top, respectively. β is a constant reduced friction $\sim \gamma/m$ and B_f is a barrier height. The factor in bracket before Γ_{BW} is so-called Kramers' factor. If $x > 1$, the motion is overdamped and if $x < 1$, underdamped. In the dynamical calculations, the fission width is time-dependent and approaches to Kramers' limit after certain time, i.e., after the transient time. During all the processes including the transient time and the time necessary for the descent from the saddle to the scission, the system evaporates particles (neutrons, protons etc) incessantly. In the present calculations, the cooling due to evaporation is calculated in the continuous limit. The validity of this treatment was studied by Delagrange *et al.*[5] Accordingly, the excitation energy and the angular momentum change continuously with time. Hence the temperature is a smooth function of time.

In the examples of calculations we assume axially symmetric and mass-symmetric nuclear shapes, which we described by Legendre-polynomial parametrization[10] of two dimensions. The potential energy surface for ^{200}Pb is shown in Fig.1. With this potential, and the mass and the friction tensor, the above Langevin equation is solved for trajectories of the order of 10^5 for each spin by iteration method with a small time step, up to the time $200\hbar$/MeV, after which the quasi-stationary fission width (which corresponds to multi-dimensional Kramers' limit) is used. This is long enough for the time-dependent width to reach the quasi-stationary value when fission barrier exists. When barrier vanishes (in higher angular momenta), this is long enough for all the trajectories to cross the scission line.

3. Results

Calculations have been made for the symmetric fission of ^{200}Pb nucleus since the following reactions have been studied experimentally: ^{19}F+^{181}Ta(E^* = 80.7MeV)[11] and ^{16}O+^{184}W(E^* = 195.8MeV).[12] Figure 2 shows the numbers of emitted particles from the compound nucleus as functions of time for the initial excitation energy $E^* = 80.7$MeV. The excitation energy of the compound nucleus (denoted by the dashed line in Fig.2) decreases with time as a result of the emission of the light particles and γ rays. Figure 3 shows the fission widths as function of time for the same initial excitation energy with the one-body friction and with $\xi = 0.014\text{MeV}^{-2}$. The fission width Γ_f is calculated as $\Gamma_f(t) = -(1/\text{N}(t))(dN(t)/dt)$, where $N(t)$ is the number of trajectories which did not escape beyond scission (saddle) at time t. One sees that the fission widths (dotted lines at saddle and solid lines at scission) approach the quasi-stationary value (dashed lines) after certain time. Note that the quasi-stationary value depends on time due to the cooling by particles evaporation. The transient time (t_{tr}) is the time up to which the dotted lines reach the dashed lines and the saddle-to-scission time (t_{ssc}) is the interval between the dotted and the solid lines. From Fig.3, both t_{tr} and t_{ssc} are about $2 \times 10^{-20}s$ for$J \leq 50\hbar$, where fission barrier exists. This calculated transient time turns out to be shorter than required by the phenomelogical analysis.

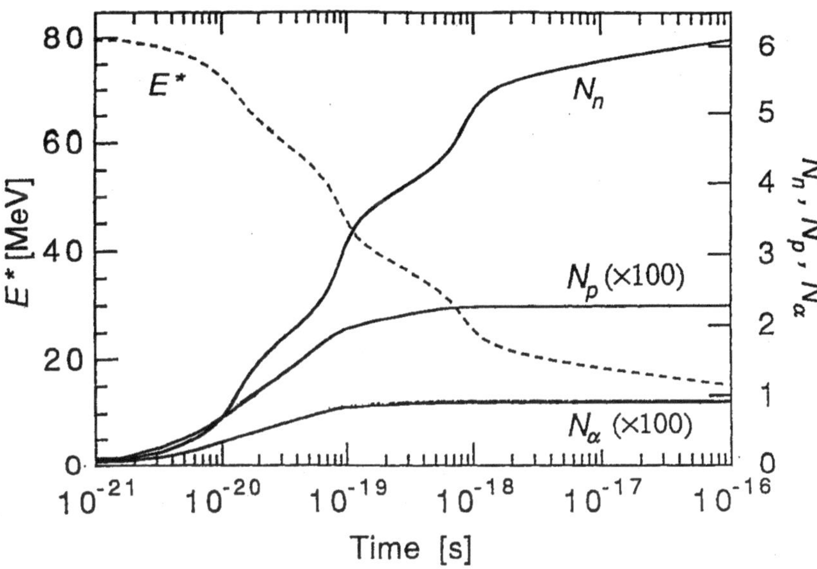

Fig.2. Numbers of emitted neutrons (N_n) protons (N_p), and α particles (N_α) as function of time for ^{200}Pb with the initial excitation energy $E^* = 80.7$MeV. Time dependence of the excitation energy (E^*) is also shown (dashed line).

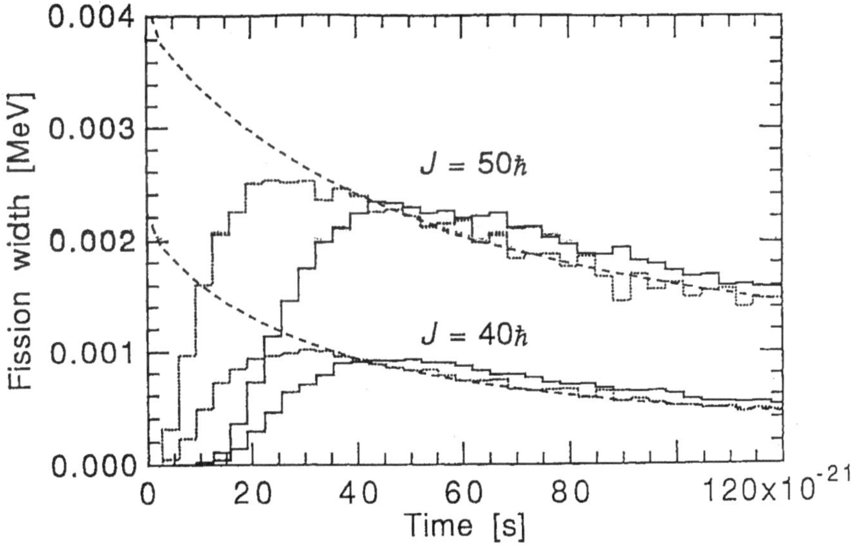

Fig.3. Time dependence of the fission width for the symmetric fission of ^{200}Pb for two values of anglular momentum $J = 40\hbar$ and $50\hbar$, and initial excitation energy $E^* = 80.7$MeV. Solid lines are the widths calculated at scission and dotted are those at saddle. Dashed lines denote quasi-stationary fission widths.

We calculated multiplicities of neutrons emitted prior to scission and kinetic energy distributions of fission fragments, averaged over the spin distribution of the compound nucleus. The spin distribution is parametrized as follows: $\sigma_{fus} = \pi\lambda^2\Sigma_J(2J+1)/[1+\exp((J-J_c)/\Delta_J)]$, where J_c and Δ_J are determined to reproduce the experimental fusion cross section (σ_{fus}). We assign an event as fusion-evaporation if the corresponding trajectory did not escape before the fission width becomes smaller than the γ decay width. Calculated results are given in Table I for the one-body dissipation case and for the two values of ξ(0.014 and 0.009 MeV^{-2}), which turned out to give not so much difference in all the results, particle multiplicities, kinetic energies, *etc.* It is remarkable that the calculated pre-scission neutron multiplicity (ν_{pre}) coincides with the experimental one within the error bar. The fusion-fission cross section is also quite well reproduced. With the same prescription, we have made calculations starting at $E^* = 195.8$ MeV. The kinetic energy distributions and the Coulomb energy contributions are shown in Fig.4 for the one-body friction. The most probable Coulomb energy indicates the most probable scission point on the scission line, which is marked with the dot on the scission line in Fig.1. The triangle on the scission line denotes that of two-body viscosity with $\mu = 0.2$ TP. The trajectories are different between the one-body friction and the two-body viscosity. The calculated variance (σ_{TKE}), however, is too small to reproduce the experiment at $E* = 195.8$ MeV. It should be noted that the experimental variance includes all masses, not only symmetric fission products. The small σ_{TKE} indicates that the present two-dimensional description of the nuclear shape is not sufficient to describe all possible scission configurations. An improvement of scission configuration description is now being made by including a mass asymmetry coordinate.

Table I. Calculated results for the case of the one-body friction. The columns contain: the excitation energy (E^*), the temperature dependence parameter (ξ), the fusion cross section (σ_{fus}), the fusion-fission cross section (σ_{fiss}), the fusion-evaporation cross section (σ_{ev}), the pre-scission multiplicities of neutron (ν_{pre}), proton (π_{pre}), α particle (α_{pre}), the average total kinetic energy of the fission fragments (TKE), and the variance (σ_{TKE}).

E^* (MeV)	ξ (MeV^{-2})	σ_{fus} (mb)	σ_{fiss} (mb)	σ_{ev} (mb)	ν_{pre}	π_{pre}	α_{pre}	TKE (MeV)	σ_{TKE} (MeV)
	0.014	1150	790	360	2.93	0.0092	0.0037	135.1	8.46
80.7	0.009	1150	758	392	3.08	0.0091	0.0036	135.5	8.33
	exp.	1150	767	383	3.2±0.3	–	–	–	–
	0.014	1400	1244	156	7.33	0.363	0.140	137.0	10.2
195.8	0.009	1400	1223	177	7.64	0.385	0.140	137.5	9.82
	exp.	1400	–	–	7.7±0.3	–	–	139	16.5

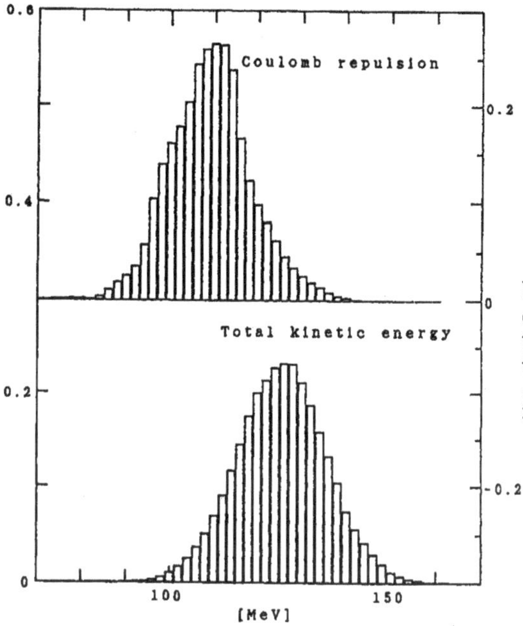

Fig.4. Distributions of Coulomb energy and total kinetic energy on the scission line for the case with the one-body friction and with $J = 60$. The ordinate is arbitrarily scaled.

Results with the hydrodynamical two-body viscosity are given in Table II. One can see that the case with $\mu = 0.06$ TP gives too small ν_{pre}. The cross sections are poorly reproduced as well. With $\mu = 0.20$ TP, ν_{pre} is closer to the experiment but *TKE* is far smaller than Viola systematics.[13] If one wants to reproduce $< \nu_{pre} >$ also of $E^* =$ 195.8MeV, one needs stronger μ and then the triangle in Fig.1 shifts upwards and *TKE* becomes still smaller. This tendency does not change when we take $\xi = 0.009$ MeV^{-2}. Therefore, one can conclude that the hydrodynamical two-body viscosity cannot give a consistent explanation of both neutron multiplicities and kinetic energies, while the one-body friction can.

Table II. Calculated results for the case of two-body viscosity and for $\xi =$ 0.014MeV^{-2}.

E^* (MeV)	μ (TP)	σ_{fus} (mb)	σ_{fiss} (mb)	σ_{ev} (mb)	ν_{pre}	π_{pre}	α_{pre}	TKE (MeV)	σ_{TKE} (MeV)
80.7	0.06	1150	928	222	2.06	0.0084	0.0035	124.9	10.2
	0.20	1150	817	333	2.84	0.0108	0.0037	108.6	9.92
	exp.	1150	767	383	3.2±0.3	–	–	–	–
195.8	0.06	1400	1338	62	4.79	0.242	0.106	123.6	13.9
	0.20	1400	1261	139	7.03	0.350	0.137	107.4	11.5
	exp.	1400	–	–	7.7±0.3	–	–	139	16.5

It is interesting to know why the calculated ν_{pre} have reproduced the experimental values, as shown in Table I, even though the calculated transient time is shorter than "delay time" in the phenomelogical analysis. The contributions from the transient time and the saddle-to-scission time are found to be fairly small, which are about 10% and 20% of the total calculated neutron multiplicities for 80.7 MeV and 195.8 MeV, respectively. They are obtained from the comparison with the case that stationary values (dashed lines in Fig.3) of scission widths are used from the beginning. (Note that we take the same procedure even in higher angular momenta with no barriers.) Therefore, reproductions of the experimental data are mainly due to the small stationary widths calculated with Langevin equation. Actually, a rough comparison of them with Bohr-Wheeler width gives factors of 1/10. In the phenomelogical analyses,[3] the fission width is set to zero during the "delay time" and it is the Bohr-Wheeler value afterwards. It should be noted that in higher excitation energies the contributions increase more and more, because the fission life gets shorter while the transient time remains rather constant. It should also be noted that a large part of fusion-fission events comes from the large values of J where fission barrier does not exist any more. In phenomenological analyses, one may take $\Gamma_{BW} = 0$ and a necessary lifetime may be provided by "delay" time. In our dynamical calculations, the average of the traveling time from the initial point to the scission line is long due to the strong friction and the distribution of this traveling time is broad due to strong random force. Therefore, the above dynamical contributions of 10% and 20% should be taken with the assumption that there are stationary limit also in higher angular momenta with no barriers. They seem to exist in actual calculations, but a physical meaning should be different from Kramers' one.

4. Summary and Discussion

We have investigated the fission dynamics of hot nuclei using the two-dimensional Langevin equation. Including the particle emission in the continuous limit, we have calculated the pre-scission multiplicities of neutrons, protons, and α particles as well as the kinetic energy distribution of fission fragments. The obtained numbers of pre-scission neutrons agree well with the experimental ones when we adopt the one-body dissipation. Unusually strong ($\mu = 0.20$ TP or larger) two-body viscosity is necessary to reproduce the observed neutron multiplicity, but this assumption is incompatible with the fission-fragment kinetic energy. This means that the fission phenomenon requires a dissipation mechanism which is very strong and makes scission configuration compact.

It is, therefore, extremely interesting to derive the one-body friction or its equivalent by microscopic theories. Recent microscopic calculations of diffusion coefficient[14] and friction constant,[15] however, appear to give too strong dissipation for nuclear collective motions. Yamaji *et al.*,[16] on the other hand, obtained a friction coefficient comparable with the wall formula using the linear response theory. The temperature dependence of friction predicted by microscopic theories is yet to investigate in comparison with experiments.

This work is performed in part (T.W.) under auspices of the Special Researcher's Basic Science Program. Y.A. is grateful for the financial support by the Grant-in-Aid for Scientific Research of Japan Ministry of Education, Science and Culture under the Grant number 03640269.

REFERENCES

1. H.A. Kramers, *Physica* 7 (1940) 284.
2. D. Hilscher and H. Rossner, preprint of a review article, Dec. 1992.
3. D.J. Hinde, D. Hilscher and H. Rossner, Nucl. Phys. **A502** (1989) 497c.
4. Y. Abe, C. Grégoire and H. Delagrange, J. Physique **47** (1986) C4-329.
5. P. Grange, Li Jun-Qing and H.A. Weidenmüller, Phys. Rev. **C27** (1983) 2063.
   H. Delagrange et al., Z. Phys. **A323** (1986) 437.
   G.-R. Tillack, Phys. Lett. **B278** (1992) 403.
   P. Fröbrich, I.I. Gonthar and N.D. Mavlitov, Nucl. Phys. **A556** (1993) 281.
   N. Carjan, T. Wada and Y. Abe, AIP Conf. Proc. **250** (1992) 230.
   T. Wada, Y. Abe and N. Carjan, Phys. Rev. Lett. **70** (1993) 3538.
   D. Cha and G.F. Bertsch, Phys. Rev. **C46** (1992) 306.
6. X. Campi and S. Stringari, Z. Phys. **A309** (1983) 239.
7. J. Töke and W.J. Swiatecki, Nucl. Phys. **A372** (1981) 141.
8. K.T.R. Davis, A.J. Sierk and J.R. Nix, Phys. Rev. **C13** (1976) 2385.
9. J. Blocki et al., Ann. Phys. **113** (1978) 330.
   J. Randrup and W.J. Swiatecki, Nucl. Phys. **429** (1984) 105.
10. S. Trentalange, S.E. Koonin and A.J. Sierk, Phys. REv. C22 (1980) 1159.
    N. Carjan, A.J. Sierk and J.R. Nix, Nucl. Phys. A452 (1986) 381.
11. D.J. Hinde et al., Nucl. Phys. A452 (1986) 550.
12. D.J. Hinde et al., Phys. Rev. C45 (1992) 1229.
13. V.E. Viola, K. Kwiatkowski and M. Walker, Phys. Rev. C31 (1985) 1550.
14. B.W. Bush, G.F. Bertsch and B.A. Brown, Phys. Rev. C45 (1992) 1709.
15. D. Boilley et al., Nucl. Phys. A556 (1993) 67.
    See also, M.C. Nemes and H.A. Weidenmüller, Phys. Rev. C24 (1981) 944 for random matrix approach.
16. S. Yamaji et al., Nucl. Phys. A475 (1988) 487.

Chaotic dynamics in the framework of the Vlasov-Nordheim equation

M. Baldo, G.F. Burgio and A. Rapisarda[1]

Universitá di Catania and INFN Sezione di Catania
Corso Italia 57, I-95129 Catania, Italy

ABSTRACT

In the framework of the time-dependent mean field theory, we solve the Vlasov-Nordheim equation and study the sensitive dependence of this highly non-linear equation on the initial conditions. We find that, when nuclear matter is inside the spinodal region, initially small differences between similar trajectories grow exponentially and produce strongly divergent paths resulting in a chaotic evolution. We focus on the fact that chaos spontaneously arises without putting any additional noise. We calculate the largest Lyapunov exponent for various initial densities. This analysis is performed for nuclear matter in two dimensions. It is argued that the mean field chaoticity can cause nuclear multifragmentation observed in intermediate energy heavy ion reactions.

1 Introduction

Intermediate energy heavy ion collisions offer a unique possibility to study nuclear matter under extreme conditions of density and temperature. In the energy range between 20 and 100 MeV/nucleon, the formation of hot nuclei and their successive decay, through emission of several intermediate mass fragments (IMF : $Z \geq 2$) with large fluctuations from event to event, has stimulated a huge debate on the main mechanism responsible of this phenomenon, commonly named "multifragmentation".

Both statistical models and dynamical simulations [1, 2] have been widely used providing suggestions on the possible paths followed by the nuclear system along the multifragmentation process. Surface instabilities or ring formation [2] have been recently considered as possible steps towards fragment formation. Quite surprisingly also the percolation model has been successfully used [3], suggesting a universal behavior typical of phase transitions and common to different fields. The onset of instability in the spinodal zone of the nuclear matter Equation of State (EOS) has been studied in ref. [4], by analyzing the corresponding linear

[1]Talk given by A. Rapisarda.

response for imaginary frequency as a function of the wave number and of the density, even including a noise of Langevin-type to the collision integral in the BUU equation [5]. However already in the 80s, in one of the first attempts to describe multifragmentation Knoll *et al.* [6] performed numerical calculations in the framework of two-dimensional Time Dependent Hartree Fock (TDHF) equation which showed many of the features described by the recent BUU calculations [2]. In particular by evolving an ensemble of Slater determinants, they found a dependence on the initial conditions and a spontaneous symmetry breaking which lead to large fluctuation from event to event and in the charge or mass distributions. This scenario suggests a strong non-linear evolution, as recently suggested also in ref.[7, 8].

In order to clarify the mechanism of multifragmentation, we study in this contribution the dynamics of the Vlasov-Nordheim (VN) equation in the spinodal zone. We find unambiguously a strong non-linear evolution which implies a sensitive dependence on the initial conditions and a spontaneous symmetry breaking [8]. This means that the dynamics in the spinodal region is chaotic, although the equation is deterministic: similar trajectories can have a completely different evolution once they enter the spinodal region. We estimate also a Lyapunov exponent, which gives the mean rate of divergence between the trajectories in density space, in line with the usual analysis of dynamical systems [9]. We find that two-body collisions do not affect the onset of chaoticity in nuclear many-body systems. Finally we draw some conclusions. Further details can be found in ref. [8].

2 Theoretical framework

2.1 *The Vlasov-Nordheim equation*

The numerical analysis is performed by solving the Vlasov-Nordheim equation

$$\frac{\partial f}{\partial t} - \{H[f], f\} = I[f] \tag{1}$$

on a two-dimensional lattice, using the same code of ref.[5]. In eq. (1) $H[f]$ is the self-consistent effective one-body Hamiltonian, while $I[f]$ is the two-body Pauli blocked collision integral. The single particle phase space is divided into cells and nuclear matter is confined in a large torus with periodic boundary conditions. The side lengths of the torus are equal to $L_x = 51\ fm$ and $L_y = 15\ fm$. In order to achieve sufficient accuracy, very small cells having $\delta x = \frac{1}{3}$ fm and $\delta p = 40$ MeV/c are employed. The time integration step was chosen equal to 0.5 fm/c. We adopt a simplified Skyrme interaction for the effective one-body field which is averaged over the y-direction [5] and convoluted over a Gaussian function whose width is $\mu = 0.43$ fm. This is equivalent to introduce a finite effective range. We employ a Fermi momentum at saturation of $P_F = 260$ MeV/c, which gives a density in two dimensions equal to $\rho_0 = 0.55\ fm^{-2}$.

2.2 *Definition of the largest Lyapunov exponent*

In the theory of dynamical systems, a measure of chaoticity is expressed in terms of Lyapunov exponents [9]. The largest Lyapunov exponent λ_∞ can be extracted by considering at a given time t the distance $d(t)$ between two trajectories, along an unstable direction in phase space, and define

$$\lambda(t) = \frac{\log(d(t)/d_0)}{t} , \tag{2}$$

where $d_0 = d(0)$. The average of this quantity along the trajectory gives the mean rate of exponential separation of neighbouring trajectories

$$\lambda_\infty = \lim_{t\to\infty} \lim_{d_0\to 0} \lambda(t) \tag{3}$$

In numerical applications one has to select a series of small values of d_0, and check that the corresponding values of $\lambda(t)$ stabilize around a definite value of $\overline{\lambda}$.

In order to characterize the system as a whole we have chosen as distance between two trajectories the difference in norm between their density profiles

$$d(t) = \sum_i |\rho_i^{(1)}(t) - \rho_i^{(2)}(t)|/N_c , \tag{4}$$

where the index i runs over the N_c cells in ordinary space, and $\rho_i^{(1)}$, $\rho_i^{(2)}$ are the densities in the cell i for the trajectories 1 and 2 respectively. The definition of eq. (4) is sufficient for the present analysis, although possible differences in momentum space are averaged out. It should include the contribution of all the unstable modes which dynamically grow up during the evolution.

3 Results and discussion

First we analyze the mean field dynamics neglecting two-body collisions, i.e. solving the left-hand side of eq.(1). We have initialized nuclear matter with a sinusoidal profile along the x-direction, characterized by a wave number $k = 2\pi(n_k/L_x)$ with $n_k = 5$ and a small amplitude equal to 1% of the local density. The local momentum distribution is assumed to be the one of a Fermi gas at a temperature $T = 3\ MeV$.

In fig.1a we show the evolution of the density profile for different times at normal density. The initial oscillation along the x coordinate is rapidly damped and, as fig.1b shows, the evolution of the Fourier transform of the density profile presents only one peak which disappears during the same time scale. Changing the initial density by a small amount, for example $d_0 = \frac{\Delta\rho}{\rho_0} = 10^{-2}$, one obtains exactly the same behavior and the evolution of the two events is indistinguishable. This is typical of the "regular" region of a dynamical system, and indicates the stability of the dynamics with respect to small perturbations.

On the contrary, if the density ρ is chosen well inside the spinodal region, $\rho = 0.5\rho_0$ in fig.2, the initial symmetry of the density profile, after a short time, is completely broken and fluctuations are rapidly enhanced.

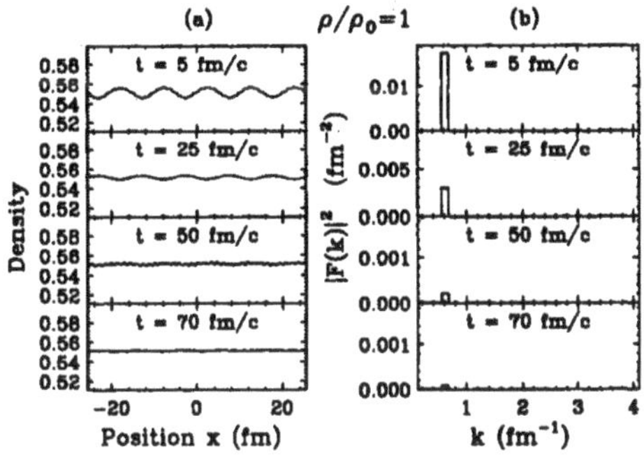

Fig.1 Evolution of the density profile (a) and its Fourier transform (b) at initial normal density, see text.

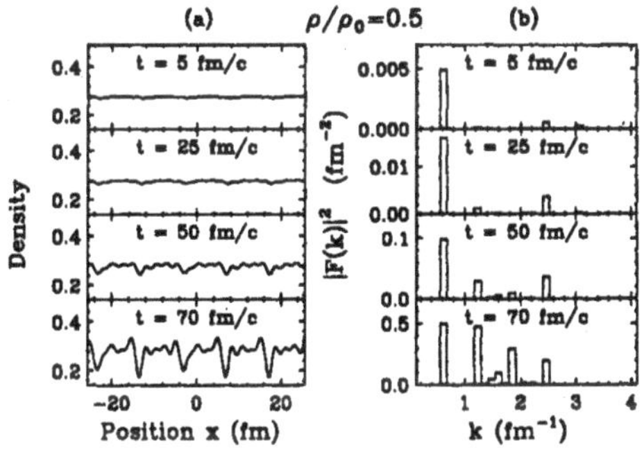

Fig.2 Evolution of the density profile (a) and its Fourier transform (b) at initial density $\rho = 0.5\rho_0$, see text.

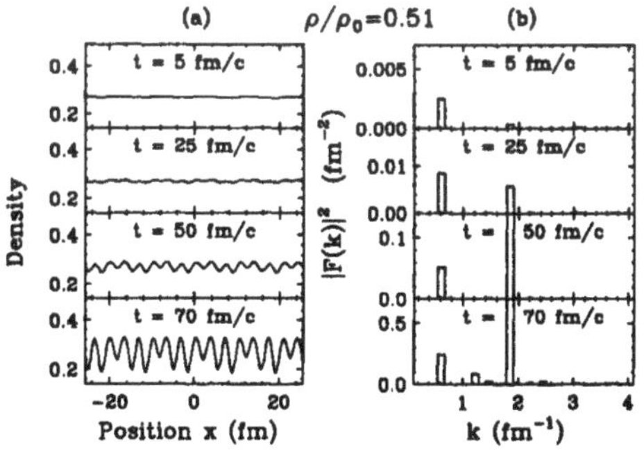

Fig.2bis Evolution of the density profile (a) and its Fourier transform (b) at initial density $\rho = 0.51\rho_0$, see text.

The Fourier transform on the other hand shows the appearance of several new modes which where not excited initially. Moreover a small change in the initial density, see fig.2bis where $\rho = 0.51\rho_0$, produces an evolution which is completely different both in shape and in amplitude from that of fig.2. This is the typical behavior of a dynamical system in a chaotic region, where due to the non-linearity of the equation, very small initial perturbations are rapidly amplified and the distance between similar trajectories diverges exponentially.

In our case this behavior occurs even if $d_0 = \frac{\Delta\rho}{\rho_0} = 10^{-4}$, see fig.4. In general one finds an increase of the populated wave numbers together with a strong mixing as the evolution proceeds, despite the fact that only one wave number k is initially occupied. The final frequencies spectrum has little resemblance with the initial one. Note however that the initial mode is not completely damped, and some memory of it is still retained. We have checked that this general trend depends neither on the initial n_k nor on the shape of the initial profile.

In fig.3a it is displayed the time evolution of the distance between trajectories, as defined in eq.(4), normalized with respect to its initial value $d_0 = 10^{-4}$. We note that, while the distance remains constant in the regular zone as time goes on, it increases exponentially when the system starts in the spinodal region. In fig.3b $\lambda(t)$ as defined in eq.(2) is plotted vs. time. As one can see, λ tends to zero when dynamics is regular, indicating that the trajectories actually do not diverge between each other. On the contrary, λ rapidly converges to a value $\overline{\lambda}$ which is not zero in the unstable regions and unambiguously defines the largest Lyapunov exponent $\overline{\lambda}$. We note that our time scale for the onset of chaos is in agreement with other numerical simulations on instability growth [7, 6, 2] and with recent experimental results on multifragmentation [10].

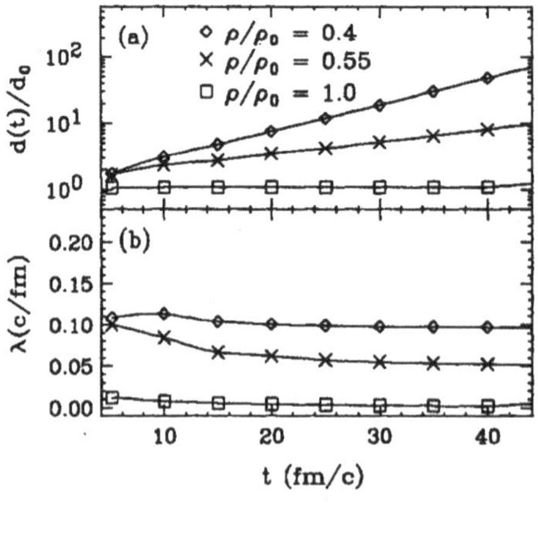

Fig.3 For three values of initial density the evolution of the distance between two trajectories (a) and the corresponding $\lambda(t)$ (b) are shown. The initial distance is $d_0 = 10^{-4}$, see text.

We have checked that the results are stable with respect to the increase of the numerical precision. This is shown in fig.4, where the time evolution of λ is plotted for different values of the time integration step and initial distance. No appreciable variation of the results is obtained using a value of d_0 smaller than 10^{-4} or decreasing the time step. Therefore the divergence of the trajectories is an intrinsic dynamical property of the nuclear mean field in this region.

The inclusion of two-body collisions, performed by solving the VN equation as done in ref.[5], but without the Langevin term, does not modify substantially the dynamics, as also found independently in ref.[7]. This can be seen in fig.5, where λ is displayed vs. time for a density well inside the spinodal region. We note only a slight variation - about 10 %- of $\overline{\lambda}$ when including the collision term. In fact, since the dynamics is chaotic, the exponential divergence between trajectories can hardly be affected by the collisions, which have a small effect at low density because of the diluteness of the gas [5, 7].

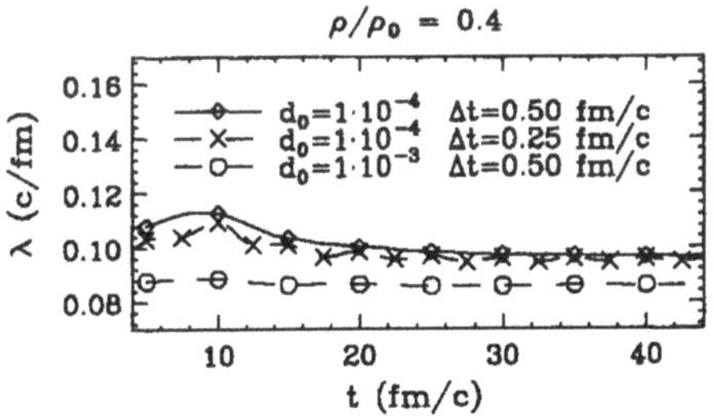

Fig.4 The quantity $\lambda(t)$ as a function of time changing d_0 and the time step. Using a value of d_0 smaller than 10^{-4} no change occurs.

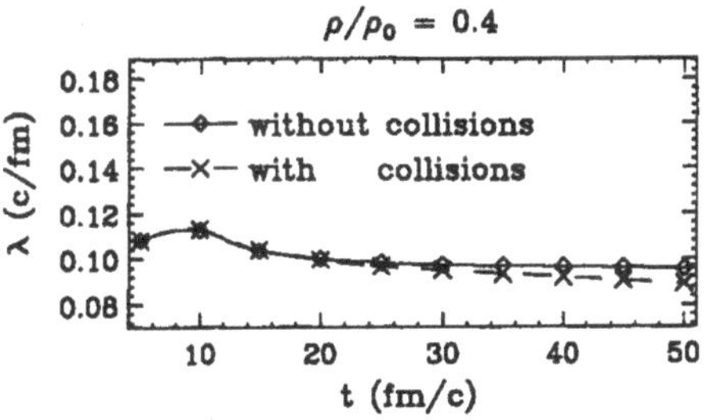

Fig.5 Behavior of $\lambda(t)$ as a function of time with and without collision term in eq.(1).

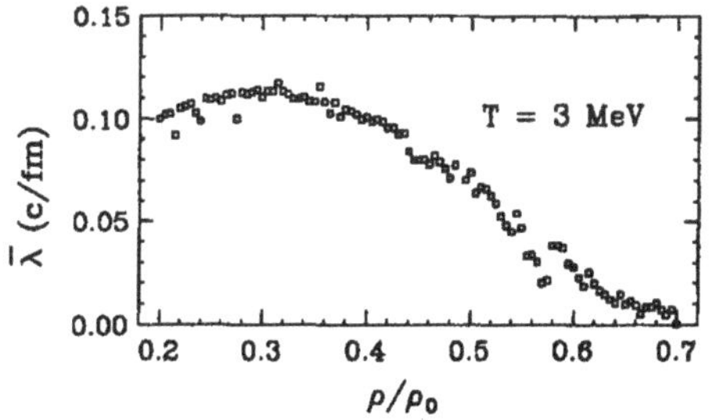

Fig.6 The Lyapunov exponent $\overline{\lambda}$ vs. ρ/ρ_0 at temperature T=3 MeV, see text.

Finally in fig.6 we show the extracted values of $\overline{\lambda}$ calculated at different densities. The values of $\overline{\lambda}$ increase from zero at $\rho = 0.66\rho_0$ to a a flat maximum around the density $\rho = 0.35\rho_0$. The curve reported in fig.6 , which is essentially independent from the initial k, clearly defines the nuclear matter spinodal region up to its upper limit.

It is worthwhile to compare the characteristic times $\tau_{MF} = \hbar/\overline{\lambda}$, which defines the time scale of the divergence between mean field trajectories, with the single particle characteristic time $\tau_{sp} = \hbar/E_F$, being E_F the Fermi energy at the given density. We have found that, for densities $\rho \leq 0.4\rho_0$, the divergence time is smaller than the single particle time, and therefore in this region the notion itself of mean field ceases to have any validity.

4 Conclusions

In the framework of a purely mean field theory, we have shown that the nuclear dynamics inside the EOS spinodal region is chaotic. Without adding any noise, and even without two-body collisions, spontaneous symmetry breaking occurs leading to large unpredictable density fluctuations. We characterize the chaoticity of the system through the largest Lyapunov exponent, for which we suggest a computational recipe. The latter does not depend on the particular choice of the initial conditions or the time integration step, but it is strongly related to the initial density. The small influence due to two-body collisions has been discussed, as well as the validity limits of the mean field theory. These results confirm quantitatively a scenario already suggested by other authors [6, 5, 7, 11] and can have important consequences for the whole picture of nuclear multifragmentation.

Generally speaking, a central collision between heavy ions at intermediate energy can be viewed as follows. In the first stage of the reaction nuclear matter gets compressed and a highly excited composite nuclear system is formed. During this phase, the dynamics is insensitive to small modifications of the initial conditions. Once the maximum compression is reached, nuclear matter starts to expand and can merge into the spinodal region. According to the results presented in this paper, at this point the dynamics becomes chaotic and the corresponding large density fluctuations dominate the cluster formation. The degree of chaoticity and fluctuations depends of course on the details of the dynamics which brings the reaction inside the spinodal region. We would like to stress that, since initially very close events follow completely different dynamical paths, an event-by-event description is strongly advised. Averaging over different theoretical paths can partially mask the non-linear evolution [12]. This scenario can support the behavior suggested by percolation model and at the same time disagree with an equilibrium statistical description since some memory of the initial mode seems to remain. Unfortunately, at the moment, the extension of these calculations to the case of collisions between heavy ions appears problematic even in two dimensions. A realistic theoretical simulation which is able to describe how the system enters into the spinodal region and how long it stays there is desirable. On the other hand an unambiguous experimental signature is needed.

References

[1] D.H.E. Gross and X.Z. Zhang , Phys Lett. **161 B** , 47 (1982); J.P.Bondorf, R.H.Donangelo, I.N.Mishustin, C.J.Pethick, H.Schultz and K.Sneppen, Nucl. Phys. **A443**, 321 (1985); G.Peilert, H.Stöcker, J.Randrup and W.Greiner, Phys. Lett. **B260**, 271 (1991); C.Dorso and J.Randrup, preprint LBL–33120; M.Colonna, M.Di Toro, V.Latora, Nucl. Phys. **545**, 111c (1992).

[2] L. Moretto, K̃. Tso, N. Colonna, G. Wozniak , Phys. Rev. Lett. **69**, 1884 (1992); W. Bauer, G. Bertsch and H. Schultz, Phys. Rev. Lett. **69**, 1888 (1992).

[3] X. Campi, Phys. Lett.**B208**, 351 (1988); M.Ploszajczak and A.Tucholski, Phys. Rev. Lett. **65**, 1539 (1990).

[4] H. Heiselberg C.J. Pethick and D.G. Ravenhall Phys. Rev. Lett. **61**, 818 (1988), and Ann. Phys. **223**, 37 (1993); M.Colonna, Ph.Chomaz and J.Randrup, GANIL preprint 1992.

[5] G.F. Burgio, Ph. Chomaz and J. Randrup, Phys. Rev. Lett. **69**, 885 (1992); G.F. Burgio, Ph. Chomaz , M. Colonna and J. Randrup, Catania University preprint 1993.

[6] J.Knoll and B.Strack, Phys. Lett. **B149**, 45 (1984).

[7] D. Idier, B. Benhassine, M. Farine, B. Remaud and F. Sébille, Phys. Rev. **C48**(1993) R498.

[8] G.F. Burgio, M. Baldo and A. Rapisarda, Catania University preprint n.93/20, submitted to Phys. Rev. Lett.

[9] A.J.Lichtenberg and M.A.Lieberman, *Regular and Stochastic motion*, Springer–Verlag (1983); M.Tabor, *Chaos and Integrability in Nonlinear Dynamics*, John Wiley (1989); Hao Bai–Lin, *Chaos*, World Scientific (1984).

[10] E. Bauge et al Phys. Rev. Lett. **70**, 3705 (1993).

[11] D.H.E.Gross, Nucl. Phys. **A553**, 175c (1993); Bao–An Li and D.H.E.Gross, Nucl. Phys. **A554**, 257 (1993);

[12] Ph. Chomaz, talk given at this conference.

DYNAMICS AND INSTABILITIES IN NUCLEAR FRAGMENTATION

M.Colonna[1,2], *M.Di Toro*[1], *A.Guarnera*[1,2], *V.Latora*[1] *and A.Smerzi*[1]

[1]*)Laboratorio Nazionale del Sud, INFN*
44, Via S.Sofia, 95125 Catania, Italy
and
Dipartimento di Fisica dell' Universitá di Catania,

[2]*)GANIL, B.P. 5027, F-14021 Caen Cedex, France*

ABSTRACT

A general procedure to identify instability regions which lead to multifragmentation events is presented. The method covers all possible sources of dynamical instabilities. Informations on the instability point, like the time when the nuclear system enters the critical region, the most unstable modes and the time constant of the exponential growing of the relative variances, are deduced without any numerical bias. Important memory effects in the fragmentation pattern are revealed. Some hints towards a fully dynamical picture of fragmentation processes are finally suggested.

1. Introduction

Heavy ion collisions at intermediate energy have been widely described through some kinetic approaches (VUU, BUU, BNV) which are able to correctly reproduce the mean trajectory of the system in phase space. However, when one is interested in unstable processes such as, for instance, multifragmentation events, these methods are not valid anymore since they do not contain fluctuations, which are very important in a not equilibrated system.

Nevertheless we will show that the use of such kinetic equations, solved within the test particle approach, can still be very useful to extract quite precise informations on physical properties of the instability regions.

Actually, since always some numerical noise is present when the systems enters an instability region, we see in the solution of these transport equations a fast growing of fluctuations, symmetry breakings and cluster formation. A lot of excitement and discussion has been recently raised around the preferred spacial geometries of fragmentation events deduced from different transport codes[1]. Indeed these geometries depend sensitively on the configuration of the system at the time of instability, on the leading unstable modes and on the time scale for cluster formation. In this paper we will show how such quantities are related to the used phase space discretization procedure (*i.e.* number of test particles per nucleon in the numerical algorithm) and we will suggest the way to extract unbiased physical informations.

Moreover we will try to go further and to follow the fragmentation pattern studying the memory effects versus the dynamical evolution of the system in different situations (*i.e.* changing the beam energy or the nuclear matter compressibility).

2. Instabilities in nuclear dynamics

As already stressed, the kinetic equation used to study heavy ion collisions describes only the mean phase space trajectory of the one-body distribution function, including two-body and Pauli correlations but averaging over the related fluctuations. It can be used only to extract mean values of one-body observables and not variances, as already known from the Time Dependent Hartree-Fock (TDHF) theory used at low energies. Since the inclusion of fluctuations is essential to form fragments in critical regions, the BNV equation seems to be useless for a fully dynamical picture of fragmentation.

However the test particle method also naturally introduces some numerical fluctuations in the initial conditions and in the Pauli blocking. These can be reduced just increasing the number of test particles and therefore they can be easily eliminated in regions of dynamical stability. In critical regime the fluctuations are growing exponentially and this numerical noise can substantially modify the relative dynamics.

The instability point, for a particular degree of freedom of the system, is reached when the corresponding RPA mode attains an imaginary frequency[2–4]. This is equivalent to say that small amplitude fluctuations in that degree of freedom will show an exponential growth. Therefore we expect that important informations on the unstable modes and on the instability point could be extracted from the study of the mean phase space trajectory if the evolution of the mean field is suitably described and, of course, if some fluctuations, from any type of source, are present. At variance, the time scale for cluster formation is very much related to the absolute value of the amount of fluctuations present in some explosive degree of freedom. We see fragments when the variances become quite large and the corresponding time evolution starts to deviate from the exponential growth, the system having reached a new region of stability. All that has been recently confirmed by some model calculation for nuclear matter[4,5]. The conclusion is that we expect any noise, in a good mean trajectory dynamics, suitable to extract informations on the instability point (most unstable modes and instability time). On the contrary, in order to get the right fragment formation, we should be able to reproduce the correct physical fluctuations at the instability point. We have checked these ideas performing a series of numerical analyses solving BNV-type transport equations for two dimensional collisions of nuclear matter slabs.

3. Dynamics and fragmentation

We have performed calculations using a simplified model. We consider a two-dimensional collision between two slabs of nuclear matter in a box of lengths 63 fm and 21 fm in order to reduce the computational time, the mean field has been averaged on the y direction. The x radius of the two nuclei is equal to 4 fm (the

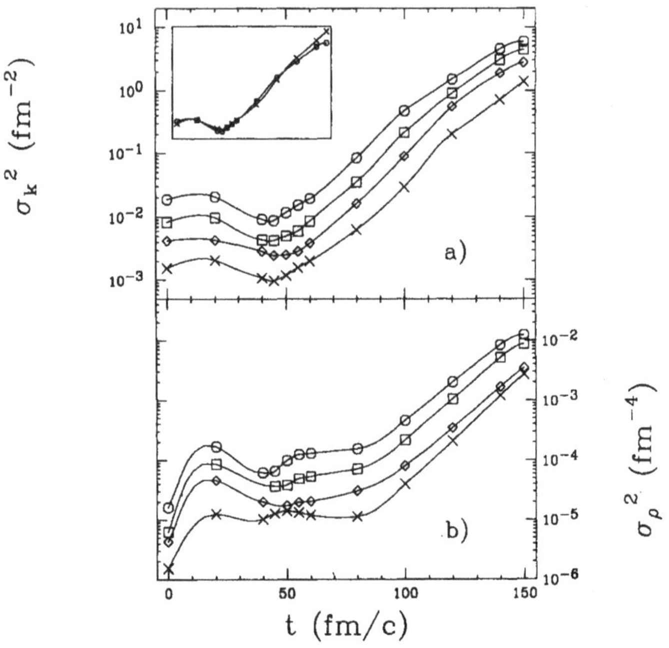

Fig. 1. Time evolution of variances for the two slab collision described in the text. a): $\sigma^2(t)$ of the most unstable modes; b) $\sigma^2(t)$ of the monopole mode. Circles: $N_{TP} = 20$; squares: $N_{TP} = 40$; diamonds: $N_{TP} = 80$; crosses: $N_{TP} = 200$.

radius of a ^{40}Ca nucleus) while the total number of nucleons is fixed requiring that the density of the two slabs is equal to the normal density, which is 0.55 fm^{-2} in two dimensions. We have simulated a central collision at a beam energy $E_{lab}/A = 40\ MeV/u$. As the time increases we see the formation of a nearly uniform region at half normal density and at a temperature around $T = 8\ MeV$. This system lies in the spinodal instability region[5]. In such a simplified situation it is possible to demonstrate that, for infinite nuclear matter, the eigenmodes of the density are plane waves, characterized by a wave number k. Even if, in our case, the system is finite, it is interesting to perform a Fourier analysis of the density fluctuation and therefore to define the following quantity:

$$\sigma_k^2 = \int_0^{L_x} dx \int_0^{L_x} dx' e^{-ik(x-x')} < \delta\rho(x)\delta\rho(x') > . \tag{1}$$

The quantity σ_k^2 represents the spectral correlator.

In fig. 1a) is shown the time behaviour of σ_k^2 for the most unstable mode ($k = 5 \div 6$), calculated using 20, 40, 80 and 200 test particles per nucleon, 50 trajectories are considered. We see that all the variances reach the minimum point at the same time $t_{inst} = 45 fm/c$; this point does not depend on the noise and it corresponds to the time when the system enters the instability region. The scaling law is preserved also in the next evolution of σ_k^2; it means that the time τ, characteristic of the exponential growing, does not depend on the number of test particles. The scaling disappears when the system enters a saturation regime, corresponding to a new equilibrium reached by cluster formation. This can be seen in the box inside fig.1a where we report two curves of this figure ($N_{TP} = 20, 200$) suitably rescaled. As expected the saturation appears earlier for the trajectories with less test-particles.

The fragmentation time, *i.e.* the time when we observe fragment formation, is therefore fully dominated, in this representation, by the value of the fluctuation σ_k^2 at $t = t_{inst}$. In this sense it would be extremely important to know the right physical amount of fluctuations that are present at that point.

We have also studied the time dependence of the density variance of the trajectories in the a circle of radius 4 fm in the overlapping zone:

$$\sigma_\rho^2 = \frac{1}{N_{events}} \sum_{i=1}^{N_{events}} \left(\rho_i(t) - \bar{\rho}(t)\right)^2 \tag{2}$$

which is related to the fluctuations of the distribution function, after some phase space averaging. In the stability region the variance follows the density as it should be in a Poisson-type distribution of test particles. Indeed if the variance on the number of test particles in the circle follows a Poisson law $\sigma_n^2 = \bar{n}$ we get for the variance on density the relation:

$$\sigma_\rho^2(t) = \frac{\bar{\rho}(t)}{S\ N_{TP}} \tag{3}$$

where S is the surface of the circle. The presence of this classical behaviour of fluctuations is not surprising since we are looking at properties of phase space averaged observables and then we expect quantum fluctuations to be washed out.

When the system enters the instability region we observe a sharp transition to an exponential behaviour:

$$\sigma_\rho^2(t) = \sigma_0^2 e^{2t/\tau} \tag{4}$$

which clearly indicates the presence of an unstable mode with associated imaginary energy $E = i\hbar/\tau$ [3,4]. σ_0^2 represents the amount of fluctuations present at the time of instability and clearly scales with the number of test particles per nucleon N_{TP}. But it is very important to remark that the time of instability, *i.e.* the time of the transition to the exponential, and the time τ, characteristic of the exponential growing of fluctuations, do not depend on the number of test particles we are using, *i.e.* on the noise introduced in the calculation. They are related just to the mean evolution and to the mean properties (density and temperature) of the system which, as we explained before, is always well described until the instability spinodal region is reached. Starting from this point, the amount of fluctuations present in the system becomes of crucial importance since the dynamics is now dominated by the exponential amplification of the initial noise and consequently calculations using a different number of test particles will lead to different results.

In fig.1b) is displayed the density variance. The scaling behaviour here is less evident because in σ_ρ^2 we have the superposition of different modes corresponding to different values of k. This superposition seems to create an uncertainty on the time when the system enters the instability region, since, using 200 test particles, the exponential growing starts later. Actually this time does not change and it is still $t = 45\ fm/c$ for all curves. When the amplitude of fluctuations is not large enough (as it happens using 200 test particles per nucleon), the system expands before exploding and therefore the growing time τ reduces, according to the reduced new density, until the explosion becomes the dominant one. In this case it happens at $t = 80\ fm/c$.

4. Memory effects in fragmentation

We focus now our attention on memory effects in the fragmentation mechanism versus the dynamical evolution of the system.

In order to study the sensitivity of fragment formation to the dynamics of the reaction, we have performed two-dimensional calculations changing the beam energy and the equation of state. The e.o.s. can be easily modified using a different value of the parameter σ in the Skyrme-like form of the mean field potential $U = A(\rho/\rho_0) + B(\rho/\rho_0)^\sigma$.

Requiring a binding energy of 16 MeV at $\rho = \rho_0$, *i.e.*

$$E/A(\rho = \rho_0) = \frac{A}{2}(\rho/\rho_0) + \frac{B}{\sigma+1}(\rho/\rho_0)^\sigma + \frac{1}{4m}\hbar^2\pi\rho = -16 \tag{5}$$

(in order to simulate three dimensional nuclear matter) and the minimum condition for this point $(\partial(E/A)/\partial\rho = 0)$, we obtain the following sets of values for the parameters A and B :

$$\sigma = 2 \qquad A = -100.3 \qquad B = 48 \qquad (I)$$

$$\sigma = 3 \qquad A = -84 \qquad B = 32 \qquad (II)$$

In the following we will refer to the parameterization (I) as "soft" equation of state, while the parameterization (II) will give the "stiff" equation of state.

Let us discuss first the effects that can be seen in the fragment formation if we change the beam energy. Increasing the beam energy, the system is allowed to perform larger oscillations in density (see fig. 2a). Because of these oscillations, a volume instability region is reached and then the formation of fragments becomes the most likely mechanism of de-excitation. Corresponding to the different initial conditions, the value of the density when the system starts to explode may be different, since the temperature in the participant zone will change. Consequently, the

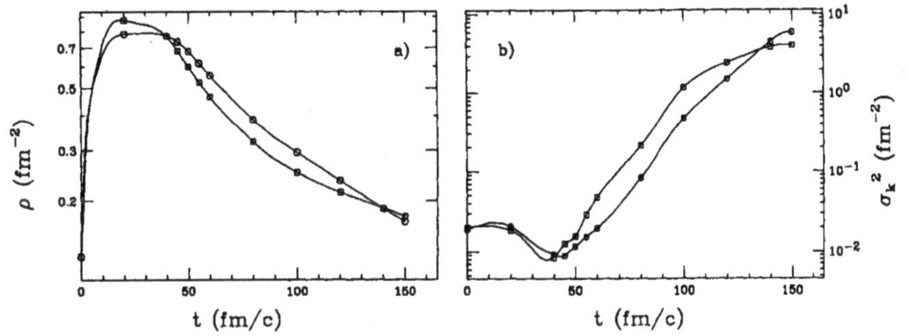

Fig. 2. Time evolution, for the two slab collision, of the mean density in the participant zone (a) and of the variance of the most unstable mode (b) for $E/A = 40\ MeV/u$ (circles) and $E/A = 70\ MeV/u$ (squares).

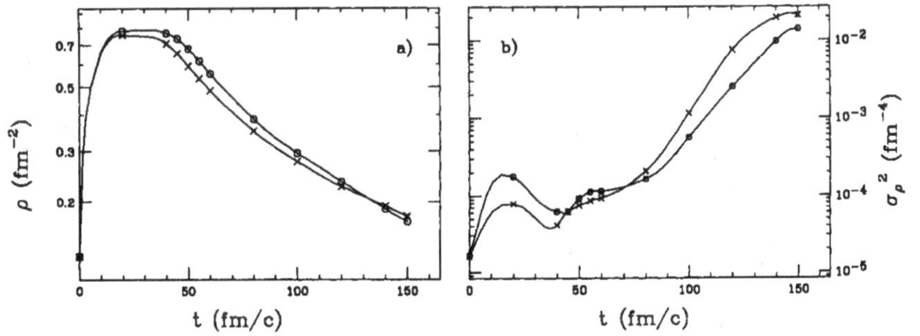

Fig. 3. Time evolution, for the two slab collision, of the mean density in the participant zone (a) and of the variance of the monopole mode (b) for a "soft" e.o.s. (circles) and a "stiff" e.o.s. (crosses).

pattern followed by the fragmentation and, in particular, the number of fragments that we obtain in the final state may also change. (Indeed this number depends both on density and temperature at the critical point).

Another effect of the increased energy is an earlier fragment formation due to the faster dynamics.

In our case we get 6 fragments for $E/A = 40\ MeV/u$ and 7 fragments for $E/A = 70\ MeV/u$. Moreover, as we can see from fig. 2b, where we plot σ_k^2 for $K = 6$ as a function of time, the instability time (*i.e.* the minimum point of σ_k^2) changes from $t \approx 50 fm/c$ to $t \approx 40 fm/c$ and correspondingly the saturation is observed before, when we pass from $E/A = 40$ to $E/A = 70\ MeV/u$. It is interesting to remark that the fragment formation time, *i.e.* the time interval between instability and saturation, seems to not depend on the beam energy. This effect has been recently observed in velocity correlation measurements[6].

Let us consider now the effects of changing the e.o.s. keeping constant the beam energy (40 MeV/u).

Smaller oscillations in the density are observed using a "stiff" equation of state (see fig. 3a). The number of fragments reduces therefore to 5.

Moreover, the expansion velocity increases, giving an effect similar to that one obtained using a larger beam energy; in fact the instability time and the saturation point are reached before (see fig. 3b). We stress here that all calculations have been done using the same number of test particles (20 per nucleon). It is very important to remind that the fluctuations present in our approach are strictly related to the used number of test particles. Thus if we increase this number (*i.e.* we correspondingly reduce the amplitude of fluctuations), the system may expand before going to fragmentation. It means that there is an interplay between the dynamical expansion and the explosion of instabilities, due to the presence of fluctuations, until the new reduced density is low enough to form fragments.

Of course, in this case, the number of fragments and the structure of the fragmentation process will depend on this new situation and consequently on the number of test particles we are using. We mean that, because of the dynamical effects, no prediction can be made about the fragmentation pattern, since it strongly depends on the amplitude of fluctuations and, in our case, on the number of test particles.

Moreover we observe that these dynamical effects are more and more important going to higher beam energies or considering a "stiff" equation of state, since in both cases we have a higher velocity of expansion.

Memory effects of the observed number of fragments are therefore lost when fluctuations are small, the fragment formation time is long and spherical symmetric dynamical expansion is dominant. In this sense, in order to correctly describe the behaviour of the system, the right physical amount of fluctuations that the system contains must be known. However, from our calculations, since the variances at the instability time should be rescaled, we predict quite large fluctuations at the critical point and therefore a corresponding short fragmentation time, of the order of 100 fm/c, is expected. This seems to be in agreement with some first experimental evaluations[6,7]. Therefore we believe that there is a good chance of observing such collective dynamical effects in the pattern of fragmentation events.

5. Conclusions

We have shown that, when one is interested in the dynamics of critical situations, it is possible to extract very important informations just solving the BNV equation within the test particle approach. In fact, if the number of test particles is large enough to avoid spurious effects, the dynamical evolution is correctly described until the system becomes unstable. Therefore the time when the spinodal region is reached and all the observables related to the mean properties (density, temperature, etc...) at that point may be calculated without any ambiguity. In particular we can extract the growing time τ, characteristic of the exponential increasing of fluctuations, which will lead to fragmentation.

We have also clearly shown the presence of memory effects in the fragmentation process. The leading unstable modes will completely fix the final fragmentation pattern (number and size of fragments, space and momentum structure and so on). This is a warning for the use of methods for fully chaotic systems, like percolation, to study nuclear fragmentation. The dynamics is playing a fundamental role. The paradox is that properties of the average phase space trajectory in the stable region will univocally determine the mean features of fragment production.

On the other hand we have also proved that time and pattern of this fragment formation depends, in a crucial way, on the noise present in the system, *i.e.* on the number of test particles. Therefore the right physical value of fluctuations is needed in order to correctly describe the dynamics. Concerning this subject, work is presently in progress.

It is finally evident that the method presented here is not limited to the study of spinodal (volume) instabilities. The recipe is to look at the time dependence of the variances σ_i^2 of various collective degrees of freedom, that can be done within the test particle approach. In this way all possible sources of instabilities (surface, Coulomb and so on) will be revealed and suitably studied.

We warmly thank Ph.Chomaz for very helpful discussions.

6. References

1. L.G.Moretto, Kin Tso, N.Colonna, and G.J.Wozniak, *Phys. Rev. Lett.* **69** (1992) 1884.
   W.Bauer, G.F.Bertsch, and H.Schulz, *Phys. Rev. Lett.* **69** (1992) 1888
   B.Borderie, B.Remaud, M.F.Rivet, and F.Sebille, *Phys.Lett.* **B302** (1993) 15
   D.H.E.Gross, Bao-An Li and A.R.De Angelis, *Ann. Phys.* **1** (1992) 467; Bao-An Li and D.H.E.Gross, HMI preprint (Sep'92), *Nucl. Phys. A* (in press)
   G.Batko and J.Randrup, preprint LBL-32722 (Sep'92), *Nucl. Phys. A* (in press)
2. L.D.Landau and E.M.Lifshits, *Statistical Physics* (Pergamon Press, New York, 1981) chap.12
   E.M.Lifshits and L.P.Pitaevskii, *Physical Kinetics* (Pergamon Press, New York, 1981)
3. M.Colonna, Ph.Chomaz, and J.Randrup, preprint LBL-33402 (Dec'92), *Nucl. Phys. A* (in press)
4. M.Colonna, G.F.Burgio, Ph.Chomaz, M.Di Toro and J.Randrup, *Phys. Rev.* **C47** (1993) 1395
5. G.F.Burgio, Ph.Chomaz, J.Randrup, *Phys. Rev. Lett.* **69** (1992) 885
6. E.Bauge et al., *Phys. Rev. Lett.* **70** (1993) 3705
7. D.R.Bowman et al., *Phys. Rev. Lett.* **70** (1993) 3534

Phase Space Density Fluctuations in Fragmenting Nuclei

C.O.Dorso

Departamento de Física, Facultad de Ciencias Exactas,
Universidad de Buenos Aires

Abstract: *We use the Quasi Classical Nuclear Model (QCNM) and the recently proposed Early Cluster Recognition Algorithm (ECRA) to investigate the dynamical evolution of the fragmentation pattern for suitably excited finite spherical spin-isospin symmetric system of 80 nucleons. We find that the asymptotic fragment structure (when clusters are far away from each other) is strongly related to the early most bound density fluctuations (MBDF) in phase space (when the system is dense).*

I. Introduction

In the last years a lot of experimental and theoretical work has been devoted to the analysis of fragmentation processes in nuclear systems. Many theoretical models have been proposed to analyze this phenomena and they can be roughly classified in two major categories. First those that assume that the fragmentation takes place at the end of a more or less complicated expansion process when the system has attained a certain degree of equilibration.[1] Secondly those that describe the full non-equilibrium dynamic evolution of the systems that eventually leads to the clusterization process. Among this second class a further classification can be established, between those that use different schemes to simulate the quantal Boltzmann kinetic equation[2] and those that rely on classical[3] and semi classical[4] microscopic descriptions in which the quantal effects are sometimes taken into account. One of the relevant differences between these two approaches is the degree to which correlations are kept. In Boltzmann like theories the main outcome is the One body distribution function and higher order correlations are systematically washed away.. In opposition the main goal of microscopic classical and semi classical models is to keep correlations of all orders at all times. In this work we use a classical model (QCNM) in which Pauli exclusion principle is simulated via a two body repulsive interaction potential in phase space.[5] Cluster analysis is performed using a cluster recognition algorithm in which the partition of the system that maximizes the total binding energy is found.[6]

II. The model

The QCNM has already been described in a series of publications (see[5] and references therein) and for the sake of completeness we briefly describe its main properties.

In this model the nuclear system is simulated via a classical interaction potential which reads:

$$V = Vn + Vp + Vc \tag{1}$$

where

$$Vp = V_{0p} \cdot \exp\left(-0.5 \cdot \left(\frac{p_{ij}^2}{p_0^2} + \frac{q_{ij}^2}{q_0^2}\right)\right) \cdot \delta\tau_{ij} \cdot \delta\sigma_{ij} \tag{2}$$

is the term responsible for simulating the Pauli exclusion principle

$$Vn = V_{0n} \cdot \left(\left(\frac{r_1}{r_{ij}}\right)^{a1} + \left(\frac{r_2}{r_{ij}}\right)^{a2}\right) \cdot \frac{1}{\exp\left(-\beta r_{ij} + \mu\right)} \tag{3}$$

it is the so called nuclear term

$$Vc = \frac{\alpha_0}{r_{ij}} \tag{4}$$

is the Coulomb term

In figure 1 we show the way in which different nuclear properties are reproduced according to this scheme

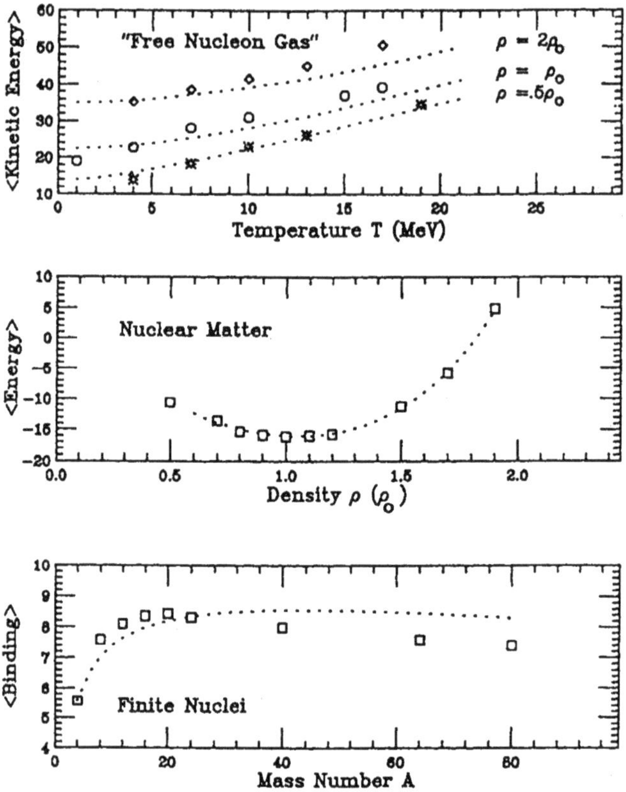

Figure 1 :The kinetic energy of a gas of nucleons that interact only via the Pauli potential (top), the binding energy of nuclear matter as a function of the density (middle), and the binding energy of finite nuclei as a function of their mass number (bottom).

III. Cluster Definition and Recognition Method

A special role in our analysis is played by the way in which we define and, consequently, recognize clusters. According to[6] a cluster is defined as that set of particles that satisfies de following condition

$$\forall i \in C, \qquad e_i = T_{iCM} + \sum_j V_{ij} < 0 \tag{5}$$

where T_{iCM} is the kinetic energy of particle i calculated in the C.M. of the cluster. And V_{ij} is the interaction energy between particles i and j belonging to the same cluster C ; i.e. all particles must be bound.

For a system of particles we define the cluster decomposition as a partition of the total system into subsets such that for each subset condition (5) is satisfied,i.e. each subset is a cluster. Because this condition can be satisfied by more that one partition we introduce the following additional condition:

cluster structure is that cluster decomposition that maximizes the total binding energy of the system when each cluster is considered as non interacting with the others.

As it can be easily seen this is a very complicated problem because the binding energy is calculated in center of momentum of each subset of the partition , as a consequence the problem is highly self consistent. To solve this problem in ref[6] an algorithm in the spirit of simulated annealing was developed. It is clear that this cluster definition is free from arbitrary "clusterization parameters", on the other side this method is statistical in nature and although we can be confident that we will be quite close to the maximizing partition, we cannot be sure.It should be emphasized at this point that being it possible to apply this method at any point in the evolution of the system there will be a stage in which what we are recognizing as clusters is the set of most bound density fluctuations in phase space. During this stage particles are close together in q space and no cluster structure can be recognized by standard configuration (percolation like) cluster recognition algorithms.At later, asymptotic, times clusters will be well separated in space and then we can call them fragments. In the following we apply this methods to the analysis of a exploding system and explore the relation between the early recognized clusters (MBDF) and the asymptotic ones (fragments).

IV. Calculations

In order to work on a rather simple problem we focus on the analysis of suitably excited spherical systems. To perform this we take as initial condition a spin-isospin symmetric system of 80 particles built via a Monte Carlo calculation in phase space at a temperature of $0.4MeV$. From the MC run configurations well separated in "time" were kept in order to be sure that they are uncorrelated both in q and p space. A ECRA calculation was performed with each of them to confirm that they were stable bound systems. In order to build excited systems the following procedure was adopted: the cold nucleus was compressed in q space and consequently expanded in p space according to the following scale transformation:

$$Q'_i = Q_i.\lambda^{-1/3}$$

$$P'_i = P_i \cdot \lambda^{1/3}$$

with $\lambda > 1$

Once this stage was accomplished the resulting time evolution was attained by solving the Hamilton equations of motion of the system using a standard self adjusting Runge Kutta scheme

We will focus our analysis in the value $\lambda = 1.8$ which leads to a typical multi fragmentation process in which light fragments up to mass 15 are found in the final mass spectra.

V. Results

The following analysis was performed over 60 configurations built according to the above mentioned prescription .

We now focus on the relation between the early MBDF spectra and the final fragment espectra of the exploding system. In fig.2 we show the total multiplicity of clusters bigger than 3 (dotted line) and the multiplicities for selected bins as a function of time All these curves are remarkably constant

To have an idea of the relevant time scales the same quantities calculated using a simple percolation-like cluster recognition algorithm, with a clusterization distance $R_{cl} = 6.0fm$ corresponding to the cut-off distance of the

$Vn(r_{ij})$ eq.(3), are displayed (thin lines). From this it can be realized that constancy of the number of clusters is attained long before spatial separation is reached.

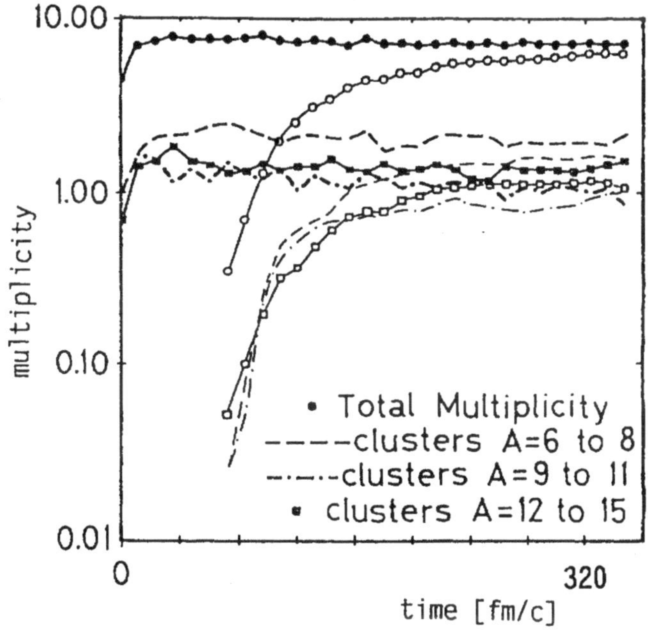

Figure 2 :Time evolution of the multiplicity. Thick lines result from the ECRA method : dots denote total multiplicity ; crosses clusters of $A = 6$ to 8; squares cluster of $A = 12$ to 15 and no symbol denote clusters of $A = 9$ to 11. Simple lines idem but resulting from a percolation like method.

Our main objective is to get information about the time evolution of the microscopic structure of the MBDF and its relation to the final fragments.

In order to explore this effect we find it convenient to define the persistence coefficient (P) :let $b(t) = 0.5 * N(N-1)$ be the number of pairs of particles in cluster C at time t .Let $a(t-\Delta t)=\sum 0.5*(n(n-1))$ the sum over all clusters present at time $t-\Delta t$, of the number of pairs of particles present in each cluster and originally belonging to cluster C, in this way

$$P=<\sum[a(t-\Delta t)/b(t)]>$$

P will be equal to 1 if all particles remain together and 0 if the cluster breaks up completely . It, then, measures the tendency of the members of a given cluster to remain together. It is easily seen that this coefficient is highly sensitive to variations in mass number reaching a value around 0.8 when just one particle is removed from sytems with 8-12 particles.

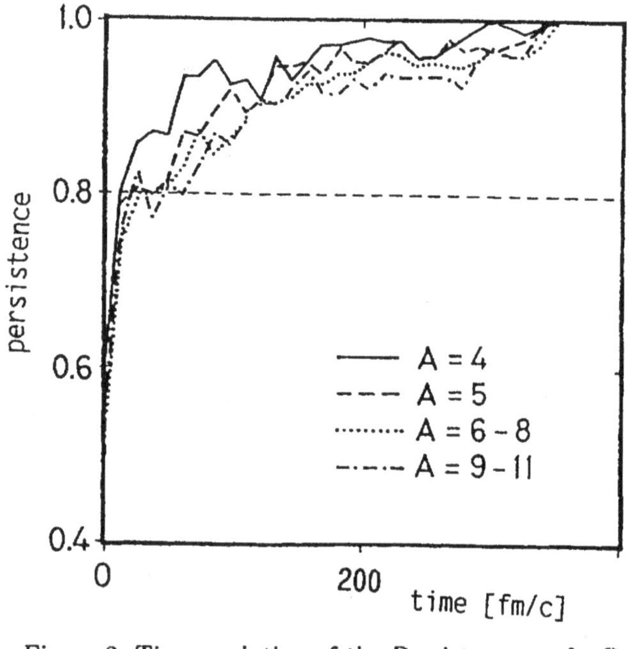

Figure 3 :Time evolution of the Persistence coef., P. Crosses indicate P for clusters of $A=4$ at the final stage, x idem for $A=5$, diamonds idem for $A=6$ to 8 and thick squares idem for $A=9$ to 11.

In Fig. 3) we show the results of calculating the coefficient P for different cluster sizes. It is clear that it reaches a value close to one very early in the evolution of the system and form comparison with the results displayed in Fig 2), this happens long before spatial separation of the clusters takes place.

From the results presented in Figs 3) and 6) it is clear that the asymptotic fragment spectra is strongly related both in the number of fragments and its microscopic structure with the clusters (MBDF) that are formed in the early, dense, stage of the explosive evolution of the excited system.

VI. Conclusions

From our calculations we conclude that the fragmentation process of excited spherical nuclei can be described as follows, as soon as the system is excited a set of density fluctuations in phase space is built. From the infinite modes that can considered a very peculiar one is extremely relevant for the analysis of the subsequent evolution of the system i.e. that one that maximize the binding energy as described previously. This set of fluctuations evolves in time in such a way that after a short time most of it's constituent particles remain together for the rest of the time evolution as can be seen from the behavior of the P coefficient. As time evolves the system reaches a volume such that clusters become physically separated but in this stage what we see as fragments are basically the time propagation of the already present MBDF. In this way the fragments, both in it's number and microscopic composition, carry information about the initial, highly excited stage of the evolution of the system. It must be noted at this point that recent analysis of experimental data confirms this picture of the time evolution of fragmenting excited systems.[7]

It is then clear that models based on the assumption of the existence of an intermediate regime in which the systems equilibrates and subsequently fragments are built up, should be critically reanalyzed (according to our microscopical calculations there is no such an equilibrated pre-fragmentation stage).

Acknowledgements
The authors gratefully acknowledge the Consejo Nacional de Investigaciones Científicas y Técnicas (CONICET) and Fundación Antorchas for partial financial support.

References

[1]J.P.Bondorf, R.Donangelo, I.N.Mishutin, C.J.Pethick, H.Schulz and K.Sneppen ; Nucl.Phys A443 (1986) 321;J.P.Bondorf, R.Donangelo, I.N.Mishutin, H.Schulz ; Nucl.Phys A444 (1986) 460;

[2]G.F.Bertsch and S.Das Gupta Phys.Rep. 160 (1988) 189 ; H.Stocker and W.Greiner Phys.Rep. 137 (1986) 277 ; P.Schuck et.al. Prog.Part. Nucl.Phys. 22(1989) 181 ; W.Bauer, C.K. Gelbke and S.Patt Annu.Rev.Nucl. Part Sci.42(1992)77

[3]T.J.Schlagel and V.R.Pandaharipande , Phys.Rev.C 36 (1987) 162

[4]J.Aichelin, Phys.Rep. 202 (1991) 233

[5]C.O.Dorso and J.Randrup, Phys.Lett.B 232 (1989) 29

[6]C.O.Dorso and J.Randrup, Phys.Lett.B 301 (1993) 328

[7]W.Bauer, J.P.Bondorf, R.Donangelo, R.Elmer,B.Jakobson, H.Schulz, F.Schussler and K.Sneppen, Phys Rev.C 47(1993)R1838

SOME RESULTS ON THE FRAGMENTATION OF HIGHLY EXCITED NUCLEI

M. Aboufirassi (1), A. Badala (1,2), B. Bilwes (3), G. Bizard (1), R. Bougault (1), R. Brou (1), J. Colin (1), F. Cosmo (3), D. Durand (1), A. Genoux-Lubain (1), T. Hamdani (1), D. Horn (1,4), J.L. Laville (1), C. Le Brun (1), J.F. Lecolley (1), O. Lopez (1), M. Louvel (1), M. Mahi (1), A. Peghaire (5), J. Péter (1), R. Regimbart (1), G. Rudolf (3), F. Saint-Laurent (5), F. Scheibling (3), J.C. Steckmeyer (1), L. Stuttgé (3), B. Tamain (1), S. Tomasevic (3)

1) Laboratoire de Physique Corpusculaire (IN2P3-CNRS/ISMRA), Bld Maréchal Juin, F-14050 CAEN cédex, FRANCE.
2) INFN Catania, Italy
3) Centre de Recherches Nucléaires (IN2P3-CNRS/Univ.Louis Pasteur), B.P. 20, F-67037 STRASBOURG cédex, FRANCE.
4) Chalk River Lab.,AECL,Canada
5) GANIL (DSM-CEA/IN2P3-CNRS), B.P. 5027, F-14021 CAEN cédex, FRANCE.

Some results about the fragmentation of heavy excited nuclei produced in heavy-ion collisions in the Fermi energy domain are presented. A transition from two- to three-body decay is observed around 3 MeV/u excitation energy. Then, fragment emission is shown to evolve from a sequential binary splitting mechanism towards a prompt simultaneous decay when the excitation energy stored in the system increases from 3 to 5 MeV/u. These results suggest a smooth transition from "standard" low-energy processes towards more exotic phenomena such as multifragmentation in this excitation energy range. Last, the question of possible compression-expansion effects in the disassembly process is adressed.

* experiments performed at Ganil

1. Introduction

The study of the decay modes of very excited nuclei produced in dissipative nucleus-nucleus collisions can help improve our understanding of the static and dynamical properties of nuclear matter under extreme conditions of temperature and pressure[1-2]. In particular, when the excitation energy is sufficiently high, many fragments can be emitted during the collision, thus leading to a complete disassembly of the excited matter. This process often referred to as multifragmentation in the literature is presently being intensively investigated with help of 4π detectors (see e.g.[3] and references therein).

In this brief communication, we present some results obtained by bombarding Au targets with Ne, Ar and Kr projectiles at incident energies from 30 to 60 MeV/u. The experiments were performed in the Nautilus scattering chamber at the GANIL facility. Fragments were detected in the Delf [4] and XYZT[5] detectors. These two detectors constitute an ensemble of 30 position sensitive parallel plate avalanche counters (PPAC), each followed by an ionisation chamber. The angular range covered was 3°-150° with a geometrical acceptance of 55%. Full detection efficiency for atomic numbers equal to or larger than 8 was obtained. The set-up had a low velocity threshold: .5 cm/ns (.13 MeV/u) for Delf and 2 cm/ns (2.07 MeV/u) for XYZT in the forward direction. The measured parameters are the atomic number Z, the fragment velocity V and the angle (with a resolution better than one degree). Light charged particles (not to be discussed here) were also detected with two complex arrays: Mur and Tonneau. The data presented in this paper concerns the fate of excited nuclei (whose mass are around 200-220) produced by a variable impact parameter-dependent transfer of linear momentum, mass and energy from the projectile to the target.

2. From binary fission to three-body decay

For moderate excitation energy (say 2 MeV/u) the decay modes of heavy nuclei are dominated either by binary fission or by evaporation residues (these are nuclei which only decay by light particle evaporation). The competition between these two processes will not be discussed here (see e.g.[6] and references therein). However, as the energy increases, new processes should take place such as three-body decay. This transition has been studied for the systems Ar+Au at 60 MeV/u and Kr+Au at 43 MeV/u[7]. The ratio between events leading respectively to two and three target-emitted fragments (fig. 1) was obtained as a function of the excitation energy per nucleon by using the Kinematical Coincidence Method. The excitation energy per nucleon E* was estimated on an event by event basis by two different methods: a first one by performing an energy balance taking into account pre-equilibrium particles and a second one relating E* to the linear momentum transferred to the target-like nucleus. Both methods lead to similar results and were confirmed by analysing the corresponding light charged particles multiplicities[8]. As shown in fig. 1 three-body decay starts to set in around E*=3 MeV/u and becomes dominant at E*=5 MeV/u.

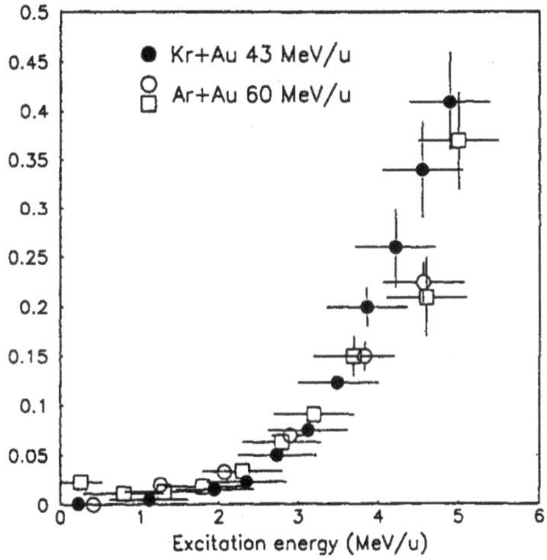

Fig. 1: Evolution of the two-fold vs three-fold fragment production as a function of the excitation energy per nucleon E. Symbols correspond to different methods for estimating E* and to two studied systems. From ref* [7]

Concerning the fragment size distribution, at moderate excitation energies, two of the three detected fragments are very similar to fission fragments while the third one has a small size thus suggesting a "natural" continuity with a "standard" low-energy binary fission process. An evolution is however observed towards more and more symmetric configurations as E* increases indicating the onset of multifragmentation.

3. Fragment emission: from long to short time-scales

The occurence of multi-body decay is however not enough to characterize properly the multifragmentation process. Indeed, one must investigate in details fragment emission time-scales since "true" multifragmentation is generally defined as a fast mechanism. This aspect of the process was studied with help of two-fragment velocity and angular correlations[9-10]. As an example, the correlation functions for three target-emitted fragments taken two by two (and sorted according to their size) for the Ne+Au system at 60 MeV/u[11] are displayed in fig. 2.

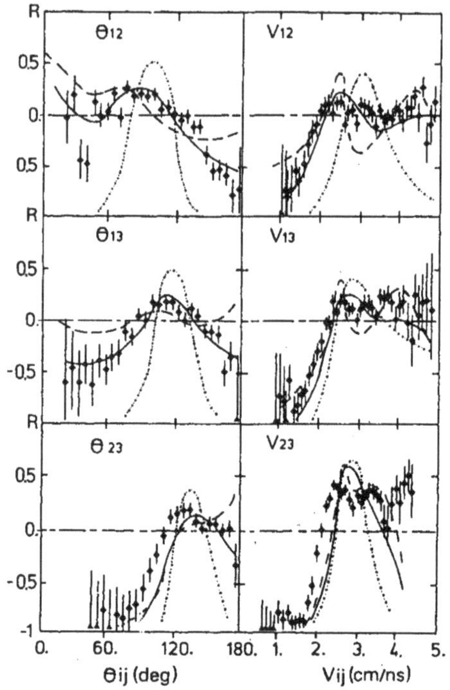

Fig. 2: Correlation function of relative velocity (right) and relative angle (left) . Black points : data. Curves are the results of simulations with the following conditions: simultaneous break-up (dotted line), sequential emission with τ=300 fm/c (full curves) and with very long τ (dashed curves). From [11].

Data are compared with the results of a simulation whose main parameter is the average time τ between successive fragment emission. A good agreement is obtained for a large value of τ around 300 fm/c, thus claiming for a slow "low-energy" process. For those events, the excitation energy was estimated around 3-4 MeV/u. Similar conclusions were drawn (although with a different method of analysis) for the same type of events in the Ar+Au case at 30 MeV/u[12] corresponding approximately to the same excitation energy range.

By increasing the incident energy or the projectile mass, larger excitation energies could be reached. Indeed, when studying three (or more) target-emitted fragments in dissipative Ar+Au or Kr+Au collisions at 60 MeV/u, excitation energies around 5 MeV/u corresponding to temperatures as high as 7.5 MeV have been observed[13-14]. The corresponding relative angle correlation function (in the Ar+Au case) is displayed in fig. 3 and compared with the above-mentionned simulation[15]. The conclusions are very different from those concerning the Ne+Au case: indeed, best agreement between the calculation and the data is obtained for a very short emission time, suggesting a fast "simultaneous" process. We therefore conclude that the onset of multifragmentation (for a nucleus whose mass is of the order of 200-220) is around E*=5 MeV/u.

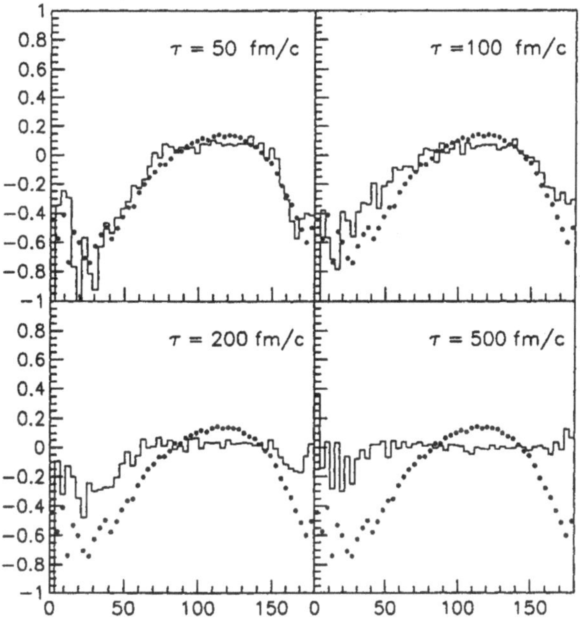

Fig 3: Correlation function for the relative angle between fragments taken two by two for three target-emitted fragments in Ar+Au at 60 MeV/u. The symbol τ has the same meaning as in fig. 2. From [15].

4. What about expansion effects in the disassembly process?

Once the simultaneous character of the decay process established, the question of a possible compression-expansion mechanism was adressed by studying fragment average kinetic energy per nucleon. These quantities were calculated in the "event frame" that is the center-of-mass frame of the detected fragments. In order to gain some quantitative insights on the existence of a possible "extra" energy in addition to Coulomb energy, we compared in fig. 4 our data with the multifragmentation model of Lopez and Randrup[16]. Details of the calculation may be found in[17]. In fig. 4 the solid line corresponds to the calculation in which no expansion energy was added to the potential (mainly Coulomb) energy of the initial configuration while the dashed line is the result of a calculation with an extra 1 MeV/u radial energy (corresponding approximately to a density of 1.3 ρ_0 for a soft equation of state). From the successful comparison of the data with the "no expansion" case, we concluded that the disassembly process is mainly driven by Coulomb forces. Nevertheless, further comparisons of our data with the results of calculations based on other multifragmentation models[18-19] would be very welcomed.

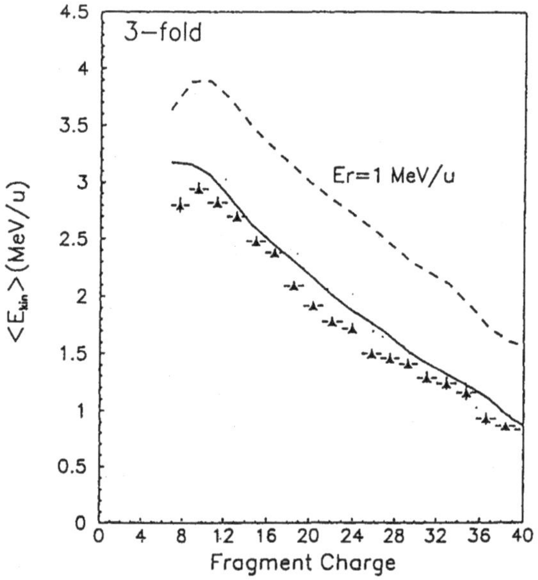

Fig. 4: Average fragment kinetic energy as a function of the atomic number for three target-emitted fragments in the reaction Kr+Au at 60 MeV/u. Black triangles: experimental points . Curves: statistical multifragmentation model (see text) with two different initial conditions; no expansion energy (solid line) and 1 MeV/u radial expansion energy (dashed line). From [17].

5. Conclusions

In this contribution, we have shown how multi-fragment emission from heavy excited nuclei was to a large extent governed by the excitation energy (or equivalently by the temperature). As soon as multi-fragment emission starts to compete with binary fission (around E*=3 MeV/u), the nucleus disassembly evolves from a slow "low energy-like" process towards a much faster (almost simultaneous) mechanism around E*= 5 MeV/u. In this energy range, it seems that compression-expansion effects play a minor role as compared with Coulomb forces in the disassembly process.

1) G. Bertsch, S. Das Gupta *Phys. Rep.* 160 (1988) 189
2) E. Suraud, B. Tamain, C. Grégoire
Prog of Nuclear and Particle Science 23 (1989) 357
3) C. A. Ogilvie et al *Phys. Rev. Lett.* 67 (1991) 1214
D. R. Bowman et al *Phys. Rev. Lett.* 67 (1991) 1527
K. Hagel et al *Phys. Rev. Lett.* 68 (1992) 2141
J. Hubele et al *Phys. Rev.* C45 N°5 (1992) R1577
4) R. Bougault et al. *NIM* A259 (1987) 473
5) G. Rudolf et al *NIM* A307 (1991) 325
6) D. Hilscher, H. Rossner *Ann. Phys. Fr.* 17 (1992) 471
7) G. Bizard et al *Phys. Letters* B302 (1993) 162
8) D. Durand et al Preprint LPCC 92-13 (submitted)
9) R. Trockel et al *Phys Rev Lett.* 59 (1987) 2844
10) Y. D. Kim et al *Phys. Rev. Lett.* 67 (1991) 14
11) R. Bougault et al *Phys Letters* B232 (1989) 291
12) G. Bizard et al *Phys. Letters* B276 (1992) 413
13) M. Louvel et al *Null. Phys.* A559 (1993) 137
14) O. Lopez et al. Preprint LPCC 93-06
15) M. Louvel et al Preprint LPCC 93-12 (accepted by *Phys. Letters*)
16) J. Lopez, J. Randrup *Nucl. Phys.* A512 (1990) 345
17) O. Lopez et al. Preprint LPCC 93-10 (accepted by *Physics Letters*)
18) D. H. Gross et al *Nucl. Phys.* A461 (1987) 641
19) J. P. Bondorf et al *Nucl. Phys.* A443 (1985) 321

PERCOLATION, INTERMITTENCY AND MULTIFRAGMENTATION

S. LERAY and S.R. SOUZA
Laboratoire National Saturne
CE Saclay, 91191 Gif-sur-Yvette Cedex, France

ABSTRACT

We show that a simple percolation model can provide informations on the mechanism responsible for projectile multifragmentation. Within the framework of this model we also investigate the possible use of intermittency to select critical events.

1. Introduction

One of the reasons for studying multifragmentation in heavy ion collisions is that it could be a manifestation of a critical phenomenon in a small size system. Recently, it has been shown by the ALADIN collaboration[1] that site-bond percolation was quite successfull in reproducing the details of the fragment size distribution observed in projectile multifragmentation induced by 600 MeV/u Au nuclei, although other approaches[2–4] seem to be also able to give a good agreement to the same data. The success of a so simple, purely geometric model in which nuclear physics is apparently absent, could appear as surprising. However, it is well known that near a critical point the details of the interaction are not important and that only the characteristic lengths of the system are relevant quantities. In Ref. 1 there was no attempt to relate percolation to nuclear physics nor to draw any conclusion on the reaction mechanism. Since we have developed, some time ago, a model[5] which is, in certain cases, similar to site-bond percolation, we want here to apply it to the ALADIN and other data in order to try to relate the parameters entering into the percolation model to nuclear physics quantities and investigate what can be learnt on the mechanism responsible for multifragmentation through this approach.

The possibility that multifragmentation could be in some cases related to a critical phenomenon raises the question of being able to experimentally select critical events. Since it has been claimed[6] that this could be realized by looking at the appearance of intermittent fluctuations in the fragment size distribution, we want also here, in the framework of our model, to study what signature can be experimentally expected.

2. The percolation model

We use here a model[5] which was originally designed to describe multifragmentation occurring in central heavy ion collisions and was called the restructured aggregation model (RAM). This model, actually, is, if the restructuration phase is ignored, equivalent to site-bond percolation but without lattice : being given a volume V representing the nucleus and A nucleons distributed at random in this volume, we suppose that two nucleons can interact and form a cluster if their centers are separated by a distance smaller that the aggregation distance, d_a. To change the volume for a given number of nucleons or the number of nucleons for a given volume is equivalent to change the number of occupied sites in percolation while the bond probability is varied by varying d_a. In order to avoid making clusters with fractal shapes, as obtained after the aggregation process, we added a restructuration phase in which the clusters are reorganized into spheres at each step of the aggregation process. In Ref. 7 the density distribution generated by a dynamical model describing the heavy ion collision was used as initial condition for the restructured aggregation model. This gave a good agreement to inclusive charge distribution data obtained in central collision between 50 and 200 MeV/u.

3. Comparison to Aladin data

The ALADIN collaboration[1] have measured the fragment charge distribution obtained after the fragmentation of 600 MeV/u gold projectile impinging on different targets. The event charge distribution was measured and high order charge correlations were plotted versus the quantity Z_{bound} defined as the sum of all charges of fragment except $Z = 1$ particles and which is supposed to be related to the impact parameter. To account for the different impact parameters present in the experimental data, we have to vary the number of occupied sites in our percolation model. There are actually two extreme ways to realize this: either we keep the volume constant and we vary the number of nucleons or we fix the number of nucleons and vary the volume. In the original model[5] we have adopted the latter prescription since we were dealing with multifragmentation arising after a hot and dense system was formed in a central collision and was then expanding to low density. Here, if this prescription is used, it is never possible to reproduce the ALADIN data, whatever the choice of the parameters of the model, with or without the restructuration phase. This, however may not be surprising as, here, the mechanism is likely different since the selected reactions are peripheral ones in which it is the projectile spectator that undergoes multifragmentation.

The other extreme possibility to vary the density of occupied sites is to keep the volume constant and vary the number of nucleons in this volume. This was tried and we found that a complete agreement to the experimental data can be obtained if the constant volume is equal to the volume of the gold projectile, $d_a = 2.0$ fm and the restructuration phase is not added. This can be seen in fig. 1 representing the charge of the largest fragment, Z_{max}, and the 3-body asymmetry between the three largest

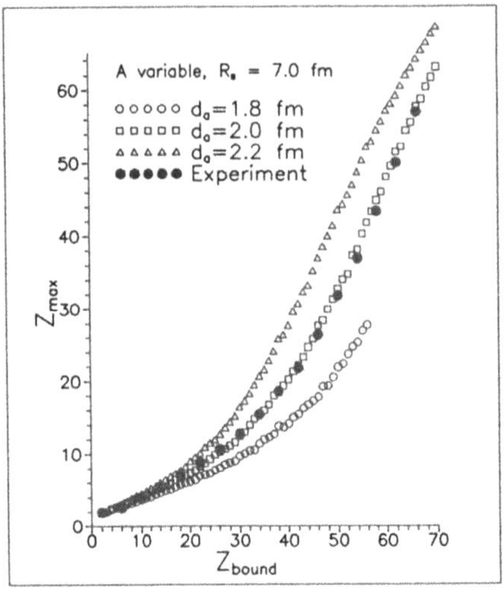

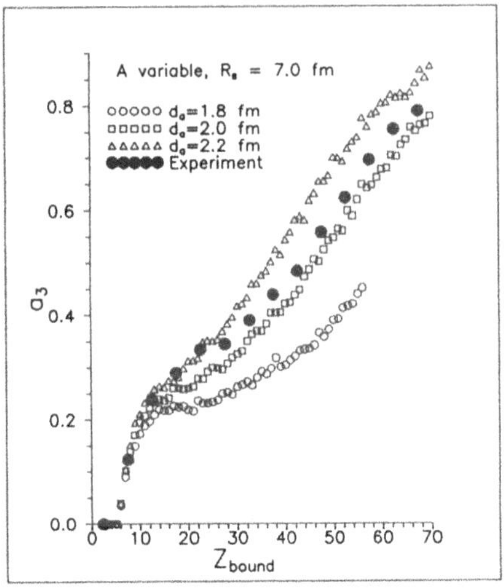

Fig. 1. *Left:* Charge of the largest fragment, *right:* 3-body assymetry between the three largest fragments versus Z_{bound}, for different values of the aggregation distance, without restructuration phase, for a variable number of nucleons in a constant volume compared to the data from Ref. 1.

fragments, A_3, versus Z_{bound}, as obtained with our model, without restructuration, for a fixed volume (equal to *Au* nucleus volume) and for different values of d_a. All the calculations presented here are corrected by the experimental acceptance. We also show in fig. 2 the combination of the moments of the charge distribution, Γ_2, as defined by Campi,[8] which is a measure of the fluctuations and is supposed to present a maximum near a critical point. It can be noticed that the results are quite sensitive to the choice of d_a. Actually, the relevant parameter of the problem is the dimensionless quantity R_S/d_a where R_S is the radius of the constant spherical volume and the agreement to the data is found for $R_S/d_a = 3.5$. However, R_S is determined by the most peripheral collisions (largest Z_{bound}) where the projectile has only very little interacted with the target and thus cannot have a radius very different from its initial value. R_S being fixed to the radius of a gold nucleus, $R_S/d_a = 3.5$ gives for d_a, 2.0 fm which is of the order of the range of the nuclear force. It appears then that the parameters entering our crude purely geometric model can be related to reasonable nuclear physics quantities as if the details of the interaction were not important and only the fundamental characteristic lengths of the system were relevant.

As above-mentioned, the data can be reproduced if we do not add the restructuration phase. In fact, we have checked that whatever d_a or R_S it is never possible

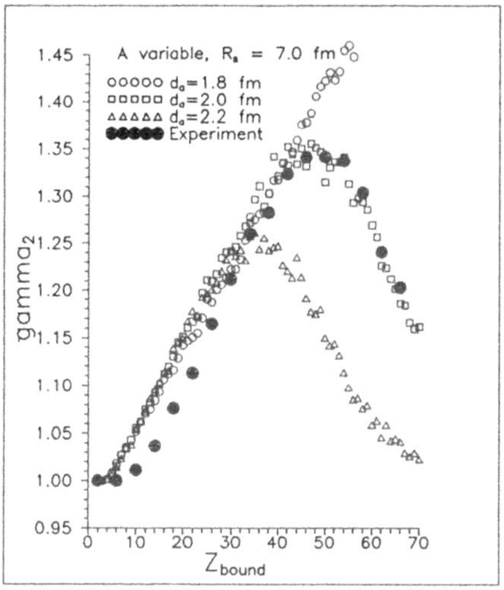

Fig. 2. Γ_2 versus Z_{bound} for different values of the aggregation distance, d_a, without restructuration phase and for a constant volume and a variable number of nucleons.

to obtain a good fit to the data with a restructuration of the fragments formed in the percolation process. This is at variance with what was found in Ref. 7 where only the model with restructuration was able to reproduce the central collision data. Here it implies that the fragments have fractal shapes at the point of breaking and that this is a necessary condition to reproduce the event size-distribution. Of course, the fragments will, in a later stage, evolve to spherical nuclei but this is out the scope of our model.

We have found that we had to vary the number of nucleons, A, while keeping the radius of the gold nucleus fixed rather than vary the volume at given A. Of course one could think of other possibilities involving simultaneous variations of V and A in a more or less complicated way. Since this would introduce new parameters without any physical justification we have not tried this. Anyhow, our results can be viewed as an indication on the multifragmentation mechanism. A possible explanation for the variable number of nucleons could be that, in a participant-spectator picture, nucleon-nucleon collisions between nucleons of the projectile-spectator and those coming from the participant zone lead to the ejection of nucleons in the spectator, whose volume appears as more or less constant because fragmentation occurs very early before equilibrium is reached and expansion has begun.

The experimental acceptance may also play a role if the quasi-projectile is more decelerated than in a pure participant-spectateur picture. Indeed, then, because only the close-to-projectile-velocity fragments can be detected, the slowest ones which are more and more numerous as the impact parameter decreases and the quasi-projectile is more decelerated, could be lost. In this case, the variable A would correspond to the variable total detected mass.

Both explanations, together with the fractal nature of the fragments at the time of breaking seem to indicate that the mechanism is completely different from the one involved in central collision multifragmentation and resembles more the fast cold breaking of glass. The difference with low energy central collisions can also be corroborated by the fact that the MSU-bond-percolation model[9] was unable to fit the $50 MeV/u$ MSU data.

4. Low energy projectile fragmentation

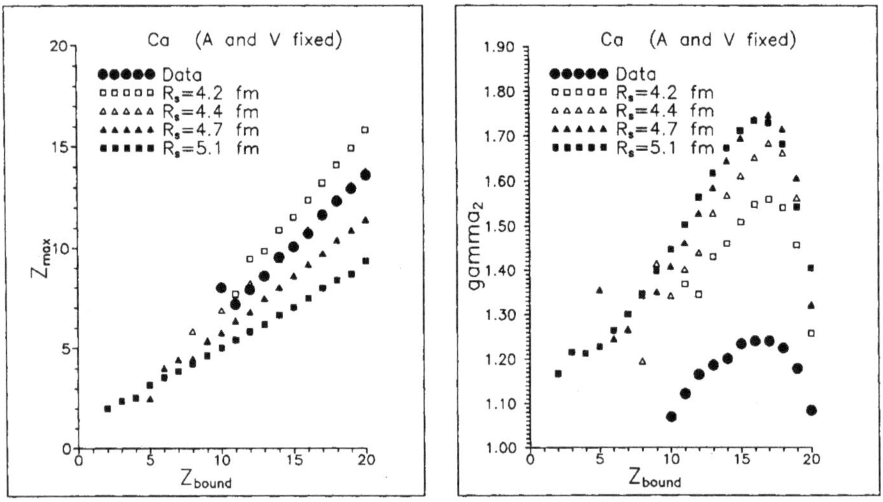

Fig. 3. *Left:* Same as fig. 1 left, *Right:* Same as fig. 2, for a Ca projectile and for a constant volume and number of nucleons. The data are from Ref. 10.

It has been said by Désesquelles et al[10] that percolation is also able to reproduce low energy data obtained in the system Ca+Cu at 35 MeV/u by the AMPHORA collaboration. However, the model used in this case was a bond-percolation model, which is different from our approach. It is therefore interesting to apply our model to this case in order to see if one can expect the same mechanism for quasi-projectile fragmentation at 35 and 600 MeV/u. We show on fig. 3 the comparison of the experimental data[10] plotted versus Z_{bound}, defined like in the ALADIN experiment, with the calculation. Here the selected events correspond to fully detected ones, i.e. with the total Z of the fragments equal to 20, so in the model both A and the volume were kept constant. It can be seen that it is never possible to reproduce simultaneously Z_{max} and Γ_2 versus Z_{bound} whatever choice of parameters of the model we make. We have checked that the fact that only fully detecting events are kept does not affect this conclusion.

The failure of the model in this case seems to indicate a different mechanism from the one responsible for fragmentation at 600 MeV/u, which is not surprising since at low energy, even for peripheral collisions, projectile and target deeply interact. This, anyway, strengthen the belief that the success found with the ALADIN data is non trivial and that this kind of simple models can tell us information on the reaction mechanism.

5. Intermittency in the percolation model

In a recent paper, Ploszajczak et al[6] have proposed to apply the formalism of the scaled factorial moments (SFM) of Bialas et al[11] to investigate the possible existence of intermittent fluctuations in the charge distribution obtained in nuclear multifragmentation. The hope was that the appearance of intermittency could be a signal of a critical behaviour and could allow to select critical events as it was found in a simple bond percolation calculation.

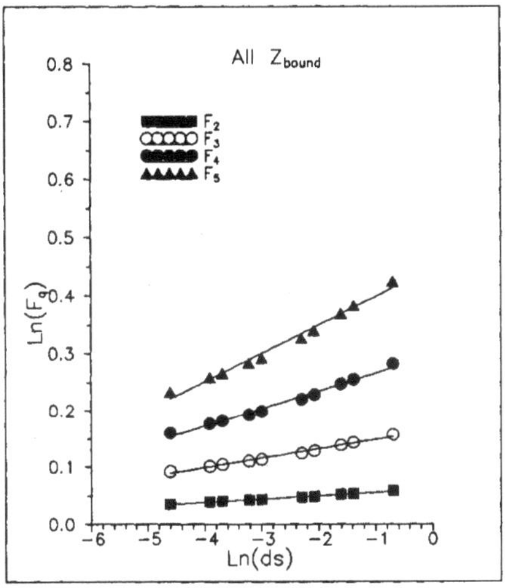

Fig. 4. Scaled factorial moments, F_2 to F_5, versus $Ln(\delta s)$ for all events generated by the model for a $A = 200$ system.

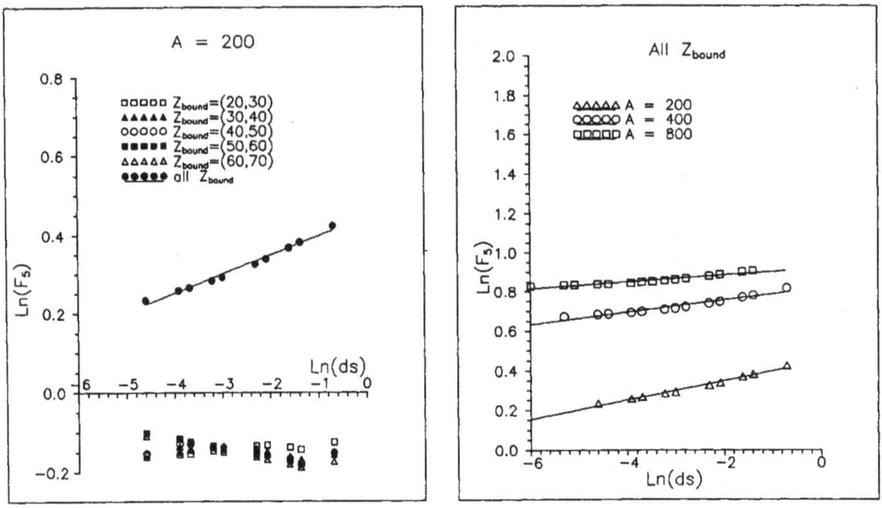

Fig. 5. Scaled factorial moment, F_5, versus $Ln(\delta s)$. *Left:* for events selected according to different Z_{bound}-bins compared to all events for a $A = 200$ system. *Right:* for all events for different sizes of the system.

Since our model reproduces perfectly the details of the event size distribution of the ALADIN experiment, it was tempting to also investigate the possible occurrence of intermittent fluctuations and what could be expected from such an analysis. The SFMs, F_i, are defined as in Ref. 6 and plotted versus δ_s, the size of the bin into which the total size of the system is divided. Fig. 4 shows F_2 to F_5, obtained for a $A = 200$ system with our model with the parameters that gave a good fit to the experimental data in the preceeding section. In this section, calculations are not corrected by experimental acceptance. The alignment of the points on a straight line with a positive slope could be viewed as an evidence of an intermittent behaviour. We have seen in fig. 2 that Γ_2 was presenting a maximum suggesting a critical point near Z_{bound} equal to 50. However, if one tries to extract critical events with the SFMs by making Z_{bound} bins, there is no longer evidence of an intermittent behaviour as the slope of the straight line becomes zero as shown in fig. 5 left for F_5. This indicates that in fact the apparent signal of intermittency seen in fig. 4 originates mostly from the mixing of events of different nature.

It was pointed out in by Bialas et al,[11] that conservation laws could influence drastically the SFMs. Since in nuclear fragmentation we are dealing with very small systems, mass conservation should be a strong constraint which can be investigated by looking at the influence of the size of the system. It can be seen in fig. 5 right that the slope of Ln (F_5) decreases when the number of nucleons in the system increases and could even be zero for an infinite system. Therefore it is not clear that what is attributed to the existence of intermittency could not be mostly due to finite size effects.

In conclusion, the study of intermittency with our model indicates that it would be extremely difficult to disentangle experimentally a true intermittent behaviour from spurious effects due to finite size and mixing of events.

6. Acknowledgments

We would like to thank P.Désesquelles for providing us with the AMPHORA data.

References

1. P. Kreutz, J.C. Adloff, M. Begemann-Blaich, P. Bouissou, J. Hubele, G. Imme, I. Iori, G.J. Kunde, S. Leray, V. Lindenstruth, Z. Liu, U. Lynen, R.J. Meijer, U. Milkau, A. Moroni, W.F.J. Müller, C. Ngô, C.A. Ogilvie, J. Pochodzalla, G. Raciti, G. Rudolf, H. Sann, A. Schüttauf, W. Seidel, L. Stugge, W. Trautmann and A. Tucholski, *Report* **GSI-92-23** (1992).
2. B.A.Li, A.R.DeAngelis and D.H.E.Gross, preprint HMI (1993)
3. A.S. Botvina and I.N. Mishustin, *Phys. Lett.* **B294** (1992) 23.
4. R. Botet and M.Ploszajczak, *preprint GANIL P92-22* (1992).
5. S. Leray, C. Ngô, M.E. Spina, B. Remaud and F. Sébille, *Nucl. Phys.* **A511** (1990) 414.
6. M.Ploszajczak and A.Tucholski, *Phys. Rev. Lett.* **65** (1990) 1539.
7. S. Leray, C. Ngô, P. Bouissou, B. Remaud and F. Sébille, *Nucl. Phys.* **A531** (1991) 177.
8. X. Campi, *J. Phys.* **A19** (1986) 917, *Phys. Lett.* **B208** (1990) 351.
9. L. Phair, W. Bauer, D.R. Bowman, N. Carlin, R.T. de Souza, C.K. Gelbke, W.B. Gong, Y.D. Kim, M.A. Lisa, W.G. Lynch, G.F. Peaslee, M.B. Tsang, C. Williams, F. Zhu, N. Colonna, K. Hanold, M.A. McMahan, G.J. Wozniak and L.G. Moretto, *preprint MSUNSCL-819* (1992) .
10. P. Désesquelles, A.J. Cole, A. Giorni, D. Heuer, A. Lleres, J.B. Viano, B. Chambon, B. Cheynis, D. Drain and C. Pastor, *preprint ISN 93-03* (1993).
11. A. Bialas and R. Peschanski, *Nucl. Phys.* **B273** (1986) 703; *Nucl. Phys.* **B308** (1988) 857.

Multifragmentation:

Surface and Coulomb Instabilities of

Sheets, Bubbles, and Donuts.

Luciano G. Moretto, Kin Tso, and Gordon J. Wozniak

Lawrence Berkeley Laboratory, Berkeley, CA 94720

Abstract: Disks, bubbles, and donuts have been observed in dynamical calculations of heavy ion collisions. These shapes are subject to a variety of surface and Coulomb instabilities. These instabilities are identified and analyzed in terms of their relevance to multifragmentation.

Introduction

Calculations of nucleus-nucleus collisions at intermediate energies performed by means of different transport codes, have shown the formation of rather peculiar shapes, like rods, disks, bubbles, and donuts, which then proceed to break up into several fragments (1, 2, 3).

These results are exciting for two reasons. On one hand they give new life to the old problem of stability for such shapes, on the other they suggest that multifragmentation may be due to the onset of easily describable static instabilities.

Much attention has been given to the spinodal instability as a possible cause of multifragmentation. Since the spinodal instability can occur in an infinite system, it can be called a bulk or volume instability. Rather strong manipulations, like compression followed by expansion, are necessary to bring about the spinodal instability. Here we are interested in the somewhat gentler physics of those instabilities that rely on surface, proximity, and Coulomb energies. We shall discuss these instabilities in the context of cylinders (rods), donuts, disks (sheets), bubbles, and spheres.

Surface Instabilities in Boltzmann-Nordheim-Vlasov Calculations

In head-on collisions of heavy-ions simulated using the Boltzmann-Nordheim-Vlasov (BNV) equation, a "disk" develops due to the side-squeezing of nuclear matter. When the disk becomes sufficiently thin, it breaks up into several fragments whose size is commensurate with the thickness of the disk.

Some of these features are shown in Fig. 1 for a head-on collisions of two ^{90}Mo nuclei. Similar calculations have been performed for a range of central impact parameters and entrance-channel mass asymmetries with similar results. At relatively small impact parameters one observes donut shapes, and at large impact parameters, rather long necks. Both donuts and necks break up into several fragments. Other calculations with different codes and for different systems show the formation of bubbles that break up in much the same manner as disks.

Metastability of a Sheet of Liquid

The characteristic way in which disk fragmentation occurs, strongly suggests that it is caused by surface instabilities. *The system appears to escape from the high surface energy of the disk by breaking up into a number of spherical fragments with less overall surface.*

One may be tempted to consider this instability as a case of the Rayleigh instability of a cylinder of liquid. The cylinder is unstable with respect to small perturbations of wavelength $\lambda \geq 2\pi R$, where R is the radius of the cylinder. However, a disk of liquid, or more generally, a sheet of liquid does not suffer the same kind of instability as a cylinder.

If we assume sharp surfaces (no surface thickness, no surface-surface interaction), a sheet can be metastable with respect to break-up into a layer of cylinders or spheres. The onset of metastability for both cases is easily calculated. On a sheet of thickness d let us identify stripes of width λ. These stripes can favorably collapse into cylinders when the surface area of a stripe (top + bottom) is greater than the surface area of the cylinder of equivalent volume.

Metastability and not necessarily instability is involved here, since there may be a barrier that prevents the sheet from reaching the more stable configurations illustrated above, and indeed there is. Clearly, any wave of infinitesimal amplitude A increases the surface area of the sheet, and thus its energy, independent of the sheet thickness, since, in the limit of infinitely sharp surfaces, the surfaces do not know of each other, until they touch. However, the systems portrayed in Fig.1 develop what appears to be a genuine instability. Perhaps, the system, which has plenty of energy, simply jumps the barrier. But, there is another, more likely possibility.

Instability of a sheet of liquid and surface-surface interactions

Nuclear surfaces are not sharp, but diffuse, and they interact with each other through an interaction of finite range called also the proximity force $\Phi(s)$, where s is the distance between surfaces. We can now calculate the incremental energy of a sheet subjected to a perturbation of wavelength λ and small amplitude A. The dimensionless proximity interaction is:

$$V_P = \frac{2}{\lambda}\int_0^{\lambda}\Phi(s)\,dx \sim \frac{2}{\lambda}\left(P(\lambda) + Q(\lambda)A^2\right) \tag{1}$$

where

$$P(\lambda) = \int_0^{\lambda}\Phi_0(x)dx \quad \text{and} \quad Q(\lambda) = \int_0^{\lambda}\Phi_2(x)dx \tag{2}$$

with $s = d + 2A \sin kx$, Φ_0 and Φ_2 being the zeroth and second order coefficients of the Taylor expansions of $\Phi(A,x)$ about $A = 0$, and $k = 2\pi/\lambda$.

The overall energy increase, including the surface energy term is:

$$\Delta V = A^2\left(\frac{2\pi^2}{\lambda^2} + \frac{Q(\lambda)}{\lambda}\right) \tag{3}$$

Instability occurs when the coefficient of A^2 is zero or negative. Thus, the critical wavelength for the onset of the instability is given by the equation:

$$\lambda_c\, Q(\lambda_c) + 2\pi^2 = 0\,. \tag{4}$$

Any perturbation with $\lambda > \lambda_c$ is then unstable, namely it will grow spontaneously and exponentially. Using for the proximity potential the expression in ref. 4, we obtain

$$\lambda_c = 1.10\, b \exp[2d/3b], \tag{5}$$

where b is the range of the proximity interaction.

When the thickness of the sheet becomes much greater than the range of the proximity interaction, the critical wavelength tends to infinity. This is the trivial result for infinitely sharp surfaces that was mentioned above. However, when the thickness of the sheet becomes comparable to the proximity range b, the critical wave-length decreases very rapidly. This result is quite interesting, because it applies in general to all liquids. Specifically, thin disks and thin bubbles can be destabilized by the sheet instability.

Stability of a bubble

Nuclear bubbles and their stability have been discussed in a variety of contexts (5). Here we shall analyse the relevant degrees of freedom, and the physical quantities that affect their stability. The bubble degrees of freedom can be divided into two classes: the radial modes and the crispation modes. The radial modes are characterized by a constant thickness throughout the bubble, while the crispation modes modulate the thickness of the bubble's wall. The physical quantities that we shall consider here are the Coulomb force, the centrifugal force, and the pressure difference across the walls of the bubble. The most important degree of freedom, which defines the bubble itself, is the radial monopole mode.

The monopole mode

Let us consider a nucleus with radius R_0, fissility parameter X, and angular momentum I; let us allow a bubble of radius R_1 to grow at the center of the nucleus with an internal pressure p. If we define $x=R_1/R_0$, the energy of the system in units of twice the surface energy E_s of the unperturbed nucleus is:

$$E = \frac{1}{2}x^2 + \frac{1}{2}\left(1+x^3\right)^{\frac{2}{3}} + 5X\left[\frac{1}{5}\left(1+x^3\right)^{\frac{5}{3}} - \frac{1}{2}x^3\left(1+x^3\right)^{\frac{2}{3}} + \frac{3}{10}x^5\right]$$

$$-x^3P + \frac{R}{\left[\left(1+x^3\right)^{\frac{5}{3}} - x^5\right]}$$

$$X = E_c^0 / 2E_s^0;$$

where: $$P = pV_0 / 2E_s^0; \qquad (6)$$

$$R = E_R^0 / 2E_s^0.$$

The first two terms of the equation give the surface energy of the inner and outer sphere, the third gives the Coulomb energy, the fourth the pressure energy, and the fifth the rotational energy. Here are the effects of each. If we neglect the Coulomb term, but we retain instead the pressure, we have a minimum at x=0, and a saddle at x=2/(3P). If the pressure does not depend on the radius, as in the case of nuclear vapor pressure, which just depends on temperature, the bubble will, for sufficiently large radii, become unstable towards indefinite expansion. On the other hand, the inclusion of both surface and Coulomb terms permits the formation of a secondary minimum for X>2.02. This is the Coulomb bubble. An example of the potential energy as a function of the inner radius is given in fig.(2). The rotational term mimics the Coulomb term, and generates a secondary minimum for R>.956. As the Coulomb (or rotational) energy increases, the secondary minimum becomes deeper and deeper, until it becomes the absolute minimum.

A word may be in order regarding the pressure. For T>0, a saturated vapor fills the bubble cavity. This vapor would have a pressure equal to the saturation pressure, which depends only upon the temperature, but not upon the volume. At first sight, one would not expect a pressure acting upon the outer surface, which is facing the vacuum. However, this is not the case. The inner surface has an outgoing flux of evaporated particles, and an ingoing flux of vapor particles. At equilibrium the two fluxes are equal: they impart the same impulse to the surface, and thus contribute equally to the pressure. The outer surface, however, has only the outgoing component, and thus feels a pressure equal to one half of the inner pressure.

<u>The higher order modes</u>

Before discussing the stability of the higher modes in a bubble, let us review the case of the liquid sphere.

<u>Liquid sphere</u> The eigen-frequencies of a liquid, incompressible non viscous sphere with irrotational flow are given by:

$$\omega'^2 = n(n-1)(n+2) \tag{7}$$

where: ω' is the frequency given in units of $(c_s/\rho R^3)^{1/2}$ (c_s is the surface energy coefficient, and ρ and R are the density and radius of the sphere respectively); and n is the order of the spherical harmonics under consideration. Since all these frequencies are real, all the modes are bound.

The introduction of a charge (uniformly distributed throughout the volume of the sphere) changes the frequencies as follows:

$$\omega'^2 = n(n-1)\,[(n+2) - 4X] \tag{8}$$

where X is the well known fissility parameter. The effect of the Coulomb field is that of destabilizing a number of modes.

For $X < 1$ all the modes are still stable. At $X = 1$, the frequency ω' goes to zero for $n = 2$. This is the onset of quadrupole instability, or of the fission instability. For $X > 1$ progressively higher modes are destabilized. The last unstable mode is:

$$n_{last} = 4X - 2.$$

One would think that when many modes are unstable, the most unstable mode would remain the lowest mode $n = 2$ or the fission mode. This is, curiously, not the case. For instance, for $X = 3$, $n_{last} = 10$, $n_{max} = 7$ and for $X = 4$, $n_{last} = 14$, $n_{max} = 10 - 9$. So, a highly charged sphere will not merely fission, but will break up in many droplets through an instability associated with a high multipole mode.

<u>Bubbles.</u> We consider first the radial modes. The stability of a Coulomb bubble towards monopole oscillations is not a sufficient condition for stability. The stability towards the higher order radial modes can be easily checked by means of a modification of eq.(8), All that one needs to do is to substitute in it for the parameter X the parameter X_{eff} containing the Coulomb and surface energy of the bubble. This is given by:

$$X_{eff}(x) = 5X\frac{\frac{1}{5}\left(1+x^3\right)^{\frac{5}{3}} - \frac{1}{2}x^3\left(1+x^3\right)^{\frac{2}{3}} + \frac{3}{10}x^5}{x^2 + \left(1+x^3\right)^{\frac{2}{3}}} \tag{9}$$

As the bubble expands, X_{eff} decreases. If the original nucleus is unstable up to the multipole of order n, as it develops into a bubble it starts stabilizing the higher multipoles. This is shown in fig.(3). At a sufficiently large value of X, the bubble will become stable even against the quadrupole deformation, and all the radial multipoles would be stable, with the possible exception of the monopole. However, at the value of x corresponding to the equilibrium of the Coulomb bubble, it turns out that the quadrupole and octupole modes are not yet stabilized. Therefore, no Coulomb bubble is absolutely stable. Of course, such bubble could exist as a transient. Radial instabilities can be escaped by a sufficiently rapid, dynamically driven expansion.

Yet, it may still be possible to have a stable nuclear bubble. If the bubble is warm, it fills up with vapor. The resulting pressure differential acts only upon the monopole mode, by displacing outwards the Coulomb minimum. The effect on the other radial modes is nil, since only changes in volume are relevant to pressure. Consequently the values of X corresponding to the stabilization of the various multipoles do not change.

Thus, as shown in fig.(2), a sufficiently large pressure can move the bubble minimum sufficiently far out, beyond the onset of quadrupole stability.

The second class of modes consists of "crispation modes" involving a thickening and thinning of the liquid layer. In the limit of sharp surfaces (no surface-surface interaction) and in the absence of the Coulomb field all these modes are stable, excepting the dipole mode which is indifferent. The "inner" sphere of the bubble in this mode is free to drift with respect to the outer sphere, leading eventually to the puncturing of the bubble. The introduction of surface-surface interaction can make all the crispation modes unstable through the "sheet instability".

On the other hand, the Coulomb field tends to stabilize these modes. This is because the Coulomb force tends to resist the attempt to concentrate the charge in "clumps" distributed on the surface of the sphere as required by the higher modes.

Bubbles and the sheet instability

A bubble behaves much like a sheet, and is subject to the sheet instability. Since a bubble, like a sheet, must rely on the proximity interaction to become unstable, it will retain its surface stability until its thickness is of the order of the surface-surface interaction range. Thus, a rather thick-walled bubble is not susceptible to surface instabilities over a broad range of its inner sphere radius. This is shown in fig.(4). At a sufficiently large inner sphere radius, a neutral bubble becomes thin enough for the onset of quadrupole instability, and it becomes unstable with respect to higher order modes when it gets thinner. The same trend is observed for a coulomb bubble; however, the curve traces the loci of the critical inner radius stops at the mode below which the bubble becomes unstable at all inner sphere radii. BNV calculations for very heavy systems at low bombarding energies show the formation of a thin bubble that seems to burst under the action of the sheet instability.

Surface Instabilities in Rods and Donuts

Rayleigh showed that a cylinder is unstable with respect to perturbations of wave-length $\lambda \geqq 2\pi R$, where R is the radius of the cylinder. Notice that this instability has a purely geometric origin, and it does not depend upon the strength of the surface tension.

The breaking of the neck, either developing in a fissioning nucleus, or forming between the partners in a heavy ion collision, has been attributed to this instability. Long necks that break up producing a few little drops, are a common observation in BNV simulations of heavy ion collisions.

A torus can be imagined as a cylinder bent so that its two bases are united. Therefore it should manifest the classical Rayleigh instability, whose critical wave-length is given by the length of the circumference of the cylinder. A torus with a length equal to the critical wave-length has a very compact shape in which the inner circumpherence degenerates into a point. Any wider torus or donut with a finite hole should be unstable , and will spontaneously pinch off. This is a serious instability that can give rise to a number of fragments.

Conclusions

Missing in the discussion of these fancy theoretical constructs is the voice of experimental evidence. The unfortunate reason for this is that there is none. Despite a great deal of experimental effort, the mechanism of multifragmentation still remains a mystery and a challenge. There are some ominous indications that statistical effects may play a rather heavy role. On the other hand, there is still plenty of room to speculate, and one should take advantage of this opportunity while there is still time to do so.

References:
1) L.G. Moretto et al.., Phys. Rev. Lett. 69, 1884, (1992).
2)W. Bauer, G.F. Bertsch, and H. Schultz, Phys. Rev. Lett. 69, 1888, (1992).
3)B. Borderie, et al. Phys. Lett. B302, 15, (1992)
4)J. Blocki, J. Randrup, W. J. Swiatecki, and C.F. Tsang, Ann. Phys.(NY) 105, 427 (1977).
5)C.Y. Wong, Phys. Rev. Lett. 55, 1973, (1985) and references therein.

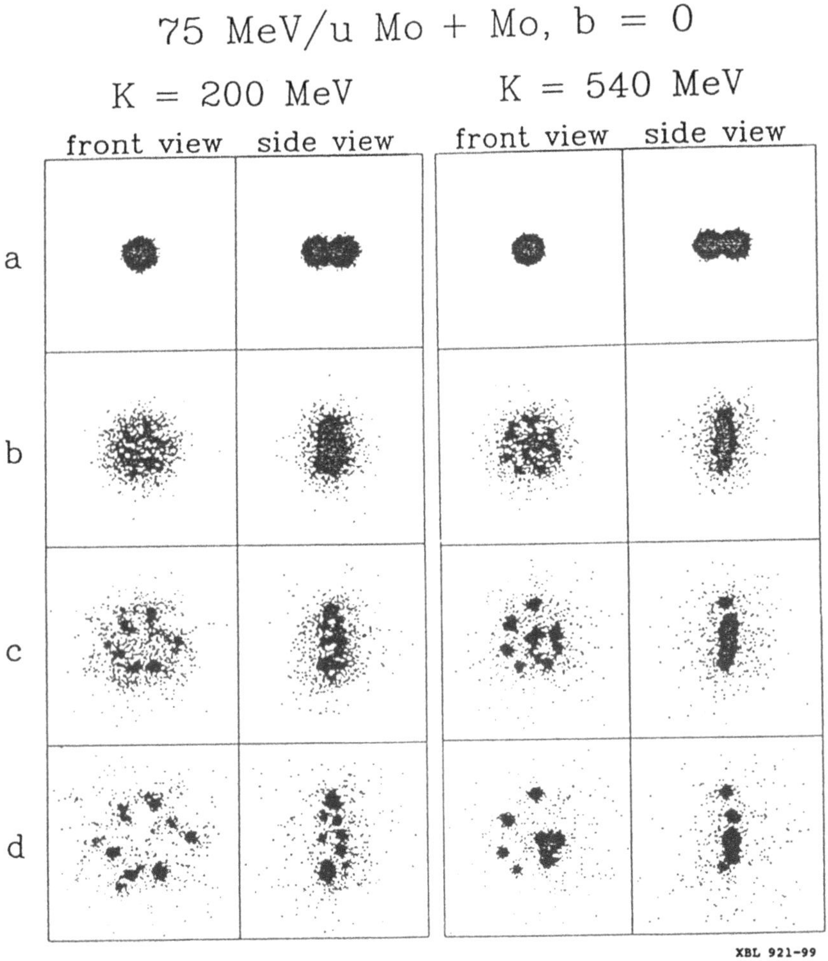

Firgure 1. BNV calculations for a head-on collision (b = 0) of the 75 MeV/u ^{90}Mo + ^{90}Mo reaction at time steps of (a) 20, (b) 60, (c) 120, and (d) 180 fm/c. The front and side-views of the colliding systems are given in columns 1 & 2, respectively for a value of the incompressibility constant, K = 200 MeV. Similar views are shown in columns 3 & 4 for K = 540 MeV.

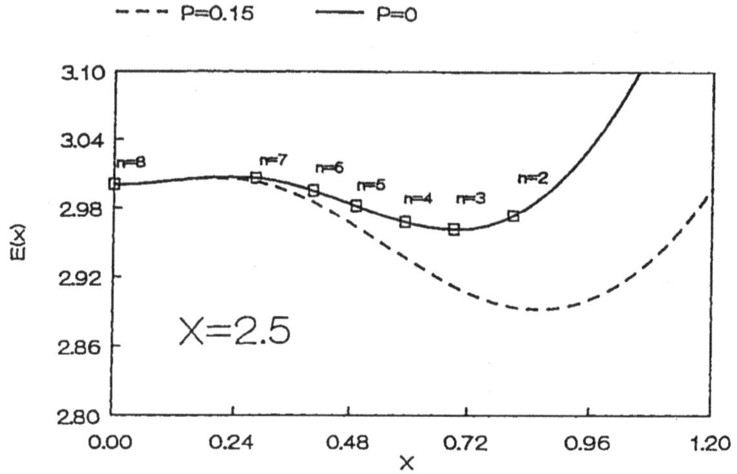

Figure 2. Potential energy of a bubble as a function of its inner radius. The fissility parameter of the equivalent sphere is X=2.5. The squares indicate the onset of instability for the various multipoles. The solid line corresponds to zero pressure, while the dashed line to the pressure that stabilizes all the multipoles.

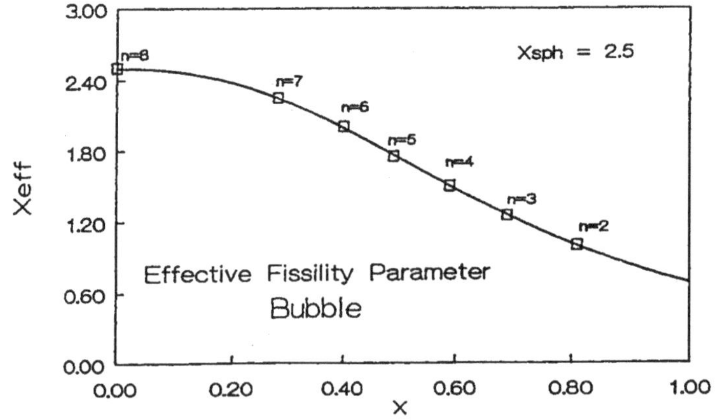

Figure 3. Effective fissility parameter of a bubble as a function of its inner radius. The fissility parameter of the equivalent sphere is X=2.5. The squares indicate the onset of instability for the various multipoles.

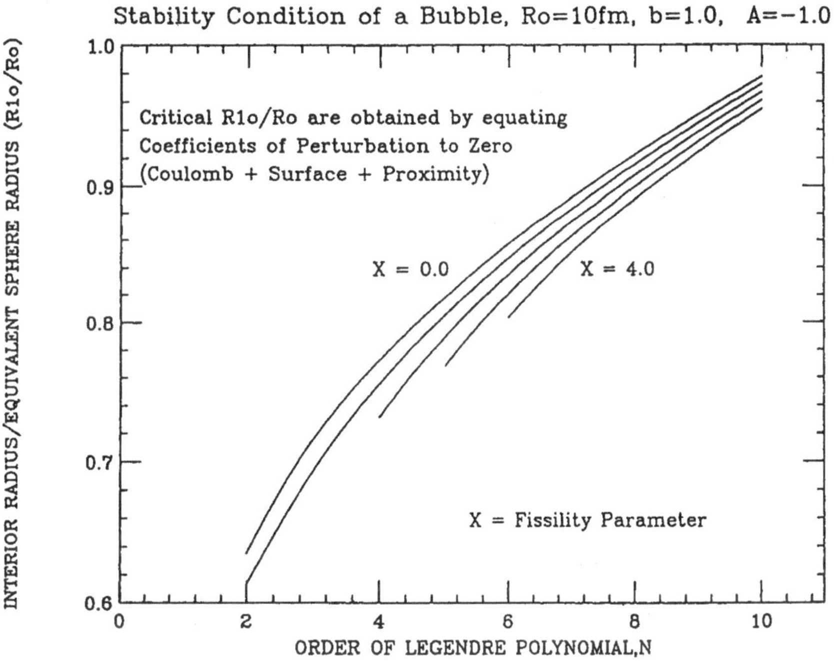

Figure 4. The critical inner radius of a bubble for the onset of the instability as a function of the order of Legendre Polynomial for various values of the fissility parameter. The perturbation on the two surfaces are equal in magnitude but opposite in phase. The radius of the equivalent sphere is Ro = 10 fm, and the range of proximity interaction is b = 1.0 fm.

Dynamical instabilities, density fluctuations and multifragmentation

J, Németh and G. Papp
Institute of Theoretical Physics, Loránd Eötvös University, Puskin 5-7 Budapest, 1088 Hungary

October 26, 1993

1 Introduction

Multifragmentation is a well-known phenomenon in heavy-ion reactions: highly excited expanding nuclei can break up into several pieces. The reason of the break up is clear: During the expansion the nucleons collide with each other, and they form clusters of different sizes in the nucleus. If the mean field is strong enough, it will smooth out this clustering of the particles. As the density decreases, the mean field gets weaker, it is not able any more to smooth out the clustering, the clusters move separately and the nucleus breaks up into parts. The resulting fragment distribution depends on several factors like nuclear masses, excitation energies, impact parameters, however dynamical effects can also play important roles in their formation.

Some theoretical models considered multifragmentation as a phase transition [1], other models tried to determine the break up with adiabatic percolation theory [2]. It is interesting to observe that both models predict the break up of the nucleus around 0.6 nuclear matter density, although the reason for the break up is very different in the two cases. In the site-bond percolation model the break up happens when there is no more percolation cluster in the system. This stage of evolution takes place whenever the number of the empty cells is big enough compared to the total number of cells.

The break up of the nucleus is determined mainly by this purely geometrical factor, and the result is only slightly modified by the strength of the bonds between the neighbouring cells, that is, by nuclear forces and excitation energies. On the other hand the density at which the phase transition appears is completely determined by the these last factors: the nucleus break up, when it reaches the so called unstable zone, where $dP/d\varrho$, the change of the pressure with the density becomes negative.

To understand the reason for the agreement of the break up densities by such very different models, we carried out detailed calculations with a simple fluid dynamical model. In this model we considered an excited nucleus under different initial conditions and examine the dynamical evolution of the system. Introducing slight fluctuations in the density distribution, we were interested in the evolution of these disturbances. If after some time the original disturbances smooth out, we consider the system to be still stable. If however, the fluctuations increase and holes are created in the nucleus, the system might be considered as unstable and will probably break up. We equate the point of multifragmentation with the transition point where the system becomes unstable. By using different excitations and different nuclear forces, we might examine the effects of these factors on the break-up density. The details of the model and the results are given in Chapter 2. The increase of bubbles and holes inside the nucleus leads to a volume multifragmentation: the instabilities will start to increase everywhere within the nucleus. There is, however, the possibility, that beside these volume effects surface instabilities can also contribute to the multifragmentation. We shall examine this possibility in Chapter 3.

2 Volume instabilities

To examine the evolution of the nuclear system with small fluctuations in the density, a simple fluid dynamical description was used. For the sake of simplicity we neglected viscosity and dissipation. The initial nuclear density has a Woods-Saxon form, with ϱ_0 central density and a surface diffuseness. In the density we created a small initial fluctuation by decreasing the density with 10-30% inside a small sphere. The calculations were carried out for a gold nucleus.

In an earlier calculation we used a different excitation energies and different nuclear forces and determined the break up as the function of the central density [3]. The nuclear forces were fitted such a way that for infinite nuclear matter they should supply the correct energy density (-16 MeV/particle) and saturation density (0.16 fm^{-3}), and we changed the type and the values of the surface forces, the Coulomb force and the excitation energies. The results for the break up density can be seen in Fig-s 1-3. It turns out that neither the excitation energy (Fig. 1) nor the Coulomb force (Fig. 2) do influence too much the value of the break up density. It depends somewhat more on the value of the surface forces used. For Yukawa force the break up density was about 20% greater, then for the corresponding Laplace force (which we get by expanding the non local Yukawa potential). If we decrease the value of the Yukawa term, the break up density further increases.

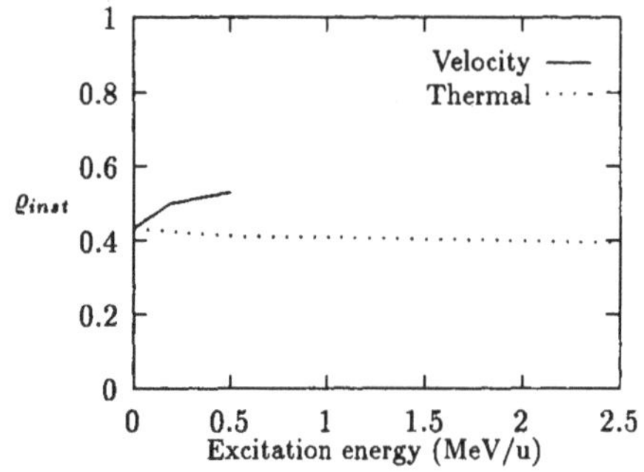

Fig.1 Dependence of the break up density on the excitation energy

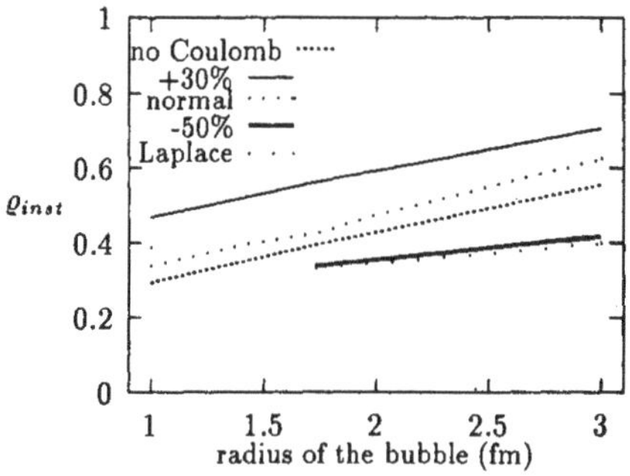

Fig.2 Dependence of the break up density on Coulomb and surface forces

In the present calculation we want to examine the effect of the changes in the equation of state (EOS) on the break up density. For this reason we simplify further our model: we consider a system with equal number of neutrons and protons, without Coulomb and surface forces. This way we do not get symmetry and Coulomb energy, and only part of the surface energy is taken into account (around half of the surface energy comes from the finite surface of the nucleus, the other halt is due to the surface forces.) For a gold nucleus we got -14.3 MeV energy/particle for a Zamick force (instead of the -16 MeV of the infinite system). For infinite system the energy per particle can be written as

$$E/A = \alpha\varrho^{2/3} - \beta\varrho + \gamma\varrho^{\sigma+1}, \tag{1}$$

with $\alpha = 23.3 MeV\ fm^2$ the infinite nuclear matter kinetic energy, and the density is expressed in nuclear matter units. At $\varrho = 1$ the expression should give -16 MeV. The saturation density equation for $\varrho = 1$ is

$$\frac{2}{3}\alpha\varrho^{2/3} - \beta\varrho + (\sigma+1)\gamma\varrho^{\sigma+1} = 0. \tag{2}$$

In the following we want to examine the instability value (the density, where for ground state energy the system becomes unstable). This can be determined from Eq. (3)

$$\frac{dP}{d\varrho} = \frac{10}{9}\alpha\varrho^{2/3} - 2\beta\varrho + (\sigma+1)(\sigma+2)\gamma\varrho^{\sigma+1} = 0. \tag{3}$$

For a stiff EOS ($\sigma = 1$) the relation to fix the saturation density to $\varrho = 1$ is

$$\beta = \frac{2}{3}\alpha\varrho^{-1/3} + 2\gamma\varrho = \frac{2}{3}\alpha + 2\gamma \tag{4}$$

From Eq. (3) one can determine the instability value. With β and γ determined from Eq. (1-2) and with $\alpha = 23.3$ $\varrho_{inst} \approx 2/3$ in nuclear matter density units. Substituting this value into Eq. (3), the relation between β and γ becomes

$$\beta = \frac{5}{9}\alpha\varrho_{inst}^{-1/3} + 3\gamma\varrho_{inst} \approx 0.64\alpha + 2\gamma \tag{5}$$

which is almost the same as Eq. (4). For a gold nucleus the instability value for the stiff EOS turned to be $\varrho_{inst} = 0.12 fm^{-3}$, or in nuclear matter density units 0.75. It is somewhat higher, than for the infinite nuclear system.

In the following we want to modify the EOS significantly, which means that either Eq. (1) or Eq. (2) will not be satisfied for infinite nuclear matter. We want to see, how will these changes influence the break up density value.

In the first case we modified the value of β and γ in such a way, that they always satisfy Eq. (1) at $\varrho = 1$. Changing the value of γ (and consequently β) the energy/particle changed only slightly, however the break up density was very different (Fig. 4 solid line). As next we were changing the $\beta - \gamma$ relation according to Eq. (4). In that case we got very different energies for different β (γ) values (Fig. 2), but the break up density did not change at all (Fig. 4 thick line).

We repeated the calculations for different σ values: the result qualitatively remained the same (Fig. 4 dotted line). If we introduce the Coulomb and surface forces, the break up density changes, but if we choose the relations between β and γ according to Eq. (2), it remains the same as the function of gamma.

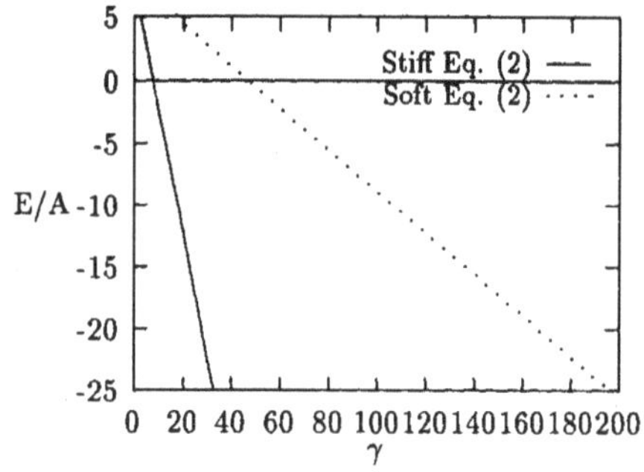

Fig.3 The dependence of ground state energy on the EOS parameters (γ)

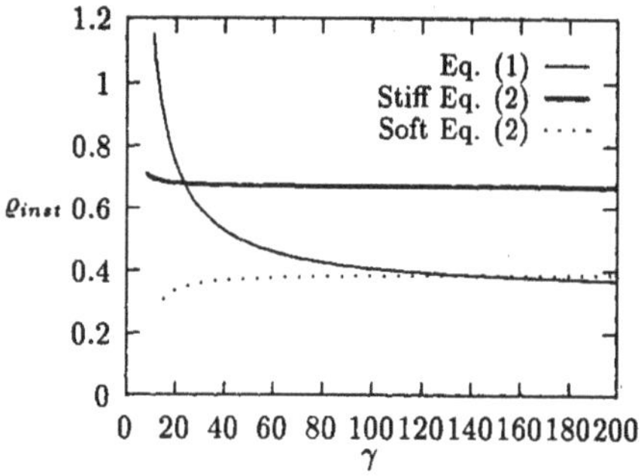

Fig.4 The break up density dependence on the EOS parameters (γ)

We can conclude from these results that even for finite nuclei the break up density is reached when the expression in Eq. (3) becomes negative, and it is not depending on the ground state energy of the system at all. If the nuclear forces were somewhat different, the phase transition break up density could be also different. The percolation break up density however does not depend on the nuclear forces, only on the total excitation energy. So it seems that it is just an accident that these two values coincide, and no deeper conclusion can be drawn from it.

It is worth while to examine, how would the change of the saturation density effect the fragment distribution in heavy ion collisions. QMD calculations to study this question are under progress.

3 Surface multifragmentation

In the macroscopic physics a well known type of growth phenomenon is the viscous fingering. It takes place when a less viscous fluid is injected into a more viscous one, for example air is pumped into glycerine. To get viscous fingering one needs a strong enough driving force, the correct viscosity difference and a noise, which cause the initial fluctuations at the surface.

A similar situation can happen in the case of heavy ion reactions too. The driving force in this case is the pressure gradient of the mean field. The fluctuations can be introduced by boundary conditions.

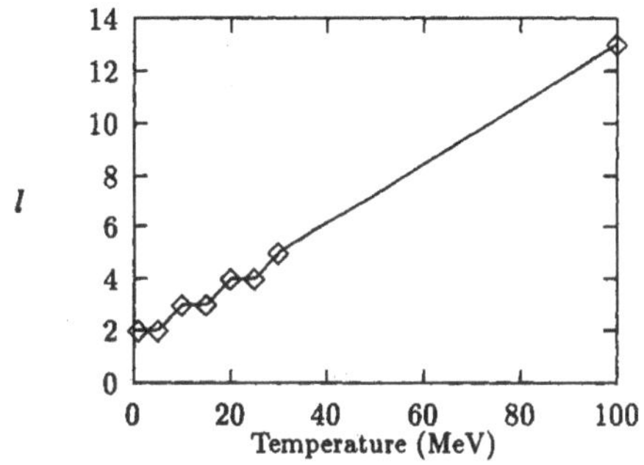

Fig.5 The critical temperature of surface instability as a function of l for gold nucleus

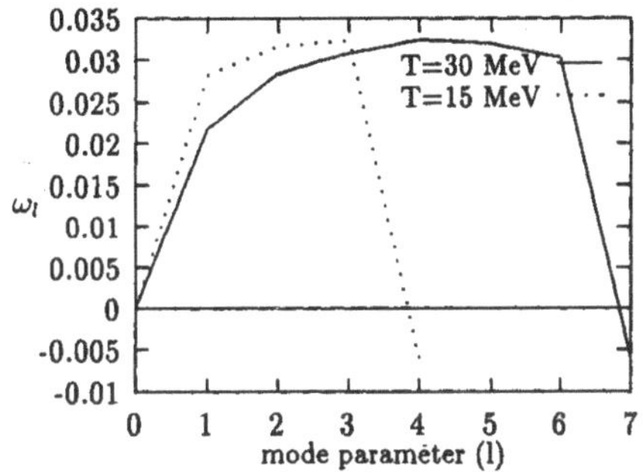

Fig.6 The critical exponents as the functions of l for two temperature values

To examine the conditions under which surface instabilities are created, we performed a model calculation. We considered a hot spherical expanding nucleus, which on the surface has some small ripples. For the sake of simplicity let us assume, that the ripples can be described by a l-th order spherical harmonics. The total density of the system can be written as

$$\varrho(\vec{r},t) = \varrho_0(r,t) + \varrho_{1,l}(r,t)Y_{lm}. \tag{6}$$

The velocities of the system can be separated into a spherical and an angle dependent part in a similar way (see [4]). At the beginning we assume that ϱ_1 is very small and differs from zero only at the surface. We integrate the fluid dynamical equations and their product with Y_{lm} for the angle. This way one can get equations for the spherical and angle dependent parts of the density and velocities. The time and radius dependent partial equations were solved on a lattice.

The results of the calculations are as follows. If the temperature (driving force) is small, the surface instabilities decrease and smooth out. For high enough driving forces however the instabilities start to increase exponentially with the time:

$$\varrho_{1,l}(r,t) = \delta_l(r)e^{\omega_l t} \tag{7}$$

In Fig. 5 we give the critical temperature (the minimum value of the temperature which is needed for the l multipole "finger" type of instability to take place) as the function of l, while on Fig. 6 the value of ω_l is given for different temperatures as the function of l. It is easy to see that the first unstable mode will always be the l=2. The fastest growing is however always the one with the biggest possible l.

In our calculations we did not take into account any viscosity. The role of the viscosity difference is played in out case by the Coulomb force. No matter, how big was the initial temperature, without Coulomb force we never got growing instability. This is represented also in Fig. 7 for a given case.

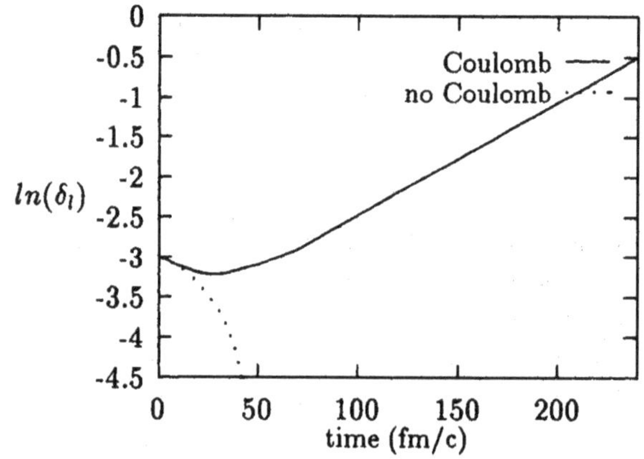

Fig.7 The development of the density perturbation on the surface

To draw further conclusions about the possibility of surface multifragmentation, we have to compare the time scale of the two instabilities. This is a further project we intend to do.

References

[1] J.P. Bondorf, R. Donangelo and K. Snepper, Phys. Lett. **B214** *No.*3 (1988) 321
C.J. Pethick and D.G. Ravenhall, Nucl. Phys. **A471** (1987)19c;
H. Heiselberg, C.J. Pethick and D.G. Ravenhall, Phys. Rev. Lett. **61** (1988) 818
G. Bertsch and P.J. Siemens, Phys. Lett. **126B** (1983) 9

[2] J. Németh, M. Barranco, J. Debois, C. Ngô, Z. Phys. **A325** (1986) 347
W. Bauer, U. Post, D.R. Dean, U. Mosel, Nucl. Phys. **A452** (1986) 699
X. Campi, J. Desbois, XIII International Winter Meeting on Nuclear Physics, Bormio, Italy

[3] G. Papp, J. Németh, J.P. Bondorf: Dynamical evolution of fluctuations in an expanding nucleus
Phys. Lett. **B278** (1992) 7-10

[4] J. Németh, G. Papp, C. Ngô, Phys. Script. Vol. T32 (1990) 160

Transport Theories and Nuclear Equation of State

COLLECTIVE EFFECTS IN BARYON AND PION EMISSION FROM CENTRAL HEAVY-ION COLLISIONS

PAWEŁ DANIELEWICZ and QIUBAO PAN

National Superconducting Cyclotron Laboratory and
Department of Physics and Astronomy, Michigan State University,
East Lansing, Michigan 48824, USA

ABSTRACT

The effects of collective expansion on baryon and pion emission from central energetic heavy-ion collisions are studied within a transport model. In the simulations, the formation of a region of dense excited matter is observed, that expands in transverse directions. Due to the expansion, baryon distributions peak at 90° in the c.m. polar angle at the lowest impact parameters, and they peak out-of-the-reaction-plane in the azimuthal angle at higher b. The mean baryon transverse-energies rise with baryon mass due to a collective contribution to these energies. The true system temperature may be read off at higher beam energies from the high-energy tail of pionic transverse spectrum. Pion yields from the collisions are significantly depressed by the expansion, while cluster yields are enhanced.

1. Historical Introduction

In the early hydrodynamic models [1] of the energetic central heavy-ion collisions, a region with dense excited matter was assumed to form between the colliding nuclei. Shock discontinuities separated that region from cold matter. With the progress of a collision, the region with dense matter expanded in the directions perpendicular to the beam axis. While the respective models were, in practise, soon discarded as overly simplistic, the experimental search to uncover traces of a collective motion set on. By the collective motion we understand one that is characterized by a correlation of the dynamic origin, between the particle positions and momenta.

Generally, the attempts to identify signatures of collective motion in the single-particle observables were plagued by ambiguities. Thus, a collective expansion in reactions was suggested [2] to be the cause of the observed differences in the transverse pion and proton slopes. However, it has been realized [3] that the production of pions in the reactions proceeds primarily through the Δ resonance excitation and decay, and thus the pion slopes could reflect the Δ-decay kinematics. As the alternatives to the collective motion as a cause of the sideward peaking of proton spectra in an asymmetric reaction [4], the Coulomb effects and shadowing were put forward [5]. Stock *et al.*[6] noted that the measured pion multiplicities in central collisions, lower than predicted within the cascade or fireball models, could be explained by assuming that a part of the available energy was used up to compress nuclear matter. Nonetheless, the dynamic calculations [7] have demonstrated a weak sensitivity of the pion yields to the compressibility of matter.

Once the 4π data became available, signatures of a different type of collective motion, than that envisioned in the head-on collisions, were discovered [8]. This motion was a sideward deflection in the reaction plane of the fragments moving forward and backward in the center of mass in semicentral collisions, quantified with a flow angle or an average momentum in the reaction plane at a given rapidity. At the midrapidity in the collisions a preference was found for the fragments to be emitted out of the reaction plane [9, 10]. The latter preference could have been of a dynamic origin or due to shadowing.

In my talk I shall discuss the effects of the collective expansion on emitted nucleons, light composites, and pions from central heavy-ion collisions as obtained within a transport model [11, 12]. Spectra of intermediate-mass fragments from the central collisions have been analyzed very recently [13, 14] and they appear to show the signs of expansion. The spectra of light fragments are being processed [15, 16]. I intend to link the results of the calculations to the past results and models.

2. Transport Description of Reactions

Calculations that I am going to discuss are carried out within a model with the explicit nucleon, deuteron, $A = 3$ cluster, pion, and delta and N^* degrees of freedom. The phase-space functions f_x of these particles satisfy transport equations which may be derived from the nonequilibrium many-body theory, in the limit of low collision rates [11],

$$\frac{\partial f_x}{\partial t} + \frac{\partial \epsilon_x}{\partial \mathbf{p}_x} \cdot \frac{\partial f_x}{\partial \mathbf{r}} - \frac{\partial \epsilon_x}{\partial \mathbf{r}} \cdot \frac{\partial f_x}{\partial \mathbf{p}_x} = \mathcal{K}_x^<(1 \mp f_x) - \mathcal{K}_x^> f_x. \quad (1)$$

In the above ϵ_x is the single-particle energy, and $\mathcal{K}_x^<$ and $\mathcal{K}_x^>$ are, respectively, the production and absorption rates of the particle x. The upper and lower signs in the statistical factor on the r.h.s are for fermions and bosons, respectively.

Deuterons within the model [11, 12] are formed in the interaction of three nucleons, and the formation rate contributing to the deuteron production rate, is given by

$$\begin{aligned} \mathcal{K}^<(\mathbf{p}) = \ & \textstyle\sum_{N=n,p} \int d\mathbf{p}_N d\mathbf{p}_p d\mathbf{p}_n d\mathbf{p}'_N \, |\mathcal{M}^{npN \to Nd}|^2 \delta(\mathbf{p} + \mathbf{p}_N - \mathbf{p}_p - \mathbf{p}_n - \mathbf{p}'_N) \\ & \times \delta(\epsilon_d + \epsilon_N - \epsilon_p - \epsilon_n - \epsilon'_N) \, f_p f_n f'_N (1 - f_N) + \cdots . \end{aligned} \quad (2)$$

Statistical and relativistic factors and powers of 2π are suppressed. Due to the microscopic reversibility the processes of deuteron formation and breakup share the matrix element squared,

$$|\mathcal{M}^{npN \to Nd}|^2 = |\mathcal{M}^{Nd \to npN}|^2, \quad (3)$$

and thus, in principle, data on deuteron breakup could be used in describing the formation. This is because the differential cross section coincides, up to a kinematic factor, with the matrix element squared,

$$d\sigma_{Nd\to npN} \propto |\mathcal{M}^{Nd\to npN}|^2. \tag{4}$$

In practise, due to the dimensionality of three-body space, the differential cross section is known only within a limited range. In the calculations, we thus use a modified impulse approximation

$$|\mathcal{M}^{Nd\to npN}|^2 \propto \sigma_{NN}\,|\phi_d|^2, \tag{5}$$

where ϕ_d is the deuteron wavefunction in momentum space. The matrix element is normalized so that the measured total deuteron breakup cross section is reproduced as a function of bombarding energy. Tritons and helions in the model are produced in the interaction of four nucleons. At low densities, in the limit of many interactions, our equations for composites yield the required law of mass action. Coalescence scaling for composite spectra is expected to emerge from the model, when production does not vary substantially with space and time.

Pion production is described as a two-step process [17]. First a Δ resonance is produced in an NN collision, and then the resonance decays into a nucleon and pion. The resonance production cross-section nearly exhausts [19] the inelastic NN cross-section at the energies of interest. The pion absorption is described with a sequence of inverse processes to those in production. The resonance formation practically exhausts the πN cross section. Microscopic reversibility yields for cross section for the resonance absorption in the interaction with a nucleon [11, 18],

$$\sigma_{N\Delta\to NN} = \frac{1}{8}\frac{p_{NN}^2}{p_{N\Delta}}\frac{\sigma_{NN\to N\Delta}}{\int^{E-m_N} dm'\,A_\Delta(m')p'_{N\Delta}}, \tag{6}$$

where A_Δ is the spectral function describing the spread of the resonance in mass. In past calculations, a detailed balance relation between the resonance production and absorption cross-sections used was such as for stable particles, leading to an excessive number of pions in equilibrium.

The single-particle energy in (1) consists of the kinetic energy and the optical potential $V_x = A_x\,V$, where

$$V = A\frac{\rho}{\rho_0} + B\left(\frac{\rho}{\rho_0}\right)^\gamma + 2D\nabla^2\rho + V_{coul}, \tag{7}$$

and ρ is baryon density. Density profiles of the initial nuclei are determined by solving the Thomas-Fermi equation.

3. Transverse Expansion

Figure 1 shows the evolution of the particle density projected onto the reaction plane from the model, in the Au+Au collisions at 250 MeV/nucleon, at three values

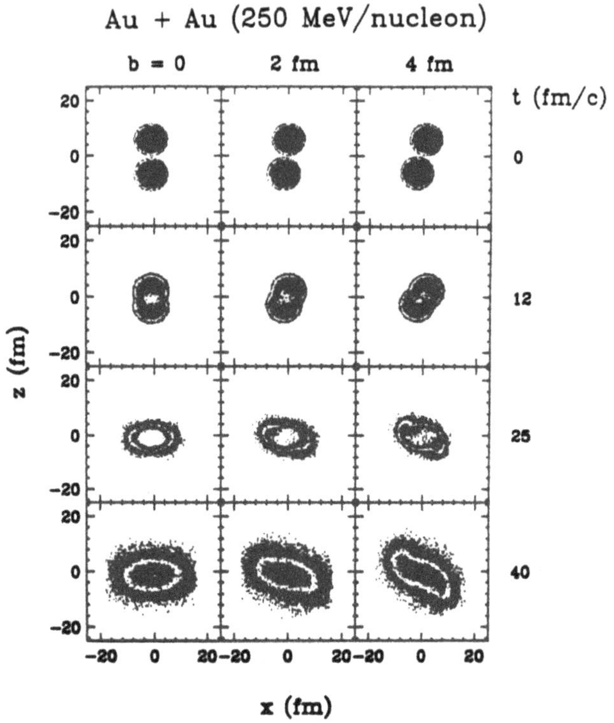

Figure 1: Contour plots of the particle density integrated over the normal to the reaction plane in Au+Au collisions at 250 MeV/nucleon.

of impact parameter. Figure 2 shows the evolution of density profiles along and perpendicular to the beam axis in a La+La head-on collision at 800 MeV/nucleon. In the center between the nuclei, at both energies, the matter gets compressed and excited. When a compression still takes place along the beam axis at $b = 0$, an expansion perpendicular to the beam axis sets in. Figure 3 shows the evolution of different quantities in the head-on collision, and in particular the growth of a transverse collective energy per nucleon. This energy is calculated for participants, from the local velocities determined using participants from surroundings. The collective energy ceases to be updated once the density of surrounding matter drops below $\rho_0/8$. The composite particles are produced late during the expansion, see Fig. 3. The expanding and approximately ellipsoidal region of matter at a late stage, which is perpendicular to the beam axis at $b = 0$, is seen in Fig. 1 to be inclined at a decreasing angle to the axis at larger impact parameters.

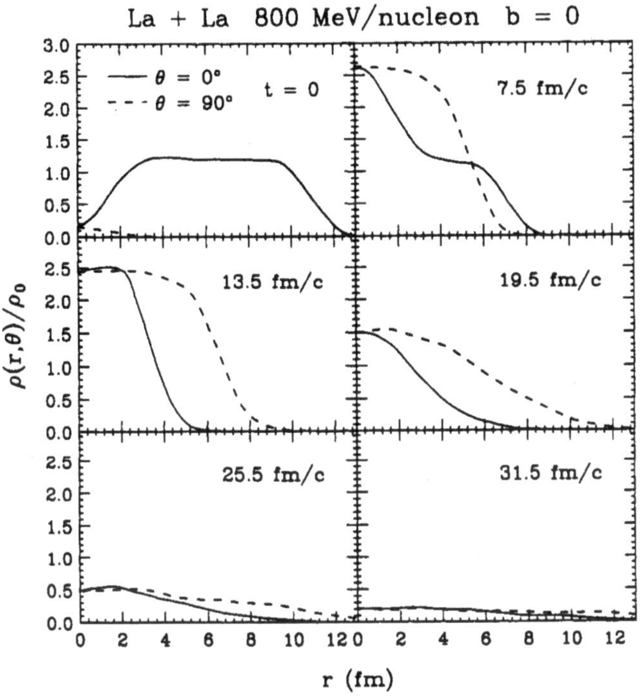

Figure 2: Baryon density as a function of the distance from the center of the system along (solid line) and perpendicular (dashed line) to the beam axis, at different times, in the $b = 0$ La+La collision at bombarding energy of 800 MeV/nucleon.

4. Angular Distributions

Judging from Figs. 1 and 2, one would expect the c.m. polar-angle distributions to peak at 90° in head-on collisions. With the change of the angle of inclination of the expanding region with respect to the beam axis, one might eventually expect the polar-angle ditribution to peak at 0° and 180° at higher b (taking into account the azimuthal-angle averaging). These results indeed emerge from the calculations, see Fig. 4. The pattern changes around $b = 3$ fm in the collisions at 250 MeV/nucleon. One could then think that traces of the collective expansion get lost unless lowest impact parameters are properly isolated. However, for finite impact parameters, orientation of the reaction plane may be determined with some accuracy. The expansion proceeding primarily within a plane inclined relative to the beam axis, should make the distributions in the azimuthal angle relative to the reaction plane peak at 90°, at the midrapidity. Amplification of the effect could be obtained by aligning the third coordinate axis with the main axis of a tensor constructed from the c.m. particle

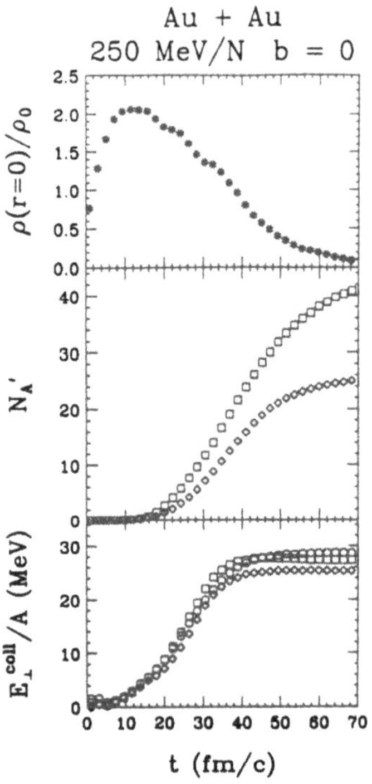

Figure 3: Time dependence of different quantities in the $b = 0$ Au+Au reaction at 250 MeV/nucleon. Top panel displays the baryon density at $r = 0$. Squares in the center panel indicate the number of deuterons, and diamonds - the number of $A = 3$ clusters, which survive intact till the end of the reaction. Diamonds, squares, and circles, in the bottom panel, indicate the collective energy per nucleon of $A = 3$ clusters, deuterons, and nucleons, respectively.

momenta. (The axis would then lie approximately within the plane of expansion.) Figure 5 displays the calculated and measured [10] azimuthal distributions that consistently show the features discussed above.

5. Mean Energies

Particle kinetic energy may be decomposed into the transverse and longitudinal components, even in a relativistic case,

$$E_{kin} = E - m = \sqrt{m^2 + p^2} - m = \frac{p_\perp^2}{m+E} + \frac{p_\parallel^2}{m+E} = E_\perp + E_\parallel. \qquad (8)$$

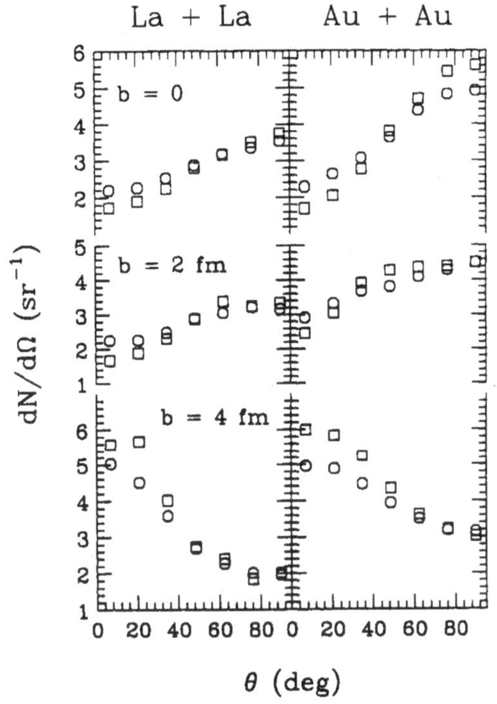

Figure 4: Number of protons (circles) and deuterons (squares) emitted per unit spherical angle, as a function of the c.m. polar-angle θ, in the La+La collisions at 246 MeV/nucleon, and in the Au+Au collisions at 250 MeV/nucleon, at different impact parameters.

In the case of isotropic emission, we should find for the mean values

$$< E_{\perp} >= 2 < E_{\parallel} >, \tag{9}$$

as there are two transverse components compared to the one longitudinal. Figure 6 shows the mean energy components for the particles emitted from $b = 0$ Au+Au collisions, in the beam energy range from 30 to 2000 MeV/nucleon. Below 70 MeV/nucleon the emission appears nearly isotropic. Between 70 and 1400 MeV/nucleon, there is a preference for the emission into the transverse over the longitudinal direction. At higher beam energies the expansion into transverse directions begins to compete with transparency.

In Fig. 6 it is seen that, generally, the mean baryon transverse energy rises with the baryon mass at any given beam energy. Thus e.g. at the beam energy of 200 MeV/nucleon, the proton, deuteron, and triton transverse energies are equal to about 50, 70, and 90 MeV, respectively. Also, a rise is seen for the mean longitudinal

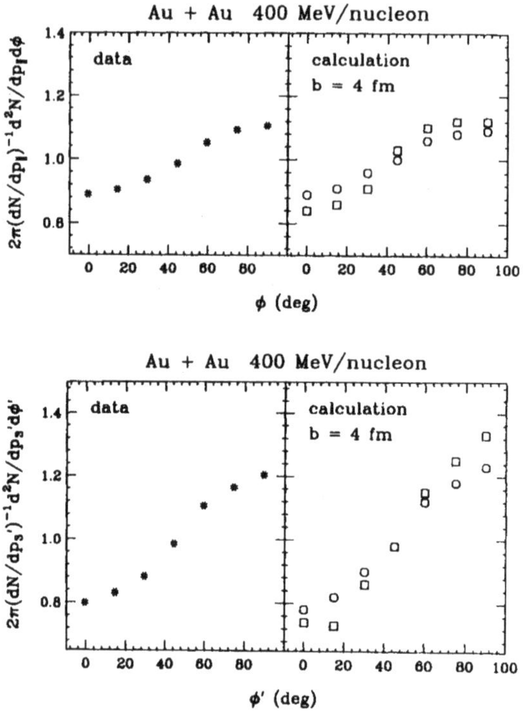

Figure 5: Distributions in the azimuthal angle ϕ about the beam axis and in the azimuthal angle ϕ' about the major axis of the sphericity tensor, for the momentum components along a respective axis approximately equal to zero. Left panels show the distributions for all particles, in the third of five multiplicity bins, as determined in the experiment of Ref. [10]. Squares and circles in the right panels indicate, respectively, the calculated deuteron and proton distributions, at $b = 4\,\text{fm}$.

energies. The mean energies at a fixed c.m. angle, particularily at $\theta = 90°$, rise too. This contradicts a naive expectation. Specifically, the proton spectrum d^3N_p/dp^3 at any one angle in a head-on collision is nearly exponential in energy,

$$\frac{d^3N_p}{dp^3} \propto \exp\left(-\frac{E_p}{T(\theta)}\right). \tag{10}$$

If the coalescence scaling were valid, then the spectra of composites should also be exponential, with the same temperature parameter as for protons,

$$\frac{d^3N_x}{dp^3} \propto \left(\frac{d^3N_p}{d(p/A)^3}\right)^A \propto \exp\left(-\frac{E_p}{T(\theta)}\right). \tag{11}$$

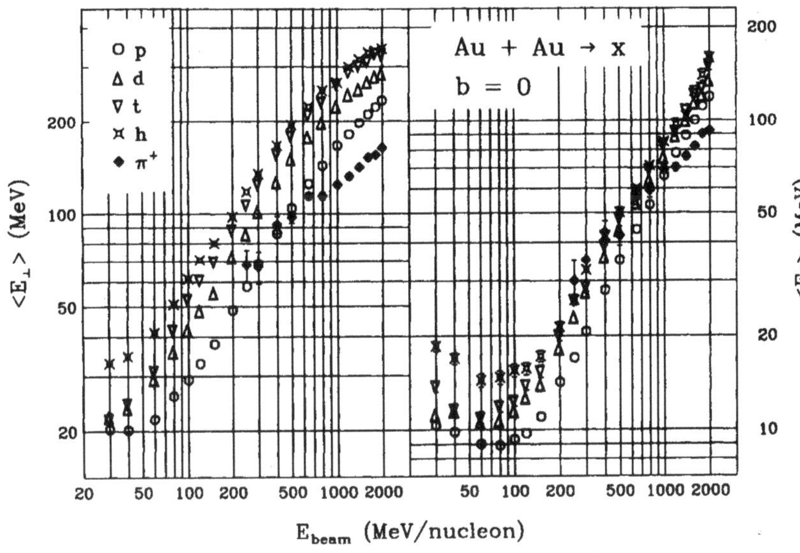

Figure 6: Mean particle energies decomposed into the longitudinal and transverse components in the $b = 0$ Au + Au collisions, as a function of bombarding energy.

In consequence, the mean energies should be independent of the fragment mass,

$$< E_x >_\theta = < E_p >_\theta = \frac{3}{2} T(\theta). \tag{12}$$

When mean energies at 90° degrees, or transverse energies averaged out over angles, are analysed as a function of the impact parameter, it is observed that the differences between the fragment energies maximize in the central reactions and vanish in peripheral (see upper panel in Fig. 7). In the central reactions particles undergo a largest number of collisions and, in principle, a global or even local equilibrium could be established. An equilibrium in a combination with the collective flow might explain the energy differences. The equilibrium phase-space functions take the form

$$f_x \approx \exp\left(-\frac{1}{T}\left(\frac{(\mathrm{p} - m_x \mathrm{v})^2}{2m_x} - B_x - \mu_x\right)\right), \tag{13}$$

and the corresponding mean energies consist of thermal and collective parts,

$$< E_x > = \frac{m_x < v^2 >}{2} + \frac{3}{2} + \frac{3}{2} T = A \frac{m_N < v^2 >}{2} + \frac{3}{2} T = A E_N^{coll} + E^*. \tag{14}$$

The thermal part is the same for different fragments, while the collective part increases linearly with the fragment mass. The decomposition analogous to (14) is valid, with (13), for separate energy components. Following the decomposition, at a beam energy of 200 MeV/nucleon the transverse collective energy could be estimated, using the

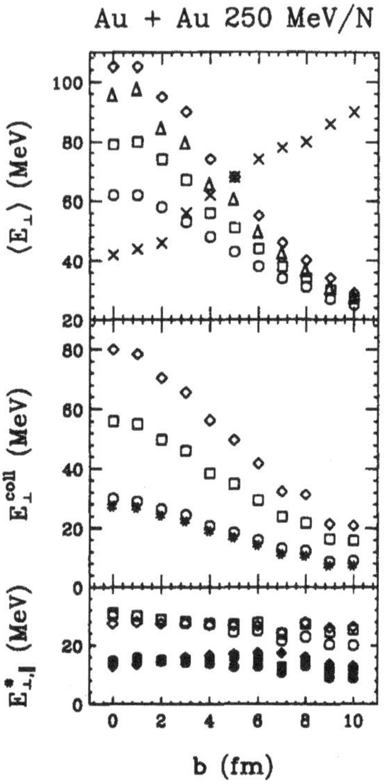

Figure 7: Different energies calculated in the Au+Au collisions at 250 MeV/nucleon, as a function of the impact parameter. The top panel shows the final transverse energies averaged over all angles for protons (circles), deuterons (squares), tritons (triangles), and helions (diamonds). Crosses in the top panel indicate twice the mean longitudinal proton energy. The center panel shows collective transverse energy obtained using local velocities for nucleons (circles), deuterons (squares), and $A = 3$ clusters (diamonds). Stars in the center panel indicate mean transverse energy per nucleon of all particles. Bottom panel shows the longitudinal (filled symbols) and transverse (open symbols) excitation energies at freeze-out, for nucleons (circles), deuterons (squares), and $A = 3$ clusters (diamonds).

aforementioned fragment-energy values, with $E_{\perp N}^{coll} \simeq < E_{\perp d} > - < E_{\perp p} > \simeq 20\,\text{MeV}$, and the thermal energy with $E_\perp^* \simeq < E_{\perp p} > - E_N^{coll} \simeq 30\,\text{MeV}$. The question, nevertheless, remains whether indeed a local equilibrium is reached in the central reactions and whether the energy differences coincide with the collective energy per nucleon that may be independently calculated within the model.

The upper and center panels in Fig. 7 display, respectively, the values of transverse energy averaged over angles for different fragments, and the values of collective transverse energy from local velocities, at a freeze-out defined with the density

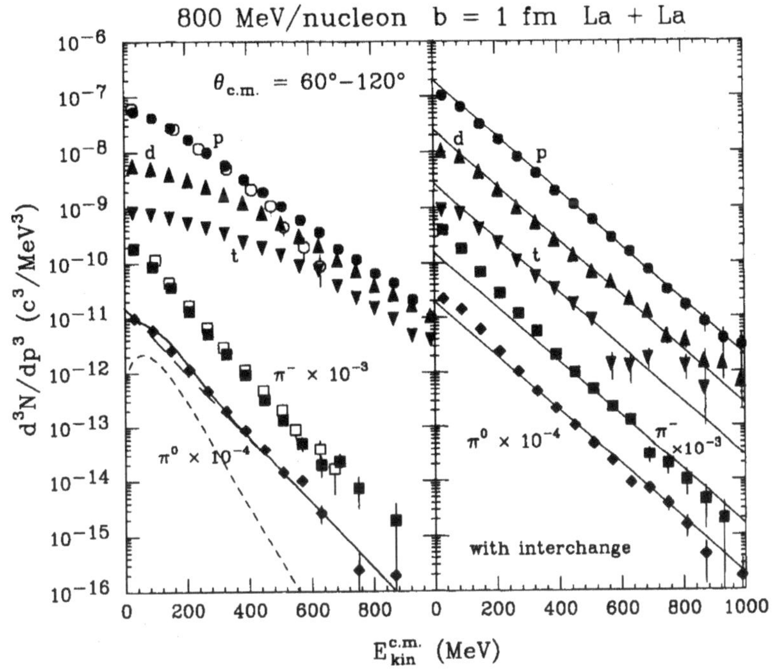

Figure 8: Transverse spectra from the central La+La reaction at 800 MeV/nucleon. Left panel shows the results of calculations at $b = 1\,\text{fm}$ (filled symbols) and data of Ref. [20]. Right panel shows the results of calculations with particle positions interchanged during evolution. Parallel lines in the right panel serve to guide the eye.

$\rho = \rho_0/8$, as functions of impact parameter in the collisions at 250 MeV/nucleon. The bottom panel displays values of transverse and longitudinal excitation energy obtained by subtracting the collective energies from the energies at the freeze-out, $E^*_{\perp,\parallel} = \langle E_{\perp,\parallel} \rangle - E^{coll}_{\perp,\parallel}$. It is seen in Fig. 7 that the transverse and longitudinal excitation energies are nearly the same for different fragments at low b, and that the transverse energy is about twice as large as the longitudinal energy, just as is expected for the thermal energy components, $E^*_\perp = T$, $E^*_\parallel = T/2$. The collective energy, calculated using local velocities, increases nearly linearly with fragment mass at any given b. Respectively, the energy differences at freezeout roughly agree with the collective energy per nucleon.

At higher beam energies, conclusions reached on the basis of the baryon spectra may be confronted with the pion spectra. The spectra from a $b = 1\,\text{fm}$ La+La reaction, in the vicinity of 90° in the center of mass, are shown in the left panel of Fig. 8. It is seen that that the pion spectra are steeper than the proton spectrum, and the latter is steeper than the deuteron or triton spectrum. The slope parameters for protons and for pions outside of the Δ-decay region are, respectively, 110 MeV and 73 MeV. From the values of mean energies for protons and deuterons at 90°, of 185 MeV and 245 MeV, respectively, we estimate the collective contribution to these energies with a difference, at about 60 MeV/nucleon. This gives an excitation energy of 125 MeV corresponding to a temperature of $T \simeq 75\,\mathrm{MeV}$ (with relativity taken into account), in an approximate agreement with the pion slope parameter. Pions as light should not be much affected by the flow.

As an ultimate test whether we indeed deal with the effects of collective expansion in the spectra or in the mean energies, we carry out calculations with a collective expansion artificially extinguished during evolution. This is done by interchanging the particle positions within pairs chosen at random, when collisions are still frequent. The procedure, while destroying the correlation between positions and momenta, leaves the spatial and momentum distributions unchanged at the instant of application. Destruction of the correlation leads to one, in practise, slope parameter for the spectra of different particles, see right panel of Fig. 8, and to close values of the mean energy for different baryons.

6. Pion Yields

Particle yields from reactions should be affected by the collective flow and local equilibrium. As I mentioned in the introduction, the measured pion yields in central reactions are lower than anticipated in the fireball model assuming a global equilibrium. Figure 9 displays the evolution of the number of pions and deltas, normalized using the baryon number, in the $b = 0$ La+La reaction at 800 MeV/nucleon. Evolution of other quantities in the reaction is shown in Fig. 10. The deltas, if left towards the end of reaction, would decay into pions. The number of pions and deltas maximizes after all the matter has entered the central hot region, cf. Fig. 2. The kinetic energy per baryon including mass at this moment, from the calculation, is 1133 MeV. If all this energy were thermalized, then, at a representative density of $2.25\rho_0$ for the compressed matter, cf. Fig. 2, the number of pions and deltas should be such as represented by the upper of the two horizontal dashed lines in Fig. 9, well above the maximum reached in collision. After the collective energy of about 30 MeV/nucleon at a given moment, cf. Fig. 10, is subtracted from the thermalized energy, the respective number gets much closer to the maximum. This number is represented by the lower of the two horizontal lines in Fig. 9. At a high density different processes, in particular those of the Δ and π production and absorption, are found fairly frequent. One then expects the system to be close to an equilibrium. Approximate location of the system when the number of pions and deltas maximizes, in the space of parameters for an equilibrated system, is shown in Fig. 11.

During the system expansion, the number of pions and deltas actually drops. This is observed in other calculations besides ours, e.g. in that reported by Fuchs at this workshop or earlier [21]. The number can in practise only follow the equilibrium value and a drop shows that the equilibrium has moved down, below the maximum. In Fig. 11 it is seen that, in an isentropic expansion, the number of pions and deltas should decrease. During the expansion more and more energy turns into collective.

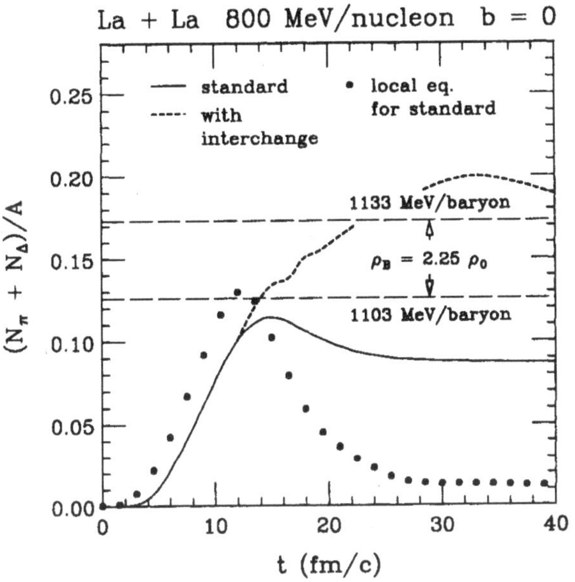

Figure 9: Time dependence of the number of pions and deltas normalized using the baryon number, in the $b = 0$ La+La reaction at 800 MeV/nucleon, from the standard calculation (solid line) and from the calculation with an interchange of particle positions (short-dashed line). The two horizontal long-dashed lines show the equilibrium number of pions and deltas for the baryon density $\rho = 2.25\rho_0$ and the kinetic energy per baryon including mass of 1133 MeV and 1103 MeV, respectively. Dots show the number obtained by integrating over space the local equilibrium values of pion and delta density in a standard calculation.

We finally calculate the equilibrium number of pions and deltas, by calculating the number locally as corresponding to a local kinetic energy, and integrating over the space. The result is shown with points in Fig. 9. After reaching a value estimated previously, the equilibrium number rapidly decreases. While the actual number of pions and deltas decreases, it freezes out, nevertheless, at comparatively high value. This freeze-out is due to a drop of the density shutting off the processes $N+N \rightleftarrows N\Delta$. The processes of pion absorption, with a much larger cross section, and the Δ decays continue longer, preserving a thermal but not chemical equilibrium of the pionic sector with nucleons (cf. also the talk by Kuehn at this workshop).

On the basis of sole information that the multiplicity of pions produced in central reactions, is lower than in the thermal model, one would (naively) expect that the time is too short to reach the equilibrium number. With the collective expansion taken into account, it is found that the equilibrium is intermittently reached. The final pion number is approached from above rather than below.

We finally illustrate the effect of collective expansion on the pion yields by com-

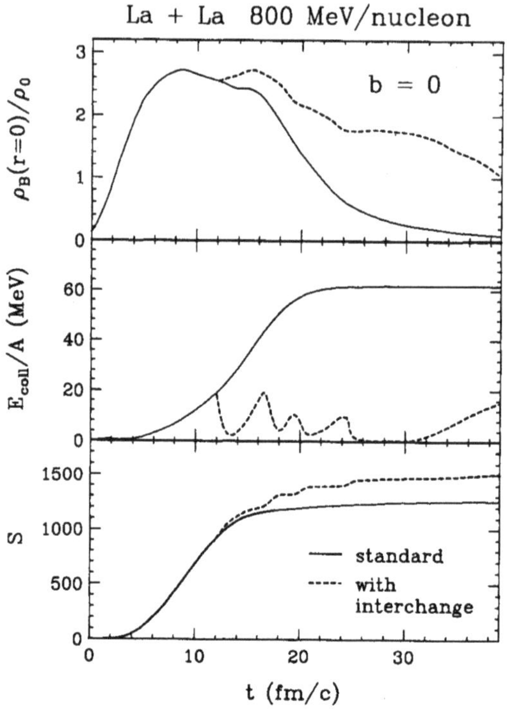

Figure 10: Time dependence of baryon density at $r = 0$ (top panel), collective energy per nucleon (center panel), and total entropy (bottom panel) in the $b = 0$ La+La reaction at 800 MeV/nucleon from the standard calculation (solid lines) and from the calculation with an interchange of particle positions (short-dashed lines).

paring the multiplicities from the standard simulations, with those from the simulations with the collective motion extinguished, and with data in Fig. 12. Standard calculations agree reasonably well with the data, while those with collective motion extinguished yield far too many pions. The latter results naturally come close to those [6] from the fireball model.

7. Cluster Yields

The collective motion in the reactions is closely related to the entropy which in turn is related [22] to the cluster yields. The collective energy expresses the work carried out by the system. By the second law of thermodynamics, less entropy should be generated and in consequence more fragments produced, when the collective motion develops, than otherwise. This connection is explored quantitatively by comparing in

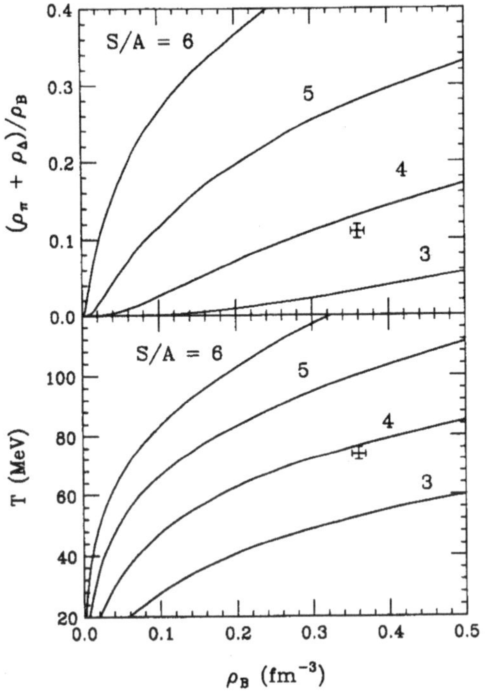

Figure 11: Isentropes in the plane of the ratio of pion-and-delta density to baryon density vs the baryon density (top panel), and in the plane of the temperature vs the baryon density (bottom panel). Approximate locations of the 800 MeV/nucleon $b = 0$ La+La system at $t = 14\,\mathrm{fm/c}$ in the two planes are marked with crosses.

Fig. 13 the yield ratios from the standard calculations with those from the calculations with a collective motion extinguished. At high beam energies, when much collective energy is generated in the standard dynamics, the ratios of the $A = 3$ cluster yields to the proton yields differ by a factor of 3 for the two types of calculation! In Fig. 13 we show also yield ratios obtained by the Plastic Ball group [23] from extrapolating their measurements beyond their detector acceptance regions (the procedure cannot be reproduced in the calculations). The experimental d/p ratios agree well with the ratios from standard calculations. The t/p and h/p ratios are generally closer to the standard results with the strong collective motion developed, than to the results with the motion extinguished.

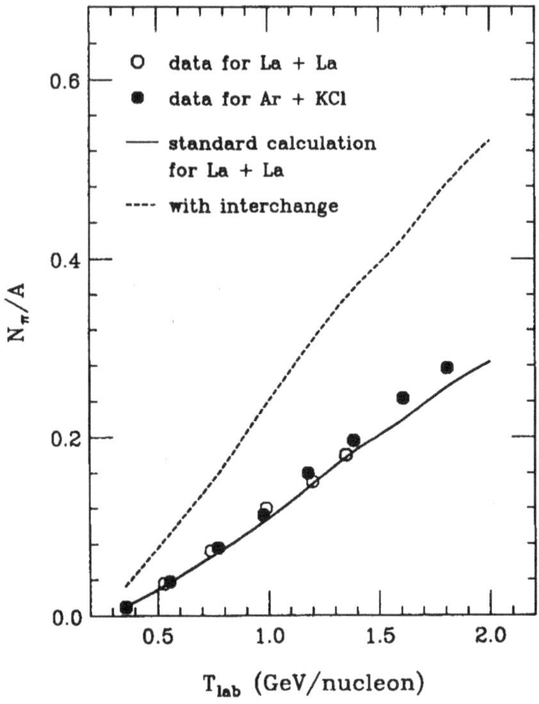

Figure 12: Ratio of the mean pion multiplicity to the number of participant nucleons in central symmetric reactions, as a function of laboratory energy. The solid and dashed lines represent, respectively, the results of the standard calculations and the calculations with particle positions interchanged, for the La+La $b = 0$ reactions. The circles represent data of Ref. [6].

8. Conclusions

In the simulations of energetic central heavy-ion collisions, a region with dense and excited matter is formed between the colliding nuclei. At the intermediate energies, 60-1400 MeV/nucleon for Au+Au, this region expands preferentially in transverse directions, as the collision progresses. The expansion is reflected in the angular distributions, mean energies, and spectra of emitted baryons. At lowest impact parameters the distributions peak at 90° in the polar angle. At higher impact parameters the distributions at midrapidity peak in the azimuthal angle out-of-the-reaction-plane, consistently with the data [10]. The mean transverse baryon energies rise with the mass number due to a collective contribution to these energies. The true system temperatures may be read off at higher beam energies from the high-energy tail of transverse pion spectrum. As far as pion yields are concerned, the energy unavailable

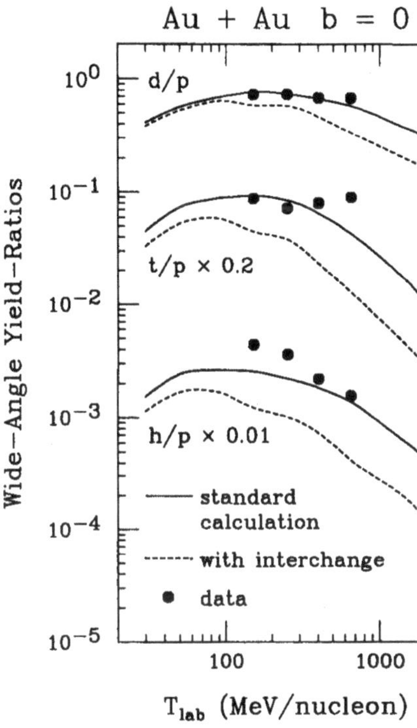

Figure 13: Ratios of cluster yields to proton yields at wide angles [11] as a function of bombarding energy in the head-on Au+Au reactions. The solid and dashed lines represent, respectively, the results of the standard calculations and the calculations with particle positions interchanged. The dots represent data of Ref. [23].

for the production, considered by Stock *et al.* [6], can be identified with the collective energy. The cluster yields are strongly affected by the expansion at the high beam energies, and to a lesser degree at the low energies. Overall, the effects of collective expansion in the energetic central heavy-ion reactions are very significant. Admittedly these effects have not been duely appreciated over the years.

This work was supported by the National Science Foundation under Grant No. PHY-9017077.

References

[1] W. Scheid, B. Müller, and W. Greiner, Phys. Rev. Lett. **32**, 741 (1974); M. I. Sobel *et al.*, Nucl. Phys. **A251**, 502 (1975).

[2] P. J. Siemens and J. O. Rasmussen, Phys. Rev. Lett. **42**, 880 (1979); H. Stöcker, A. A. Ogloblin, and W. Greiner, Z. Phys. A **303**, 253 (1981).

[3] R. Brockmann *et al.*, Phys. Rev. Lett. **53**, 2012 (1984).

[4] R. Stock *et al.*, Phys. Rev. Lett. **44**, 1243 (1980).

[5] O. Scholten, H. Kruse, and W. F. Friedman, Phys. Rev. C **26**, 1339 (1982).

[6] R. Stock *et al.*, Phys. Rev. Lett. **49**, 1236 (1982); J. W. Harris *et al.*, Phys. Rev. Lett. **58**, 463 (1987).

[7] G. F. Bertsch, S. Das Gupta, and H. Kruse, Phys. Rev. C **29**, 673 (1984).

[8] H. A. Gustafsson *et al.*, Phys. Rev. Lett. **52**, 1590 (1984); R. E. Renfordt *et al.*, *ibid.* **53**, 763 (1984).

[9] P. Danielewicz *et al.*, in *Proc. XV Workshop on Gross Properties of Nuclei and Nuclear Excitations*, Hirschegg, 1987, ed. H. Feldmeier, p. 91; Phys. Rev. C **38**, 120 (1988).

[10] H. H. Gutbrod *et al.*, Phys. Lett. B **216**, 267 (1989).

[11] P. Danielewicz and G. F. Bertsch, Nucl. Phys. **A533**, 712 (1991); P. Danielewicz, Nucl. Phys. **A545**, 21c (1992).

[12] P. Danielewicz and Q. Pan, Phys. Rev. C **46**, 2002 (1992).

[13] H. W. Barz *et al.*, Nucl. Phys. **A531**, 453 (1991); W. Bauer *et al.*, Phys. Rev. C **47**, R1838 (1993).

[14] S.C. Jeong *et al.*, report GSI-93-38, 1993.

[15] R. Lacey, private communication.

[16] G. Rai, talk at this Workshop; G. Poggi, private communication.

[17] Y. Yariv and Z. Fraenkel, Phys. Rev. C **20**, 2227 (1979).

[18] B.-A. Li, Nucl. Phys. **A552**, 605 (1993).

[19] B. J. VerWest and R. A. Arndt, Phys. Rev. C **25**, 1979 (1982).

[20] S. Hayashi *et al.*, Phys. Rev. C **38**, 1229 (1988).

[21] J. Cugnon, D. Kinet, and J. Vandermeulen, Nucl. Phys. **A379**, 553 (1982).

[22] P. J. Siemens and J. I. Kapusta, Phys. Rev. Lett. **43**, 1486 (1979); G. Bertsch and J. Cugnon, Phys. Rev. C **24**, 2514 (1981).

[23] K. G. R. Doss *et al.*, Phys. Rev. C **37**, 163 (1988).

COLLECTIVE MATTER FLOW IN RELATIVISTIC HEAVY ION COLLISIONS

J.P. ALARD
Laboratoire de Physique Corpusculaire
IN2P3-CNRS, 63177 Aubière Cedex

for the FOPI Collaboration [1]
and the DIOGENE Collaboration [2]

ABSTRACT

Measurements of nuclear collective flow in relativistic nuclear collisions are presented for symmetric and asymmetric systems. The presented data are obtained using two experimental setups, and we show results concerning proton like, pions and fragments. Nuclear collective flow is observed in the reaction plane and out of plane. We show the existence of a new component of flow for very central collisions.

We give a brief review of some experimental data on nuclear collective flow in relativistic heavy ion collisions. We present some characteristics results obtained with the Diogene detector [3] (Saturne) and the FOPI detector [4] (GSI Darmstadt).

The first experiments performed with the Plastic Ball/Wall [5] at LBL confirmed the predictions of the theoreticians [6]: in central collisions of relativistic heavy ions (RHIC) we observe a stopping of the nuclear matter and the formation of a source of energetic ejectiles which is centered at the c.m. velocity for symmetric systems. Experiments also showed the existence of nuclear flow in the reaction plane (sidewards flow) and out of reaction plane (squeeze out). Recently, the first experiments performed with the FOPI detector at GSI [4] showed the existence of a radial flow for very central collisions. These dynamical aspects of collisions are believed to be strongly correlated with the nuclear Equation of State [7](EOS) and the Momentum Dependant Interaction [8].

In the first part of this talk,we describe briefly the two experimental setups used at Saturne and at GSI Darmstadt. Then we discuss on the determination of the centrality of events; finally we present experimental results concerning observation of flow, both in reaction plane and out of plane, for asymmetrical and symmetrical systems.

1-Experimental setup

1-1- The Diogene experimental setup [3] has been built at the Saturne Synchrotron in Saclay. It is made of a pictorial drift chamber (PDC) associated with a multiplicity trigger, a plastic wall and triggering scintillators. The PDC is a cylindrical drift chamber, covering the angular range 25 to 140^0 lab, with its axis oriented parallel to the beam, and placed inside a 1 Tesla magnetic field which is parallel to the beam. Particle identification is achieved by the determination of magnetic rigidity and energy loss in the gas of the chamber.

The plastic wall consists of arrays of scintillators. It covers the angular range 0-6^0 and permits measurements of both charge and velocity of fragments. The trigger is composed of a barrel of scintillator slabs around the chamber and a set of beam counters. This barrel is a multiplicity counter which can help for selection of central events.

1-2- *The FOPI detector at GSI Darmstadt* .

A complete description of the newly built FOPI detector has been given elsewhere [4]. The first measurements used the Au + Au system with the so-called Phase I of FOPI: it consists of a highly granular plastic scintillator wall with altogether 764 elements, each of them providing an energy loss and TOF signal, and spanning laboratory polar angles between 1 and 30^0 with full azimuthal coverage. Low energy fragments are detected by a shell of energy loss detectors (188 elements) mounted in front of the scintillators. The system allows element identification of charges up to 15, with detector thresholds increasing from 14 to 50 A.MeV as Z increases from 1 to 15. The Phase II detectors will include a time projection chamber inside a longitudinal magnetic field, a frontal drift chamber and barrels of scintillators around the TPC. Fig 1 shows a global view of the detector.

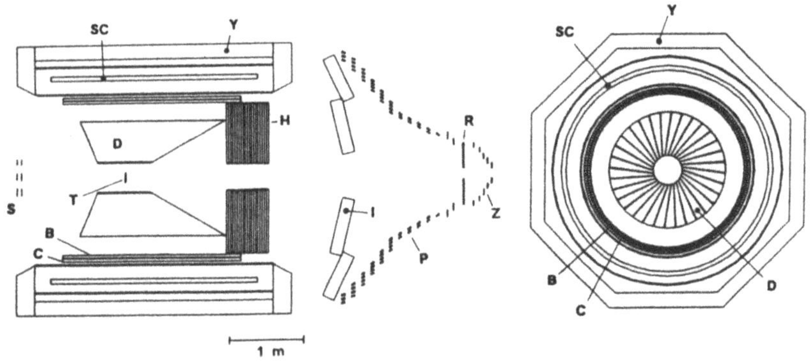

Fig 1. Complete FOPI facility in its final layout. S Start/Beam counters; T target; Y magnet yoke; SC superconducting coil; D central drift chamber; H forward chamber; B,C scintillator and Cerenkov barrels; P,Z plastic walls; I,R cluster detectors.

2- Determination of the centrality of events

A trivial method currently used for an evaluation of the centrality of events is the multiplicity of charged particles emitted in a given acceptance. Such a criterion has been used both for Diogene and FOPI experiments. As an example, Fig 2 shows the multiplicity distribution for Au + Au collisions at 150 MeV, from FOPI data. The full line corresponds to distributions obtained after background elimination. We observe the usual flat plateau and a fall off at higher multiplicities. The cut called PM5 starts at half of the plateau value [9], corresponding to a cross section of 350 mb. The remaining multiplicity range is then divided into 4 equally spaced bins PM 1 to 4.

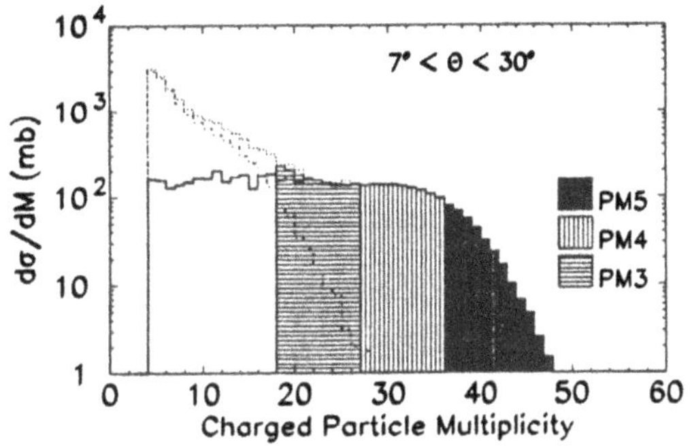

Fig 2. Charged-particle multiplicity distribution for target-in (dotted line) and target-out (dash-dotted line) measurements. The solid histogram gives the background-subtracted distribution with the selection cuts indicated by differently hatched areas (PM3–PM5).

The second step in the determination of centrality is the use of the transverse momentum directivity, defined as:

$$D = |\sum_i \vec{p_i}| / \sum_i |\vec{p_i}| \quad for \quad y \geq y_{cm}$$

The sum extends over all charged particles with $Z \geq 15$, and the momenta have been calculated using A = 2Z.

Another way for the estimation of the centrality is the use of the ratio E_{rat} defined by:

$$E_{rat} = \sum_i E^i / \sum_i E^i \quad for \quad y \geq y_{cm},$$

central events corresponding to values of E_{rat} larger than a specific limit.

In our analysis, the multiplicity criterion has been used both for Diogene and FOPI data; The D and E_{rat} criteria have been used only for FOPI experiments. Let us note that it has been shown clearly [9] that the multiplicity cuts alone do not permit a good selection of central events, especially for symmetrical collisions.

3-Some experimental results

We present some experimental results concerning nuclear matter flow, obtained with the two experimental setups described above.

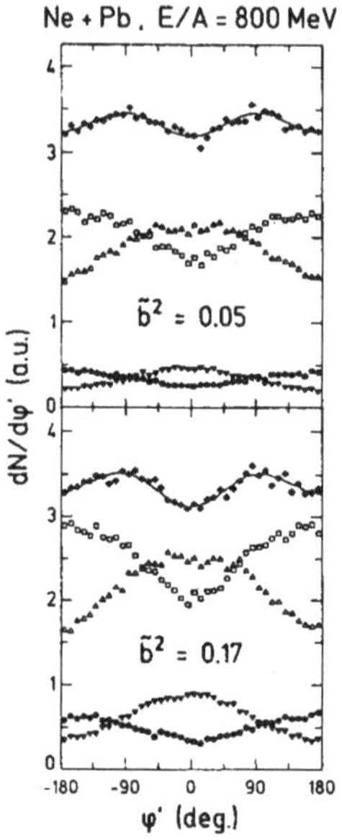

Fig 3. Azimuthal distributions of proton like particles emitted in different rapidity bins (Ne + Pb 800 A.MeV) The lines correspond to Fourier series fits (cf text) Symbols: • $y < y_{cm} - 0.3$; □ $y_{cm} - 0.3 < y < y_{cm} - 0.1$ ♦ $y_{cm} - 0.1 < y < y_{cm} + 0.1$; △ $y_{cm} + 0.1 < y < y_{cm} + 0.4$ ▲ $y_{cm} + 0.4 < y$
The statistical error bars are approximately as large as the symbols sizes. Points are not corrected for the estimated reaction plane azimuthal fluctuations due to finite number effects.

3-1-Flow measurements with Diogene

Fig 3 shows azimuthal distributions of proton like particles [10] emitted in different rapidity bins for Ne + Pb collisions at 800 A.MeV. Note that the points are not corrected for the estimated reaction plane fluctuations (finite number effects). The two parts correspond to the two multiplicity cuts (≥ 18) and (13-17) in the angular range 20 to 140^0. We observe sidewards flow in the reaction plane for the two backward rapidity bins (minimum at 0^0) and for the forward rapidities (maximum at 0^0). For intermediate rapidity bins we observe the squeeze out component (out of the reaction plane). Note that we can summarize the ϕ' distribution by a fit using a second order Fourier serie:

$$dN/d\phi' \quad = \quad 1 + a'_1.cos\phi' + a'_2.cos2\phi'$$

a'_1 being related to the usual sideward flow observable and a'_2 containing new information related to the $\pm 90^0$ values vs the 0 and 180^0 ones:

$$r = (1 - a_2')/(1 + a_2')$$

a maximum at $\pm 90^0$ corresponding to negative values of a_2'.

Other similar results have been obtained [10] for Ne + Nb and Ne + NaF. Let us note that these two components of the nuclear flow are observed even with small projectiles. It is necessary now to make comparisons with the actual improved models describing the dynamics of nuclear collisions, in order to check the sensitivity of these observables to the nuclear EOS and in medium effects.

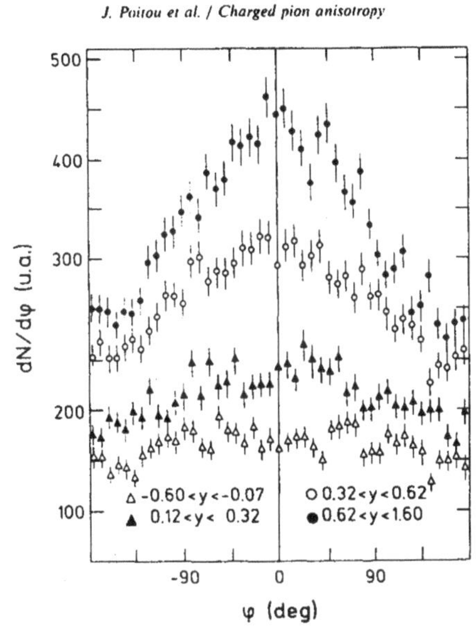

Fig 4. Pion azimuthal distributions with respect to the reaction plane for rather central collisions, at four rapidity bins; $b^2 = 0.17$ (Ne + Pb at 800 A.MeV)

Another interesting aspect is given by the flow of charged pions [11], especially for asymmetric systems. Fig 4 shows the azimuthal distribution of pions with respect to the reaction plane. The data are for Ne + Pb at 800 A.MeV, for rather central collisions ($b^2 = 0.17$) and for four different rapidity bins. This can be due either to an excess of pions emitted towards positive p_x, or the average transverse momentum is larger for pions emitted towards positive p_x. On the other hand, fig 5 shows the pion average in-plane transverse momentum vs system asymmetry, in the same impact parameter window; note that for asymmetrical systems $< q_x >$ is incompatible with zero values. On the other hand, we observe that the squeeze out component is present only for symmetrical systems and large impact parameters. From these analysis, it appears that the pion absorption in nuclear matter is probably an important cause for the aspect of pion flow. So the pion flow data represent a new important constraint on the theoretical models. They can help to probe the pion dynamics in nucleus nucleus collisions.

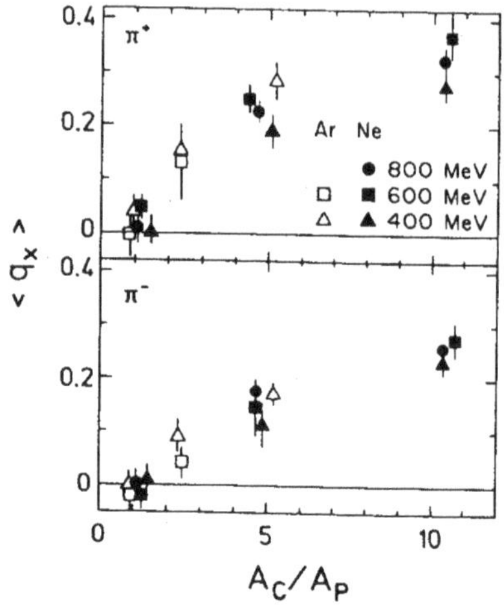

Fig 5. Pion average in plane transverse momentum versus system asymmetry for the same reduced impact parameter window; The top is for π^+,the bottom for π^-

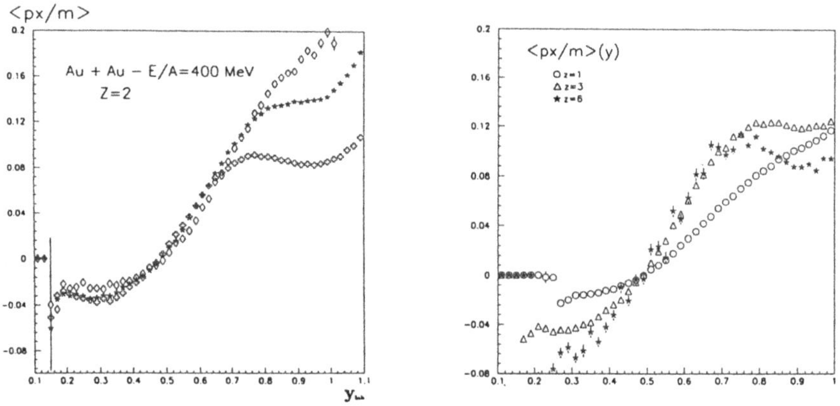

Fig 6. Transverse momentum analysis as a function of the centrality for Z = 2 particles (left).Open diamonds: central ($E_{rat} > 0.72$); Full stars: semi-central ($0.43 < E_{rat} < 0.72$); Open crosses: Peripheral events ($0.14 < E_{rat} < 0.43$).

On the right, the same transverse momentum analysis is represented for Z = 1 (circles) Z = 3 (triangles) and Z = 6 (stars).

3-2-Brief survey of some of the first FOPI data.

Let us focus on Au + Au collisions studied with the Phase I of the FOPI setup. Fig 6 shows a transverse momentum analysis [12] obtained for Z=2 particles, as a function of the centrality. We observe a weak sensitivity of the flow parameter (slope at $< p_x/m > = 0$) which is probably more reduced by the strong FOPI angular cuts; the strongest differences occur in the spectator region. On the other hand, a transverse momentum analysis as a function of Z shows clearly that the flow parameter is larger for fragments as for Z = 1 particles (see fig 6). Such an observable, using intermediary mass fragments, can be a good probe for the study of the dynamics of the collisions.

A more global representation of the sidewards flow can be obtained through the parameter F_s given by:

$$F_s \quad = \quad K.[(\sum \vec{p_t})^2 - \sum p_t^2]$$

where K is a normalization factor. Fig 7 shows that F_s presents a maximum for intermediate values of the centrality and vanishes almost for events with a high degree of stopping [13].On the other hand, values of F_s can be compared with QMD model, as a function of the incident energy. The rise observed for the data seems to be reproduced only if we use the MDI version. However, all different parametrizations in QMD are not able to reproduce the magnitude of the signal above 250 A.MeV.

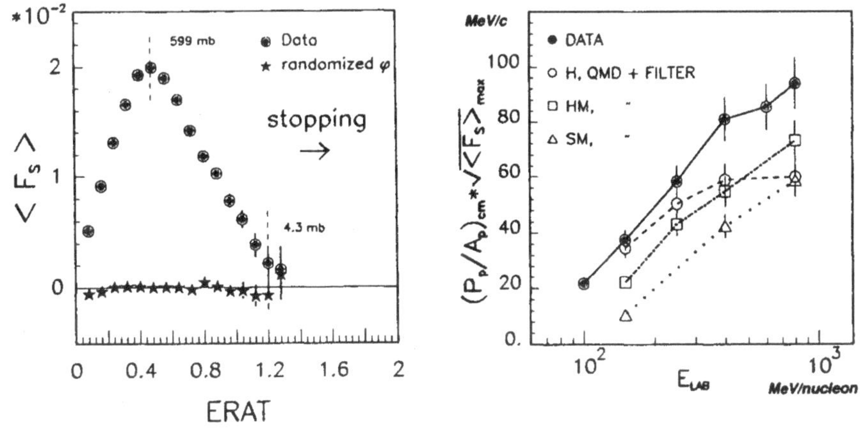

Fig 7. Mean value of the flow quantity F_s (left) versus the value of E_{rat}; on the right, the maximum of the directed sidewards flow as a function of the beam energy.

On the other hand, out of plane collective flow (squeeze out) can be observed on fig 8 as a function of the charge of emitted fragments [14]. We note that the extracted signal increases with the charge; the solid lines correspond to the usual fit (§3 – 1).

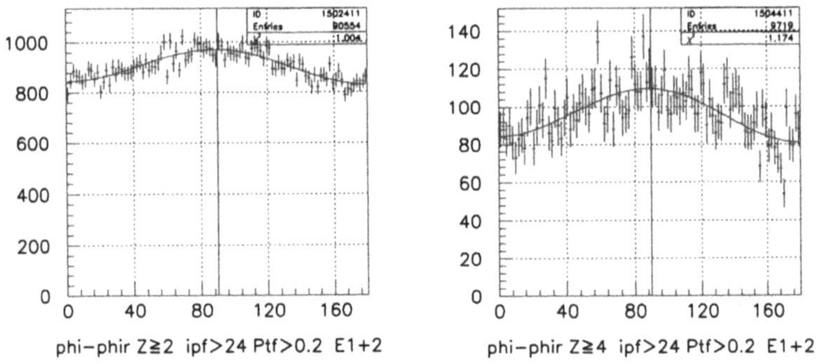

Fig 8. Azimuthal distribution of charged fragments in Au + Au collisions at 150 A.MeV. The left correspond to $Z \geq 2$ particles and the right corresponds to $Z \geq 4$ particles.

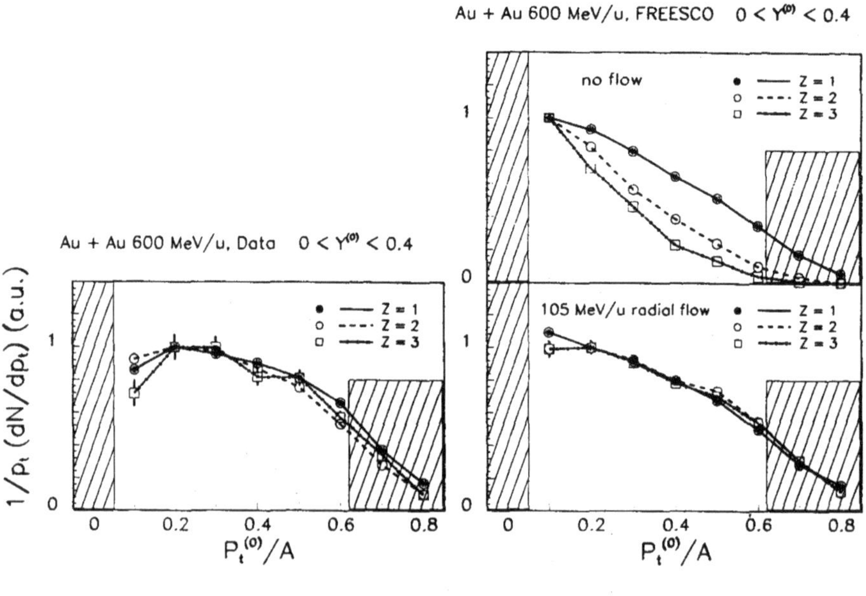

Fig 9. Transverse momentun distributions in a region close to midrapidity for Au + Au at 600 A.MeV. In the hatched area the distributions are distorted by the detector. Left: experimental data; Right: the top correspond to FREESCO calculations with complete thermalization, the bottom includes a collective part of 105 A.MeV.

Finally we present some data concerning collective expansion in highly central collisions. The FOPI data experiments allows to study the small cross sections of truly central events. Head on collisions provide an ideal scenario where the reaction pattern has to be azimuthally symmetric: this allows an integration on ϕ. We present on fig 8 preliminary results [15] on the ϕ integrated p_t distributions of the midrapidity source for very central events (PM5 + D1, see §2) at 600 A.MeV. These data can be compared with predictions of the improved FREESCO model [16,17] used in the single source mode. We can see clearly that the extreme case where all the available energy is converted into heat do not describe correctly the experimental data. On the other hand, a radial flow component of 105 A.MeV has been included in the calculations, associated with a thermal component equal to 30 A.MeV. In this case we observe a rather good qualitative description of the data. This result is in agreement with the existence of a radial component of the flow.

The kinetic energy spectra of the fragments emitted from the isolated midrapidity source has been investigated [18] for center of mass angles between 25 and 45^0, in order to minimize the effect of the experimental cuts of the FOPI facility (fig 9). The mean kinetic energy $< E_{kin} > /A$ has been compared with the predictions of the improved FREESCO code for very central collisions ($E_{rat} > 0.65$). We observe a strong discrepancy (fig 9) between FREESCO and experimental data for intermediary mass fragments [18]: in this case, FREESCO predicts a small thermal component. If we include into the FREESCO calculation an isotropic flow with values (4,4,4) for (E_x, E_y, E_z),we obtain the correct slope for fragments. So a simple "thermal + Coulomb" calculation underestimates the measured value by $\sim 20MeV$. Such a value can be an estimate of the collective flow, it represents $\sim$ 54% of the initial available kinetic energy of 37 A.MeV.

Fig 10. The mean kinetic energy compared with the predictions of the FREESCO model for very central collisions.

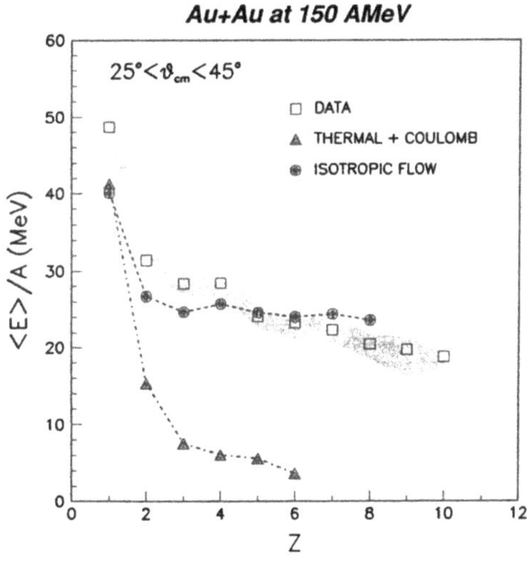

Conclusion

Experimental data clearly put in evidence the existence of a nuclear collective flow in relativistic heavy ion collisions, both in the reaction plane and out of the reaction plane. It is now important to compare results obtained from the different experiments, as a function of the size of the systems and of the incident energies, in order to check the sensitivity of these observables to the models and finaly to the nuclear equation of state. Observations with different species of emitted particles (pions, protons, fragments, neutral particles) are of a great interest. We have also shown the existence of a new component of the flow for very central collisions of symmetric systems. This radial flow can be an interesting observable and another constraint on the theoretical models.

References

1. The FOPI Collaboration: Bucharest, Budapest, Clermont, Darmstadt, Florence, Heidelberg, Mainz, Moscow, Rossendorf, Strasbourg, Warsaw, Zagreb.
2. The Diogene Collaboration : Clermont, Saclay, Strasbourg.
3. J.P. Alard et al, *Nucl. Instr. Meth.* **A261** (1987) 379.
4. A. Gobbi et al, *Nucl. Instr. Meth.* **A324** (1993) 156.
5. A. Baden et al, *Nucl. Instr. Meth.* **203** (1982) 189.
6. W. Scheid et al, *Phys. Rev. Lett.* **32** (1974) 741.
7. G. Peilert et al, *Phys. Rev.* **C39** (1990) 1402.
8. K. Weber et al, *Nucl. Phys.* **A515** (1990) 747.
9. J.P. Alard et al, *Phys. Rev. Lett.* **69** (1992) 889.
10. M. Demoulins et al, *Phys. Lett.* **B241** (1990) 476.
11. J. Poitou et al, *Nucl. Phys.* **A536** (1992) 767.
12. J.P. Alard et al, *GSI Rep.* **93-1** (1992) 39.
13. T. Wienold et al, *GSI Rep.* **93-1** (1992) 33.
14. J.P. Alard et al, *GSI Rep.* **93-1** (1992) 37.
15. T. Wienold et al, *GSI Rep.* **93-1** (1992) 34.
16. J. Randrup and G. Fai, *Phys. Lett.* **115B** (1982) 281.
17. J. Randrup, *LBL Rep.***33865**, to be published *Comp. Phy. Com.*
18. S.C. Jeong et al, *GSI Rep.* **93-38** (1993), to be published.

Relativistic Transport Calculations in the Local Phase Space Configuration Approximation *

C. Fuchs, L. Sehn+, H. H. Wolter
Sektion Physik, Universität München,
85748 Garching, Germany

Abstract
In relativistic transport calculations for heavy ion collisions we use self energies which are based on a parametrisation of Dirac-Brueckner G-Matrix results and which are determined for the scenario of colliding nuclear matter in a local phase space approximation in order to take into account non-equlibrium effects in the effective interaction. We calculate flow observables and Δ and π production rates and show that non-equlibrium effects are important in the determination of the equation of state of nuclear matter.

1 Introduction

The investigation of heavy ion collisions (*hic*) at energies of about 1 to 2 GeV/A bombarding energy is largely motivated by the search for the equation of state (*eos*) of nuclear matter (*nm*), i.e. for the properties of equilibrated nuclear matter at densities away from saturation at non-zero temperatures. In a *hic* the *eos* enters via the mean field in the participant zone of the colliding nuclei. However, through most of the reaction the phase space is far from equilibrium. Thus to determine the *eos* in a *hic* involves the problem of describing realistically the many body system in non-equilibrium situations. Then the same theory can be used to determine the equilibrium *eos* for the ground state.

A well established framework for the description of heavy ion collisions are transport theories of the Boltzmann-Uehling-Uhlenbeck (BUU) type with a self consitent mean field and a collision term [1,2]. In relativistic theories the mean field is replaced by the self energy, whose density and temperatur behavior for ground state nuclear matter has extensively been studied in Dirac Brueckner (DB) Hartree Fock theory [3,4,5]. In many applications, however,the Walecka model [6] or nonlinear modifications [7] of it were utilized. Some of the nonlinear versions have been quite sucessful in reproducing experimental flow data

[7]. However, these approaches are mostly carried out in a local density approximation with the local phase space distribution characterized only by the density. Thus no account is taken of the non-equlibrium situation occuring in heavy ion collisions. To improve on this approximation, we treat some charactaristic features of the non-eqilibrium phase space configuration in an approach which will be called *local phase space configuration approximation* (LCA). A typical such configuration in the early stage of the collision is that of two separated Fermi spheres, corresponding to two interpenetrating streams of nuclear matter (called colliding *nm* in the following). Such configurations were included in the work of the Tübingen group [8], where the G-matrix calculated for colliding *nm* was used as an effective interaction in a non-relativistic formulation. In the work described here a similar attempt is made in a relativistic framework [9,10]. We are, however, not actually solving the DB theory for non-equlibrium situations, e.g. for colliding *nm*, but we extrapolate the ground state DB results using covariance and selfconsistency arguments. We can also view this as realistically taking into account the momentum dependence, in particular the momentum *space* dependence, of the mean fields. In related work of the Giessen group scalar and vector fields were parametrized on a Hartree-Fock level in order to reproduce the momentum dependence of the optical potential [11,12].

2 The Local Phase Space Configuration Approximation

In the Walecka model [6] including nucleon and σ and ω meson fields in the classical mean field approximation the self energies are given as

$$\Sigma_S = g_S^2/m_S^2 \rho_S = \Gamma_S \rho_S \ , \ \Sigma^\mu = g_V^2/m_V^2 j^\mu = \Gamma_0 j^\mu, \tag{1}$$

where the coupling constants and masses have been combined in coupling parameters Γ_S and Γ_0 , which are constants adjusted to the ground state properties. ρ_S and j^μ are the scalar density and fermion current, respectively. It has been shown in ref. [13] that the self energies in nuclear matter obtained in Dirac-Brueckner calculations can be parametrized in a similar form

$$\Sigma_i(k, k_F) = \Gamma_i(k, k_F)\rho_i(k_F, m^*) \ , \tag{2}$$

where the index $i = S, 0$ corresponds to the scalar and the timelike vector component of the self energy. $\rho_0 = j_0$ is the baryon density. The DB results, which we took from ref. [3], are then extrapolated in a self consistent way to the configuration of colliding *nm*, i.e. to two Fermi ellipsoids. To include the full momentum dependence of the resulting self energies in a transport calculation is at the moment unfeasable. Therefore the self energies Σ_i^{12} of the colliding *nm* system were finally averaged over the momentum configuration[1] and expressed in mean field form as functions of the Fermi momenta and the relative velocity of two currents

$$\bar{\Sigma}_S^{12} = \bar{\Gamma}_S^{12}(k_{F_1}, k_{F_2}, v_{rel})\rho_S^{12}, \quad \bar{\Sigma}_0^{12} = \bar{\Gamma}_0^{12}(k_{F_1}, k_{F_2}, v_{rel})j_\mu^{12} \quad . \tag{3}$$

The construction of the self energies is described in more detail in refs [10,13]. The coupling functions $\bar{\Gamma}_S^{12}$ and $\bar{\Gamma}_0^{12}$ contain in an effective way the correlation and exchange effects of the DB calculations results. In fig. 1 the $\bar{\Gamma}_i^{12}$ are shown as functions of the subsystems densities and for different relative velocities. Since for the low density regime ($\rho < 0.5\rho_0$) no values from Brueckner calculations were available, we had to extrapolate to that region. It is seen that scalar and vector coupling functions show a similar behavior and decrease generally with density. The velocity dependence is more complex due to effects from the momentum dependence of the G-matrix and effects, stemming from the Pauli principle in the colliding *nm* system.

In fig. 2 the different *eos* are shown: for ground state *nm* in the DB model [3], for colliding *nm* with a relative velocity of 0.88 c which corresponds to an incident lab energy of about 1 GeV/A, with DB selfenergies as discussed above, and for a nonlinear version of the Walecka model, NL2 from ref. [7]. The extraction of the *eos* in a system of colliding *nm*, which is not an a priori evident procedure, is described in more detail in ref. [13]. As already shown there the *eos* is obviously more repulsive, i.e. stiffer, in the colliding system than for *nm* at rest.

3 The Transport Equation

The relativistic BUU-equation has the form [2]

$$\begin{gathered}\left[k^{*\mu}\partial_\mu + (k^*_\nu F^{\mu\nu} + m^*\partial_\mu m^*)\partial_\mu^{k^*}\right] f(x,k^*) = \\ \frac{1}{2}\int d^4k_2^* d^4k_3^* d^4k_4^* \frac{d\sigma_{eff}}{d\Omega}(k^* + k_2^*)^2 \delta^4(k^* + k_2^* - k_3^* - k_4^*) \\ \left[f_3 f_4 (1-f)(1-f_2) - f f_2 (1-f_3)(1-f_4)\right] \quad .\end{gathered} \tag{4}$$

The physical input of the RBUU-equation are the self energies, which determine the effective momenta $k_\mu^* = k_\mu - \bar{\Sigma}_\mu^{12}(x)$ and the effective mass $m^* = M - \bar{\Sigma}_S^{12}(x)$ of the nuclei as well as the field strength tensor $F^{\mu\nu} = \partial^\mu\bar{\Sigma}^{12\nu} - \partial^\nu\bar{\Sigma}^{12\mu}$.

[1]The superscript ^{12} refers to the colliding system.

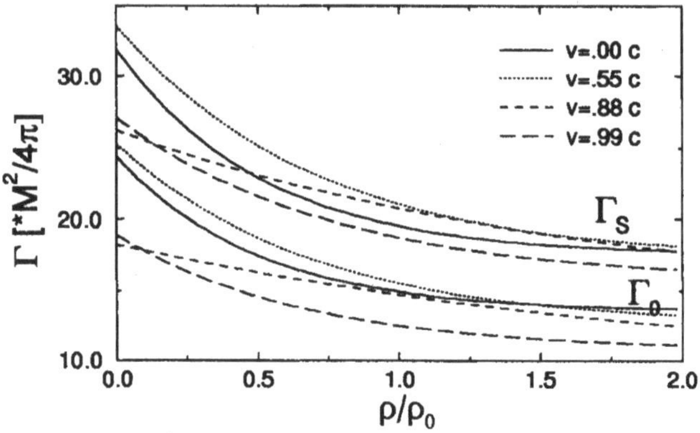

Figure 1: Effective scalar and vector coupling functions $\bar{\Gamma}_S^{12}$ and $\bar{\Gamma}_0^{12}$ in dimensionless units for colliding nuclear matter with different streaming velocities of the subsystems.

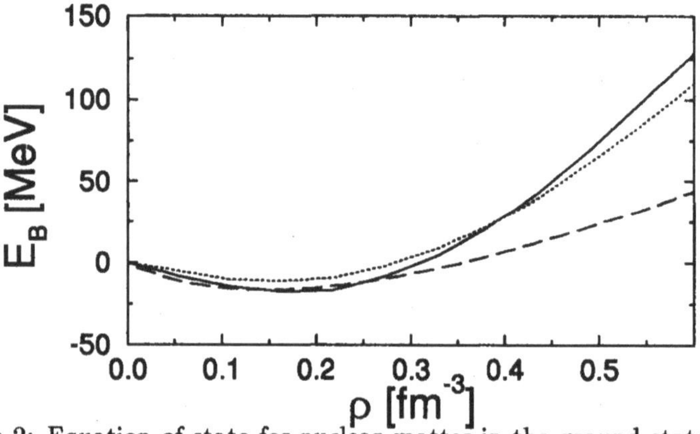

Figure 2: Equation of state for nuclear matter in the ground state for the DB model [3] (dotted), and for a mean field calculation with NL2 (dashed), and for colliding nuclear matter with relative velocity v=0.88 c with extrapolated self energies (solid).

In the LCA approach we take the self energies in the form of eq. (3) and parametrize the momentum configuration in a cell in coordinate space in the overlap region by two separated Fermi ellipsoids. The parameters of the actual configuration in a *hic* collision, the Fermi momenta and the relative velocity of the subsystems, are extracted from the phase space distribution function $f(x, k^*)$. In the moment we assume a locally symmetric configuration $k_{F_1} = k_{F_2} = k_F$, but this is no restriction in principle.

To have a fully consistent description of the transport equation it would be necessary to determine also the in medium cross section σ_{eff}, which enters in the collision integral, on the same level as the mean field. One method to do this is by the use of a generalized optical theorem connecting the G-matrix with the imaginary part of the self energy which leads to a proportionality $\sigma_{eff} \propto \Im\Sigma/v\rho$. In the present calculation, however, we adopt the usual Cugnon parametrisation [16] for the NN-cross section.

We also included Δ-degrees of freedom since they play an important role in the dynamics and are the source for the pion production. Deltas are created and annihilated through inelastic collisions ($NN \longleftrightarrow N\Delta$). Additionly deltas can decay to nucleons and pions ($\Delta \rightarrow N + \pi$) with an exponential decay law $e^{-\Gamma t}$, $\Gamma = 112 MeV$, in the delta rest frame. For the propagation of the deltas we use the same field as for the nucleons, i.e. $m^*_\Delta = M_\Delta - \bar{\Sigma}^{12}_S(x)$. The pions propagate freely with a mean free path $\lambda_\pi = v\rho/4\rho_0$ until they are reabsorbed through the process $NN\pi \rightarrow N\Delta$.

4 Results

In fig. 3 the transverse flow and rapidity distribution for the reaction $^{40}Ca + ^{40}Ca$ at 0.8 GeV/A lab energy and impact parameter $b = 1.62$ fm are compared for nonlinear versions NL1-NL3 of the Walecka model, all taken from ref. [7], and our LCA appproach. As can be seen here the apparent increase of the stiffness of the *eos* in the LCA (see fig. 2) does not necessarily lead to more flow. This is because there is a complicated interplay between the density and velocity dependence in the mean fields.

No comparison to experiment is shown here, but in ref. [7] it was found that NL1 and NL2 are able to reproduce experimental flow data quite well, whereas NL3 fails because of its small effective mass. It is seen that the DB model in the LCA, i.e. with proper account for the non-equilibrium effects, is in good agreement with the NL1 and NL2 models and therefore should reproduce experiments as well. If, on the other hand, the coupling functions in the DB model are kept *fixed* at their values at the saturation point during the entire reaction, this strongly overestimates the flow (not shown here). Such a procedure, which we will abbreviate as DB0, is equvalent to the original Walecka model with coupling coefficients ajusted to the ground state of DB calculatons.

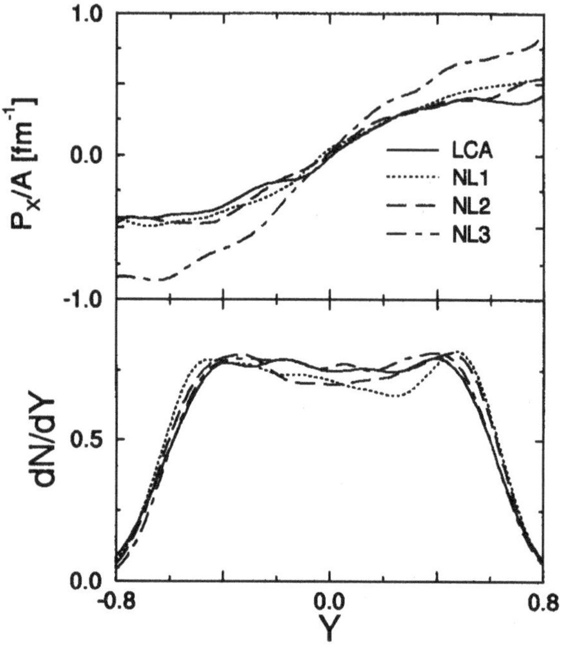

Figure 3: Transverse flow and rapidity distribution for the models NL1-NL3 and the DB model in the LCA for the reaction $^{40}Ca +^{40} Ca$, E=0.8 GeV/A, b=1.62 fm.

Differences between the NL2 and the LCA calculations are found, however, in the pion production rate. In fig. 4 the time evolution of the delta and pion multiplicities is displayed for a central collision of ^{40}Ca on ^{40}Ca at 1.8 GeV/A in the LCA, NL2 and DB0 approaches. The time evolution of the deltas is quite similar for the LCA and NL2 but substantially differs in the DB0, where the deltas have disappeared at an earlier time and consequentely less pions are produced. The amount of deltas is determined by the balance of delta production through inelastic scattering, delta decay, and pion reabsorption. As can be seen in fig. 4 a high Δ-density survives slightly longer in the LCA than in the NL2 approach and gives rise to a significant increase of the pion yield. We see here that the pion production - and possibly particle production in general - is a better probe for the *eos* than flow observables.

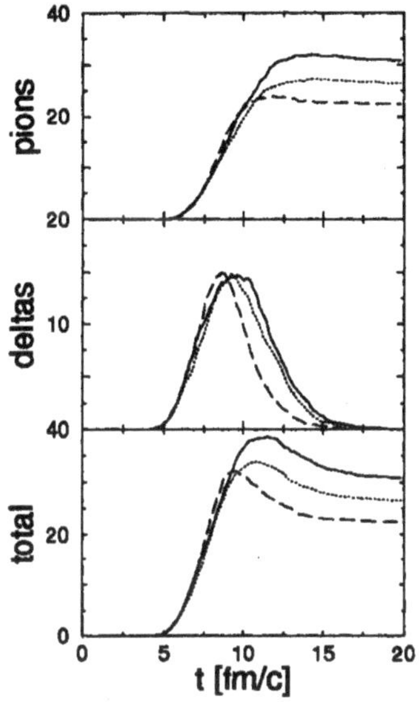

Fig. 4: Pion and delta production for a central collision of ^{40}Ca on ^{40}Ca at E=1.8 GeV/A, LCA (solid), NL2 (dotted) and DB0 (dashed).

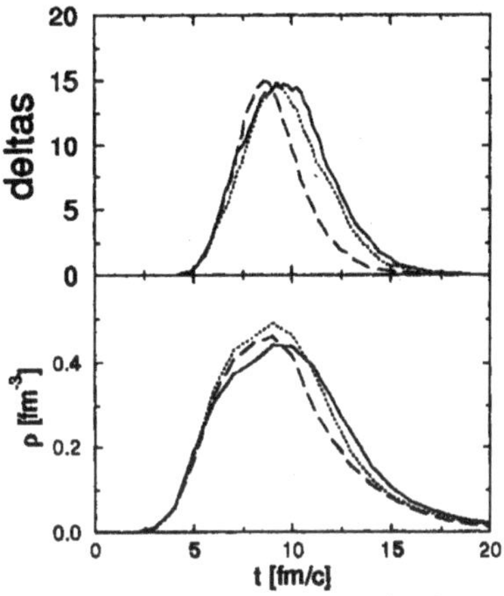

Fig. 5: Delta production (top) and central density (bottom) for the same reaction as fig. 4, LCA (solid), NL2 (dotted) and DB0 (dashed).

In fig. 5 the number of deltas (same as in fig. 4) is shown parallel to the central density in the reaction. As can be seen there, not the maximum value of the density reached but the duration of a sufficiently high compression is responsible for the number of deltas in the system and the resulting pion yield.

To summarize, we have included realistic forces in relativistic BUU-calculation which are based on the Dirac-Brueckner theory and which contain non-equilibrium features of the phase space. To do this we have approximated the phase space locally by colliding *nm*. We found a good agreement of flow data in the LCA approach with the NL1 and NL2 parametrisations of the Walecka model. Differences, however, can be seen in the pion yields showing that particle production provides a more sensitive probe for the *eos* than flow observables. Finally it is seen that the inclusion of non-equilibrium effects is important for the description of heavy ion collisions, especially if the *eos* of *nm* is to be reliably extracted from such processes.

References

* Supported by the German Minister for Research and Technology, grant 06LM352/4 and by GSI.

+ Present address: University of Nantes, France.

[1] H.T. Elze, M. Gyulassy, D Vasak, H. Heinz, H. Stöcker, and W. Greiner, *Mod. Phys. Lett.* **A2**(1987)451

[2] B. Blättel, V. Koch, W. Cassing, and U. Mosel, *Phys. Rev.***C38**(1988)1767

[3] B. ter Haar, R. Malfliet, *Phys. Rep.* **149**(1987)207

[4] C. J. Horowitz, B. D. Serot, *Nucl. Phys.* **A464**(1987)613

[5] R. Brockmann, R. Machleidt, *Phys. Rev.* **C42**(1990)1965

[6] J.W. Walecka, *Ann. Phys.* **C25**(1974)497

[7] B. Blättel, V. Koch and U. Mosel, *Rep. Phys. Phys.* **56**(1993)1

[8] J. Jänicke, J. Aichelin, N. Othsuka, R. Linden, A. Faessler, *Nucl. Phys.* **A536**(1992)201

[9] C. Fuchs, L. Sehn, H. H. Wolter, *Nucl. Phys.* **A545**(1992)151c

[10] C. Fuchs, L. Sehn, H. H. Wolter, *Prog. Part. Nucl. Phys.* **30**(1993)247

[11] K. Weber, B. Blättel, W. Cassing, H.-C. Dönges, V.Koch, A. Lang and U. Mosel *Nucl. Phys.* **A539**(1992)713

[12] T. Maruyama, B. Blättel, W. Cassing, A. Lang, U. Mosel and K. Weber, *Phys. Lett.* **B297**(1992)228

[13] L. Sehn, H. H. Wolter, *Nucl. Phys.* **A519**(1990)289c

[14] H. Elsenhans, L. Sehn, A. Faessler, H. Müther, N. Othsuka, H.H. Wolter, *Nucl. Phys.* **A536**(1992)750

[15] C. Gregoire, B. Remaud, F. Sebille, L. Vincet, and Y. Raffay, *Nucl. Phys.* **A465**(1987)317

[16] J. Cugnon, T. Mizutani, and J. Vandermeulen, *Nucl. Phys.* **A352**(1981)505

Flow and nuclear compressibility with non-local forces

F. Haddad[1], M. Farine[3], P. Schuck[2],

F. Sébille[1] and V. de la Mota[1]

[1]Laboratoire de Physique Nucléaire, Université de Nantes-IN_2P_3. F-44072 Nantes Cedex 03. France.

[2]Institut des Sciences Nucléaires. F-38026 Grenoble Cedex. France.

[3]Ecole Navale, Lanvéoc Poulmic. 29240 Brest Marine. France.

Abstract

A study of collective flow in heavy-ion reactions at intermediate energies is performed in the framework of the Landau-Vlasov model. The influence of the effective mass and the compressibility modulus is analyzed through a non-local nucleon optical potential over a wide range of energies.

Current experimental facilities offer the opportunity to perform Heavy-Ion collisions that allow the formation of very excited nuclear systems through various dynamical processes depending on the bombarding energies as well as on the impact parameters. Then a large variety of phenomena provides the possibility to investigate the behaviour of strongly interacting finite fermionic systems in the vicinity or far from their ground states. A prime objective of this Heavy-Ion physics is to extract informations on the properties of the nuclear matter in different conditions of density and excitation. To achieve this goal, detailed analysis of experimental observables are needed due to strong dynamical effects inherent in violent collisions between microscopic objects of finite size.

Among the different attempts to track down the nuclear matter properties, special attention has been devoted to the appearance of collective flow of nuclear matter in a wide range of beam energies [1]. A considerable effort has already been undertaken to investigate this phenomenon both from the experimental and theoretical point of view. Nevertheless, it turns out that the determination of fundamental characteristics, like those related to the nuclear Equation of State, is more difficult than expected. As a matter of fact, the build up of the nucleon collective motion is a dynamical process, and a straightfoward connection with usual equilibrium concepts is not obvious. In particular, the attempts to describe observables using 4π detectors with either the cascade model [2] or the hydrodynamical model [3] demonstrate that the formation of sidewards flow is sensitive to the mean-field felt by the nucleons during the different stages of the reaction.

Due to the strong influence of the dynamics on experimentally observable signatures, transport calculations constitute actually the most convenient way to disentangle the effects of the main features of the mean nuclear potential, and probably provide the opportunity to gain quantitative informations on their strength. For beam energies ranging from 100 to 400 MeV, where experimental data are currently available, the non locality of the optical potential is expected to play a significant role. The purpose of this contribution is to adress its influence on the construction of collective nucleon motion.

The investigation of this problem has been performed in a semi-classical transport model, referred to as the Landau-Vlasov model [4], which describes the dynamics of nuclear collisions through the time evolution of the one-body phase-space distribution function $f(\vec{r},\vec{p})$ which is driven by the following kinetic equation

$$\frac{\partial f}{\partial t} + \{f, H\} = I_{coll}(f) \tag{1}$$

where $\{\ ,\ \}$ stands for the Poisson bracket, H is the one-body Hamiltonian and $I_{coll}(f)$ is the Uehling-Uhlenbeck collision term. In this approach, the statics as well as the dynamics are treated selfconsistently, utilizing a Lagrangian point of view where the one-body distribution function $f(\vec{r},\vec{p})$ is projected on a moving basis of coherent states:

$$f(\vec{r},\vec{p}) = \frac{A}{N} \sum_i \omega_i \; g_\chi(\vec{r}-\vec{r}_i) \; g_\phi(\vec{p}-\vec{p}_i) \tag{2}$$

where ω_i are the projection coefficients determined by the initial conditions and g_χ and g_ϕ symbolise the moving elements of the basis.

The non-local feature of the mean-field potential has been introduced in this model through the phenomenological effective Gogny interaction [5] which has proven to be able to reproduce many nuclear properties at low energy. In the phase space representation, the corresponding mean-field potential without spin orbit term reads:

$$V_q(\vec{r},\vec{p}) = [\frac{3}{4}(\gamma+2)\rho^{\gamma+1}(\vec{r}) - \frac{3}{4}\gamma\rho^{\gamma-1}(\vec{r})(\rho_n^2(\vec{r})+\rho_p^2(\vec{r})) - \frac{3}{2}\rho^{\gamma}(\vec{r})\rho_q(\vec{r})]t_3$$
$$+\sum_{i=1}^{2}\int d\vec{r}'[(W_i+\frac{B_i}{2})\rho(\vec{r}') - (H_i+\frac{M_i}{2})\rho_q(\vec{r}')]exp - \frac{(\vec{r}-r')^2}{\eta_i}$$
$$-\frac{2(\sqrt{\pi}\eta_i)^3}{(2\pi)^3}\sum_{i=1}^{2}\int d\vec{k}' \quad [(\frac{W_i}{2}+B_i)f_q(\vec{r},\vec{k}')-(\frac{H_i}{2}+M_i)f(\vec{r},\vec{k}')]exp-[\frac{(\vec{k}-k')^2\eta_i^2}{4}]$$

where $\rho_n(\vec{r})$ and $\rho_p(\vec{r})$ represent respectively the neutron and proton local density, $\rho(\vec{r})$ is the sum of the neutron and proton densities. The usual parametrization referred to as the D1-G1 can be found in ref [5]. The resulting effective mass is $m^*/m = 0.67$, and the incompressibility modulus is $K_\infty = 228MeV$.

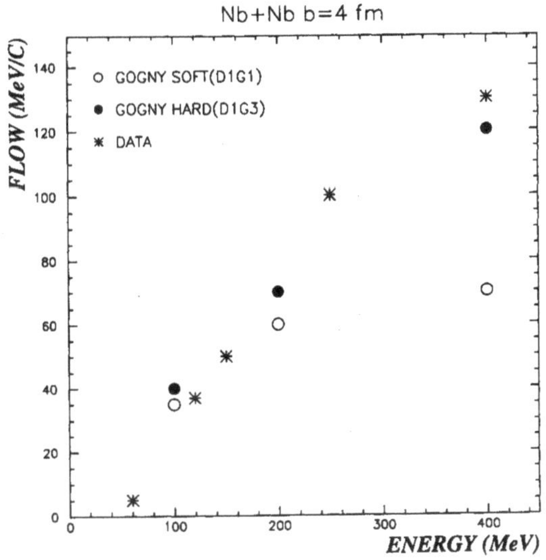

Figure 1: *Transverse collective flow as a function of the incident energy for the Gogny D1-G1 and D1-G3 forces compared with the experimental data from [6].*

In the present study the non-locality of the nuclear mean-field is addressed in connection with the flow parameter commonly used to evaluate the transverse collective motion projected in the reaction plane. The flow parameter is defined as follows:

$$F = \left(\frac{d < p_x >}{d(y/y_p)}\right)_{y/y_p=0} \tag{3}$$

where $< p_x >$ is as usual the mean value of the transverse momentum projected into the reaction plane, and y and y_p are the particle rapidity and the projectile rapidity in the center of mass frame respectively.

In addition to the normal Gogny force we also have constructed one with K_∞=360 MeV with all other characteristics, in particular m^*/m=0.67, unchanged. In Fig. 1 we show the flow values obtained with these two forces. The cross section used for the collision term was the free one with correct isospin and energy dependences. One observes that the experimental values are rather well reproduced by both forces up to beam energies of around 200 MeV per nucleon. This weak dependence on K_∞ for these energies has already been reported in Ref. [7]. Above 200 MeV per nucleon, the behavior changes drastically, the high incompressibility reproducing experiment far better than the original Gogny force (K_∞=228 MeV).

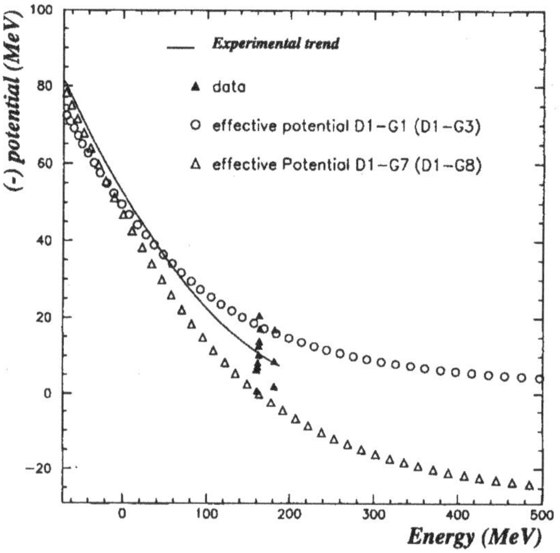

Figure 2: *Optical potential as a function of the energy for two parametrizations of the Gogny force and the experimental data from [9].*

However the low flow values of the latter can be traced back to a incorrect behavior of the optical potential at higher energies ($\simeq$ 200 MeV) (see Fig 2). Indeed the data [9] shows that the nucleon nucleus optical potential turns positive at around 200 MeV for nuclear matter at saturation. We therefore have constructed new "Gogny forces" with a more repulsive optical potential behavior(see Fig 2).

We call these forces D1-G7 and D1-G8 having respectively incompressibility moduli of 228 MeV and 300 MeV whereas their effective mass is $m^*/m = 0.56$. The parameter values for the D1-G7 are given in table 1. The asymptotic value of the optical potential for higher energies is 40 MeV in the new parametrisation and 0 MeV in the old one. Our new potential is quite similar to the one of Wiringa [10] as a function of density and energies.

Table 1	i	η_i	W_i	B_i	H_i	M_i	T_3	γ
D1-G7	1	0.7	-402.4	-70	-496.2	-23.56	916.723	0.35
	2	1.2	61.087	-83.963	95.463	-165.197	916.723	0.35

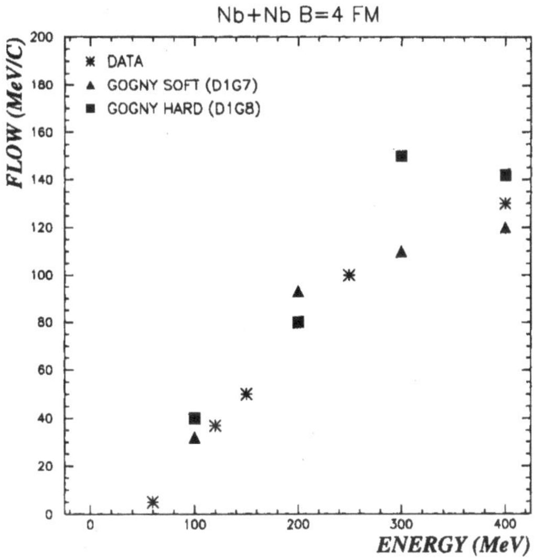

Figure 3: *Same as figure 1 for the Gogny D1-G7 and D1-G8 interactions.*

The flow values corresponding to D1-G7 and D1-G8 are shown in Fig. 3. Below 200 MeV per nucleon again the theoretical flow does not discriminate between a soft and hard equation of state. This corresponds to an energy range where the variation of the real part of the optical potential versus the energy is the steepest, and exhibits similar behavior for all parametrizations used. For beam energies above 200 MeV per nucleon the calculation shows a clear dependence on K_∞ even though it is somewhat reduced with repect to Fig 1. We observe from Fig.3 what a non local interaction having the right optical potential and a $K_\infty \simeq$ 200 MeV produces a flow up to 400 MeV per nucleon which is in rather good agreement with experiment. The corresponding stiff equation of state(K_∞= 300 MeV) clearly gives flow values too high.

These results confirm an earlier findings of Pan and Danielewicz [8] where they however used an approximation to calculate the equation of state probably resulting in comparatively more flow. It is also worth mentioning that our results agree with experiments for the flow of the Ar+Pb reaction at 400 MeV per nucleon where experimental cuts permit a clear distinction between local and non local forces in favor of the latter ones.

In conclusion we have shown here that non-local forces which couply with constraint from optical potential data give a real sensitivity of the flow above beam energies of 200 MeV per nucleon on the nuclear equation of state. Our findings are in agreement with studies by Pan and Danielewicz and a value $K_\infty \simeq$ 200 MeV seems to emerge from the analysis. This confirm older values extracted from giant monopole resonance [11] but contradicts recent claims where a much higher incompressibility value is advocated [12].

The authors are indebted to P. Danielewicz for discussion and helpful exchanges of scientific results.

References

[1] H. Gutbrod et al., Rep. Prog. Phys. **52** (1989) 1267; J. Peter et al., Nucl. Phys. **A519** (1990) 611 J. Krofcheck et al., Phys. Rev. Lett. **63** (1989) 2028 C.A. Ogilvie et al., Phys. Rev. **C42** (1990) 10

[2] J. Cugnon, Phys. Rev. **C22** (1990) 1885

[3] R.B. Clare and D. Strotman, Phys. Rep. **141** (1986) 16

[4] B. Remaud, F. Sébille, C. Grégoire. L. Vinet and Y. Raffray, Nucl. Phys. **A447** (1985) 555c; C. Grégoire, B. Remaud, F. Sébille, L. Vinet and Y. Raffray, Nucl. Phys. **A465** (1987) 317; B. Remaud, C. Grégoire, F. Sébille, and P. Schuck, Nucl. Phys. **A488** (1988) 423

[5] J. Dechargé and D. Gogny, Phys. Rev. **C21** (1980) 1568

[6] K.G.R. Doss et al., Phys. Rev. Lett. **57** (1986) 302; D. Krofcheck et al., Phys. Rev. **C46** (1992) 1416

[7] V. de la Mota, F. Sébille, B. Remaud and P. Schuck, Phys. Rev. **C46** 667 (1992)

[8] G. Pan and P. Danielewicz, Phys. Rev. Lett **70** (1993) 2062

[9] M. Bauer et al., J. Phys. G: Nucl. Part. Phys. **8** (1982) 525

[10] R.B. Wiringa, Phys. Rev. **C38** 2967 (1988)

[11] J.P. Blaizot, Phys. Rep. **64** 171 (1980)

[12] M.M. Sharma, Phys. Rev. **C38** 2562 (1988);

Collective Flow and Azimuthal Correlations in Nucleus-Nucleus Collisions at the Bevalac

G. Rai and the EOS Collaboration:

S. Albergo,(7) F. Bieser,(1) F. P. Brady,(4) Z. Caccia,(7) D. Cebra,(4)
A. D. Chacon,(5) J. Chance,(4) Y. Choi,(8) S. Costa,(7) J. Elliott,(3)
M. Gilkes,(3) A. Hauger,(3) A. Hirsch,(3) E. Hjort,(3) A. Insolia,(7)
M. Justice,(2) D. Keane,(2) V. Lindenstruth,(6) H. S. Matis,(1) M. McMahan,(1)
C. McParland,(1) W. F. J. Müller,(6) D. L. Olson,(1) M. D. Partlan,(4) N. Porile,(3)
R. Potenza,(7) G. Rai,(1) J. Rasmussen,(1) H. G. Ritter,(1) J. Romanski,(7)
J. L. Romero,(4) G. V. Russo,(7) H. Sann,(6) R. Scharenberg,(3) A. Scott,(2)
Y. Shao,(2) B. Srivastava,(3) T. J. M. Symons,(1) M. Tincknell,(3) C. Tuve,(7)
S. Wang,(2) P. Warren,(3) H. Wieman,(1) K. Wolf (5)

(1)Lawrence Berkeley Laboratory, Berkeley, California, 94720
(2)Kent State University, Kent, Ohio 44242
(3)Purdue University, West Lafayette, Indiana, 47907-1396
(4)University of California, Davis, California, 95616
(5)Texas A&M, College Station, Texas, 77843
(6)Gesellschaft für Schwerionenforschung, Darmstadt, Germany, D-64220
(7)Universita' di Catania & INFN-Sezione di Catania, Catania, Italy, 95129
(8)Sung Kwun Kwan University, Suwon, Rep. of Korea, 440-746

Abstract

The EOS experiment at the Bevalac has recently carried out exclusive event-by-event measurements of relativistic heavy ion collisions with a variety of projectile, target and beam energy combinations. The data was obtained using the EOS Time Projection Chamber. We present preliminary results on inclusive spectra, collective flow and azimuthal correlations obtained from a study of Au+Au reactions with beam energies covering 0.6 - 1.2 A GeV.

1 Introduction

Past experiments [1, 2] have proven collective flow exists in high energy heavy ion collisions. Different aspects have been identified (bounce-off, side-splash, and squeeze-out) and their mass and energy dependence have been studied. Flow effects resulting from nuclear compression are known to depend on both the static, that is, the equation of state (EOS), and the dynamic properties of the collision. Progress towards an understanding of the nuclear matter EOS has been slow. This is due in part to the uncertainties in the theoretical description of the collision dynamics, for example, the pronounced influence of in-medium effects and momentum-dependent interactions on the EOS. Experience has shown that it might be possible to resolve ambiguities by requiring models to simultaneously describe a comprehensive set of observables that are sensitive to the underlying EOS. Previous experiments lacked the systematics and did not measure most of the signal in a collision event because of detector limitations.

New measurements on relativistic heavy ion collisions have been made at the Bevalac using a 4π detector system, with particle indentification and momentum measurement. High statistics data was obtained with a variety of projectile, target and beam energy combinations listed in table 1. In the first phase of the analysis we have concentrated on the mass and energy dependence of flow in symmetric Au+Au collisions. Our preliminary results are presented in this paper.

Beam Projectile	Incident Energy	Target
Au	0.25 - 1.2 A GeV	Be, C, Al, Ca, Cu, Nb, La, Au, U
La	0.4 - 1.3 A GeV	C, Al
Ni	0.25 - 2.0 A GeV	Be, Cu, Au
Kr	1.0 A GeV	C, Al
Ca	0.8 - 2.1 A GeV	Be, Ca, Au

Table 1: Summary of the data obtained by the EOS Experiment

2 The EOS Experimental Arrangement

The arrangement of detectors are shown schematically in Fig. 1. The apparatus provides substantial phase space coverage in the center of mass system (close-to-4π) with simultaneous particle identification. The principal tracking detector is a sophisticated heavy ion Time Projection Chamber called the EOS TPC [3] which provides continuous three dimensional tracking of charged particles and most of the event characterization. The TPC is located between the pole tips of a large aperture, 1.3T, superconducting dipole magnet, and therefore, particle identification (up to Z $\sim$ 5-6) is obtained from the magnetic rigidity and multiple energy loss measurements along the path of an individual track.

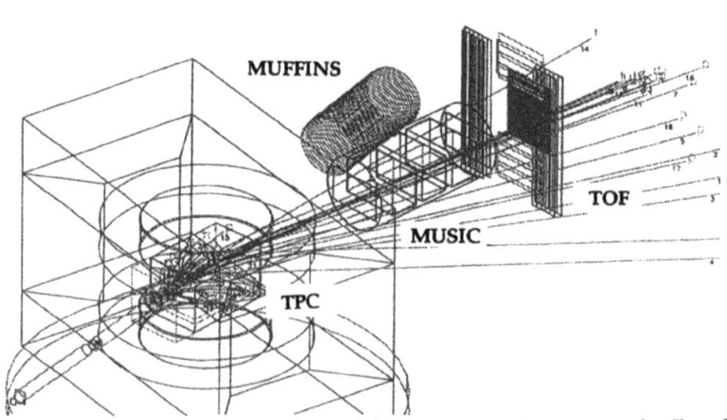

Figure 1: A perspective view of the EOS experimental setup at the Bevalac.

Two detectors called MUSIC and the TOF wall are located downstream of the EOS TPC. MUSIC is a large volume, multiple sampling ionization chamber which measures the trajectory and identifies the charge of heavy fragments ranging from Carbon to beam particles. The TOF wall consists of a movable array of scintillator slats which measures the time of flight of light mass fragments and identifies their element number. The combination of MUSIC and the TOF wall were used to identify the forward going projectile fragments which are beyond the dynamic range of the TPC and thus provide full acceptance for the reverse kinematic experiments. The MUFFINS detector recorded the energy and angular position of the neutrons emitted in the beam direction. The data from all the detectors were combined and used in the offline analysis to reconstruct the collision event-by-event. In the following sections we discuss the results obtained from the EOS TPC detector alone.

3 Inclusive Spectra

The inclusive spectra of midrapidity (-0.1<y<0.1) pions, protons, deuterons and alphas emitted in midcentral, 1.2 A GeV, Au+Au collisions are plotted in Fig. 2. In terms of a single isotropic fireball model, the momentum distribution for emitted particles at midrapidity (y=0) is given in the center of mass by:

$$\frac{dn}{dydm_t} = const. \times \frac{m_t^2}{e^{m_t/T} + d} \tag{1}$$

where d = +1,-1,0 for a Fermi, Bose, or Boltzmann distribution, m_t is the transverse energy equal to $\sqrt{P^{\perp 2} + m^2}$ and T is the temperature.

The solid lines in Fig. 2 represent the fit to the exponential part of Eqn. 1. The inverse slope parameters 1/T are plotted in Fig. 3 as function of particle type and multiplicity bin. In general the slope of the distributions are larger for the more central collisions and heavier mass fragments. The discussion of spectra in terms of a thermal distribution is a gross oversimplication since several complicating factors occur. Any equilibration reached is at best transient and the radial expansion of the system confuses the interpretation of source temperature. The effect of radial hydrodynamic flow on inclusive spectra can be indentified in two extreme cases. If thermal equilibrium is reached, the average kinetic energy of the fragments must be constant and independent of their mass. In the case of hydrodynamic flow, one might assume that each fragment is in motion with the flow velocity. If so, the average kinetic energy is linearly proportional to the mass of the fragment and the source would appear to be hotter. However, an apparently hotter source could also arise from the final state clustering of the nucleons and any collective motion, directed or radial, would merely enhance the effect. A detailed comparison of the data with microscopic models which include dynamical effects, for example Quantum Molecular Dynamics, is in progress.

4 Collective Flow

The global transverse momentum analysis technique developed by Danielwicz and Odyniec [4] was used to measure collective flow. In this method the reaction plane is constructed for each event by summing the transverse momentum components of all particles in the center of mass forward hemisphere and likewise in the backward hemisphere. The difference vector between the two hemispheres, called the Q vector, is an estimate of the reaction

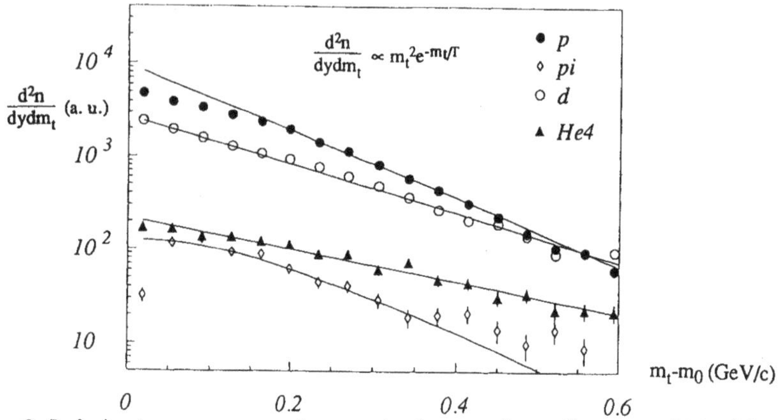

Figure 2: Inclusive transverse momentum spectra for several mass fragments obtained from 1.2 A GeV Au+Au collisions.

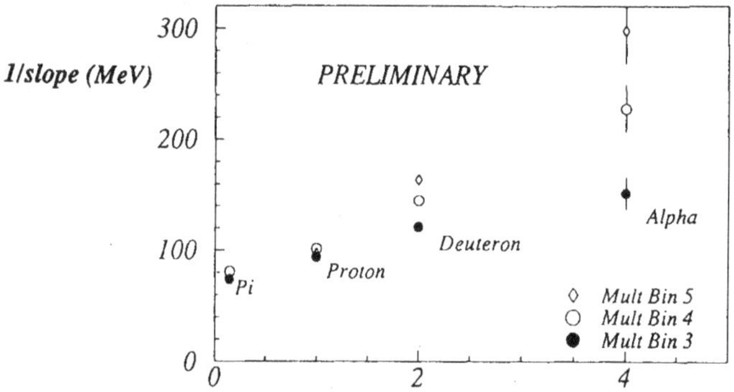

Figure 3: Inverse slope parameters plotted for several mass fragments and multiplicity bins.

plane. Auto–correlation is removed by determining the reaction plane for every fragment by excluding that fragment from the sum:

$$Q = \sum_{i=1}^{n} \mathrm{w(i)}\mathbf{P}_{\perp}(\mathrm{i}) \tag{2}$$

The weights w(i) varied linearly from -1 at target rapidities to $+1$ at projectile rapidities to allow for the fact that fragments near midrapidity are less correlated with the reaction plane.

The transverse momentum vector of each particle is projected onto the reaction plane, yielding the in-plane transverse momentum per nucleon P_x/A, which is subsequently averaged over all events satisfying a multiplicity cut. The rapidity dependence of $\langle P_x/A\rangle$ is shown in Fig. 4 as a function of beam energy. The characteristic S-shape of the emerging curve is a clear sign of nuclear matter flow. The general features of the distributions are: a) particles close to the beam rapidity show maximum directed flow, b) the peak value of $\langle P_x/A\rangle$ is a constant 0.1 GeV/c/A, independent of beam energy above 0.6 A GeV, and c) for the stopped, thermalized particles at mid-rapidity, the transverse momentum $\langle P_x/A\rangle$ vanishes. The projectile and spectator regions are observed to behave differently at 1.2 A GeV than at 0.6 A GeV. This effect is not fully understood.

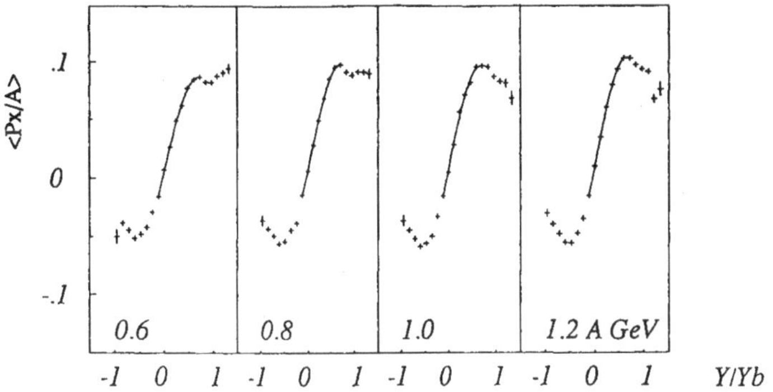

Figure 4: Transverse momentum $\langle P_x/A \rangle$ versus rapidity for Au+Au collisions at several beam energies.

The slope of the "S-shaped" $\langle P_x/A \rangle >$ curve at midrapidity is given by

$$P_x^{slope} = \left(\frac{d < P_x/A >}{d(y/y_{proj})} \right)_{y=y_{cm}} \tag{3}$$

This quantity is unaffected by the spectator fragments and is often used in the literature as a good indicator of collective flow. We have determined P_x^{slope}, for mid-central (Mul3) Au+Au collisions at 0.6, 0.8 1.0 and 1.2 A GeV beam energies. The excitation function is shown in Fig. 5 for fragments with identified charge up to Z=2, albeit the results are dominated by the protons. For comparison with earlier work, the corresponding data from the Plastic Ball group is also illustrated. The EOS results join smoothly with the Plastic Ball data at 0.6 A GeV and extend the measurements to higher energies. The values of P_x^{slope} increase slowly as function of beam energy whereas the 0.8 A GeV Plastic Ball data point might have suggested the onset of saturation.

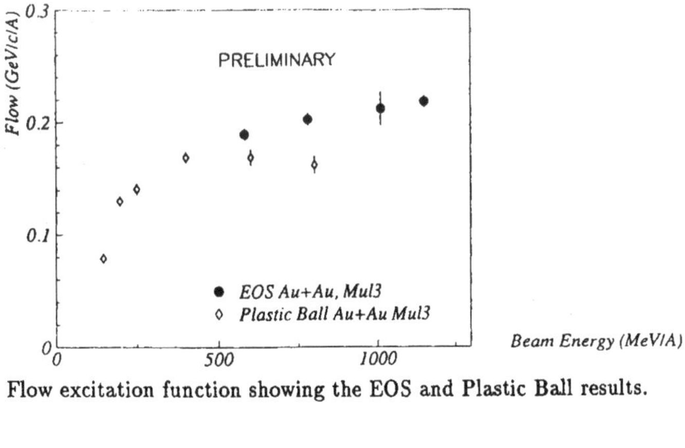

Figure 5: Flow excitation function showing the EOS and Plastic Ball results.

5 Correlations

The azimuthal particle-pair correlation method of Wang et al [5]. was used to study in detail the emission pattern of collective flow. This approach is independent of the reaction plane and, in principle, has the advantage of separating the directed flow correlations into their geometrical and momentum components. The technique relies on the azimuthal angle difference for each fragment pair, defined as $\Phi = |\phi_i - \phi_j|$, where ϕ_i represents the azimuthal angle of the ith fragment. Next one constructs the histogram $dN/d\Phi$ whose functional form should obey:

$$P(\Phi) = \frac{dN}{d\Phi} = A(1 + 0.5\lambda^2 \cos\Phi) \tag{4}$$

where A is a constant normalization factor and λ is related to the azimuthal anisotropy and represents the strength of the collective flow. Adapting the approach of interferometry analysis we define the azimuthal correlation function as:

$$C(\Phi) = \frac{P_{corr}(\Phi)}{P_{uncorr}(\Phi)} \tag{5}$$

where $P_{corr}(\Phi)$ is the measured Φ distribution of pairs in which both fragments are selected from the same event and $P_{uncorr}(\Phi)$ is the Φ distribution for uncorrelated pairs generated by event mixing. Collective flow appears as $C(\Phi) > 1$ at small Φ and as $C(\Phi) < 1$ al large Φ. The magnitude of the observed flow can be quantified by the value of λ in Eqn. 4 that best fits the data for $C(\Phi)$.

In Fig. 6, the correlation function is plotted for 1.2 A GeV Au+Au collisions selected for mid-to-central impact parameters. The data is represented in five rapidity intervals covering the mid-rapidity and beam projectile region. The correlation function develops a stronger $\cos\Phi$ dependence as the intervals approach projectile rapidity. The solid line show the fit used to extract the strength parameter λ, which in turn is plotted in Fig. 7 (left picture) vs. scaled rapidity. Similarly, the values obtained at beam energies of 0.6, 0.8 and 1.0 A GeV are also shown on the same figure. In general, we observe the strength of directed flow is zero at mid-rapidity and increases almost linearly with rapidity. This feature is consistent with the "S-shaped" flow curves shown in the previous section. However, the slope of the strength parameter versus rapidity is essentially independent of beam energy which therefore suggests the azimuthal emission pattern of fragments due to collective flow is also independent of incident beam energy.

We have extended the correlation formalism to investigate the radial part of the directed flow. Specifically, we consider a model in which the directed flow momentum is presumed to be constant. In that case, the measured transverse momentum vector for each fragment can be decomposed into a thermal component, $P^{\perp}_{therm}$, and a part which comes from the directed flow momentum vector, $P^{\perp}_{flow}$, projected along the thermal component. This is not to be confused with the radial or cylindrical flow which might also exist but cannot be studied using the technique described here. Thus, we can write for transverse momentum of the ith fragment:

$$|P^{\perp}_i| = |P^{\perp}_{therm}| + |P^{\perp}_{flow}| \tag{6}$$

Again for each fragment pair we define the sum transverse momentum as

$$|P^{\perp}_{sum}|_{i,j} = |P^{\perp}_i| + |P^{\perp}_j| \tag{7}$$

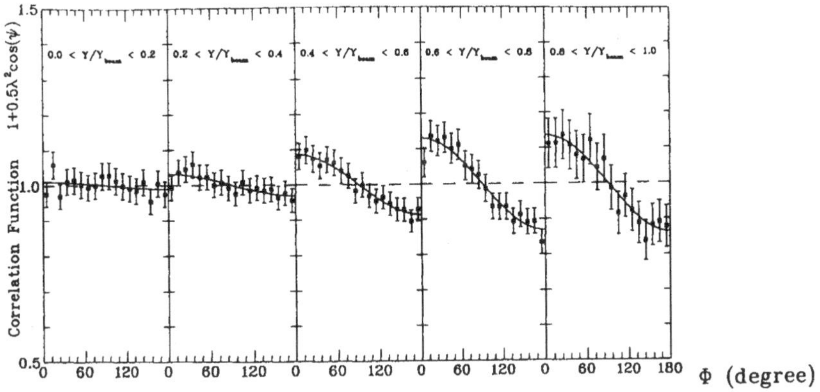

Figure 6: Azimuthal correlation function in several rapidity intervals

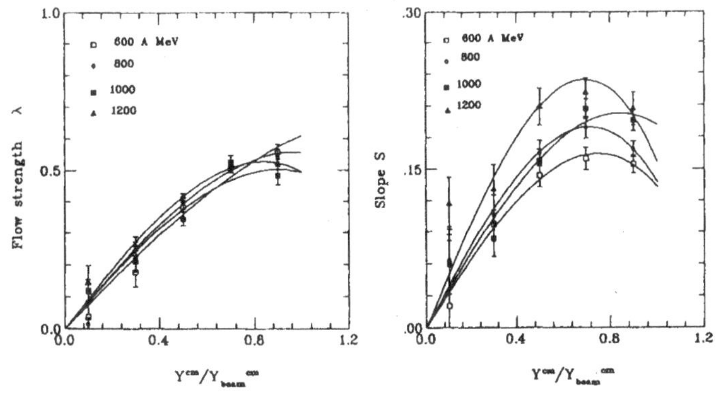

Figure 7: The flow strength parameter, λ, (left) and the slope parameter, S, (right) plotted as a function of rapidity and beam energy.

and its standard deviation by

$$\sigma^2(\Phi) = \langle P_{sum}^{\perp 2} \rangle_{(\Phi)} - \langle P_{sum}^{\perp} \rangle_{(\Phi)}^2 \tag{8}$$

for a given angle difference bin Φ. The averaging is done within a bin. In the absence of flow, a plot of σ^2 vs. Φ should be constant and reflect the width of the thermal background. Collective flow will appear to increase the values of σ^2 by approximately $2(P_{flow}^{\perp})^2$, $P_{flow}^{\perp}$, and 0 for Φ equal to 0°, 90°, and 180° respectively. Thus the slope of σ^2 vs. Φ is measure of the magnitude of directed flow momentum.

In Fig. 8 the value of σ^2 vs. Φ is plotted in five rapidity intervals for Au+Au data at 1.2 A GeV. The slope, S, of the fitted line is steeper for the higher rapidity bins which suggest fragments closer to the projectile rapidity receive a stronger momentum kick from sideward flow. The value of the slopes are plotted in Fig. 7 (right picture) versus rapidity together with the data obtained at the lower beam energies. Unlike the azimuthal correlation function, we observe the gradient of S versus rapidity is enhanced at the higher beam energies, that is, the directed sideward momentum of fragments becomes stronger with increasing incident beam energy.

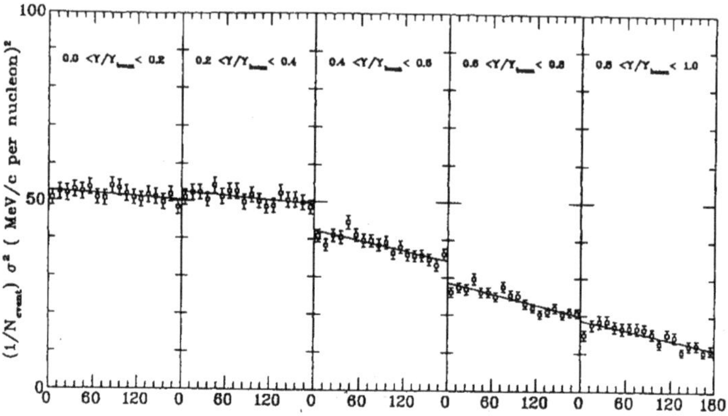

Figure 8: σ^2 versus. Φ in several rapidity intervals

6 Summary

The EOS experiment has obtained extensive data on heavy ion collisions at the Bevalac using the new Time Projection Chamber. We have measured the inclusive spectra for different mass fragments. We find a non-saturating excitation function for sideward directed flow. Azimuthal correlations suggest the emission pattern for directed sideward flow is independent of beam energy while the average flow momentum per nucleon appears to increase at higher beam energies.

This work was supported by the Director, Office of Energy Research, Office of High Energy and Nuclear Physics, Division of Nuclear Physics of the U.S. Department of Energy under Contract DE-AC03-76SF00098.

References

[1] H.H. Gutbrod, et al., Phys. Rev. C42, 640 (1990).
H.A. Gustafsson, et al., Phys. Rev. Lett. 52, 1590 (1984).

[2] D. Beavis et al., Phys. Rev. C45, 299 (1992).
A. Sandoval et al., Nucl. Phys. A400, 365c (1983).

[3] G. Rai, et al., IEEE Trans. Nucl. Sci. 37, 56 (1990).

[4] P. Danielwicz and G. Odyniec, Phys. Lett. 157B, 146, (1985)

[5] S. Wang et al., Phys. Rev. C44, 1091, (1991)

LANDAU PARAMETERS AT FINITE TEMPERATURE WITH THE GOGNY FORCE

J. Ventura, A. Polls, X. Viñas
Dept. E.C.M., Facultat de Física, Diagonal, 647, 08028-Barcelona, Spain

and

E. S. Hernández
Dept. Física, Facultad de Ciencias Exactas y Naturales
Univ. Buenos Aires, RA-1428 Buenos Aires. Argentina

ABSTRACT

An extension of the Landau Fermi liquid theory at finite temperature is discussed in the context of symmetric nuclear matter. Special attention is devoted to the meaning of the Landau parameters and their stability requirements at finite temperature. Explicit calculations are performed in a framework of a finite temperature density dependent Hartree-Fock method using the D1-Gogny finite range effective interaction. Comparisons with zero range effective interactions are also made.

1. Introduction

Heavy-ion collisions offer the oportunity to probe the nuclear equation of state (EOS) in a very wide range of densities and temperatures. In particular, the multifragmentation phenomena observed at intermediate energy heavy-ion reactions have been related to the liquid-gas phase transition predicted by the theoretical study of EOS.[1,2]

By construction, the EOS expresses the thermodynamical properties of infinite nuclear matter and therefore it is only in an indirect way that one can predict features of the multifragmentation processes. The basic assumption is that when the many nucleon system is placed sufficiently deep within the liquid-gas coexistence region, predicted by the EOS, the value of the speed of sound becomes imaginary, density fluctuations grow rapidly and fragmentation can occur.[3,4] Of course, a precise analysis requires to face a full dynamical description.

Recently, there have been some analytical dynamic studies on the growth of instabilities in hot expanding nuclear matter which have been carried out in the framework of Landau theory, extended to non zero temperatures, using a zero range effective interaction.[5,6]

The ideal program for a theoretical approach to static and dynamic properties of nuclear matter would be to start from the bare interaction between its constituents. However, the strong short range repulsion present in any realistic nucleon-nucleon interaction makes the microscopical calculations very difficult. To avoid those difficulties, most of the existing finite temperature nuclear matter calculations have been carried out using zero range effective interactions of the Skyrme type in the framework of the finite temperature Hartree-Fock (HF) theory.[7] In spite of their success, using Skyrme forces supposes a too crude approximation, specially if one studies dynamical properties of the system. To better simulate the intermediate and long range parts of the interaction, Gogny et.al. constructed an effective force containing finite range terms[8] which is also very successful in reproducing average properties of nuclei and has been recently used to study thermodynamic aspects of the EOS.[9,10]

The purpose of the present paper is to discuss an approximation scheme to extend Landau theory to non zero temperatures for finite range forces along the lines suggested by Heiselberg et.al. in Ref. 6. The key point is to take properly into account the smearing of the Fermi surface at finite temperatures which implies, as discussed in section. 3, performing appropriate averages of the Landau fields.

2. Particle-hole effective interaction

Let us consider a symmetric nuclear matter system interacting through a $D1$-Gogny effective force[8] of the form

$$\begin{aligned} V(\vec{r}) &= \sum_{i=1,2}[W_i + B_iP_\sigma - H_iP_\tau - M_iP_\sigma P_\tau]\exp(-r^2/\mu_i^2) \\ &+ t_0(1+x_0P_\sigma)\rho^\alpha\delta(\vec{r}), \end{aligned} \tag{1}$$

where P_σ, P_τ are the spin and isospin exchange operators, respectively. The parameters which define the force are given in Ref. 8. As we are dealing with nuclear matter, the spin-orbit term has not been considered.

According to the Landau Theory of Fermi liquids,[11,12] the excitations in the system are treated as quasiparticles which have, at zero temperature, an infinite lifetime on the Fermi surface. Therefore, it is assumed that the properties of the system may be described in terms of the quasiparticle (qp) occupation numbers n_i, i.e., the total energy density E is a functional of n_i. One can define the particle-hole (ph) effective interaction as its second functional derivative with respect to the distribution function

$$f_{\vec{k}\sigma\tau,\vec{k}'\sigma'\tau'}(\vec{q}) = \frac{\delta^2 E}{\delta n_{\vec{k}+\vec{q}/2}\,\delta n_{\vec{k}'-\vec{q}/2}}, \tag{2}$$

where the single particle (sp) labels are assigned according to Fig.1.

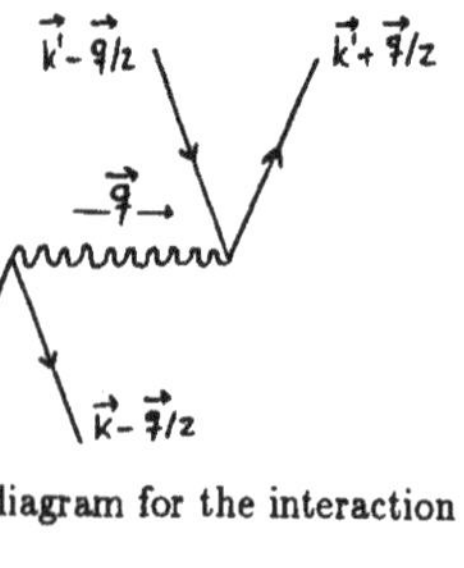

Fig 1. Feynman diagram for the interaction in the ph channel.

Some applications of Landau theory require the knowledge of the effective interaction away from the Fermi surface. Then, one must take into account the limitations of the theory due to the finite lifetime of quasiparticles in this regime. Including the temperature in the system spreads the Fermi surface in such a way that all quasiparticles have finite lifetime. In this sense, any extension of the Landau theory to finite temperatures should be regarded as a model.

Obviously, the effective interaction is also a function of spin and isospin. For symmetric nuclear matter one can separate f in particle-particle (pp) channels of well defined total spin and isospin. Using transformation relationships among the pp and the ph channels one can write the effective ph interaction in channels of (anti)symmetric (a)s spin and isospin in the following way:

$$f_{\vec{k}\vec{k}'}(\vec{q}) = f^{ss}_{\vec{k}\vec{k}'}(\vec{q}) + f^{as}_{\vec{k}\vec{k}'}(\vec{q})\vec{\sigma}\cdot\vec{\sigma}' + (f^{sa}_{\vec{k}\vec{k}'}(\vec{q}) + f^{aa}_{\vec{k}\vec{k}'}(\vec{q})\vec{\sigma}\cdot\vec{\sigma}')\vec{\tau}\cdot\vec{\tau}', \qquad (3)$$

where σ and τ are the Pauli matrices acting on spin and isospin spaces respectively. The fisrt (second) superscript stands for the spin (isospin). Due to rotational invariance, the qp effective interaction depends only upon the modulus of q, k, k' and the Landau angle θ_L between $\vec{k}$ and $\vec{k}'$. For the specific case of the Gogny force, we obtain an effective interaction with the following structure:

$$f^{\sigma\tau}_{kk'}(\vec{q}, cos\theta_L) = \varphi^{\sigma\tau}\rho^{\alpha} + \sum_{m=1}^{2}\left[\phi^{\sigma\tau}e^{-\mu_m^2 q^2/4} + \psi^{\sigma\tau}e^{-\mu_m^2(k^2+k'^2-2kk'cos\theta_L)/4}\right], \qquad (4)$$

where $\varphi^{\sigma\tau}$, $\phi^{\sigma\tau}$ and $\psi^{\sigma\tau}$ are certain combinations of the parameters of the force.[13] In a next step, a multipole expansion in terms of Legendre polinomials $P_l(cos\theta_L)$ give rise to the multipolar amplitudes $f_l^{\sigma\tau}(k, k', q)$.

Similarly, the effective interaction for Skyrme forces,[14] can be written as

$$f^{\sigma\tau}_{kk'}(\vec{q}, cos\theta_L) = A^{\sigma\tau} + B^{\sigma\tau}q^2 + C^{\sigma\tau}(k^2 + k'^2) + D^{\sigma\tau}\vec{k}\cdot\vec{k}'. \qquad (5)$$

The relevant Landau amplitudes correspond to the values of $f_l(k, k', q)$ at the Fermi surface, so that for fixed transferred momentum q, k^2 and k'^2 are required to be equal to $k_F^2 - q^2/4$. Usually one considers the dimensionless Landau coefficients F_l, multiplying f_l by the density of sp states at the Fermi level $N(0) = 2m^*(k_F)k_F/\pi^2\hbar^2$.

Fig.2 shows $F_0^{ss}(q)$ and $F_1^{ss}(q)$ at the Fermi surface as a function of q, for the Gogny force at saturation density. It is worthwhile to recall that for $q = 0$ one obtains the so-called Landau parameters. The same amplitudes for the SkM* force[15] are shown in Fig.3. One can clearly appreciate the finite range effects. In fact, the two forces present a similar behaviour only at low q. The different values at $q = 0$ are related to the different values of the incompressibility at the saturation density.

3. The effective interaction strengths at finite temperatures

The low energy excitations of a Fermi liquid at zero temperatures lie on the Fermi surface, so that the standard Landau coefficients are a good characterization of the qp effective interaction for zero transferred momentum. The situation is not

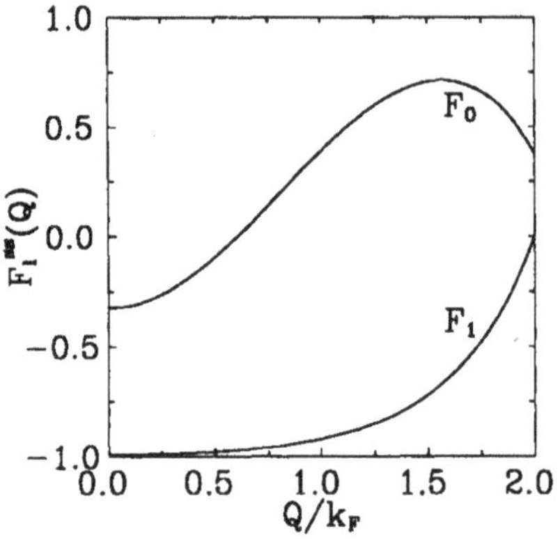

Fig 2. Dimensionless monopolar and dipolar Landau coefficients as a function of q at the Fermi surface, for the Gogny force at saturation density.

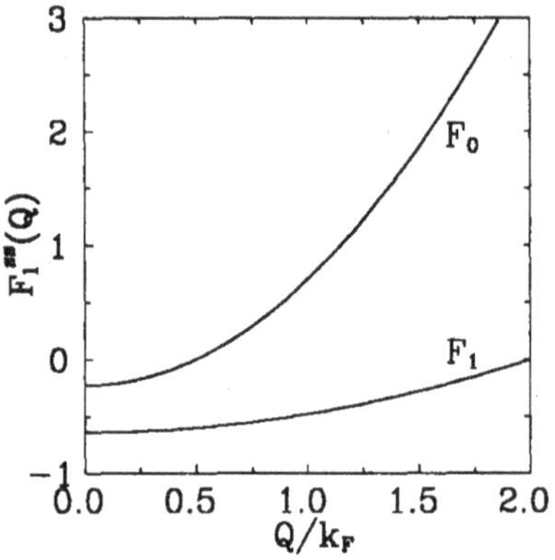

Fig 3. Same as Fig.2 for the SkM* force.

so clear at finite temperatures, since in such a case, one can also get low energy excitations promoting quasiparticles whose momenta lie inside the Fermi sphere.

Heiselberg, Pethick and Ravenhall[5,6] have outlined a possible generalization to finite temperatures of the monopolar and dipolar Landau parameters, valid for ph strenghts $f_l^{\sigma\tau}(k,k',0)$ with a quadratic dependence on both k and k', or with no dependence at all. In what follows, we present a brief derivation of the monopolar ss-amplitude that generalizes the zero temperature one and illustrates the outcome for a finite range force.

In order to compute the isothermal incompressibility given by

$$K(T) = 9\rho \left.\frac{\partial \mu}{\partial \rho}\right)_T , \tag{6}$$

one needs the derivatives $\partial n^0_{\vec{k}}/\partial\mu)_T$. Taking into account the variations in the qp energies[11,12] in terms of the ph effective interaction, the following integral equation is established:

$$\left.\frac{\partial n^0_{\vec{k}}}{\partial \mu}\right)_T = -\frac{\partial n^0_{\vec{k}}}{\partial \epsilon_{\vec{k}}}\left[1 - \sum_{\vec{k}'} f_{\vec{k}\vec{k}'} \left.\frac{\partial n^0_{\vec{k}'}}{\partial \mu}\right)_T\right] . \tag{7}$$

Performing a summation over the momentum label $\vec{k}$ one gets

$$\frac{\partial \rho}{\partial \mu} = N_0(T) - \sum_{\vec{k}} F_0^{ss}(k,T)\frac{\partial n_{\vec{k}}}{\partial \mu} , \tag{8}$$

where

$$N_n(T) = \sum_{\vec{k}} k^n \left(-\frac{\partial n^0_{\vec{k}}}{\partial \varepsilon_{\vec{k}}}\right), \tag{9}$$

$$F_0^{ss}(k,T) = \sum_{\vec{k}'} \left(-\frac{\partial n^0_{\vec{k}'}}{\partial \varepsilon_{\vec{k}'}}\right) f_0^{ss}(k,k'). \tag{10}$$

If f_0^{ss} does not depend on (k,k'), $F_0^{ss}(k,T)$ is just $N_0(T)f_0^{ss}$ and Eq.(8) can be easily solved:

$$\left.\frac{\partial \rho}{\partial \mu}\right)_T = \frac{N_0(T)}{1+F_0^{ss}(T)}. \tag{11}$$

This expression leads to an incompressibility modulus

$$K(T) = \frac{9\rho}{N_0(T)}\left[1+F_0^{ss}(T)\right], \tag{12}$$

which is formally identical to the zero temperature case,[11,12] with $N_0(T)$ being the generalization of the density of states at the Fermi surface. Now, if f_0^{ss} is a function of the qp momentum, as it actually happens with the Gogny force, Eqs.(11) and (12) do not hold. However, the kernel $-\partial n^0_{\vec{k}}/\partial \varepsilon_{\vec{k}}$ is a narrow peaked function at $\varepsilon_{\vec{k}_\mu} \equiv \mu$ with a density-independent energy width of $\sim 2.63T$. For sufficiently low temperatures, and provided that $f_0^{ss}(k,k')$ is a smooth function of its arguments, a linear expansion of this function in terms of k' gives rise to

$$F_0^{ss}(k,T) \sim f_0^{ss}(k,k_\mu)N_0(T) + \left.\frac{\partial f_0^{ss}(k,k')}{\partial k'}\right|_{k_\mu} m_1(T)N_0(T), \tag{13}$$

where $m_1(T) = N_1(T)/N_0(T) - k_\mu$.

The quality of this approximation can be tested through Fig.4, where we plot the exact field $F_0^{ss}(k,T)$ given by Eq.(10) as a function of the momentum k, together with the approximated one taking only the first term of Eq.(13), for densities $1/2\rho_s$, ρ_s and $2\rho_s$ and temperature equal to 9 MeV. The inclusion of the linear term produces a curve which is practically indistinguishable from the exact one within the scale of the plot. It is seen that the agreement is remarkably good. Under these conditions, we intend a solution of Eq.(8) based on treating the variable k in $F_0^{ss}(k,T)$ on an equal footing as k'. This means that on the linear expansion on k of expression (13)

$$F_0^{ss}(k,T) \sim N_0(T)\left[f_0^{ss}(k_\mu,k_\mu) + \left.\frac{\partial f_0^{ss}}{\partial k}\right|_{k_\mu,k_\mu}(k-k_\mu) + \left.\frac{\partial f_0^{ss}}{\partial k'}\right|_{k_\mu,k_\mu} m_1(T)\right], \tag{14}$$

we assume that the characteristic value of $k-k_\mu$ inside the integral in (8) is identical to $m_1(T)$, whose non-zero value is due to the fact that the distribution $-\partial n_k/\partial \varepsilon_k$ is not symmetrical with respect to k_μ. With such an assumption, Eq.(8) can be solved leading to an incompressibility modulus

$$K(T) = 9\rho \frac{1 + N_0(T)\left[f_0^{ss}(k_\mu, k_\mu) + 2\,(\partial f_0/\partial k)|_{k_\mu}\, m_1(T)\right]}{N_0(T)}, \tag{15}$$

which has the same structure as Eq.(12). The role of $F_0^{ss}(T)$ in Eq.(12) is here played by $N_0(T)$ times the square bracket, which in turn corresponds to the linear expansion of $F_0^{ss}(k,T)$ (taken from (14)) evaluated at $k_\mu + m_1$, labelled hereafter as $\bar{F}_0^{ss}(k_\mu + m_1, T)$. As it is shown in Fig.5, the exact incompressibility modulus computed on the $\mu\rho$-isotherms and the approximation (15) are in very good agreement up to moderate high temperatures. Fig.5 also includes the calculation performed taking only $f_0^{ss}(k_\mu, k_\mu)$, i.e., the first term in the square bracket of Eq.(15). This was a reasonable approximation to calculate $F_0^{ss}(k,T)$ but turns out to be rather poor to compute the incompressibility. This effect indicates the necessity of averaging the effective interaction over the smeared Fermi surface.

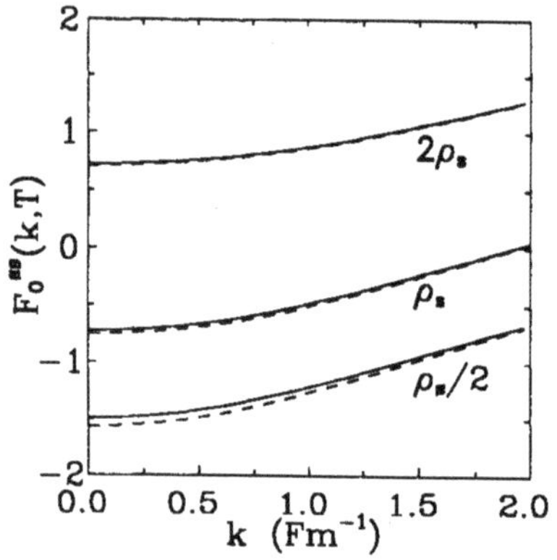

Fig 4. Comparison of the exact field strength $F_0^{ss}(k,T)$ (solid lines) with the first term of the r.h.s of Eq.(13) (dashed lines), for the Gogny force as a function of k at several densities and $T = 9$ MeV.

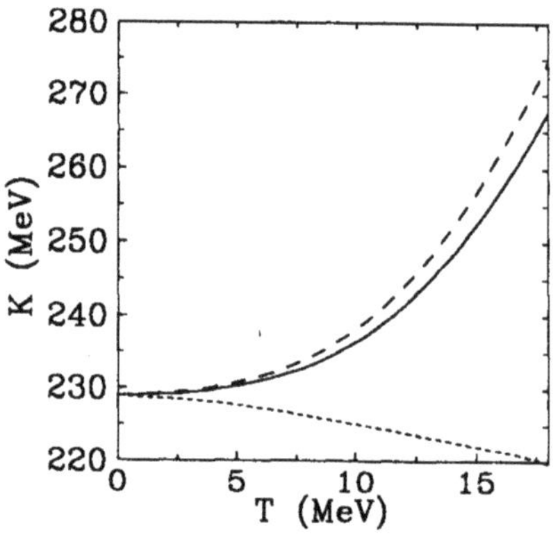

Fig 5. Incompressibility modulus (solid line), compared with the aproximate ones: (a) Eq.(15) (long-dashed line), (b) taking only the first term in the square bracket of Eq.(15) (short-dashed line).

In Fig.6, $\bar{F}_0^{ss}(k_\mu + m_1, T)$ is reported as a function of the density for several temperatures. Eq.(15) also provides a thermodynamical stability criterium, namely $\bar{F}_0^{ss}(k_\mu + m_1, T)$ must be larger than -1 at any given temperature. As far as $\bar{F}_0^{ss}(k_\mu + m_1, T)$ is a good approximation to the exact field, the previous condition should correspond to the spinodal points defining the threshold for unstable density fluctuations. As it is shown in Fig.7, this stability criteria gives rise to spinodal densities which are in very good agreement with the exact ones.[9]

Furthermore, one may consider perturbations acting on the spin-isospin channels - i.e., the system is subjected to polarization in a magnetic field, or to neutron-proton asymmetry. In either situation, the natural finite temperature expressions of the Landau parameters seem to be

$$F_0^{\sigma\tau}(k,T) = \sum_{\vec{k}'} \left(-\frac{\partial n^0_{\vec{k}'}}{\partial \epsilon_{k'}}\right) f_0^{\sigma\tau}(k,k',0). \tag{16}$$

In terms of these Landau parameters and under the same approximation criteria that give rise to Eq.(13), the corresponding thermodynamic response parameters can be computed using the zero temperature formulae with $N_0(T)$ and $\bar{F}_0^{\sigma\tau}(k_\mu + m_1, T)$.

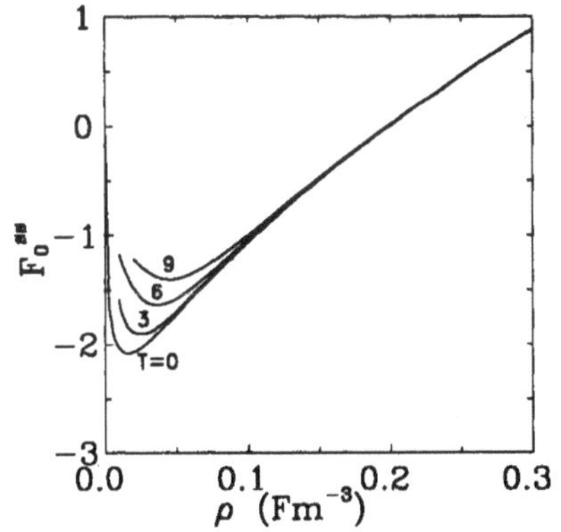

Fig 6. Dimensionless Landau parameter as a function of the density, $\bar{F}_0^{ss}(k_\mu + m_1, T)$ for the Gogny force at several temperatures.

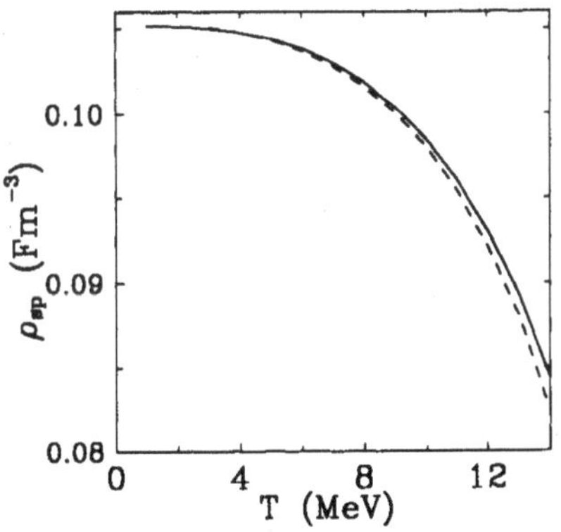

Fig 7. Spinodal densities as a function of the temperature calculated thermodynamically (solid line) and using the generalized stability criteria (dashed line).

4. Conclusions

The finite range effects of an effective two nucleon interaction, here represented by a D1-Gogny force, have been investigated. Opposite to the bulk features, both the *qp* properties and the *ph* effective interaction are rather sensitive to the finite range. This brings as a consequence, that the dynamical response of nuclear matter together with the two *qp*-scattering properties will be affected by this short scale.

We have also investigated the possibility of linking the macroscopic response as measured by thermodynamic magnitudes like compressibility, to the corresponding effective microscopic field. Since the Landau theory of Fermi liquids yields an afirmative answer to the above question at zero temperature, we have examined a temptative extension to finite temperature of these major ideas, along the lines suggested in Ref. 6 for zero range forces. We have shown that the Landau zero temperature relationship among the computed isothermal incompressibility and the microscopic fields can be appropriately generalized to finite temperatures. For the particular case of the Gogny force, the agreement is good up to 10 MeV, the discrepancies being smaller than 2%.

This work has been supported by DGICYT, PB92/0761, (Spain).

References

1. *The Nuclear Equation of State*, Eds. W. Greiner and H. Stöcker, New York, plenum 1982.
2. Proceedings of the International Workshop on Dynamical Fluctuations and Correlations in Nuclear Collisions, *Nucl. Phys.* **A545**, Aussois, France, 1992
3. H. Heiselberg, C.J. Pethick and D.G. Ravenhall, *Phys. Rev. Lett.* **61** (1988)
4. S. Pratt, *Phys. Rev.* **A42** (1990) 7447
5. H. Heiselberg, C.J. Pethick and D.G. Ravenhall, *Nucl. Phys.* **A519** (1990) 279c
6. H. Heiselberg, C.J. Pethick and D.G. Ravenhall, *Ann. Phys.* **223** (1993) 37-74
7. E. Suraud, Ch. Gregoire and B. Tamain in *Progress in Particle and Nuclear Physics*, Ed. A. Faessler, Vol.**23** (1989) 357
8. J. Dechargé and D. Gogny, *Phys. Rev.* **C21** (1980) 1568
9. J. Ventura, A. Polls, X. Viñas, M. Pi and S. Hernandez, *Nucl. Phys.* **A545** (1992) 247c
10. H. Song, G. Zheng and R. Su, *J. Phys.* **G16** (1990) 1861.
11. D. Pines and P. Nozières, *The Theory of Quantum Liquids*, Benjamin, New York, 1966
12. G. Baym and C.J. Pethick, *Landau Fermi-Liquid Theory*, Wiley, New York, 1991
13. J. Ventura, A. Polls, X. Viñas and S. Hernandez, *Submitted to Nucl. Phys.*
14. T.H.R. Skyrme, *Phil. Mag.* **1** (1956) 1043
15. J. Bartel, P. Quentin, M. Brack, C. Guet and H.B. Håkansson, *Nucl. Phys.* **A386** (1982) 79

Particle Production

CLASSICAL AND QUANTUM COHERENCE IN BREMSSTRAHLUNG †

Jörn Knoll

Gesellschaft für Schwerionenforschung (GSI)
Postfach 110 552
D-94220 Darmstadt, Germany

Abstract

Recent considerations about the importance of coherence effects for bremsstrahlung in dense matter (Landau - Pomerančuk - Migdal - effect) are presented and extended for the application to QCD transport and parton kinetic models. Low energy (soft limit) theorems are discussed and pocket correction formulae for the in-matter radiation cross sections are derived using the experience from an exactly solvable hard-scattering model. In contrast to recent derivations by others we require a stronger in-medium suppression factor of the form $((\omega\tau)^4/(1+(\omega\tau)^2)^2$ where ω is the energy of the radiated quantum and τ is the mean free time between successive collisions. This stronger suppression of e.g. gluon radiation implies an even weaker stopping power of high energy partons than discussed by Wang and Gyulassy.

1. Introduction

With the growing interest in relativistic nucleus-nucleus collisions considerable effort has been put into the understanding of the properties of hot and dense matter. Of immediate interest were the properties of quasi-particles in dense matter, their propagation and scattering formulated through self energies, spectral functions and in-medium collision terms. Comparatively little has been done to account for changes of radiative processes in dense matter. Rather most of the dynamical schemes employ

†also presented at the conference on "Strong Interactions at Finite Temperature", ITP Santa Babara, California, Aug. 1993, and at the "Workshop on Pre–equilibrium Parton Dynamics in Heavy Ion Collisions", LBL Bekeley, California, Aug. 1993

the *incoherent quasi-free collision limit* (IQF), with on-shell particles and incoherently added contributions from different microscopic collisions on the basis of the free production cross section. That is, one adds intensities rather than amplitudes [1, 2]. Among others the IQF–picture ignores coherence effects like those related to multiple scattering, i.e. related to the finite *mean free time* τ or *mean free path* λ between successive collisions.

Over the last years it became quite apparent, that a proper understanding of QCD transport and therefore a justification of parton kinetics requires careful investigations and estimates of the soft part of all processes. These non-perturbative effects require either particular resummation schemes or guidance from other appropriate limits as for instance the classical limit of field theory. Well known is the Rutherford type of singularity in scattering cross-sections when the force is mediated by the exchange of a space-like zero mass quanta (photon/gluon). In dense matter the infinite range of the force is screened (Debey-screening) inducing a finite real mass of this quantum. An other case where singularities occur are radiation or absorption processes. Induced by free scattering (bremsstrahlung) they are singular in the soft limit, i.e. in the limit of zero four-momentum $q = (\omega, \vec{q}) = 0$ of the radiated quanta, due to the infinite time available before and after scattering. In dense matter only *finite free* propagation times τ are available and the corresponding rates become regular. This effect (LPM) has first been discussed by Landau, Pomerančuk, Migdal[3, 4] (and many others later) for bremsstrahlung from fast electrons traversing matter. They found a reduction of the radiation amplitude for the contributions from intermediate scatterings such that the terms from the initial and final state propagation dominate (factorization limit). This effect which is entirely in the scope of classical electrodynamics is the subject of this paper. While the LPE–effect discusses how to cure the infra–red 'catastrophe', we like to discard contributions from the asymptotic state propagation (initial and final scattering states) and concentrate on systems where the radiation entirely arise from acceleration or scattering processes in matter. This applies for example for the radiation of photons from a dense and confined gas (e.g. hadron gas) or plasma, for the neutrino cooling of neutron stars, or, since there are no colored asymptotic states in QCD, for the production (or absorption) of gluons in a quark-gluon plasma.

Apart from the study of Migdal [4], who gave an analytical result for the infra–red problem, most of the papers deal with the coherence–time effect only on a qualitative level in terms of inequalities or estimates. Due to time dilation this effect becomes important with increasing energy of the scattered charged particle. Its comprehension is vital for a proper description of the space–time behaviour of high energy hadronic [5, 6] and nuclear collisions [7, 8, 9], and concepts of history–dependent cross sections, creation – and formation times were introduced. While the formation time refers to the time needed to form composite particles like hadrons from the 'elementary' partons, a quite complex issue on its own, we are here concerned with the coherent creation of elementary fields and the corresponding coherence time problem.

With this contribution I like to present some simple considerations that clarify the nature of bremsstrahlung in dense matter. In section 2 we discuss the classical and quantum radiation formulae for some classes of physical systems and arrive at analytical constraints for the in–matter radiation. Subsequently we use the experience from an exactly solvable quantum system and derive a pocket formula that appropriately corrects the false *quasi–free* prescription. We conclude with some remarks for the applicability to the QCD inspired parton picture and compare to results that have recently been given by others [10, 11].

2. Bremsstrahlung and its limitations

The basic approximation employed throughout this paper is that the radiation field couples only perturbatively (to lowest order) to the object (source) system, while the source can interact in any non-perturbative way. We further assume the classical interaction time $\tau_{\rm int}$ to be short (we take it as zero) relative to the typical time τ between successive scattering. With the interest in the production of vector particles we take the example of electrodynamics, but when appropriate comment on the possible applications to QCD.

The number of photons with energy ω, momentum $\vec{q}$ and polarization $\vec{\epsilon}$ can be given by the square of an invariant amplitude (or matrix elements in the quantum case)

$$d\mathcal{N} = \frac{1}{4\pi^2} \left|M(\omega, \vec{q}, \vec{\epsilon})\right|^2 2\delta(\omega^2 - \vec{q}^{\,2})\, d\omega d^3 q. \tag{1}$$

For off-shell photons, e.g. dilepton production, the same invariant amplitude enters, however at off-shell $\omega, \vec{q}$ together with the intermediate photon propagator and the decay matrix elements. In classical electrodynamics the radiation results from the time integration over the entire classical path $\vec{x}(t)$ with velocity $\vec{v}(t)$ of the accelerated charge (assumed to be point-like)[12]

$$M_{\rm cl}(\omega, \vec{q}, \vec{\epsilon}) = e \int_{-\infty}^{\infty} dt\, \vec{\epsilon}\vec{v}(t) \exp\{i\omega t - i\vec{q}\vec{x}(t)\}. \tag{2}$$

This classical expression has been used by many others so far to study the LPM effect. We use it only for some illustrative examples, since in the general case it is quite problematic and has to be evaluated with considerable care †.

Acceleration during limited time

It is instructive to look at classical scattering cases, where a charge incident with four-momentum p_i is accelerated only during a limited time between space-time points x_i and x_f, to the final four momentum p_f, c.f. fig. 1, left part. At large ω and/or $\vec{q}$ all individual accelerations are resolved and the production rate approaches the incoherent sum of the bremsstrahlung from the individual accelerations.

†First, the integration extends over the whole time interval from $-\infty$ to $+\infty$ and has to be taken with a convergence prescription of the form $\exp[-|t/t_0|]$ with $t_0 \to \infty$. Further on, the invariance of this expression is not obvious at first glance, since it is written in a special frame dependent gauge, the radiation gauge, where $\{\epsilon_\mu\} = (0, \vec{\epsilon})$. The point is that all non-invariant parts of the integrand conspire to a delta-function at $q = 0$ for the integral which does not contribute to finite ω.

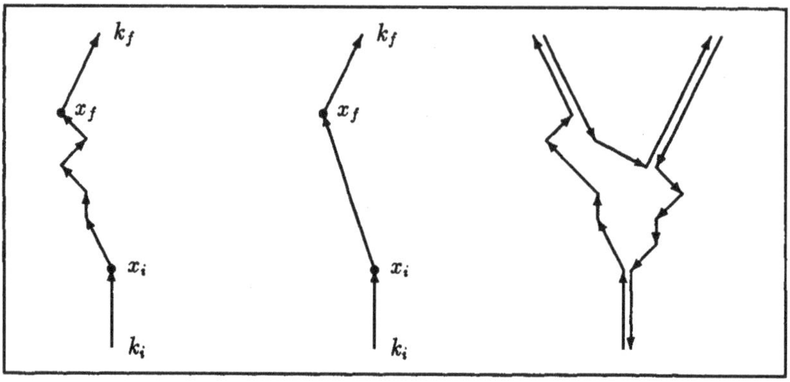

Figure 1: Examples of classical paths in space–time with accelerations during a limited time. Left and middle part for a single charge; right part for a neutral particle that dissociates for a finite time and finally forms two neutral particles (the line sense (up– or downwards) denotes the sign of the charge)

Using four-vector notation and Einstein's summation convention for the Lorentz indices (in greek), we decompose the amplitude into three parts

$$M(q,\epsilon) = e\left\{\frac{\epsilon p_i}{qp_i}e^{iqx_i} + \int_{x_i}^{x_f} \epsilon_\mu dx^\mu e^{iqx} - \frac{\epsilon p_f}{qp_f}e^{iqx_f}\right\}. \tag{3}$$

In the soft limit, where $|(x_f^\mu - x_i^\mu)q_\mu| \ll 1$ one safely can ignore the intermediate integral together with all phase differences and obtains the factorization limit where the amplitude merges that of single scattering process with the common infra-red divergence. While the LPM studies focuses on this infra-red problem, we are interested in the radiation that arises just from the passage of the charge at intermediate time steps†. Apart from an overall phase factor $\exp(iq\bar{x})$ with $\bar{x} = (x_i + x_f)/2$ the intermediate integral can be expanded into powers of q due to the finite time interval. The zero order term $\int_{x_i}^{x_f} dx = x_f - x_i$ is path independent and suggests a decomposition of M into two invariant parts (omitting the arguments q and ϵ) $M = M_2 + \Delta M_{\rm path}$, with the idea that M_2 depends only on the initial and final paths, i.e. on (p_i, x_i, p_f, x_f). M_2 can simply be given by the straight-line connection (fig. 1, middle part) between x_i, while x_f, $\Delta M_{\rm path}$ corrects for the particular intermediate path. About this decomposition we can state the following theorem:

All terms singular and constant in q of the $M_{\rm cl}$ are contained in M_2 and thus depend only on (p_i, x_i, p_f, x_f). The path dependent part ΔM is a regular function in photon four-momentum q and vanishes linearly with $q \to 0$.

†At this point one may be tempted to assign each term of the integral (3) to the radiation amplitude arising from the segments of propagation. This however is not permitted since each term by itself is non-invariant!

Since the path-independent parts survive all stochastic treatments, they require a coherent evaluation in any case, there is no way out. However, there is the hope that under certain kinetic and stochastic conditions the averaging of the path-dependent parts leads to an *incoherent* result such that mutual interferences cancel. This is what one is looking for, since it opens a kinetic description even for the softer, τ-dependent part of bremsstrahlung.

Therefore we focus on cases where all asymptotic and path-independent parts cancel. For scattering problems this occurs only accidentally. Luckily there are some relevant cases where this cancellation happens strictly. One of them is when all incident and leaving particles are neutral, i.e. they dissociate only for a finite time, fig. 1, right part; thus:

The bremsstrahlung-amplitude of a scattering problem with only neutral initial and final states is regular in q and vanishes linear for $q \to 0$.

This is a very important result for the application to QCD, where all asymptotic states are *color-neutral.* Gluon bremsstrahlung, rather being infra-red divergent is *regular* and its rate requires a regularisation of *four powers* in q relative to the IQF limit.

Acceleration during all times

Systems of finite size, where all charges experience accelerations at all times, cannot be studied as above due to the infinite time integration of the classical law (2)†. Examples are systems in statistical equilibrium such as plasmas, either self-bound or enclosed in a vessel. Void of asymptotic free motion there is already no infra-red problem. Rigorous conclusions about the low q behavior in this case can be drawn from the corresponding quantum transition rate M_{QM} as shown by R. Lenk and myself [13]

$$
\begin{aligned}
|M_{QM}|^2 &= \sum_{\mathrm{f}} |M_{\mathrm{if}}|^2 \, 2\pi\, \delta(E_{\mathrm{i}} - E_{\mathrm{f}} - \omega) \mathrm{d}t \quad \text{with} &(4)\\
M_{\mathrm{if}} &= \int \mathrm{d}^3 r\, e^{i\vec{q}\vec{r}}\, \vec{\epsilon}\, \vec{j}_{\mathrm{if}}(\vec{r}) \quad \text{and} \quad \vec{j}_{\mathrm{if}}(\vec{r}) = \left\langle \mathrm{f}\, | \vec{j}(\vec{r}) |\, \mathrm{i} \right\rangle, &(5)
\end{aligned}
$$

expressed by the matrix-elements of the current density $\vec{j}(r)$ between the exact stationary eigenstates $|\,\mathrm{i}\,\rangle$ and $|\,\mathrm{f}\,\rangle$ of the system. Due to charge conservation Siegert's theorem transforms M_{if} into

$$M_{\mathrm{if}} = -iq_\mu \int \mathrm{d}^3 r\, e^{i\vec{q}\vec{r}} (\vec{\epsilon}\vec{r})\, j^{\mu}_{\mathrm{if}}, \tag{6}$$

which displays the desired property:

†The exponential function in (2) cannot be expanded in powers of ω since for any finite ω there are t such that $|\omega t| \gg 1$. Note also that (1) is infinite for trivial reasons, since it gives the total number of photons. This can be cured normalizing (1) by $1/t_0$ to arrive at a rate, where $\exp(-|t/t_0|)$ is the convergence factor for (2)

Steady–state systems of finite size have a photon spectrum, where $\|M_{QM}\|^2$ is regular and vanishes at least quadratically with $q \to 0$.

In ref. [13] we went even further and discussed sum-rules which follow from either form of $\vec{j}_{if}$ as a quantitative tool to control approximation schemes for the rate [14, 13]. These sum rules are entirely classical in nature and assure all above considerations be valid also in the classical limit for systems of finite size governed by time-independent hamiltonians.

Experience from an exactly solvable model

While the above theorems only give some analytical constraints for the rate, the derivation of its explicit form requires a coherent evaluation of an increasing number of collisions towards the soft limit.

Recently we have presented a clarifying study [13] of an exactly solvable quantum mechanical model for which the many–body wave functions are known in closed form [15, 16]. It is a non–relativistic hard scattering model in one space dimension, where distinguishable constituents interact mutually via a repulsive hard–core potential of infinite strength. For this model the quantum transition rates could be brought into simple closed forms in the limit of large quantum numbers (semi-classical limit) even for a large number of interacting particles. For details I refer to our paper [13].

Only the essential experience and results are summarized here. Ignoring coherence effects, that arise from the collective motion of many source particles, the radiation of a single charge due to its restricted zig–zag motion in a neutral collision environment is studied. The analysis of the resulting quantum mechanical rates leads to the following simple and expected picture. The finite lifetime for the particle to stay in its state, either before or after scattering due to other interactions implies a *damping* of the quantum mechanical wave function with the mean free time or path. The point is that damping has to be accounted for on the level of amplitudes rather than on the level of probabilities. This is well known in scattering theory, where optical waves are introduced to account for the damping, giving rise to distorted wave transition matrix elements. The same is true here. Quite generally this effect can be accounted for by the corresponding imaginary parts of the self energy and medium–dependent form factors of the source particle. Thus wave vectors become complex and one has to evaluate the production T–matrix with correspondingly complex momenta. Transferred to the classical amplitude (2) this means, that for each single scattering it has to be evaluated with corresponding exponential damping factor before and after scattering.

Two effects are observed: The first is a genuine quantum effect (which is not seen on the level of the classical amplitude), namely that conservation laws for the binary processes are softened and can be violated in energy and momentum on scales of $1/\tau$ and $1/\lambda^{\dagger}$. The second refers to the above discussed reduction of the in-medium

†For the model, this is seen for momentum conservation only, since it has quasi–particles of sharp energy.

production rate relative to the IQF–rate. We express this by a quenching factor C by which the IQF–rate has to multiplied in order to obtain the proper in–medium contribution from each collision. The correction factor can be quite simply determined in that frame which is the most appropriate for scattering, namely the *Brick–wall frame.* In this frame the scattered charge just reverses its velocity such that $\vec{v}_{\rm i} = -\vec{v}_{\rm f} = \vec{v}$. With all quantities defined in that frame one obtains

$$C(\omega, \vec{q}) = \frac{\left(\omega^2 - (\vec{v}\,\vec{q})^2\right)^2 \tau^4}{\left(1 + \left(\omega^2 - (\vec{v}\,\vec{q})^2\right)\tau^2\right)^2 + 4(\vec{v}\,\vec{q})^2\tau^2}, \tag{7}$$

which for vanishing photon momentum $\vec{q} = 0$ simplifies to

$$C(\omega\tau) = \left(\frac{(\omega\tau)^2}{1 + (\omega\tau)^2}\right)^2. \tag{8}$$

These expressions are valid in the semi–classical limit. An other condition is that the mean free path before and after scattering is the same in that brick–wall frame, which may be fulfilled in many cases. So far we did not succeed in a more general derivation†. We also like to mention that the soft behaviour of the correction factor C is general, yet its specific analytic form at finite q is tied to the zero range interaction considered here.

3. Discussion and perspectives

Altogether the correction factor (7) provides an extension of the *incoherent quasi–free* approximation from the ultra violet limit, where $C \to 1$, down to the soft limit with appropriately modified production cross sections. It does not only regularize the infra–red divergence of the free rate, but it also produces the right q dependence in the soft limit as required from the analytical considerations of section 2. For simplicity we refer to eq. (8)) for the subsequent discussion. With decreasing mean free time τ an increasing part of the low energy spectrum is affected by this reduction such that the former $1/\omega$ singularity of the quasi–free rate changes to a regular ω^3 behaviour in matter. The factor C reflects the duality between frequency and time. The lower the photon frequency ω, the longer is the corresponding coherence time for this process and therefore the longer the mean free time τ has to be in order that the process occurs quasi–free. The result is quite intuitive and shows that part of the coherence phenomena (not all) can be accounted for by a medium corrected production cross section, just accounting for quenching factor C. This is true at least for the low density limit for which the results have been derived. Since in this limit the radiating system behaves nearly classically, these corrections of the *productive* parts of the cross section occur already even though the corresponding *scattering* part of the cross section may still be unchanged in the medium and be still the free one. The result illustrates in

†The problem is that this condition is fulfilled for the solvable model, and the naive extension to more general cases leads to path–independent terms, which we know should cancel on the long time average.

particular that for zero mass fields, like the gluons in a quark–gluon plasma, there are no infra–red divergences in matter, since the required time scales are not available. Such singularities only occur for asymptotic states as they exist in QED but not in QCD.

There is no problem to extend our QED results to QCD–inspired parton kinetics. The essential condition used in our demonstration, the conservation of charge, is also fulfilled for color in QCD. I see presently no complication due to the non–abelian character of QCD. The only point to observe is, that in terms of diagrams, for example, only those different diagrams add coherently which are identical in their asymptotic states. This implies that interference occurs only in case that the gluon is radiated off the 'continuous' line of one color. In this respect gluon–exchange interactions have to be treated like charge–exchange interactions in QED (soft interactions lead to strong accelerations of the color charges and the brick–wall frame is the mid–rapidity frame), while pommeron exchange behaves like a 'normal' (i.e. neutral) interaction in QED. Since in a plasma the gluon acquires a finite screening mass, one may argue that the effect discussed here is of no importance because for real gluons one never reaches the soft point. One sees however, that the correction factor C is general and applies to any mass of the radiated quantum and it is a matter of the kinematical conditions to see whether damping is not equally important. For illustration we discuss some cases on scales relevant for nuclear collision systems at beam energies of a few hundred MeV per nucleon where nucleons are the constituents of the system. For a mean free path of $\lambda = 2$ fm we compare the incoherent quasi–free production rate with the medium corrected rates which include the factor C from eq. (8) in fig. 2. The dashed curves

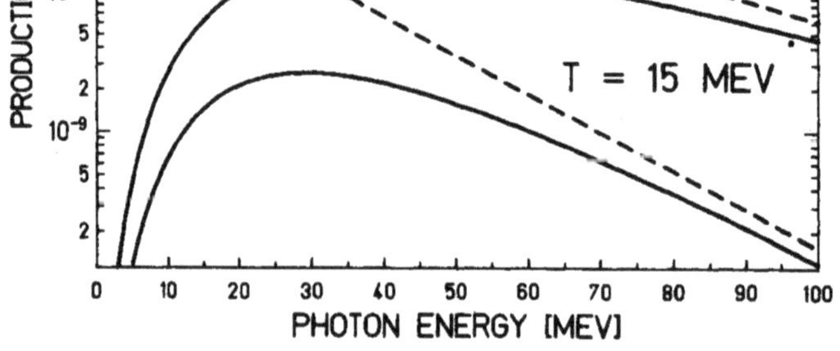

Figure 2: Production rate for the binary scattering of the charged particle with its neutral neighbour for T = 15 and 30 MeV. Coherent result: full lines, incoherent quasi–free approximation: dashed lines.

give the quasi–free result. We see the importance of the medium correction (full lines) for low energetic photons. Corrections of more than a factor two occur for energies up to 60 MeV. For typical cases in parton kinetics the corresponding time scales are about a factor 10 shorter so that the corresponding suppression effect can be read off increasing the energy scale by a factor 10.

Recently some studies have been presented addressing to the same problem by Cleymans et.al. [10] and by Wang and Gyulassy [11]. Their results deviate from ours, in that both obtain a suppression factor which is essentially the square root of ours. Since this contradicts the analytic requirements discussed in section 2, I have to comment on their derivations. Cleymans et. al. studied the radiation of photons from an equilibrated hadron gas. In their stochastic treatment of the classical production rate they essentially kept the terms which are *path-independent* and which have to be treated coherently *for all times.* As shown for the quantum rates these parts of the amplitude entirely cancel for equilibrated systems. To see this cancellation on the basis of the classical formulas is very subtile. Wang and Gyulassy considered a fast parton that randomly changes its color due to scattering. The problem in their derivation is that they considered this parton with its random charge isolated from the rest of the system. Thus gauge invariance is violated and their results are not at all invariant. As color is conserved in QCD, complementary color changes occur in the target with corresponding contributions to the gluon amplitude. In this respect it is very important to realize that *the soft limit* $q \to 0$ *coincides in all frames.* This implies that the projectile part of the gluon amplitude interferes with that from the target in the soft limit, and this interference is destructive due to charge conservation! The interesting aspect in this connection is that the authors have used their estimate of the gluon bremsstrahlung for the stopping power of high energy partons. The account of the correct reduction of gluon bremsstrahlung as given here implies an ever weaker stopping power than predicted by Wand and Gyulassy.

One of the major questions is, how the medium effects discussed here can be implemented in kinetic simulation models of nuclear collisions at high and ultra–relativistic energies for the production of hadrons or partons. The central problem is to devise a kinetic scheme that includes the possible *off–shell* propagation of the constituents. Due to time dilation this problem becomes important with increasing beam energy. Therefore for multiple collision processes in high energy physics theoretical and phenomenological concepts in the form of eikonalized multiple collision amplitudes, modified rescattering cross sections and the account of creation and formation times have been introduced and studied. I am not expert enough to estimate the details of the different particular models and to judge how properly this effect is already included in the present schemes of hadron and parton cascades models in particular for the application to the collision of two nuclei. The study presented here opens the perspective for an analytically based concept rather than just employing sharp cut–offs as done many times so far.

Estimates of collision times in actual calculations for the collision of two nuclei at high energies show that for many production processes these coherence effects are significant and it would be of great importance to find a dynamical scheme that improves the presently available kinetic models in that respect in particular in a regime where the relevant degrees of freedom change from hadronic to partonic ones.

Acknowledgments

I thank G. Bertsch, P. Danielewicz, B. Friman, M. Herrmann, J. Kapusta, P. Henning, and V. Toneev. I enjoyed stimulating and clarifying discussions with J. Cleymans, K. Geiger, M. Gyulassy, B. Müller and X.N. Wang during the summer '93 studies at the ITP in Santa Babara, at the LBL Berkeley and at the INT in Seattle in particular about the differences in the results at that time.

References

[1] G. Bertsch, Prog. Part. Nucl. Phys. **4** (1980) 483; G. Bertsch and S. Das Gupta, Phys. Rep **160** (1988) 189

[2] C. Cassing, V. Metag, U. Mosel and K. Niita, Phys. Rep. **188** (1990) 363

[3] L. D. Landau and I. Pomerančuk, Dokl. Akad. Nauk SSSR **92** (1953) 553, 735; also in Collected Papers of Landau, ed. Ter Haar (Gordon & Breach, 1965) papers 75 - 77; V. M. Galitsky and L. I. Gurevich, Nuovo Cim. 32 (1964) 396

[4] A. B. Migdal, Sov. Phys. JETP **5** (1957) 527

[5] N. N. Nikolaev, Sov. Phys. Usp. **24** (1981) 531

[6] F. Niedermeyer, Phys. Rev. **D34** (1986 3494

[7] A. Białas and M. Gyulassy, Nucl. Phys. **B291** (1987) 793

[8] K. Geiger, Phys. Rev. **D46** (1992 4965 and 4986

[9] K. Werner, Habil–thesis, University of Heidelberg, to be published in Phys. Rep.

[10] J. Cleymans, V.V. Goloviznin, and K. Redlich, Phys. Rev.**D47** (1993) 173 and 989; and Z. Phys **C** to be published

[11] X.N. Wang and M. Gyullassy, Phys. Rev.**D** to be published

[12] J. D. Jackson, Classical Electrodynamics (Wiley, New York, 2nd ed., 1975), sect. 14.5

[13] J. Knoll and R. Lenk, GSI–preprint 93–13, submitted to Nucl. Phys. **A**

[14] J. Knoll and C. Guet, Nucl. Phys. **A494** (1989) 334;
M. Durand and J. Knoll, Nucl. Phys. **A496** (1989) 539

[15] H. A. Bethe, Z. Phys. **71** (1931) 205

[16] E. H. Lieb and W. Liniger, Phys. Rev. **130** (1963) 1605

PARTICLE PRODUCTION IN RELATIVISTIC HEAVY ION COLLISIONS

W. Kühn
II. Physikalisches Institut
Universität Gießen, Germany

for the TAPS collaboration

Abstract

Particle production in relativistic heavy ion collisions in the BEVALAC/SIS energy regime is discussed . Meson production probabilities follow simple scaling laws with energy and number of participants. In contrast, the shape of pion transverse momentum distributions exhibits strong variations with system size, pointing out the importance of multi-step production processes and final state interaction.

1. Introduction

One of the main goals of relativistic heavy ion Physics in the BEVALAC/SIS energy regime is the study of the equation of state of hadronic matter. This field has received strong attention both theoretically[1] as well as from the experimental point of view[2], since it is of fundamental importance not only for nuclear Physics but also in astrophysics. Despite considerable progress, the problem is still open and requires further work. Information on the EOS can basically be derived in two complementary ways, involving the investigation of collective flow and related phenomena[3] as well as the study of particle production, which is the subject of this article.

Of particular interest are particles which are created during the collision, since they carry information on the dynamics of the dense, hot hadronic medium where they are produced. For energies up to 2 AGeV, light pseudoscalar mesons[4] (π,η,K) as well as antiprotons[5] have been investigated. At lower incident energies, below the π^0 - threshold in the nucleon-nucleon system, direct photon production (bremsstrahlung) has been observed. In contrast to all other probes, photons and to some extent K^- undergo no significant final state interaction and permit direct experimental access to the production dynamics. In contrast, the information obtained from pions is influenced by hadronic final state interaction. As a consequence, the analysis is more complicated but provides additional information about the expansion phase. Density-dependent effects are expected to be sizeable, since large compression factors (3 times normal nuclear matter density for central Au + Au collisions at 1 AGeV) are predicted[6] .

In the following, we shall concentrate on neutral pion - production at SIS/GSI studied with the TAPS, the two-arm photon spectrometer[7] . The two-photon decay of neutral mesons offers the unique advantage of a high quality measurement down to zero transverse momentum.

Experimentally, measurable quantities include production probabilities, transverse momentum spectra and rapidity distributions as a function of bombarding energy, target/projectile mass and impact parameter.

2. Experimental Techniques

TAPS is a modular detector system consisting of 256 BaF_2 - detectors arranged in 4 blocks with individual charged - particle plastic veto detectors. The blocks are arranged in two movable towers positioned symmetrically with respect to the beam direction to detect the decay of neutral mesons emitted near mid-rapidity. Meson identification is done via invariant mass analysis from the measured angles and energies of coincident photons. Photon / hadron discrimination is done via time-of-flight with respect to a start detector as well as by pulse shape analysis. The combinatorial background from uncorrelated photons is obtained from event mixing and subtracted to derive transverse momentum (p_T) - spectra. The overall mass resolution is 15 % FWHM.

3. Experimental Results

The inclusive production cross sections can be factorized into the reaction cross section, the average number of participant nucleons and the production probability per participant nucleon. Fig. 1 shows the production probability per participant nucleon for a large number of different systems as well as for different particles as a function of the incident energy.

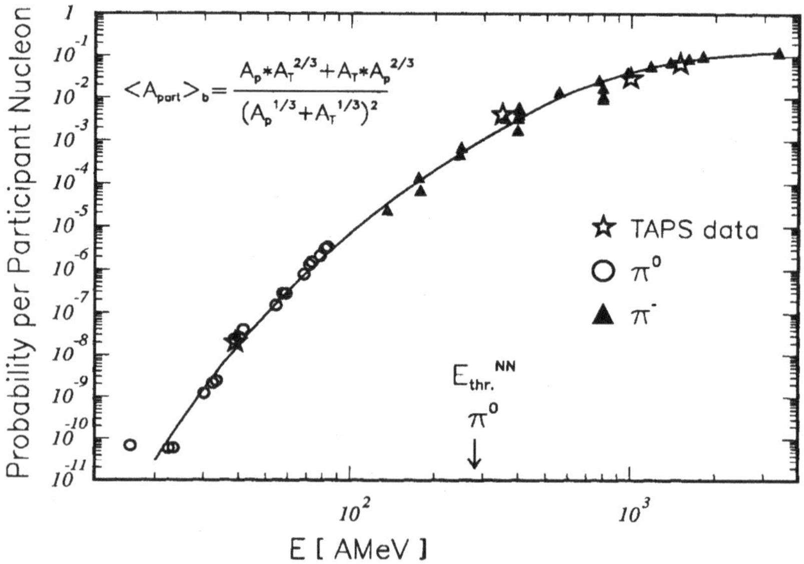

Fig. 1:
Pion production probability per participant nucleon, as a function of the beam energy per nucleon.

This systematics includes inclusive data from systems with different target/projectile combinations.

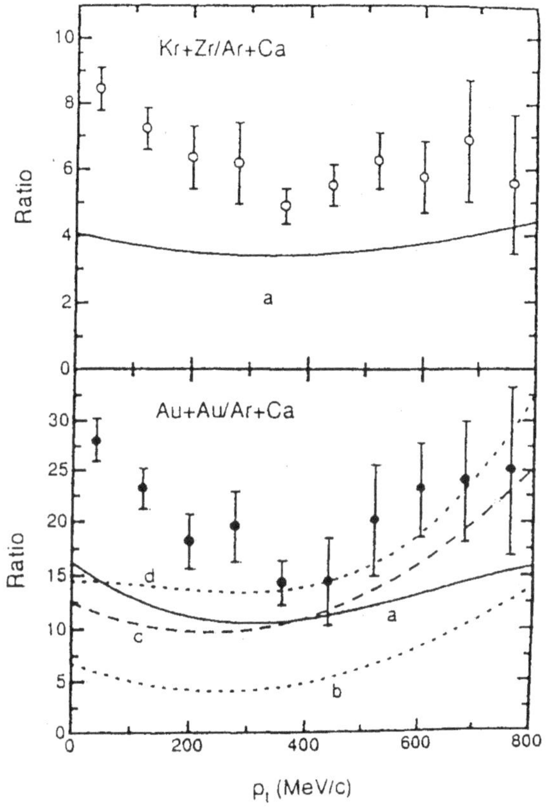

Fig. 2:
Ratio of transverse momentum spectra of neutral pions from Au+Au/Ar+Ca and Kr+Zr/Ar+Ca collisions at 1 AGeV. Also shown are various model predictions: a=BUU(W.Ehehalt et al.), b=IQMD (S. Bass et al.) c=RQMD(H.Sorge et al.) d=QGSM(N.S.Amelin et al.)

At first glance, the scaling with the number of participant nucleons is striking, since it seems to suggest that absorption effects which depend strongly on the size of the system should not play a role. This observation would clearly contradict the known experimental results from pion absorption on cold nuclei[8]. To resolve this question, a closer look to the spectral distribution of the pions is necessary. Fig. 2 shows the ratio of measured transverse momentum π^0 distributions of Au + Au / Ar + Ca and Kr + Zr / Ar + Ca collisions at 1 AGev. For the heavier systems, there are enhancements of the ratio at small and large transverse momenta . Also shown are predictions from various model calculations, which all fail to reproduce the experimental result, in particular for the low transverse momentum region.

4. Discussion

Several effects need to be considered which might influence the shape of the transverse momentum distribution as a function of system size. First of all, the initial conditions will be different, when comparing large and small collision systems. The size, density and lifetime of the central highly compressed zone will be different. Since most of the pions are assumed to originate from resonance decays, the relative population of nucleon resonances will affect the shape of the spectra.

Most important, absorption effects should play an important role. From pion scattering experiments, it is well known that pions are strongly absorbed in cold nuclear matter. Except for radiative capture, pion absorption does not occur on an isolated nucleon, since this process violates energy/momentum conservation. The lowest order absorption process is shown in fig. 3.

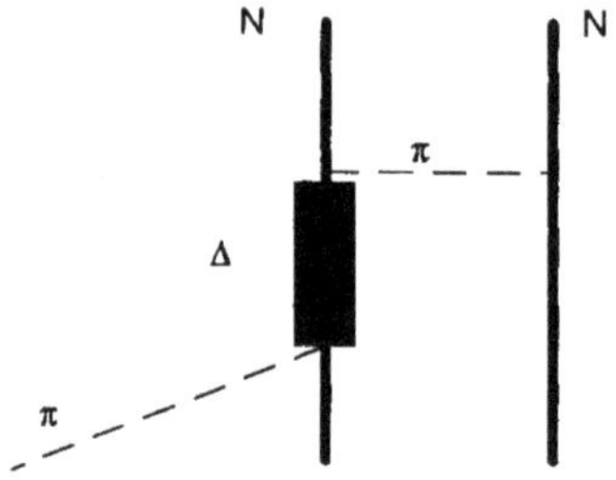

Fig. 3:
Lowest order process for pion absorption. Absorption on a single nucleon is impossible, except for radiative capture.

As a consequence, the pion absorption cross section should be strongly (at least quadratically) dependent on the nuclear matter density. This has an important consequence : since the density of nuclear matter decreases rapidly during the expansion phase of the collision, pion absorption will take place only on a time scale which is sufficiently short to allow a considerable fraction of the primarily produced pions to escape. Moreover, since pion absorption is momentum dependent - with a maximum in the region of the delta resonance - a quantitative study of the shape of transverse momentum spectra of high mass collision systems might provide direct access to the time scale of the expansion phase, which itself is related to the EOS. Considering a scenario with slow expansion, a strong quenching of pions with momenta near the absorption maximum would be expected. In contrast, a stiff EOS resulting in faster expansion would lead to a much more rapid decrease of the nuclear matter density with smaller absorption .

It is obvious that a detailed analysis of such effects requires models featuring a realistic description of the nuclear matter distribution during the expansion phase. This includes also the distribution of cold spectator matter in semi - central collisions. From the experimental point of view, more exclusive experiments along the line of ref. 9 are required, as well as a comparison of probes with different absorption properties such as eta and K mesons.

Part of the data shown in this article is taken from the PhD thesis of O.Schwalb[10]. The author would like to acknowledge fruitful discussions with W.Cassing, H.Löhner, V.Metag, U.Mosel and Gy. Wolf. This work was supported in part by GSI under contract GI Met K, BMFT under contract 06GI174I and by FOM.

5. References

1. U. Mosel, Ann. Rev. Nucl. Part. Sci. **41**(1991)29
   J. Aichelin, Phys.Rep. **202**(1991)233
2. R.Stock, Phys.Rep. **135**(1986)259
3. H.A. Gustafson et al., Phys.Rev.Lett. **52**(1984)1590
   H.H. Gutbrod et al., Phys.Rev. **C42**(1990)640
4. J.W. Harris et al., Phys.Rev.Lett. **58**(1987)377
   F.D.Berg et al., Z.Physik **A340**(1991)297
   D.Brill et al., Phys.Rev.Lett. **71**(1993)336
5. A.Schröter et al., Nucl.Phys. **A553**(1993)775c
6. W.Cassing et al. Phys.Rep. **188**(1990)363
7. R.Novotny, IEEE Trans. Nucl. Sci. **38**(1991)379
8. D.Ashery and J.P.Schiffer, Ann. Rev. Nucl. Part. Sci. **36**(1986)207
9. L.Venema et al., Phys.Rev.Lett. **71**(1993)835
10. O.Schwalb, PhD Thesis Universität Gießen 1993

Subthreshold Meson and Dilepton Production - Consequences of Resonance Matter Formation*†

U. Mosel

Institut fuer Theoretische Physik

Universitaet Giessen, D–35392 Giessen, Germany‡

Abstract

Based on results of semiclassical transport theories that incorporate explicitly π and η mesons and three nucleon resonance states I review the results obtained for pion and eta production in heavy-ion collisions. I then present the theoretical evidence for formation of *resonance matter* in such collisions and discuss its observable consequences, first, the enhancement of far-subthreshold particle production cross-sections, and, second, the presence of a pion-annihilation component in dilepton spectra. In the last part of my talk I briefly discuss electromagnetic formfactors of baryons and mesons, their in-medium changes and observable consequences.

1 Pion and Eta Production

In heavy-ion reactions in the GeV/u energy regime pions are copiously produced [1]. Since it is well known that at higher energies nearly all of the pions originate from

*Based on work with W. Cassing, W. Ehehalt, M. Schäfer and G. Wolf

†Work supported by BMFT and GSI Darmstadt

‡Email: mosel@piggy.physik.uni-giessen.de

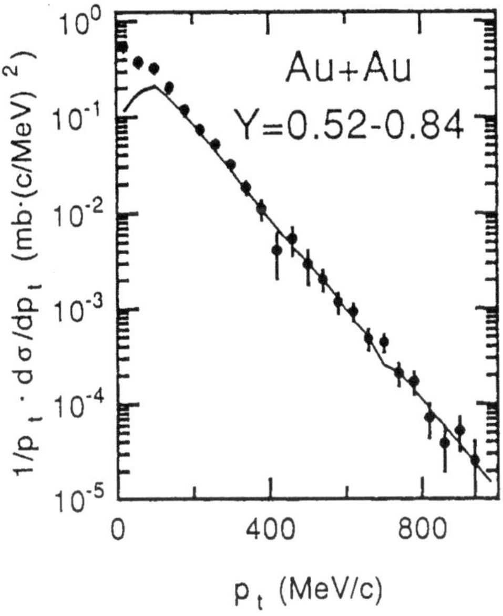

Figure 1: Transverse-momentum spectrum of neutral pions produced in a ^{197}Au + ^{197}Au collision at 1 GeV/u. The data were obtained by the TAPS collaboration at SIS [7]. The curve gives the results of a BUU calculation, using the method of Wolf et al. [3].

the decay of the Δ resonance and are frequently reabsorbed and reemitted on their way to the detector, a realistic theoretical treatment of such processes requires the simultaneous, coupled description of nucleons, Δ's and pions [2]. Such descriptions, based on semiclassical transport theories, are nowadays available [3, 4, 5]. These descriptions all use the BUU-equation [6] as a starting point and propagate explicitly mesons and baryons which are coupled through inelastic collision terms in addition to the usual Boltzmann-like collision term of the BUU-equation. In Fig. 1 I show a typical result obtained with the method of Wolf et al. [3] for the spectrum of neutral pions in the collision $Au + Au$ at 1 GeV/u. It is seen that the spectral shape is very

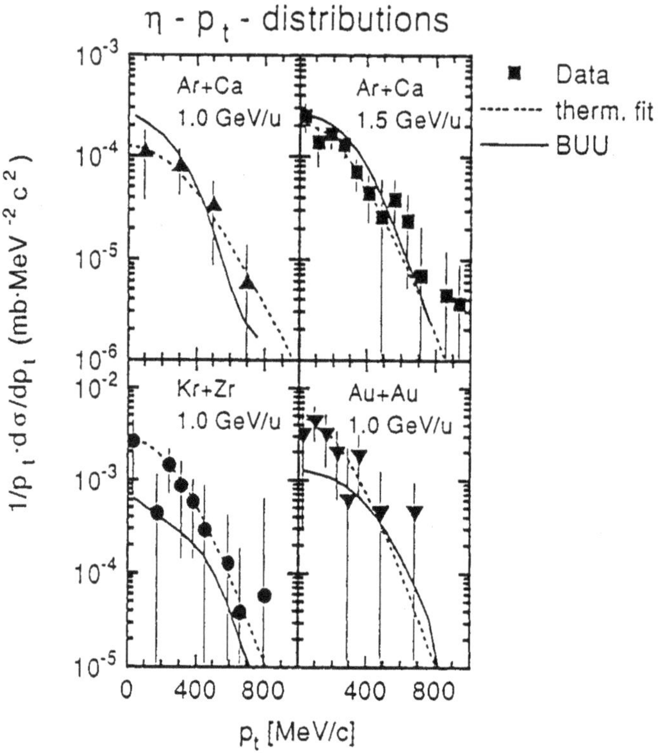

Figure 2: Transverse-momentum spectra for η-mesons in comparison to data obtained by the TAPS collaboration [8] for the systems and energies indicated.

well reproduced, except for very small transverse momenta p_t (note that the plotted spectrum is weighted by the factor $1/p_t$ and thus accentuates this discrepancy).

Besides the Δ resonance and the $N(1440)$ resonance, that are important for pion production and absorption, the calculations by Wolf et al. also include the $N(1535)$ resonance that at these energies is the most important source for η mesons. The η meson yield has therefore also been calculated; the results for the spectra are shown in Fig. 2. The overall agreement is very good. The remaining discrepancies may be due to our insufficient knowledge of the energy-dependence of the elementary η production

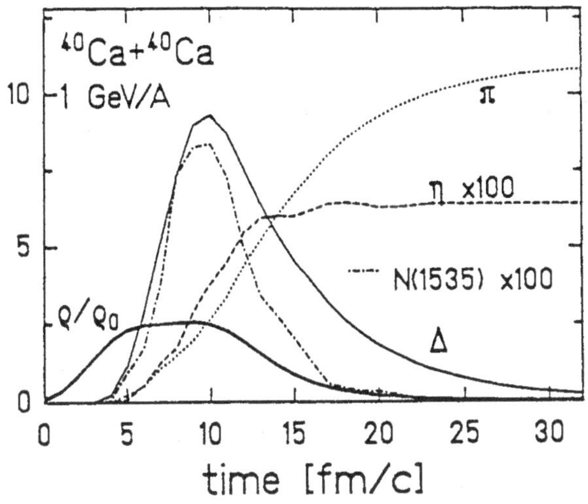

Figure 3: Time-dependence of the Δ and $N(1535)$ resonances and the pion and eta yields as a function of time during the course of the collision (from [3]).

cross section close to threshold and to a still rather rough description of the $N(1535)$ resonance properties in the calculations used. η-mesons are interesting because they freeze out earlier than pions and thus may still carry some information about the dense collision zone outwards to the detector. This can be seen in Fig. 3 which shows that after the initial collisions first the Δ and the $N(1535)$ resonances are populated on roughly the same time-scale. The deexcitation, however, is quite different for both resonances. The Δ resonance can be reexcited through pion-absorption on a nucleon even in the later, cooler stages of the collision because of its relatively small excitation energy. The $N(1535)$ resonance, on the contrary, requires rather high-energetic pions and etas for its reexcitation which exist only in the early preequilibrium phase of the collision. Thus, the $N(1535)$ resonance cannot be significantly reexcited, the η mesons consequently experience much less reabsorption and thus freeze out earlier. Note that this phenomenon is not connected to any special physical properties of the

η meson, but only to its higher mass; high-energy pions would show a very similar behavior.

2 Resonance Matter

Since in the calculations pions, η's and Δ's are – besides the nucleons – explicitly propagated, we can analyze their space-time behavior during the collision. The relative population of the baryon species in the central reaction volume is shown in Fig. 4 as a function of bombarding energy. At about 2 GeV/u about 30% of all baryons are Δ's. Since at this energy the baryon-density reaches a value of about 2 - 3 ρ_0, we have for the Δ-density

$$\rho_\Delta = 0.3\rho_B = 0.3(2-3)\rho_0 \approx (0.6-0.9)\rho_0 \tag{1}$$

The density of the Δ resonances is thus nearly that of the nucleons at the nuclear saturation density. In a heavy system, like $Au + Au$ this system of dense Δ's lives for about 10 fm/c, i.e. for about 6 lifetimes of a free Δ [9]. This allows us to speak of the formation of *Resonance Matter*, a new form of nuclear matter, in such reactions [9].

2.1 Observable consequence: subthreshold particle production

An observable consequence of the formation of resonance matter in relativistic heavy-ion collisions is an enhancement of far-subthreshold particle production in such collisions. It was, for example, shown by Lang et al. [10] and by Batko et al. [11] that the cross-sections for the subthreshold production of K^+ mesons and antiprotons in the energy-regime of 1 - 2 GeV/u were dominated by ΔN (K^+) and $\Delta\Delta$ processes ($\bar{p}$.) These are typical many-body processes. For example, in the case of K^+ production two nucleons first collide and form a Δ which then collides with a third nucleon. In

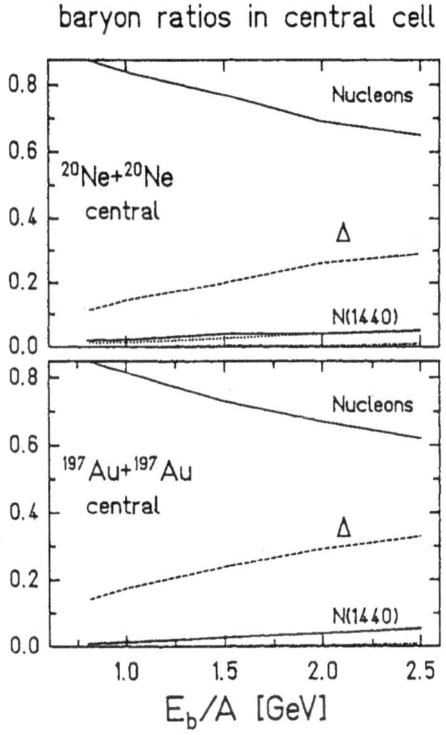

Figure 4: Relative population of nucleons and nucleon resonances in the central reaction cell as a function of bombarding energy (from [9]).

this subsequent collision not only the kinetic energy of the reaction partners, but also the mass-difference between Δ and nucleon is available for meson production: the Δ has acted as an energy storage device. In total, three nucleons participate in the production process. Because the Δ is produced with a rather small kinetic energy, its collision will take place very close to the location of the first NN collision, so that the production process goes like ρ^3. Similarly, in the case of $\bar{p}$ production two Δ's are produced in two NN collisions, and these two Δ's then collide with each other to form the antiproton; in this process two times the Δ-nucleon mass-difference is available for the production. In this case the production process goes like ρ^4. Both subthreshold processes thus sample selectively the high-density region of the collision volume.

Obviously, such processes can play a major role only if enough resonances are created in the first collisions and if the high resonance density survives long enough for secondary collisions to take place. The observed magnitude of the subthreshold particle production cross sections is thus a direct indication of the formation of resonance matter. This is reminiscent of a well-known phenomenon in low-energy nuclear physics: the observed enhancement of subtreshold heavy-ion fusion cross-sections over the values expected from simple one-dimensional barrier penetration models found its explanation in the coupling to other (shape) degrees of freedom. In both cases the physical system thus exploits internal degrees of freedom.

3 Dilepton Production

We have seen in the preceding section that Δ resonances are abundantly produced in relativistic heavy-ion collisions. Besides the enhancement of subthreshold particle production processes a direct consequence of this formation of resonance matter is the high multiplicity of pions produced at these energies. This high pion multiplicity actually allows one to study pion-pion interactions and, in particular, pion annihilation reactions. In this section I will briefly review what we have learned about dilepton production in heavy-ion collisions and will show that the pion-pion annihilation process dominates the high-energy part of the observed dilepton spectra.

Pioneering experiments by the DLS group at the BEVALAC in Berkeley have proven that dilepton spectra can be measured even at the relatively low bombarding energies at that accelerator [12]. Data have been obtained both for heavy-ion induced reactions and for proton-induced reactions in the energy range from 1 GeV/u up to nearly 5 GeV/u. They are given in the form of invariant-mass spectra where the invariant mass of the electron-positron pair is just that of the (virtual) photon before pair production. Obviously, the electron and the positron can be emitted back-to-back, so that events with total momentum zero, but a nonzero invariant mass are possible. Because of four-momentum conservation at the electromagnetic photon-dilepton vertex also the photon carries no momentum, but energy; it is therefore a time-like photon that is involved in dilepton production.

Several groups [3, 5] have performed theoretical analyses of the measured dilepton spectra. There is consensus now, that although there are many dilepton-sources, like, e.g., nucleon-nucleon bremsstrahlung, pion-nucleon bremsstrahlung, Δ decay, etc, the heavy-ion spectra are really dominated by only two components: up to an invariant mass of about 500 MeV the Dalitz decay of the η meson dominates. Since the η yield has now been measured by the TAPS collaboration (see discussion above), the dilepton yield from η Dalitz decay can be calculated in a completely model-independent way; it agrees with the theoretical predictions. Above invariant masses of about 500 MeV, the measured spectra indicate a shoulder-like structure. In all the simulations cited [5, 3] this shoulder is assigned to the $\pi^+ - \pi^-$ annihilation process. That there is a measurable probability for this process to take place is due to the high pion-multiplicity in heavy-ion collisions and is thus a consequence of the formation of resonance matter, as discussed above.

A comprehensive comparison of the dilepton invariant-mass spectra for both $p+A$

and $A + A$ collisions for 1 and 2 GeV/u is shown in Fig. 5. Included in this figure are also the results for the heavy system $Au + Au$ for which no measurements exist. It is seen that the overall agreement between theory and experiment is quite good, but that the experimental uncertainties prohibit any more detailed comparison. In particular, the shoulder-hump like structure at invariant masses above about 600 MeV in the theoretical calculations cannot unambigously be identified in the data because of the rather large error bars.

This shoulder is caused by the pion-annihilation component; all other components show an exponential decrease with mass. Its origin lies in the elementary $\pi^+\pi^- \rightarrow e^+e^-$ cross section which exhibits a well-defined peak at an invariant mass of about 750 MeV. This peak is caused by a corresponding structure in the pion's electromagnetic formfactor [13] and reflects the phenomenon of Vector Meson Dominance (VMD) [14]. According to the original VMD hypothesis photons couple to hadrons only via intermediate vector mesons; this hypothesis finds its explanation in the coupling of the photon ($J^\pi = 1-$) to virtual quark-antiquark ($q\bar{q}$ pairs with the same quantum numbers, i.e. to vector mesons. The electromagnetic formfactor of the pion is experimentally well-known and indeed exhibits a strong peak just at about the ρ meson mass [13].

This fact opens the possibility to study the properties of the ρ meson in dense matter, since the pion annihilation component receives most of its contributions from regions where the pion density is high, i.e. from the high-density phase of the heavy-ion collision. There are recent theoretical predictions that the mass of the ρ meson should drop in dense matter as a precursor phenomenon to the restoration of chiral symmetry expected at very high densities; other calculations predict that the mass should stay constant, but that the width of the ρ meson should increase significantly. The only known experimental consequence of such phenomena is a change in the dilepton spectra in the region around the ρ meson mass.

We have, therefore, taken the predictions of one particular theory, namely that

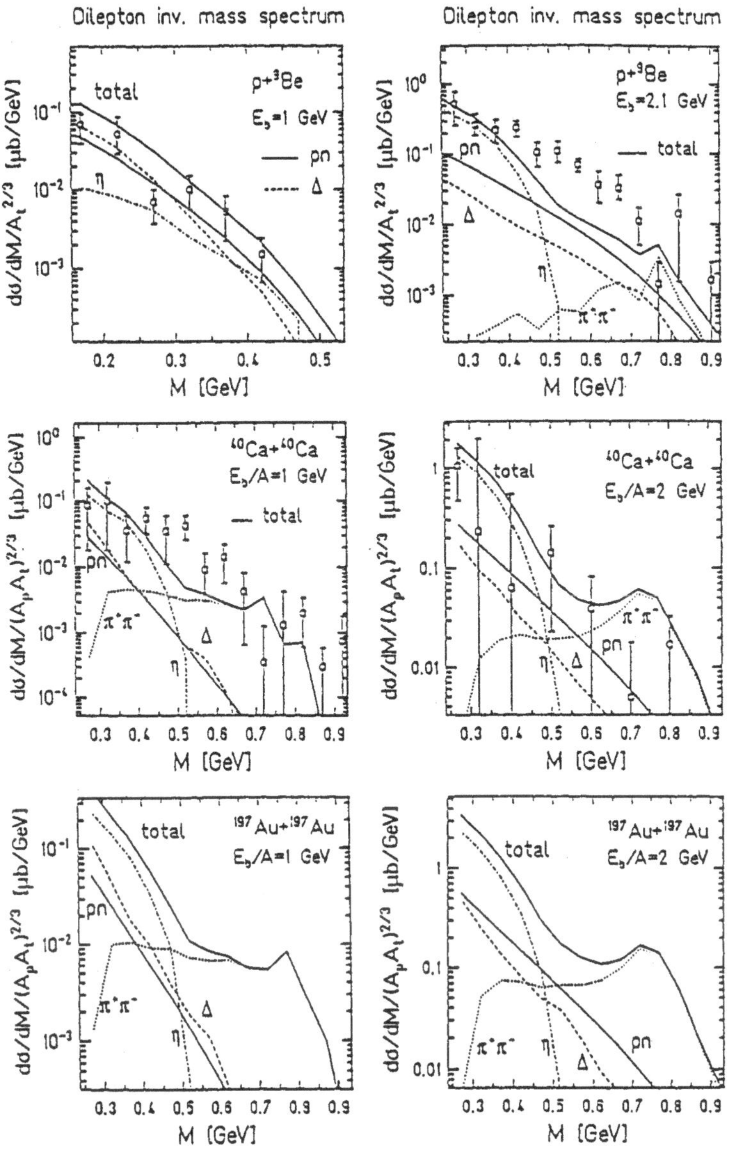

Figure 5: Invariant mass spectra for dileptons for various systems and energies. The data are from the DLS collaboration [12] whereas the lines represent results of calculations of Wolf et al. [3].

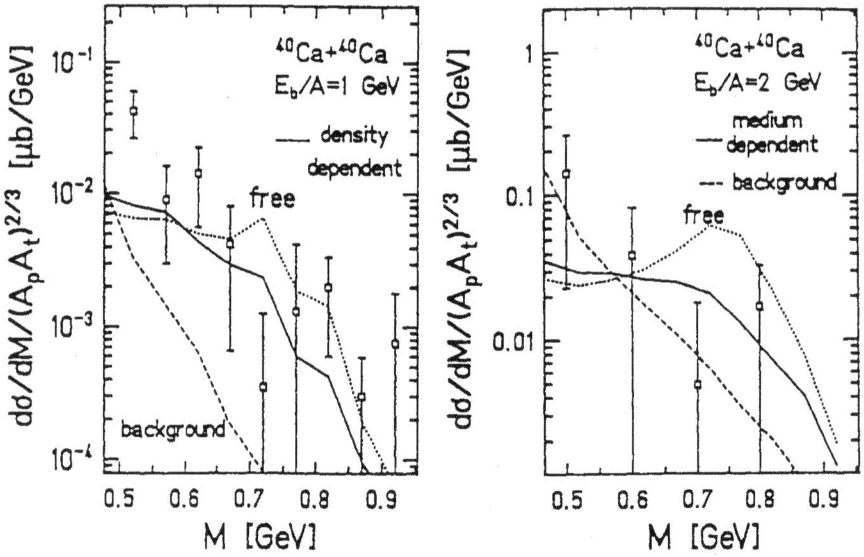

Figure 6: Observable consequences of the in-medium changes of the ρ-meson on dilepton spectra. The two curves represent results of calculations [3] with the free and with the medium-corrected formfactor of Herrmann et al.[15]

of Herrmann et al. [15], which predicts a strong broadening of the ρ even at density ρ_0, and have used this density-dependent formfactor at each annihilation vertex; this is possible since the calculations yield the precise space-time point of, and thus also the density at, each pion-annihilation process. The results of this calculation are shown in Fig. 6. It is seen that around the ρ meson mass the dilepton cross section changes by about a factor of 3 – 4; the shoulder-like structure becomes flatter. Using predictions of other theories like that of Ko et al. [16] which predict a lowering of the ρ meson mass gives an even stronger disappearance of the shoulder. In all of these results one has to remember, of course, that the medium effects on the electromagnetic formfactors have been calculated under the assumption that the ρ meson is embedded in homogenous, static nuclear matter with infinite extensions in space and time. In a real heavy-ion collision, the high-density zone lives, however, only for about 10 fm/c and is localized in the central reaction cell with rather large inhomogenieties at its surface. The effects of this transient nature of the high-density phase on in-medium properties have not yet been investigated. Nevertheless, the results shown in Fig. 6 serve to define the requirements on a next generation dilepton spectrometer, such as HADES.

Dilepton spectroscopy could also give interesting information on the electromagnetic formfactor for baryons where the situation is more complicated since for these the time-like sector is not accessible below an invariant mass of $2m_p$; the formfactor in the space-like sector is well described by the so-called dipole fit [17] whereas a naive VMD model would lead to a monopole fit since the propagator of the vector mesons appears only linearly in the formfactor. Also other evidence points to the fact that the VMD hypothesis is for nucleons not so well established as it is for mesons. Fig. 6 shows that in the proton-induced reaction $p + B$ at 2 GeV bombarding energy there is a clear window for proton-bremsstrahlung which is sensitive just to this formfactor (half-off-shell). Experiments on the dilepton spectroscopy for the elementary pn reaction for which some exploratory data by the DLS group exist now, could thus give information on the timelike part of the nucleon's electromagnetic formfactor.

4 Summary

In this talk I have described how transport calculations predict a high density of Δ resonances in the central reaction volume of a relativistic heavy-ion collision. While a detailed study of the properties of these Δ's and their decay in the future may yield interesting information on in-medium changes of nucleon resonances, the consequences of this resonance matter formation are already now observable in the production cross sections of far-subthreshold particles. I have described how the mass of the Δ resonance may serve as an energy reservoir and that reactions involving one or even two Δ resonances in the entrance channel become the more important for particle production the farther the bombarding energy is below the free production threshold.

A particular manifestation of resonance matter is the high pion multiplicity in heavy-ion reactions and the ensuing possiblity to study pion annihilation in a nuclear-matter environment. I have shown that BUU calculations can well describe the experimental dilepton spectra, but that the search for in-medium effects requires a second-generation dilepton spectrometer with a much higher efficiency.

I gratefully acknowledge many helpful and extremely stimulating discussions with V. Metag. Most of the material covered in this talk is based on work with G. Wolf.

References

[1] R. Stock, Phys. Rep. **135** (1986) 259

[2] U. Mosel, Ann. Rev. Nucl. Part. Sci **41** (1991) 29

[3] G. Wolf, W. Cassing and U. Mosel, Nucl. Phys. **A552** (1993) 549 , and references therein

[4] Bao-An Li and W. Bauer, Phys. Rev. **C44** (1991) 450

[5] L. Xiong, Z.G. Wu, C.M. Ko and J.Q. Wu, Nucl. Phys. **A512** (1990) 772, and references therein

[6] G.F. Bertsch and S. Das Gupta, Phys. Rep. **160** (1988) 189

[7] O. Schwalb et al., Giessen preprint (1993), to be published

[8] F.D. Berg et al., Giessen preprint (1993), to be published

[9] W. Ehehalt, W. Cassing, A. Engel, U. Mosel, Gy. Wolf, Phys. Rev. **C47** (1993) R2467

[10] A. Lang, W. Cassing, U. Mosel and K. Weber, Nucl. Phys. **A541** (1992) 507

[11] G. Batko, W. Cassing, U. Mosel, K. Niita and Gy. Wolf, Phys. Lett. **256B** (1991) 331

[12] G. Roche et al., Phys. Rev. Lett. **61** (1988) 1069 C. Naudet et al., Phys. Rev. Lett. **62** (1989) 2652 G. Roche et al., Phys. Lett. **B226** (1989) 228

[13] T. Ericson and W. Weise, *Pions and Nuclei*, Clarendon Press, Oxford, 1988

[14] J. J. Sakurai, *Currents and Mesons*, University of Chicago Press, 1969

[15] M. Herrmann, B. Friman and W. Nörenberg, Nucl. Phys. **A546** (1992) 267c

[16] M. Asakawa and C.M. Ko, Phys. Rev. C (1993), in press

[17] D.H. Perkins, *Introduction to High Energy Physics*, Addison-Wesley, Reading, 1982

COHERENT PION PRODUCTION INDUCED BY PHOTONS, ELECTRONS, PROTONS AND IONS

E. Oset, P. Fernández de Córdoba, M. J. Vicente-Vacas and J. Nieves

Departamento de Física Teórica and IFIC Centro Mixto
Universidad de Valencia-CSIC 46100 Burjassot (Valencia) Spain.

Abstract

Some processes in which pions are produced in particle nucleus collisions leaving the nucleus in its ground state are discussed. Some features of the coherent production are independent of the projectile used. The fraction of pions produced coherently is about 1/4 of the total production but it is qualitatively very important to explain the shift of the delta peak observed in nuclear reactions induced by protons or light ions.

1 Introduction

Quite a few talks in this workshop have dealt with the question of pion production in heavy ion collisions. Pions are produced as soon as there is enough energy and in their way out some of then are reabsorbed, hence giving their energy to pairs or trios of nucleons and helping redistribute the energy among the nucleons. One interesting topic in pion production is the subthreshold pion production, which requires a certain coherence, or collaboration of several nucleons, in order to proceed [1, 2]. Although with similar qualitative features to the mechanism of [1, 2], our coherent production displays clear qualitative and quantitative features which make it rather different from the mechanisms considered in subthreshold production. We will study coherent pion production in subthreshold production. We will study coherent pion production in (p,n) or (p,p'), $(^3He,t)$ reactions and also in reactions induced by photons and electrons.

The peak of the pion energy distribution appears around the delta resonance region, hence the proper study of this process and its emplacement in a broader context passes through the study of the delta excitation in nuclei. We devote next section to this topic.

2 Δ excitation on nucleus and nuclei induced by protons or light ions.

Experiments on Δ excitation with (p,n) reactions are done in [3] and with the $(^3He,t)$ reaction in [4]. We show some relevant features of this latter experiment.

An overall view to the ($^3He,t$) induced Δ excitation on proton, deuteron and ^{12}C targets is shown in fig. 1

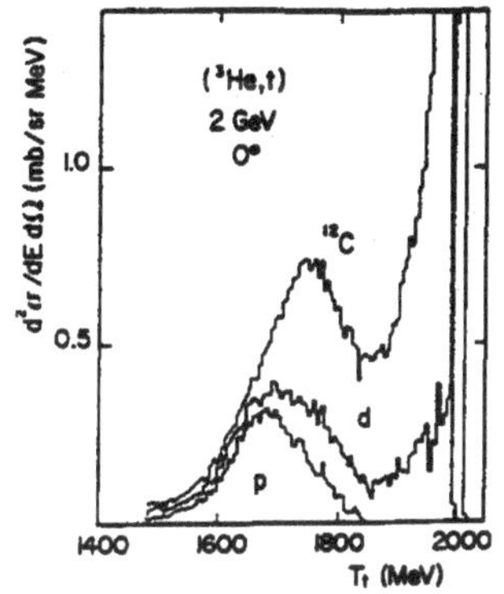

Fig. 1: Double differential cross section for (^{3}He, t) on different layer nuclei as a function of the t kinetic energy.

Two features are worth noting with respect to this figure. The deuteron spectrum is clearly shifted towards higher t energies with respect to the spectrum for p targets. The peak is not much shifted but lots of strength accumulates now at lower excitation energies. On the other hand at the high side of the t spectrum some strength accumulates which is clearly attributable to quasielastic collisions of the type $He\,n \rightarrow t\,p$ and which do not occur with the p targets. A smooth extrapolation of this part of the spectrum below the Δ excitation peak also shows that the new strength of the d excitation function with respect to the proton can not be attributed to these quasielastic collisions and one must find a different interpretation.

On the other hand the peak of the distribution in ^{12}C is clearly shifted toward higher t energies and a considerable amount of strength is also shifted. However, contrary to the case of the deuteron, the quasielasticrpeak is so large that one immediately realizes that its extrapolation below the apparent delta peak must have much to do with the strength of this peak. Hence any serious attempt to explain the inclusive spectra shown in fig. 1 must take the quasielastic channels into consideration.

3 Projectile and target Δ excitation

Let us begin with the deuteron. The deuteron spectrum finds a nice interpretation by considering the delta excitation in the projectile in addition to the excitation in the target [5].

Most approaches [6] [7] neglect the delta excitation in the projectile. The two mechanisms are depicted in fig. 2.

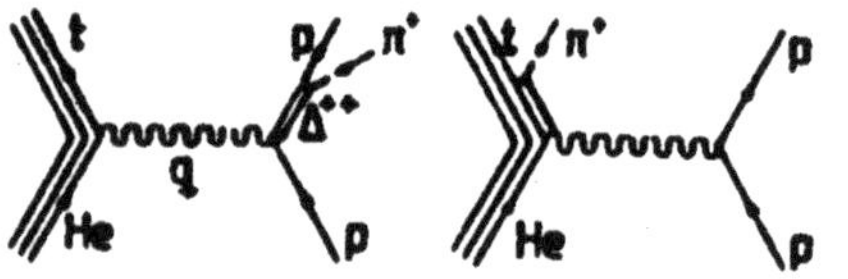

(a) (b)

Fig. 2

a) Delta excitation in the target (DET).

b) Delta excitation in the projectile (DEP).

One of the reasons for neglecting the DEP mechanism is that its strength is small in the $(^3He, t)$ reaction on p targets. This is seen in table 1 which shows a strength 9 times bigger for the DET mechanism than for the DEP one, based only on the counting of isospin coefficients. However, in the same reaction on neutron targets the weight of the DET mechanism becomes three times smaller and the weight of the DEP mechanism three times larger and thus both have now the same weight. This is important to note because the shape of the energy distribution for both mechanisms is quite different. There is no mystery in this latter statement since both shapes are simply related by a Lorentz boost. One can evaluate the DEP mechanism as if it were a DET mechanism for a (p, p') collision on a ^3He target at rest and then make a boost to the frame where the original proton is at rest and the ^3He moving.

The different energy shapes of the DET and DEP mechanisms can be appreciated in fig. 3. The DEP mechanism peaks at higher t energies than the DET mechanism.

In the $(^3He, t)$ reaction on the proton target the DEP mechanism introduces some corrections in the high energy side of the t spectrum but the cross section on a neutron target has a very different shape to the one on the proton target because of the different weights of the two mechanisms.

It is also interesting to see that the simple sum of the predicted $(^3He, t)$ cross section on p and n targets reproduces fairly well the cross section on the deuteron as can be seen in fig. 4, by shifting strength at high t energies but without moving the peak position.

Another way to assess the validity of the ideas exposed above is to carry out experiments with different probes where the weights of the DET and DEP mechanisms are very different. In table I we show the weight of the DET and DEP mechanisms for several reactions. It is remarkable to see that the $(^3He, ^3He)$ reaction is largely dominated by the DEP mechanism and the

(^4He,^4He) reaction is given exclusively by it.

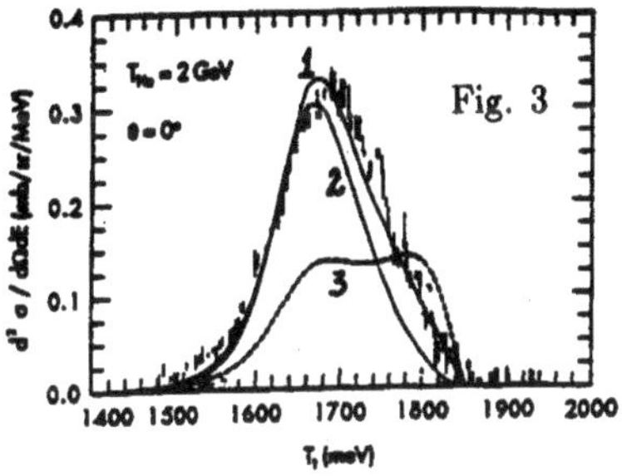

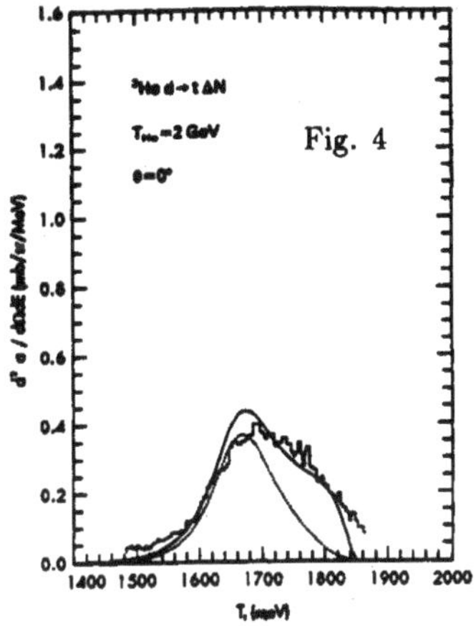

Fig. 3: Double differential cross section in the $p(^3He,t)p\pi^+$ reaction in the p lab system (from ref. [6]).

1. Results including the DET + DEP + s-Wave mechanisms.
2. Results with the DET mechanism alone.
3. Results for the $n(^3He,t)N\pi$ reaction, including the DET + DEP + s-wave mechanism.

Fig. 4: Double differential cross section for ($^3He,t$) on deuteron target at fixed angle as a function of the kinetic energy. Dotted curve: DET + DEP + s-wave.

It is easy to see this last point. As shown in fig. 5, the DET mechanism, a) is forbidden in this latter reaction while the Δ excitation can proceed through Δ excitation in the projectile, b).

Fig. 5

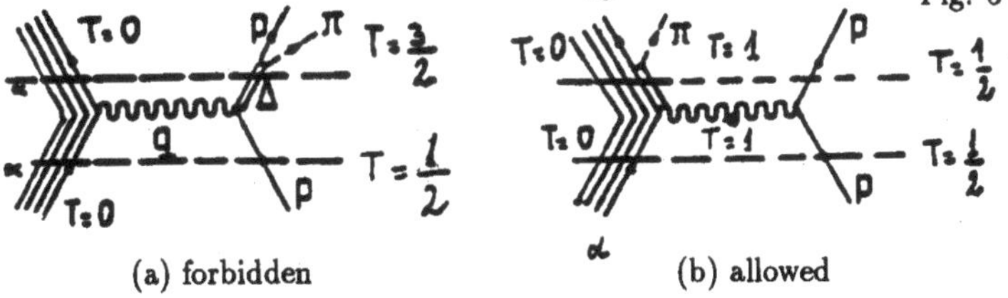

(a) forbidden (b) allowed

TABLE I

	target	projectile
$p(^3He,t)$	2	2/9
$n(^3He,t)$	2/3	2/3
$p(^3He,^3He)$	6/9	134/9
$n(^3He,^3He)$	6/9	86/9
$p(^4He,^4He)$	0	64/3
$n(^4He,^4He)$	0	64/3

In fig. 6 we show the results for the $(^3He,^3He)$ reaction on a proton target calculated in [8]

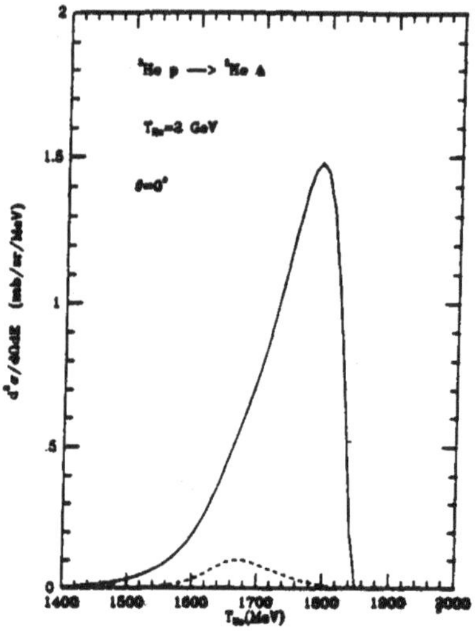

Fig. 6

Fig. 6: Double-differential cross section for $(^3He,^3He)$ on a proton target at fixed angle as a function of the outgoing ^3He kinetic energy. Dashed line:DET mechanism. Full line: DET + DEP + s-wave.

We observe that the DEP mechanism is very important while the DET mechanism provides only a small contribution.

The shapes are however completely different and the peak of the DEP mechanism is shifted by about 140 MeV at higher energies of the outgoing ^3He with respect to the peak of the DET mechanism. The total strength is also much bigger than in the $(^3He,t)$ reaction. It is curious that so much experimental work has been done for the $(^3He,t)$ reaction and none for the $(^3He,^3He)$ which does not require major changes in the experimental techniques. The theoretical results of [8] indicate that the study of this latter reaction is an important complement for the global understanding of the mechanism of delta excitation on nucleons and nuclei.

The curious shape of the DEP mechanism has been confirmed in a recent experiment at Saclay [9]. Although the form factors of the ^3He and ^4He nuclei are a bit different, the experimental shape found in the $(^4He,^4He)$ reaction [9] is remarkably similar to the one shown in fig. 6. We show in fig. 7 the experimental cross section of [9] together with our results with arbitrary normalization.

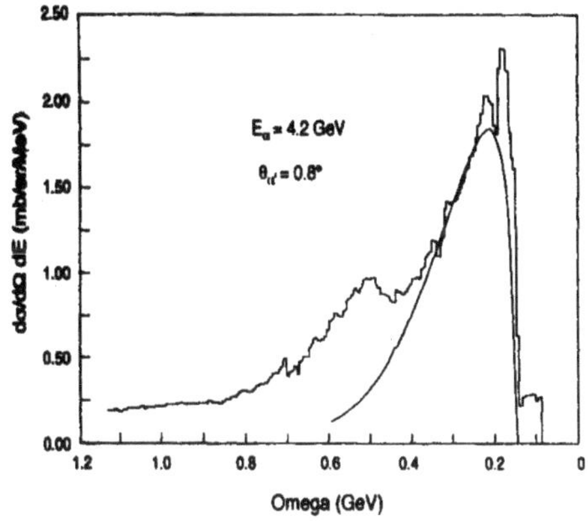

Fig. 7: Cross section for the (α, α') reaction on a proton target. Exp. from [9]. The results are our calculation with the DEP mechanism alone.

It is interesting to see that the shape of the energy distribution is nicely reproduced by the DEP mechanism for delta excitation. The strength and angular distribution are approximately reproduced with the same interaction as for $(^3He, t)$ and considering distortion of the proton and pion waves. Since no distortions were considered in the $(^3He, t)$ reaction on the proton because distortion effects were effectively included in the transition amplitude, the setting of the $(^3He, t)$ and (α, α') reaction in the same framework, with the same elementary interaction, and distortion corrections properly included in both cases, becomes advisable. Thus we have reevaluated simultaneously the $(^3He, t)$ and the (α, α') reactions using the same interaction and considering the proper distortions. The results [10] give a similar shape as in fig. 7 and an angular distribution in the angle α' which we show in fig. 8 and compares favorably with the experimental data within 20%.

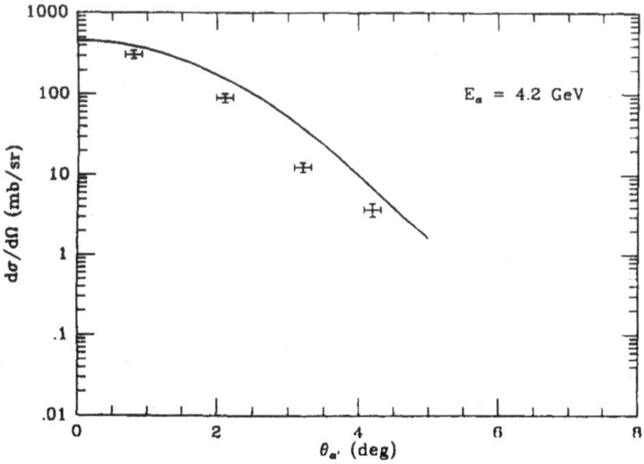

Fig. 8: $\frac{d\sigma}{d\Omega_{\alpha'}}$ for the (α, α') reaction on the proton. Experimental data from [9].

4 The shift of the delta peak in coherent processes.

Now we turn to the old problem of the shift of the delta peak in nuclei [11, 12]. It is instructive to see what happens in analogous reactions in nuclei. In fig. 9 we show the total πA and γA cross sections for ^{12}C compared with the elementary πN and γN cross sections. We observe that the total $\pi\,^{12}C$ cross section (and so is for the elastic channel) has the peak shifted at lower excitation energies by about 60 MeV with respect to the position of the free delta peak. However, there is no shift of the peak for the $\gamma\,^{12}C$ cross section. Although sometimes one hears that there is a trivial explanation for it, I would like to say that these is nothing trivial in this feature. The complexity of the multiple reaction channels has been made evident in a recent work where all the channels are studied microscopically in detail [13]. Furthermore, to say that there is no shift in the photonuclear or electronuclear reactions is misleading, because, as we shall see below, there is a channel, the coherent pion production channel, which has a clear shift, as large as in the πA scattering or in the $(^3He, t)$ reaction.

In order to illustrate this point let us concentrate in πA elastic scattering. One finds a shift of the Δ peak like in the total cross section of fig. 9. Although one can find different explanations for the shift in technical terms I would like to show here qualitatively the reasons for this shift. They come from two sources: the nuclear form factor and the pion distortion. If one has a quasielastic π scattering in the nucleus the reaction probability per unit length is roughly proportional to the πN elementary cross section since there are no constraints on the outgoing nucleon momenta (leaving apart Pauli blocking) and the outgoing nucleon can compensate for the momentum transfer taken by the pion.

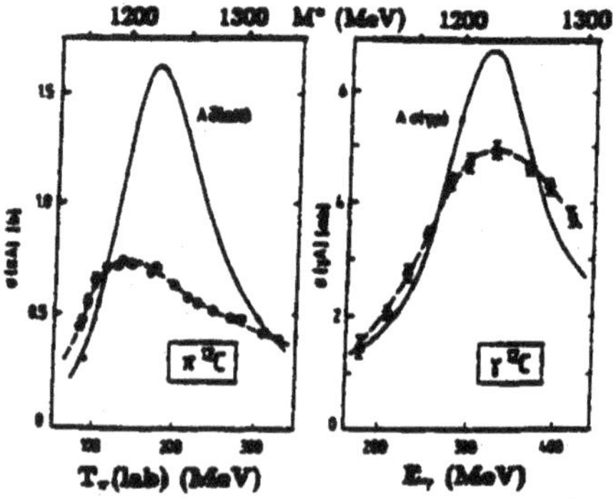

Figure 9: The total cross section for $\pi +^{12} C$ and $\gamma +^{12} C$ reactions are compared. Also shown are the corresponding cross sections on the nucleon times $A(= 12)$. the figure is adapted from the book by Ericson & Weise [28].

In the nuclear elastic collision one finds an extra element which is the nuclear form factor, $F(\vec{k}-\vec{k'})$, since we go from the ground state of the nucleus to the same ground state and the pion momentum transfer has to be taken from the momentum distribution of the nuclear wave function. Now, as the energy of the pion grows, $(\vec{k} - \vec{k'})$ is bigger for the same angle and the form factor smaller. Hence the presence of the form factor tends to decrease the elastic cross section as the energy increases. More important than this is the effect of distortion on the pion waves at energies close to the resonance . Since we are interested in the elastic channel (nucleus in the ground state) every time the pion undergoes a quasielastic collision, or is absorbed, it is removed from the elastic channel and hence the pion wave suffers an attenuation as it passes through the nucleus. Because both the quasielastic collisions and pion absorption have its maximum strength at resonance, the pion loss of flux is maximum at this energy. The net elastic cross section combines the product of the elementary cross section, the distortion and the form factor and the net effect is a shift (and an appreciable one) at lower energies than resonance. Technically this can be envisaged in the eikonal approximation where

$$t_{\pi A}(\vec{k}, \vec{k'}) \propto \int d^3 r e^{i(\vec{k}-\vec{k'})\vec{r}} t_{\pi N}(\vec{k}, \vec{k'}) \rho(\vec{r})$$
$$e^{\frac{-i}{2}\int_{-\infty}^{\infty} \frac{1}{k} t_{\pi N}(\vec{k},\vec{k}) \rho(\vec{b},z') dz'} \tag{1}$$

One finds then an atenuation factor

$$e^{-1/2 \int_{-\infty}^{\infty} \frac{1}{k}(-Im t_{\pi N}(\vec{k},\vec{k})) \rho(\vec{b},z') dz'}$$
$$\equiv e^{-1/2 \int_{-\infty}^{\infty} \sigma_{\pi N} \rho(\vec{b},z') dz'} \tag{2}$$

The last step given by virtue of the optical theorem. (Note that this standard eikonal form does not contain the effect of pion absorption which has to be put in addition).

As $t_{\pi N}$ approaches the resonance, the attenuation factor becomes smaller and $t_{\pi N}$ has a resonant structure. As a consequence, the peak position in $t_{\pi A}$ moves to lower pion energies.

Hence the coherence of the process is responsible for this shift. We can hence look for coherent pion production in electromagnetic processes which is the analogous reaction to π elastic scattering. The shift is also present in coherent (γ, π^0) photoproduction, a feature noticed both in experiments [14] and calculations [15, 16, 17]. In fig. 10 we show the results for the angle integrated (γ, π^0) coherent cross section as a function of the γ energy. We can see that there is a big shift of the peak of the energy distribution with respect to the free delta peak.

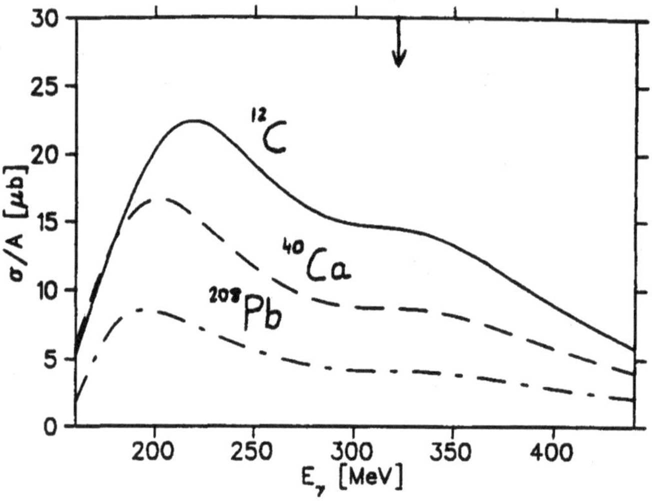

Fig. 10: Cross section per nucleon for (γ, π^0) coherent photoproduction as a function of the γ energy for ^{12}C, ^{40}C and ^{208}Pb. The arrow indicates the position of the free Δ peak. The free γN cross section is of the order of 650 μb.

The coherent (γ, π^0) process has a special feature. Note from fig. 11, that the coupling of the photon to the $N\Delta$ vertex is transverse $\vec{S}^+ \times \vec{k}$, while the coupling of the pion to the $N\Delta$ vertex is longitudinal, $\vec{S}.\vec{q}$. On the other hand, energy is conserved in the process and hence $k = \omega(q)$ which implies $k > q$.

Fig. 11

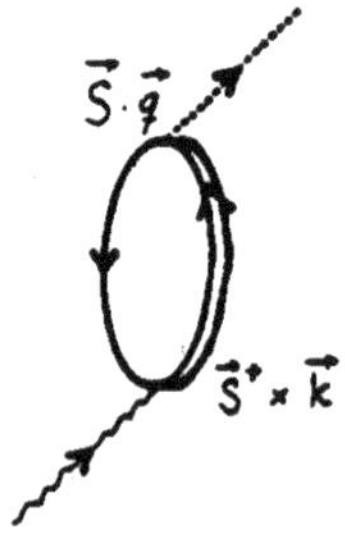

The cross section goes then as:

$$\frac{d\sigma}{d\Omega} = |\vec{k} \times \vec{q}\,|^2 F(\vec{k} - \vec{q}\,)^2 = k^2 q^2 sin^2\theta F(\vec{k} - \vec{q}\,)^2 \tag{3}$$

Now one has to go to finite angles to have non zero cross section but then $|\vec{k} - \vec{q}\,|$ becomes bigger and the nuclear form factor reduces appreciably the cross section. The coherent pion photoproduction cross section is definitely shifted, however, for the reasons given above its strength is small compared to the incoherent channels and can not produce a shift in the total photonuclear cross section.

The same feature can be observed in coherent $(e, e'\pi^0)$ electroproduction. We show the results in fig. 12, from a recent calculation [18]. We observe that the impulse approximation, which contains the effect of the form factor but not the pion distortion, already shows a shift of about 30 MeV with respect to the peak position of a free Δ. On the other hand the distortion of the pion wave induces a further shift of 40 MeV. The strength however is small enough compared to the incoherent (e, e') channels and does not produce a shift in the total (e, e') cross section.

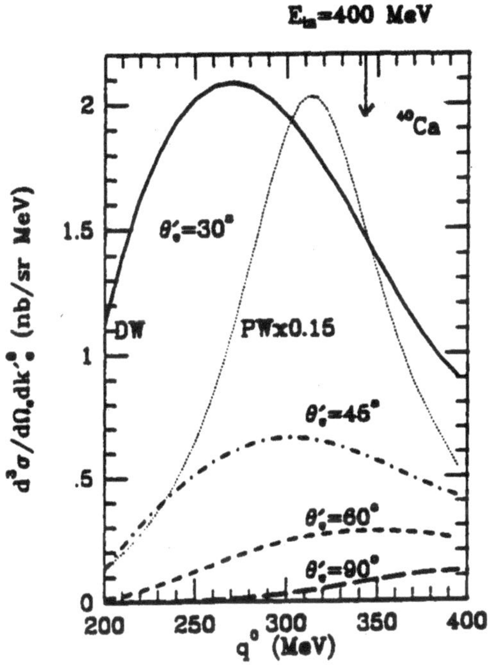

Fig. 12: $d^3\sigma/d\Omega_e' k_e'^0$ for the coherent $(e, e'\pi^0)$ in ^{40}Ca at several electron angles. Solid line : $\theta_e' = 30^0$; dash-dotted line: $\theta_e' = 45^0$; short-dashed line: $\theta_e' = 60^0$; long-dashed line: $\theta_e' = 90^0$; dotted line: $\theta_e' = 30^0$, plane wave calculation (with no Fermi motion and no medium effects on the Δ-resonance) scaled by a factor 0.15. The arrow indicates the position of the free Δ peak.

We can now turn our attention to coherent pion production in the $(^3He, t)$ reaction. The process is depicted in fig. 13

Fig. 13

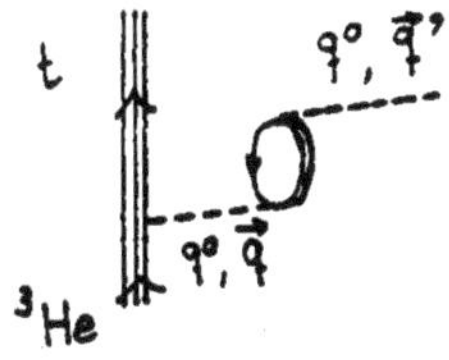

We have there:

$$q^0 = E_{He} - E_t$$

$$q = p_{He} - p_t \quad (assume\ forward\ direction)$$

$$q' = \sqrt{q^{02} - m_\pi^2} < q \tag{4}$$

Then the coherent pion production cross section goes roughly (neglecting the distorsion of the final pion) as

$$\frac{d\sigma}{d\Omega} \propto \left| \frac{1}{\sqrt{s} - M_\Delta + i\frac{\Gamma}{2}} \right|^2 F_A^2(\vec{q} - \vec{q}')A(q, q')(\frac{-q^2}{\vec{q}^{\,2}}) \tag{5}$$

In a model for the $NN \rightarrow N\Delta$ transition which appears in fig. 13 one can put π and ρ exchange evaluated within correlated wave functions. The interaction splits into a longitudinal and a transverse part and one gets,:

$$A(q, q') = [V_l \hat{q}_i \hat{q}_j + V_t(\delta_{ij} - \hat{q}_i \hat{q}_j)]\hat{q}'_j . [V_l \hat{q}_i \hat{q}_m + V_t(\delta_{im} - \hat{q}_i \hat{q}_m)]\hat{q}'_m$$

$$= V_l^2(\hat{q}.\hat{q}')^2 + V_t^2[1 - (\hat{q}.\hat{q}')^2]$$

$$= V_l^2 cos^2\theta + V_t^2 sin^2\theta$$

$$V_l \simeq \frac{\vec{q}^{\,2}}{q^2 - \mu^2} F(q)^2 + g' \quad , \quad V_t \approx \frac{\vec{q}^{\,2}}{q^2 - m_\rho^2} F_\rho(q)^2 C_\rho + g' \tag{6}$$

Hence one finds that the form factor will also tend to reduce the cross section. However in the electromagnetic processes we always found a $sin^2\theta$ factor which forced the processes to take place at finite angles where the form factor was small. Now, by the contrary, we have the longitudinal part of the interaction which goes with $cos^2\theta$ and hence the form factor in the forward direction will not be so drastically reduced as it happens in the electromagnetic reactions. Thus we should expect a larger relative weight of the coherent pion production process in the $(^3He, t)$ or (p, n) reactions with respect to the incoherent processes than in the electromagnetic processes. Since this coherent pion production process, following the general trend of the coherent processes, is also shifted because of the form factor in (5) plus the additional effect of the pion distortion, we might hope that its increased relative weight might be large enough to produce a global shift in the inclusive cross section shown in fig. 1. This will be shown to be the case although other channels are also responsible for the total shift of the peak and the strength.

There is another point worth making. The form factor will force the contribution of small angles between q and q' and hence all the weight of the coherent pion production comes from the longitudinal part of the interaction essentially. The process is thus selective to the longitudinal part of the $NN \rightarrow N\Delta$ transition.

On the other hand the elementary $(^3He, t)$ reaction on a proton target is selective to the combination $V_l^2 + 2V_t^2$, and the transverse part has a bigger weight than in the coherent process. It is thus clear that a combined study of the elementary reaction and the coherent pion production process will lead to a better understanding of the ingredients of the interaction than the elementary reaction alone.

Another interesting comment is worth at this point. Fig. 13 makes a very specific assumption on how coherent pion production proceeds. It requires the renormalization of the pion in the medium. The ^3He, and t waves obviously are distorted waves when one studies this reaction. Hence, this mechanism of pion production differs substantially from the ordinary way of producing coherent photons in electron nucleus collisions: bremsstrahlung. We might think that coherent pion production could proceed in the same way, in which case the mechanism for the production would be depicted in fig. 14.

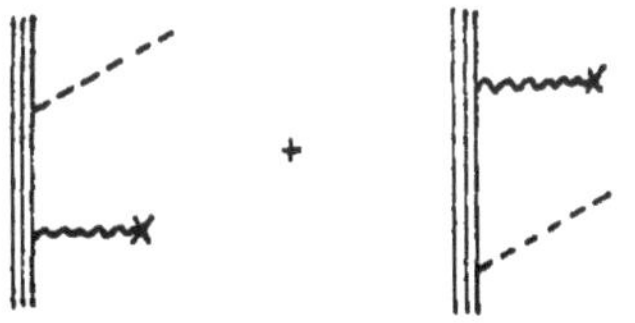

Fig. 14

where the interaction line and the cross symbolize the interaction of the He or t with the nucleus producing distorted waves. The nucleus acts as a source of momentum (and not energy) which could make possible the pion production process, otherwise forbidden by energy conservation in the absence of external fields.

However, one can show that the process, at least for forward or small t angles, does not proceed with the Bremsstrahlung mechanism of fig. 14. This can be easily proved if one uses the Glauber eikonal approximation for the ^3He or t wave functions in the forward direction. The wave function picks up some transverse momentum components but the free longitudinal momenta are not modified. Hence the $^3He \rightarrow t + \pi^+$ transition in this case is as forbidden as if the ^3He or t did not interact with the nucleus. For all these reasons the process proceeds as in fig. 13 and it qualifies as virtual pion production followed by elastic scattering of this virtual pion with the nucleus till it becomes real. Hence the study of this channel is bound to offer valuable information about the pion nucleus interaction, complementary to the one obtained from the scattering of real pions.

5 Incoherent channels in the $(^3He, t)$ reaction in nuclei

We list here the incoherent reactions which contribute to the inclusive $(^3He, t)$ cross section.

1) Quasielastic collisions which include one step or two step processes:

$He + N \rightarrow t + N$

$He + N \rightarrow He(t) + N$ followed by $He(t) + N \rightarrow t + N$

$He + N \rightarrow He^*(t^*) + N$ followed by $He^*(t^*) + N \rightarrow t + N$

where in the last process there is a quasielastic collision with breakup of the He or t in the first step and recombination in the second one.

2) Pion production, for which the model of fig. 2 is used including delta excitation in the target and the projectile. Two step processes in which a quasielastic collision appears before or after pion production are also considered. In all these processes the nucleons are ejected to the continuum or into ^{12}C excited states and the process accounts for incoherent pion production.

3) Virtual pion absorption. These are processes of the type of fig. 15a where the dashed circle represents the $\pi N \rightarrow \pi N$ amplitude which contains s and p-waves. A particular case of the p-wave piece is shown in fig. 15b corresponding to delta excitation followed by the non mesonic Δ decay, $\Delta N \rightarrow NN$. It must be said, however, that at low virtual pion energies where this mechanism peaks, the mechanism of fig. 15b only accounts for about 1/3 of the strength of the p-wave part, the rest coming from nucleon pole or crossed nucleon and delta pole terms and the interference with the other diagrams of fig. 15a [19]. The pion in fig. 15 is in practice substituted by the whole spin-isospin interaction, like in fig. 2, and eqs. (6).

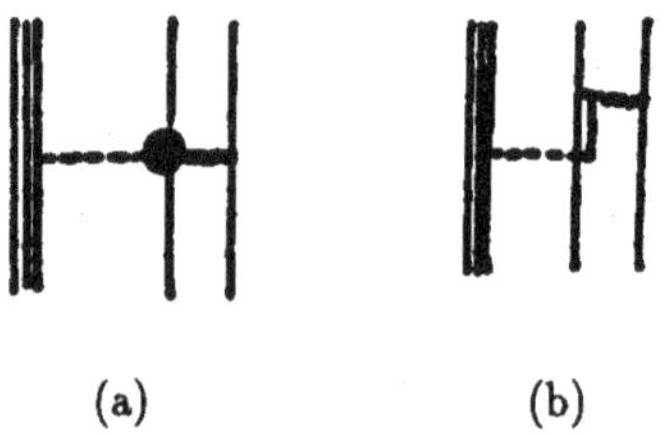

Fig. 15

The multiple channels and their ramification in the scattering process are treated by means of a Montecarlo simulation with the individual probabilities either taken from experiment or from models if experiments do not provide enough information. The coherent pion production channel is calculated fully quantummechanically.

In fig. 16 we show the contribution of all the channels discussed above classified by the first step which takes place in the simulation. Final state interaction would redistribute the strength between the different channels but does not change the cross section. As we can see, the one step quasielastic collisions fill the region of very higt t energy while the two step quasielastic incoherent pion production peaks at the same position as the elementary $(^3He, t)$ reaction on the p, something confirmed experimentally [4]. The most interesting feature, however, is that the peak of the coherent pion production is considerably shifted (about 70 MeV) with respect to the incoherent one. This feature is responsible for the ultimate shift of the peak in the total inclusive $(^3He, t)$ cross section. We can observe in the figure that the sum of all the incoherent channels shifts considerably the strength at higher t energies, but the position of the peak is not altered. When one then adds the coherent pion production cross section the peak is moved and one finds agreement with experiment.

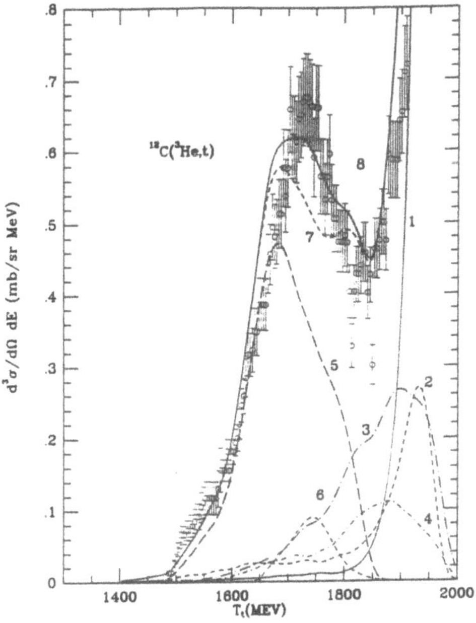

Fig. 16: Different contributions to the cross section for the $(^3He, t)$ inclusive reaction on ^{12}C. 1) Quasielastic $He + N \rightarrow t + N$. 2) Two step quasielastic $He + N \rightarrow He(t) + N$ followed by $He(t) + N' \rightarrow t + N'$. 3) Two step quasielastic with intermediate $He(t)$ break up $He + N \rightarrow He^*(t^*) + N$ followed by $He^*(t^*) + N \rightarrow t + N$. 4) Virtual pion production followed by pion absorption. 5) Incoherent pion production. 6) Coherent π^+ production. 7) Sum of incoherent processes 1-5. 8) Total sum of coherent and incoherent processes.

6 Coherent pion production

Coming back to coherent pion production, another feature of the coherent channel is that the pions are created in a very narrow cone around the direction of the (He, t) momentum transfer, particularly in heavy nuclei. This property and the fact that the nucleus recoil energy is very small, and well known if all pions come in the same direction, converts the coherent pion production in a tool to produce beams of highly monochromatic and unidirectional pions. Its potential practical use, similar to the tagging of highly monochromatic photons, should be investigated.

We have now some data for the coherent channel extracted from the SATURNE measurements [20]. These measurements are still partial and more devoted experiments are still needed. However they confirm our theoretical findings about the position of the peak. On the other hand we can also perform calculations for the (p, n) reaction and we can compare our results for coherent π^+ production in the analogous reaction (p, n) on ^{12}C with the data of Chiba et al. [3] when they separate the π^+ channel. Even here they have a background in this channel of neutral particles coming from $pn \rightarrow nn\pi^+$. We can see our results in fig. 17 compared to the experiment. We have adapted our results to the experimental cuts and resolution.

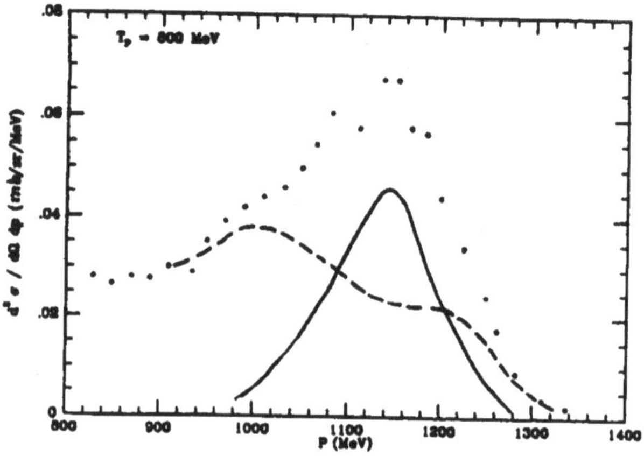

Fig. 17: Dots experimental data from [3]. Continuous line: Coherent π^+ production. Dashed line: Difference between experiment and the calculated coherent part.

One observes that our coherent production peaks at the place of the experimental peak, which is appreciably shifted with respect to the peak of the incoherent pion production. On the other hand when our coherent pion production is subtracted from the experimental data one obtains a smooth background which looks very much like the cross section of the analogous $^3He\,n \rightarrow tn\pi^+$ which is obtained in the $(^3He, t)$ cross section on the neutron target of fig. 3. The bigger weight at lower t energies (in the region of the Δ resonance) for the background of the $pn \rightarrow nn\pi^+$ reaction is understandable because of the absence of the $(^3He, t)$ transition form factor that reduces the cross section more strongly at lower t energies in this latter reaction.

As one can see, the information obtained from our theoretical analysis matches so far the available experimental information. However, experiments devoted to the measurements of the coherent channel should definitely be pursuit. We should also point out that Chiba [21] has made an analysis of the data of [3] placing some cuts in order to separate the pion coherent channels. His results agree qualitatively with our theoretical findings.

As mentioned before the angular distribution of the coherent production is very forwad peaked. We show in fig. 18 the differential cross section for coherent π^+ production in ^{12}C induced by the $(^3He, t)$ reaction. The shape of the distribution is very similar to the one measured in [20] which is however integrated over different t energies.

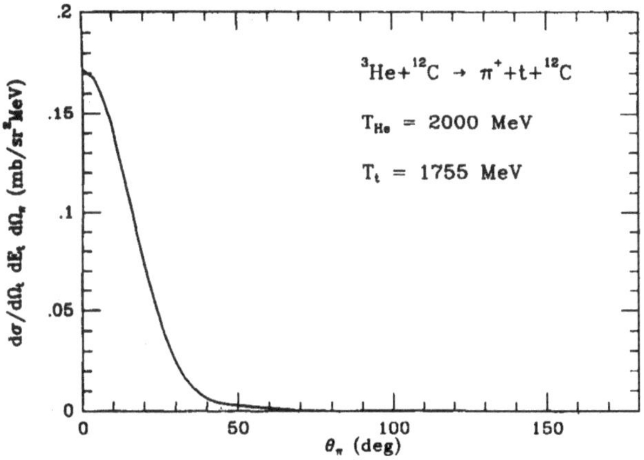

Fig. 18

On the theoretical side the idea of coherent pion production has also stimulated work [22, 23, 24]. Although the qualitative features like the position of the peak or the shape of the angular distributions are similar, there are still discrepances regarding the strength of the cross section. For instance, the cross section of ref. [24] for coherent pion production is about a factor of two bigger than the one we show here or in ref. [23]. It is clear that precise experimental values are needed to help us understand the process better. On the other hand the (α, α') reaction qualifies as coherent pion production induced by the (p, n) or (p, p) reactions on the ^4He nucleus and with the pionic variables integrated. The fact that we reproduce fairly well this reaction gives us confidence about our results in the coherent channel.

7 Summary

Summarizing, the study done here has shown that the shift of the Δ peak in the $(^3He, t)$ (or (p, n)) reaction is due to a collaboration of many channels, amongst which the coherent pion production plays a crucial qualitative role as responsible for the shift of the peak position. This channel was implicitly included in the calculation of refs. [25, 26] and is one of the reasons, in our opinion, of the shift found there, although other reasons were given in those papers. More recent work on the topic [22, 27] shows a convergence of the different theoretical groups towards the ideas using this talk, although expressed in different formalisms.

Finally, the coherent pion production channel shows up as an interesting tool to learn about the pion nuclear interaction. Also, its properties as a source of highly monochromatic and unidirectional pions are bound to have some practical use in the future.

References

[1] P. A. Deutchman, J. W. Norbury and L. W. Townsend, Nucl. Phys. A454 (1986)733

[2] A. H. Blin, C. Guet and B. Hiller, Nucl. Phys. A454(1986) 746.

[3] J. Chiba et al. , Phys. Rev. Lett. 67 (1991) 1982.

[4] T. Hennino et al., Phys. Lett. B283 (1992) 42.

[5] E. Oset, E. Shiino and H. Toki, Phys. Lett. B224 (1989) 249.

[6] T. Udagawa, S. H. Hong and F. Osterfed. Phys. Lett. B245 (1990) 1.

[7] J. Delorme and P.A.M. Guichon, Phys. Lett. B263 (1991) 157.

[8] P. Fernández de Córdoba and E. Oset. Nucl. Phys. A544 (1992) 793.

[9] P. Morsch et al., Phys. Rev. Lett. 69 (1992) 1336.

[10] P. Fernández de Córdoba, Yu. Ratis, E. Oset, J. Nieves, M. J. Vicente-Vacas, B. López-Alvaredo and F. Gareev, University of Valencia preprint.

[11] D. Contardo et al., Phys. Lett. B168 (1986)331.

[12] V. G. Ableev et al., Sov. Phys. JETP Lett. 40 (1984) 763; Phys. Lett. B 264 (1991) 264.

[13] R. C. Carrasco and E. Oset, Nucl. Phys. A536 (1992) 445.

[14] J. Arends et al., Z. Phys. A311 (1983) 311.

[15] E. Oset and W. Weise, Nucl. Phys. A368 (1981) 375.

[16] J. M. Koch and E. Moniz, Phys. Rev. C27 (1983) 751.

[17] R. C. Carrasco, E. Oset, and J. Nieves, Nucl. Phys. to be published

[18] S. Hirenzaki, J. Nieves. E. Oset and M. J. Vicente Vacas, Phys. Lett. B304 (1993) 198.

[19] J. Nieves, E. Oset and C. García Recio, Nucl. Phys. A554 (1993) 309; ibid, page 554.

[20] T. Hennino et al. Phys. Lett B303 (1993)236.

[21] J. Chiba, invited talk in the International Workshop on deltas in nuclei, Tokyo, May 1993.

[22] P. Oltmanns, F. Osterfeld and T. Udagawa, Phys. Lett. B299 (1993) 194.

[23] E. Oset, P. Fernández de Córdoba, J. Nieves and M. J. Vicente- Vacas, Phys. Scr. 47 (1993).

[24] V. F. Dimitriev, Phys. Rev. C48 (1993) 357.

[25] T. Udagawa, S. H. Hong and F. Osterfeld, Phys. Lett. B245 (1990)1.

[26] J. Delorme and P. A. M. Guichon, Phys. Lett. B263 (1991) 157.

[27] T. Udagawa, invited talk in the International Workshop on deltas in nuclei, Tokyo, May 1993.

[28] T.E.O. Ericson and W. Weise, Pions in nuclei, Clarendon Press, 1989.

'SIGMA'-AND-RHO-MESONS IN DENSE MATTER

Z. AOUISSAT[1], G. CHANFRAY[2] and P. SCHUCK[1]

[1]Institut des Sciences Nucléaires, 53 Avenue des Martyrs, 38026 GRENOBLE CEDEX
[2]Institut de Physique Nucléaire, 43 Bd du 11 Novembre 1918, 69622 VILLEURBANNE CEDEX

Abstract

A review of recent theoretical results on pion-pion correlations in nuclear matter is given. We concentrate on the sigma (l = I = 0) and rho (l = I = 1) - meson channels. Extremely strong reshaping of σ-meson mass distribution in matter due to Δ-h and ph self energies of the pions is observed. For instance a strong resonance below the $2m_\pi$ threshold for slightly compressed matter is exhibited. In the rho-meson channel an upward shift, quenching and broadening of the rho-meson peak results from our calculation. No lowering of the bare rho meson mass is taken into account at this point. Due to constraints from gauge invariance a new structure at ≈ $3m_\pi$ is predicted. The influence of our results on the dilepton mass spectrum is estimated.

Introduction

Today, it is well belived that QCD is the basic theory for hadrons and thus in principle also for nuclei. Three quark bags are then interacting via $q\bar{q}$ excitations (Fig. 1).

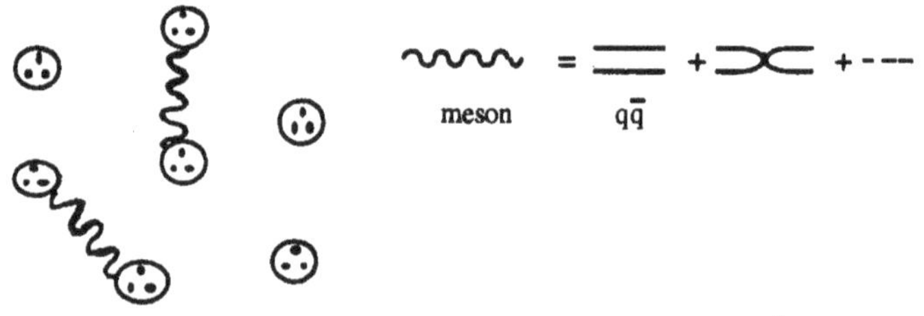

Fig. 1. Nucleons viewed as quark bags interacting via elementary ($q\bar{q}$) excitations (the mesons).

However, between the two extremes : pure QCD and phenomenological NN force, one can imagine that a semi phenomenological approach based on nucleons, deltas,...and mesons could be a reasonable intermediate step. In this picture we view the NN, N-Δ, etc. forces as mediated by meson exchanges. The short range repulsion is mainly due to the ω-exchange (780 MeV) whereas one attributes the medium range attraction to the fictitious 'σ-meson' (Fig.2).

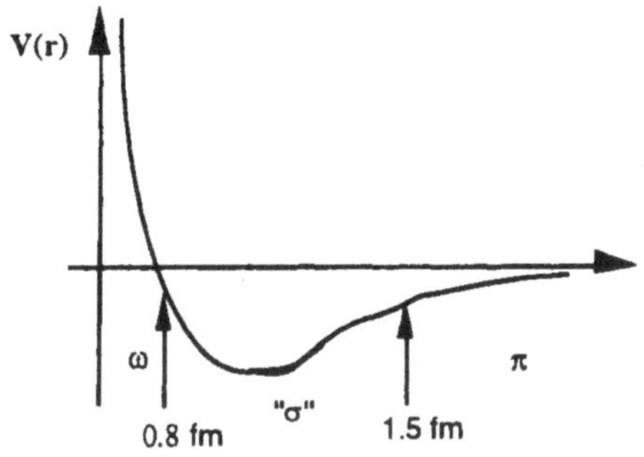

Fig. 2. Schematic view of the NN force divided into a short range repulsion (ω-meson), a medium range attraction ("σ"meson) and a long range tail (π-meson).

To a large extent (Paris-Bonn approach) one views this meson as being representative of a correlated two pion exchange as indicated in Fig.3.

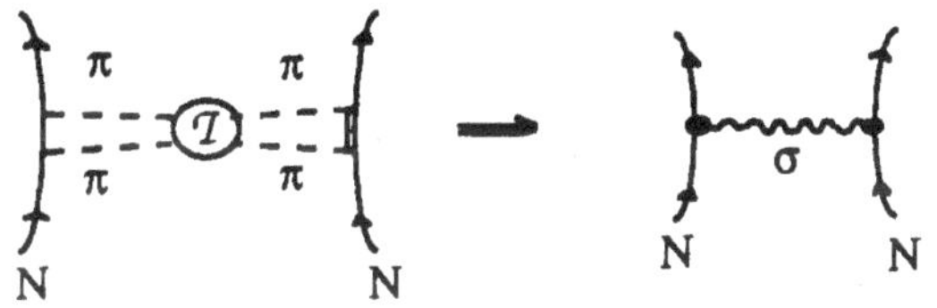

Fig. 3. Correlated π–π exchange replaced by an effective σ-meson exchange

When nucleons scatter inside matter they are renormalized. Also pions are supposed to be strongly renormalised by medium effects mainly due to its coupling to Δh states (Fig.4).

Fig. 4. Coupling of the pion to delta-hole states in the medium.

The pion propagator then takes the following form

$$D_\pi(\omega,\mathbf{q},\rho)=\frac{1}{\omega^2-\omega_\mathbf{q}^2-\mathbf{q}^2\tilde{\Pi}_\Delta(\omega,\mathbf{q},\rho)} \tag{1}$$

with $\qquad \omega_q=\sqrt{m_\pi^2+\mathbf{q}^2}$

and $$\tilde{\Pi}_\Delta(\omega,\mathbf{q}) = \frac{\Pi^0_{\Delta h}(\omega,\mathbf{q})}{1-g'\,\Pi^0_{\Delta h}(\omega,\mathbf{q})}\ ; \quad \Pi^0_{\Delta h}(\omega,\mathbf{q}) = \frac{c(q)}{\omega^2-\varepsilon^2_{\Delta h}(q)+i\eta} \tag{2}$$

$$\varepsilon_{\Delta h}(q) = M_\Delta - M_N + \frac{q^2}{2M_\Delta}$$
$$c(q) = \frac{8}{9}\left(\frac{f_{\pi N\Delta}}{m_\pi}\right)^2 \varepsilon_{\Delta h}\,\Gamma^2_{\pi N\Delta}(q)\rho \tag{3}$$

where ρ is the density of the system and g' (≈ 0.5) takes care of short range correlations. Inspite of its simplicity this kind of pion mass operator leads to a pion-nucleus optical potential for elastic pion nucleus scattering which is quite realistic. In Fig.5 we show schematicaly typical pion dispersion relations obtained from

$$\omega^2(\mathbf{q}) = m_\pi^2 + \mathbf{q}^2 + \mathbf{q}^2\tilde{\Pi}_\Delta \tag{4}$$

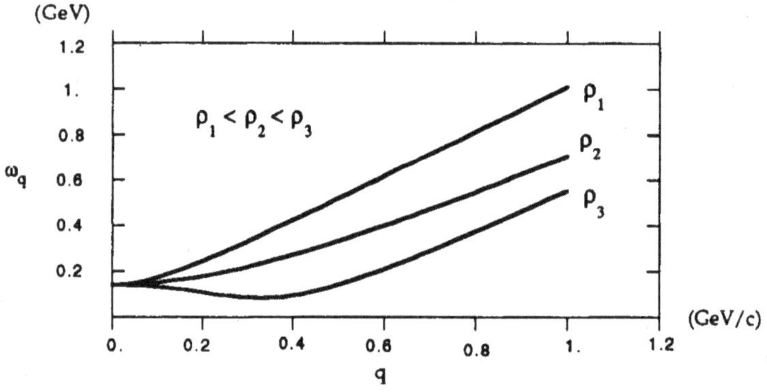

Fig. 5. Medium modified pion dispersion relation for ρ = 0, ρ = ρ0 and ρ = 2 ρ0. (schematic)

The 'sigma - meson ' in matter

We now turn our interest to pion-pion scattering. First we want to consider a model which reproduces the measured phase shifts in the L = I = 0, i.e. σ-meson channel. To this end we take a separable one pole π–π interaction similar to the one used by Johnstone and Lee [1] :

$$V(q,q',E) = 2\pi^2 g^2\, 2\omega_{\mathbf{q}}\, 2\omega_{\mathbf{q}'}\; v(q)\frac{1}{E^2-M_\sigma^2}v(q') \tag{5}$$

with

$$v(q) = \frac{\sqrt{2M_\sigma}}{1+q^2/q_c^2} \tag{6}$$

In Fig. 6 this interaction is represented graphically.

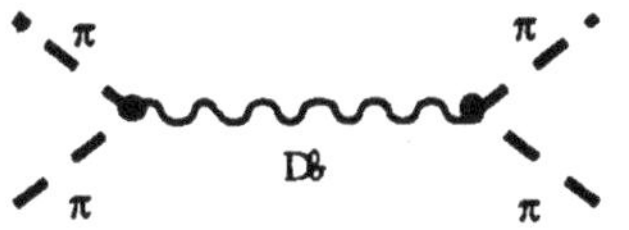

Fig. 6. Pion-pion interaction mediated by a "σ-meson".

The resonant state $D_\sigma{}^0$ stands for a free σ-meson propagator such as it could for example be calculated within the Nambu-Jona-Lasinio model. M_σ then represents the $q\bar{q}$ bound state energy. Indeed we will see in a moment that the fit to the phase shifts yields $M_\sigma \approx 1$GeV a value which is compatible with a $q\bar{q}$ interpretation of this mass. However, on the other hand there is also precisely in this region the f_0 resonance which is sometimes interpreted as a $K\bar{K}$ bound state (molecule). Whether we now view M_σ as $q\bar{q}$ or as $K\bar{K}$ does not really matter as long as we take (5, 6) simply as a phenomenological ansatz. With such a separable force the T-matrix is immediately found analytically from the Lippman Schwinger equation (Fig. 7) and one

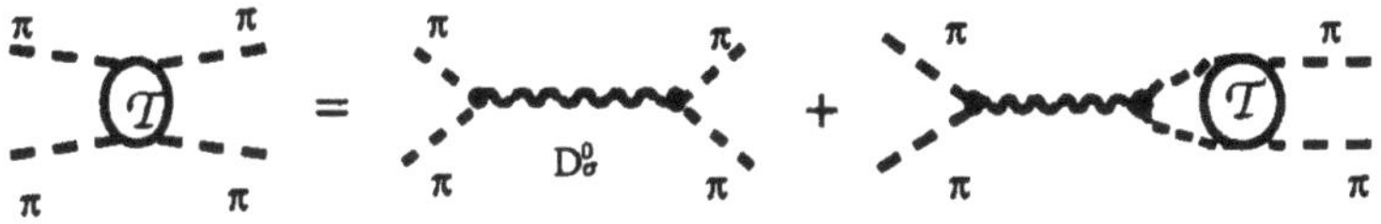

Fig. 7. Lippman-Schwinger equation for π–π scattering.

obtains

$$\mathcal{T}(q,q',E) = 2\pi^2 g^2 2\omega_q 2\omega_{q'}\, v(q)\, D_\sigma(E)\, v(q') \tag{7}$$

with

$$D_\sigma(E) = \frac{1}{E^2 - M_\sigma^2 - \Sigma_\sigma(E)} \tag{8}$$

$$\Sigma_\sigma(E) = g^2 \int dq\, q^2 4\omega_q^2\, v^2(q)\, G_{\pi\pi}(E,q) \tag{9}$$

The σ-meson propagator (8) can now be interpreted as the physical σ-meson, since the σ-meson mass operator couples the σ-meson to two pion intermediate states represented by the propagator

$$\begin{aligned} G_{\pi\pi}(E,\mathbf{q}) &= \int \frac{idq_0}{2\pi} D_\pi(q_0,\mathbf{q}) D_\pi(E-q_0,-\mathbf{q}) \\ &= \frac{1}{\omega_\mathbf{q}} \frac{1}{E^2 - 4\omega_\mathbf{q}^2 + i\eta} \end{aligned} \tag{10}$$

From (7) we can calculate the phase shifts δ_{00} and determine the parameters g, M_σ and q_C. This yields

$$g = 0.533\, m_\pi^{-0.5}\ ; \quad M_\sigma = 1.GeV\ ; \qquad q_c = 3.20\, m_\pi . \tag{11}$$

and the corresponding fit to the experimental phase shifts is shown in Fig. 8. We see that the δ_{00} phase shifts

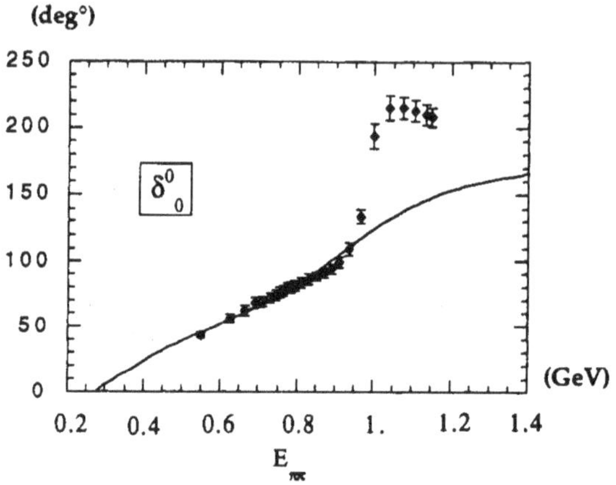

Fig. 8. Fit to the experimental phase shifts [2] in the scalar isoscalar channel.

are reproduced quite satisfactorily in the whole domain below the f_0 resonance. To reproduce also the resonance would need a more sophisticated coupled channel model which will not be considered here because we will be concerned with energies rather close to the two pion threshold. In Fig. 9 we show the structure function $S(q) = -\frac{1}{\pi}\,\mathrm{Im}\,D_\sigma$ of the σ-meson in free space corresponding to the model we just presented.

We see that the σ-meson comes out close to expectation, since the σ-meson is believed to be a very broad resonance such that its experimental detection is quasi impossible. Indeed looking at Fig. 9 we see that its width is ≈ 500 MeV (position at ≈ $6m_\pi$) which is in agreement with the estimate of other authors [3]. Having at hand a rather satisfying

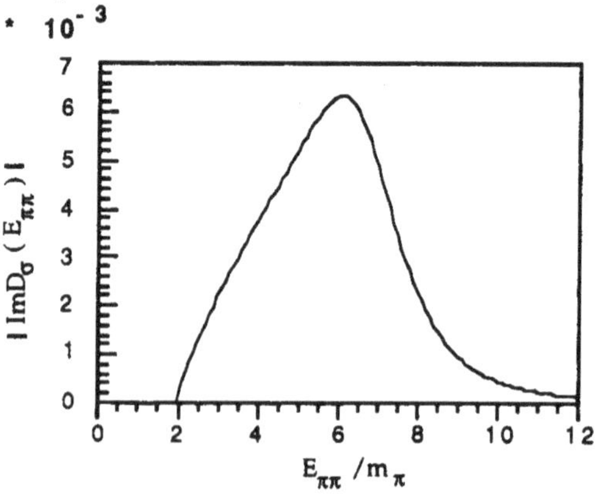

Fig. 9. "Sigma"-meson structure function in free space.

theory for the σ-meson in free space we are now interested in what happens to it in the nuclear medium. To have a quick idea it will be sufficient to parametrise the in medium pion dispersion relation by

$$\omega(\mathbf{q}) \rightarrow \omega_\gamma(\mathbf{q}) = \sqrt{m_\pi^2 + \gamma q^2} \tag{12}$$

where at saturation density γ lies in the range 0.2 to 0.4. It can indeed be checked that the pion dispersion curves of Fig.5 are rather well reproduced for not too high momenta. Before coming to the explicit results let us discuss what we are qualitatively expecting. In the limit of low q we have

$$\omega_\gamma(\mathbf{q}) \approx m_\pi + \frac{q^2}{2m_\pi^*} \tag{13}$$

with $m_\pi^*/m_\pi \approx 2\text{-}3$ for $\gamma = 0.2 - 0.4$. A small γ therefore suppresses kinetic energy. On the other hand one sees from the δ_{00} phase shifts that there exists a rather strong attraction of the two pions already in free space. However, it is too weak to bind the two pions. Suppression of kinetic energy will favor binding. Indeed we see on Fig.10 that in lowering γ (increasing m_π^*) one first accumulates strength around the 2 m_π threshold, whereas the original σ-meson is pushed upwards. At $\gamma = 0.2$ a bound state at about 10MeV below threshold of the two pions is formed.

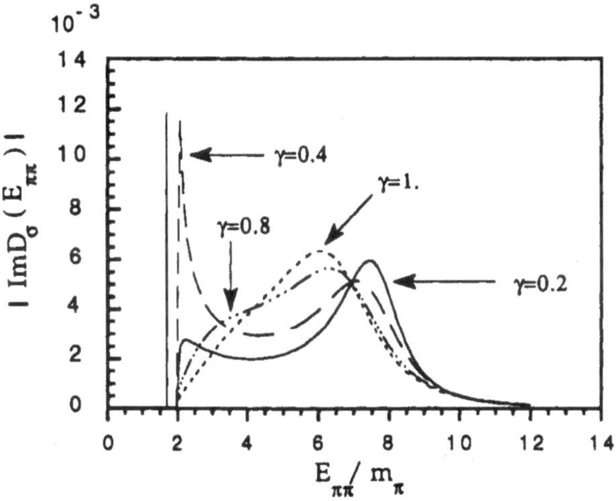

Fig. 10. Structure function of 'sigma' meson in medium (schematic model).

On Fig.11 we show that this spectacular behavior is reproduced in a calculation using in (10) the full in medium pion propagator (1). We should emphasize that this dramatic reshaping of the "σ-meson" mass distribution in nuclear matter has been obtained with rather conventional ingredients and it has in the meanwhile been confirmed by other authors [4].

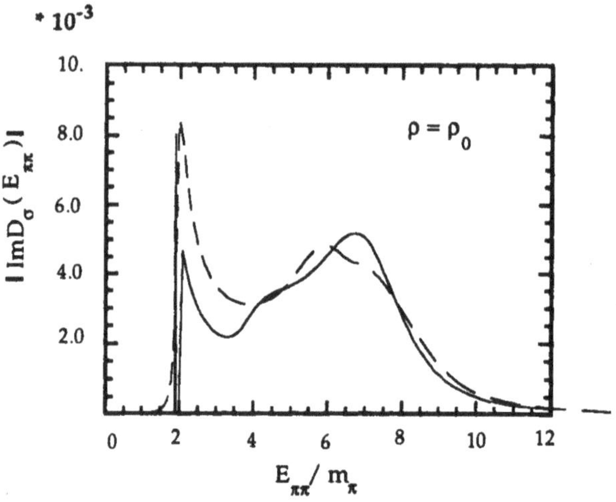

Fig. 11. Structure function of 'sigma' meson in medium. Dashed curve for the full model and solid curve for the two level model.

Can we hope to measure such things ? One could think of measuring two γ's in coincidence. We can imagine that this in medium σ-meson decays into two γ's in the way shown in Fig.12a. However in a hot medium quantum numbers are not conserved and the σ-meson can also decay in a detectable dilepton pair. (Fig.12b) [5].

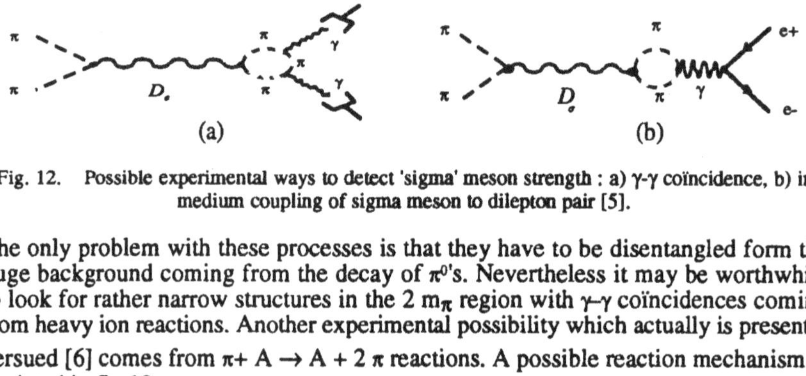

Fig. 12. Possible experimental ways to detect 'sigma' meson strength : a) γ-γ coïncidence, b) in medium coupling of sigma meson to dilepton pair [5].

The only problem with these processes is that they have to be disentangled form the huge background coming from the decay of π^0's. Nevertheless it may be worthwhile to look for rather narrow structures in the 2 m_π region with γ-γ coïncidences coming from heavy ion reactions. Another experimental possibility which actually is presently persued [6] comes from π+ A $\rightarrow$ A + 2 π reactions. A possible reaction mechanism is depicted in fig.13.

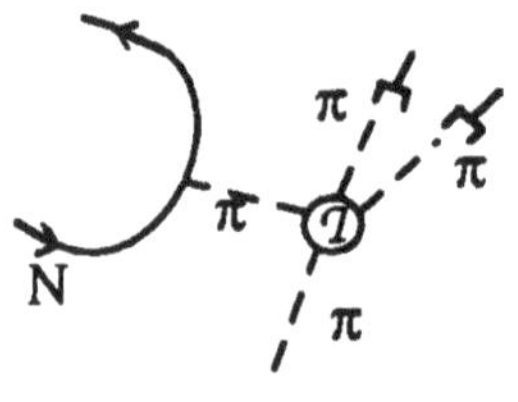

Fig. 13. π, 2π knock out reaction with possibility for in medium π–π correlations.

The incoming pion catches a pion in flight and after (in medium) rescatterings (T-matrix) the two pions leave the nucleus. Such reactions are shown in Fig. 14 where the invariant mass of the two pions is plotted for various target masses [7].

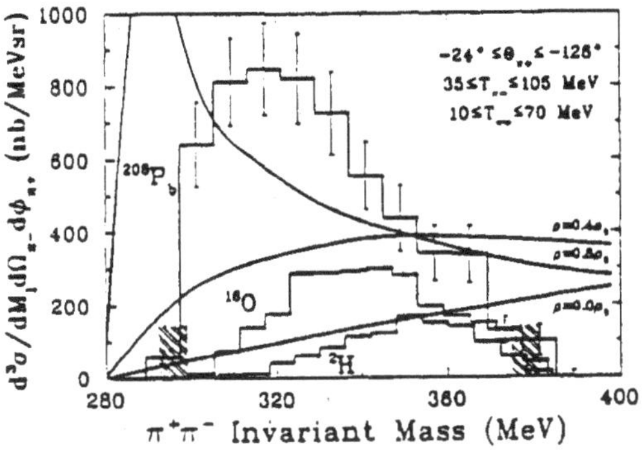

Fig. 14. Invariant π^+–π^- mass spectra from π, 2π knock out reaction on various tragets [7].

It is intriguing that one sees indeed a progression with increasing target mass which is similar to the one shown in Fig.10 as a function of density (theoretical curves coresponding to Fig.10 are shown as continuous lines in Fig.14). In this respect, one could attribute a lower mean density to lighter nuclei than to heavier ones. However, on the one hand more exhaustive measurements have to be performed (presently under way [8]) before anything definite can be said. On the other hand also the theory needs elaboration. One ingredient which is missing so far and which will become important once a $\pi\pi$ bound state goes appreciably below threshold, is the coupling of the pion to the particle-hole continuum (so far only Δ-h states have been considered). This introduces a further component into our model which consists now of three configurations : i) the pion branch, ii) the Δ-h branch, iii) the p-h branch. In Fig.15 we show the dispersion of the three branches.

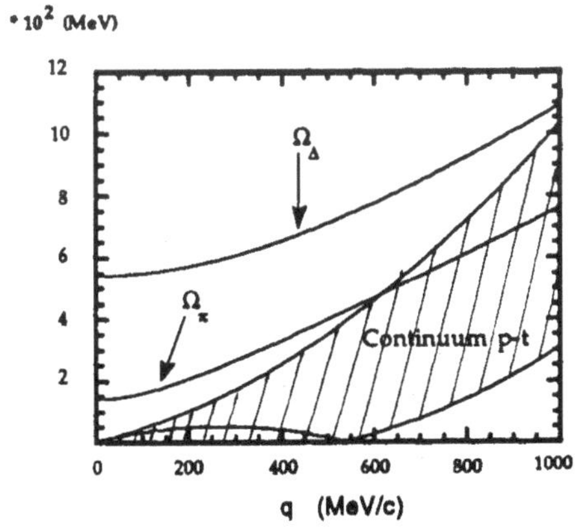

Fig. 15. Schematic representation of the three interacting branches : upper branch : Δh; middle branch : pion lower branch : p-h continuum.

The p-h branch is easily taken care of, since the corresponding polarisation propagator is known analytically (Lindhardt-function [9]). Also the width of the delta-resonance can be included. The combined effect gives a width to the π–π bound state but more importantly it still shifts the resonance further down in energy (Fig.16a). At $\rho = 1.3\ \rho_0$ a very collective and narrow resonance close to zero energy develops as can be seen on Fig.16b and further compressing the system the resonance goes through zero energy indicating instability of matter with respect to this kind of σ-meson condensation [10]. In fact this instability can be viewed as a superposition of two interacting single "pion condensation"- modes (with center of mass momentum $\vec{q}$ and - $\vec{q}$ each) which attract one another via the strong π–π interaction in the L = I = 0 channel. Of course such extremely spectacular behavior must be carefully checked with respect to other processes not considered so far and which may counter balance the effects of collectivity.

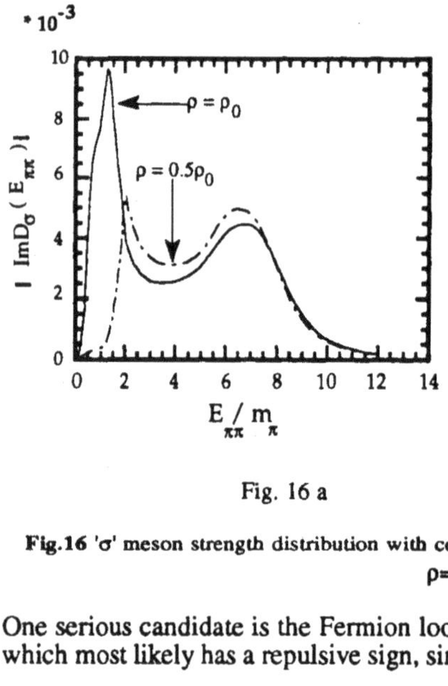

Fig. 16 a

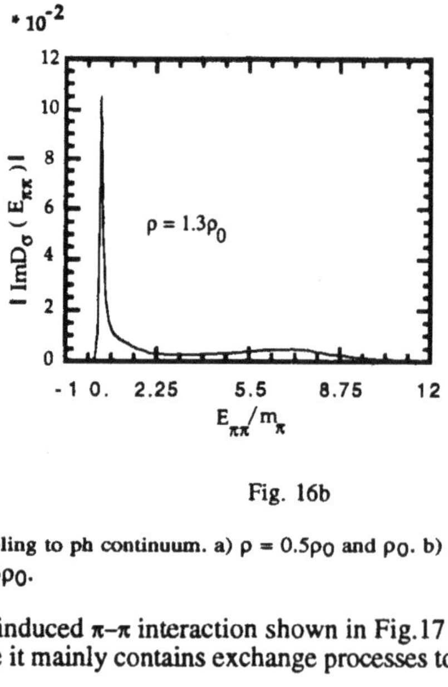

Fig. 16b

Fig.16 'σ' meson strength distribution with coupling to ph continuum. a) $\rho = 0.5\rho_0$ and ρ_0. b) $\rho = 1.3\rho_0$.

One serious candidate is the Fermion loop induced π–π interaction shown in Fig.17 a which most likely has a repulsive sign, since it mainly contains exchange processes to

Fig.17.a

Fig.17.b

Fig. 17. a) Induced π–π interaction through a Fermion loop. b) Cross channel exchange of the "collective" two pion propagator.

the ones treated so far when dressing the pions with ph bubbles [11]. Also the t-channel exchange of the correlated in medium $\pi\pi$ propagator could have a profound

influence on the σ-meson strength (Fig.17.b). At any rate we see from the above studies that due to a quite strong π–π attraction in the σ-meson channel strong reshaping of the σ-meson structure function can occur in nuclei.
On the experimental side we have to be very careful not to misinterpret structures found in π–π correlation function measurements. For example recently the π^+–π^- correlation function as obtained in heavy ion reactions has been reported [12]. This correlation function shows a peak at invariant mass close to $2m_\pi$ which could be interpreted in the sense of the features discussed above. However numerical simulations [13] showed that this peak is almost entirely due to shadowing effects. Indeed pions have rather short mean free path in nuclear matter and therefore when a projectile nucleus hits a target at a certain impact parameter the pions all leave the impact region to the same side. This means that the pions have on the mean small relative momenta which favors low invariant mass.
An effect which we have completely neglected so far and which plays a certain role even at HIC's with energies around 1GeV/n is temperature. In as much as we can attribute to the creation of about 50 pions a temperature, we have to modify the two pion propagator of eq. (10) to

$$G_{\pi\pi}(E,q)=\frac{1}{\omega_q}\frac{1+2f_q}{E^2-4\omega_q^2+i\eta} \tag{14}$$

where f_q is the Bose-Einstein occupation factor. This factor may reinforce the π–π interaction and again favor the appearance of such structures at low energy as shown in Fig.16. On the other hand we have also solved the Bethe-Goldstone equation, i.e. the $\pi\pi$ T-matrix equation with (14) for the two pion propagator, in a gas of hot pions such as produced in URHIC's. Finite temperature and finite chemical potential act again in favor of π–π bound state formation. These "Cooper pairs" may then give raise to pairing like phenomena in a hot pion gas [14].
In this section we treated exclusively the σ-meson channel. The rho meson channel may be less spectacular but easier detectable experimentally and we will turn to it in the next section.

The rho meson in matter and dilepton production:

The rho meson decays practically 100% into two pions. Its energy in free space is m_ρ=768.3 ± 0.5 MeV, its width Γ_{total} = 149.1 ± 2.9 MeV and $\Gamma_{\pi+\pi-}/\Gamma_{tot} \approx 1$.
The neutral rho meson is particularly interesting, since it possesses the same quantum numbers as the photon and can therefore decay into a e^+-e^- pair (Fig.18).

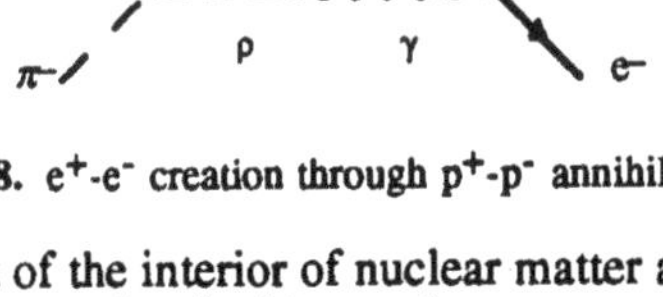

Fig.18. e^+-e^- creation through p^+-p^- annihilation.

The photon can come out of the interior of nuclear matter and can therefore teach us about the rho meson properties inside nuclear matter which may be hot and compressed when created by heavy ion reactions. As the σ-meson, the rho-meson can be viewed as a π^+–π^- resonance

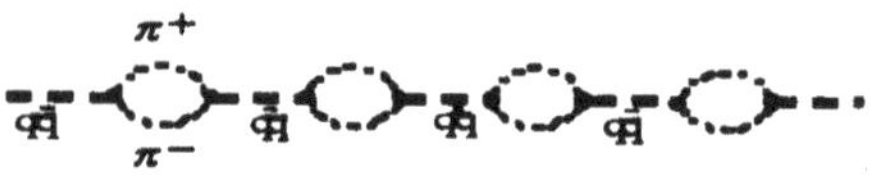

Fig.19. Free rho-meson (viewed as $q\bar{q}$ bound state) renormalized from pion bubbles.

where the $q\bar{q}$ is supposed to be the "bare" ρ-meson propagator,

$$\frac{1}{E^2 - M_\rho^2}. \tag{15}$$

The form factors at the vertices are again parametrised by phenomenological expressions where care has to be taken that the ρ-meson has L = 1 :

$$v_\rho(q) = g_\rho\sqrt{2M_\rho}\,\frac{q/q_\rho}{\left(1+q^2/q_\rho^2\right)^2} \tag{16}$$

Again the integral equation for the T-matrix can be easily solved and the three parameters g_ρ, M_ρ, q_ρ adjusted such that the ensuing phase shifts very well fit the experimental values.

$$g_\rho = 0.668\, m_\pi^{-0.5}\,; \quad M_\rho = 826.7.MeV\,; \quad q_\rho = 3.34\, m_\pi$$

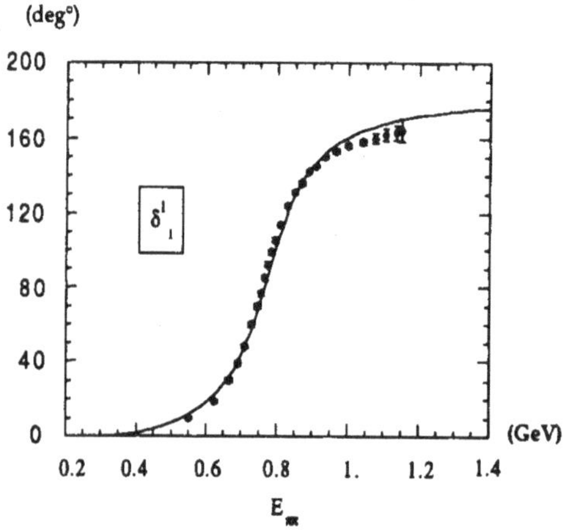

Fig. 20. Fit to the δ_{11} experimental phase shifts [1].

Another experimentally measured quantity is the electromagnetic form factor of the rho meson

$$\left|F_\pi(q^2)\right|^2 = \left|\frac{m_\rho^2}{q^2 - m_\rho^2 - \Sigma(q^2)}\right|^2 \tag{17}$$

where Σ is the rho-meson mass operator entering the rho-meson propagator

$$G^{\mu\nu}(q) = -\left(g^{\mu\nu} - \frac{q^\mu q^\nu}{q^2}\right)G(q^2) - \frac{q^\mu q^\nu}{m_\rho^2 q^2} \tag{18}$$

with

$$G(q^2) = \frac{1}{q^2 - m_\rho^2 - \Sigma(q^2)} \tag{19}$$

This form factor together with the experimental values is shown on Fig. 21

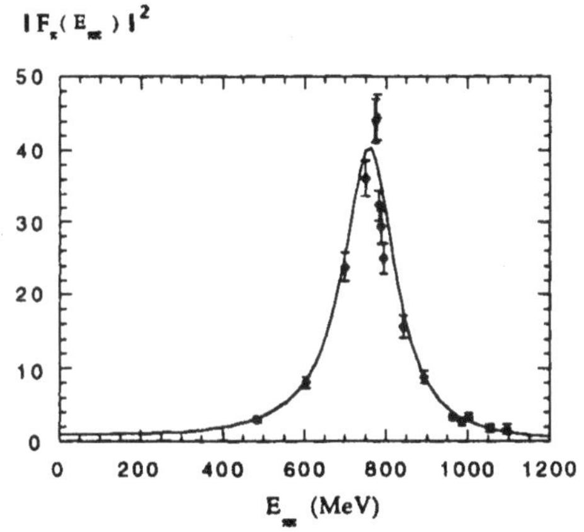

Fig. 21. The rho-meson form factor and the fit of the model.

The phase-shifts (Fig. 20) are related to the mass operator by

$$\delta(E) = arctg\left[\frac{-\operatorname{Im}\Sigma(E)}{E^2 - m_\rho^2 - \operatorname{Re}\Sigma(E)}\right]$$

Since the rho-meson directly couples to the photon, which is conform with the so-called Vector-Dominance Model (VDM), one has to care about gauge invariance. For the moment let us, however, postpone this somewhat subtle question and let us first introduce the most naive model for Σ treating it exactly as in the case of the σ-meson, i.e. as a pure π^+–π^- bubble shown in Fig.22. To get the in medium ρ-rmeson we again take the renormalised pion propagator of eq.(1) and calculate the ρ-meson structure function Im G

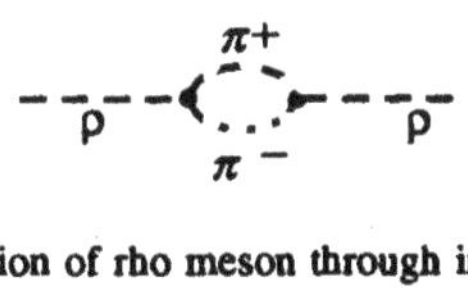

Fig. 22. Renormalization of rho meson through in medium pion bubble.

We then obtain the mass distribution shown in Fig. 23 which is analogous to Fig. 11 for the σ-meson. From this figure we see already the main features for the ρ-meson ρ-meson resonance proper which more or less will stay unaltered even when we later on take gauge invariance into account : in matter the rho meson is slightly quenched, broadened, and shifted upwards in energy (some tens of MeV). Of course other processes unconsidered so far still can change the situation dramatically and we will come to a short discussion of this later. Around threshold we see a similar feature as for the σ-meson : a build up of strength which in this case is, however, much attenuated

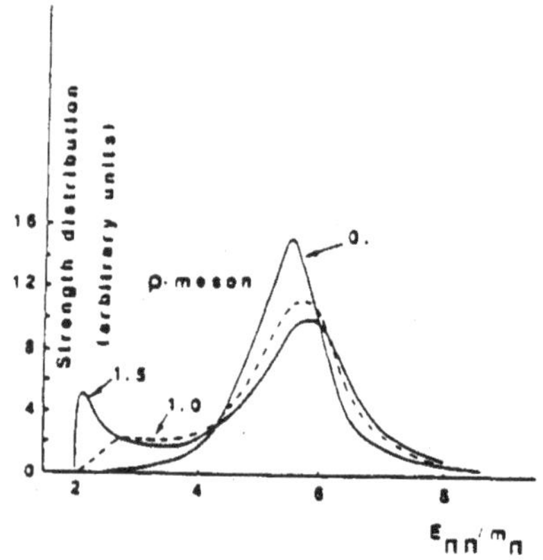

Fig. 23. Rho meson mass distribution in the naïve model for densities ρ/ρ_0 = 0., 1.0, 1.5 from [15].

because in the rho-meson channel there is less attraction between the pions around threshold. Anyway, this threshold enhancement will be suppressed by inclusion of gauge invariance as we will see now.

As we already said the rho meson acts like a photon. According to the VDM hypothesis the rho meson has therefore to obey gauge invariance in the same way as the photon. Gauge invariance means that one has to replace momenta by their canonical momenta :

$$\begin{aligned} \vec{p} &\rightarrow \vec{p} + \frac{e}{c}\vec{A} && \text{photon} \\ \partial_\mu &\rightarrow \partial_\mu + ig\rho_\mu && \text{rho meson} \end{aligned} \tag{20}$$

Therefore, e.g. the kinetic energy of the pions $(\partial_\mu \pi)^*(\partial^\mu \pi)$ now becomes

$$\left[\left(\partial_\mu + ig\rho_\mu\right)\pi\right]^*\left[\left(\partial^\mu + ig\rho^\mu\right)\pi\right] \tag{21}$$

and we see that this generates an energy independent $\rho\rho$–$\pi\pi$ scattering vertex. It leads to an energy independent contribution to the ρ-meson mass operator (the so-called seagull term) (see Fig.24) which has to be added to the graph of Fig.22.

Fig. 24. Seagull graph

The evaluation of the sum of the Feynman diagrams of Figs.22 and 24 leads to the following expression of the ρ-mass operator [16]

$$\bar{\Sigma}(q^2)=\frac{4g^2}{3}\int\frac{d^3t}{(2\pi)^3}\frac{v^2(t)}{\omega(t)}\frac{t^2}{q^2-4\omega^2(t)+i\eta}-g^2\int\frac{d^3t}{(2\pi)^3}\frac{v^2(t)}{\omega(t)} \tag{22}$$

where the first term is seen to involve the dynamic two pion propagator whereas the last term corresponds to the seagull contribution. The form factor is given by :

$$v(t)=\left(1+\frac{t^2}{m_\pi^2}\right)^{1/2}\left(1+\frac{t^2}{q_\rho^2}\right)^{-2} \tag{23}$$

where the constants g_ρ, q_ρ and M_ρ are again adjusted to reproduce phase shifts and form factor. Before doing so we have to realize, however, that with the form (9) of the mass operator we have $\Sigma(0) \neq 0$ and thus the normalization condition $F(0) = 1$ is violated. We therefore have to slightly modify our effective theory in redefining the physical mass operator as

$$\Sigma(q^2)=\bar{\Sigma}(q^2)-\bar{\Sigma}(0) \tag{24}$$

This leads to

$$\Sigma(q^2)=\frac{4g^2}{3}\int\frac{d^3t}{(2\pi)^3}v^2(t)\frac{q^2}{4\omega^2(t)}\frac{1}{\omega(t)}\frac{1}{q^2-4\omega^2(t)+i\eta} \tag{25}$$

and it can be shown that our theory for the rho meson in free space is explicitly gauge invariant [16]. This completes our theory for the free case.

In matter we have to recall that the renormalization of the rho-meson comes from the renormalization of the pion. In turn the renormalization of the pions comes from their coupling to delta-hole states. Relativistically the coupling of the pion to delta-nucleon states is described by the coupling

$$\mathcal{L}_{\pi N\Delta}=\frac{f_{\pi N\Delta}}{m_\pi}\bar{\Psi}_N\vec{T}\,\Psi_\Delta^\mu\partial_\mu\vec{\Phi}_\pi+\text{h.c} \tag{26}$$

where T is a isospin transition operator. Again, due to gauge invariance we have to make in (26) the minimal coupling substitution (20) introducing the vertex shown in Fig.25.

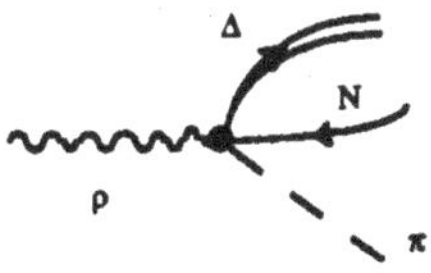

Fig. 25. Coupling of the ρ-meson to Δ-h states.

It is easily imagined that this vertex leads to a series of additional diagrams in the rho meson mass operator. They are shown in Fig. 26

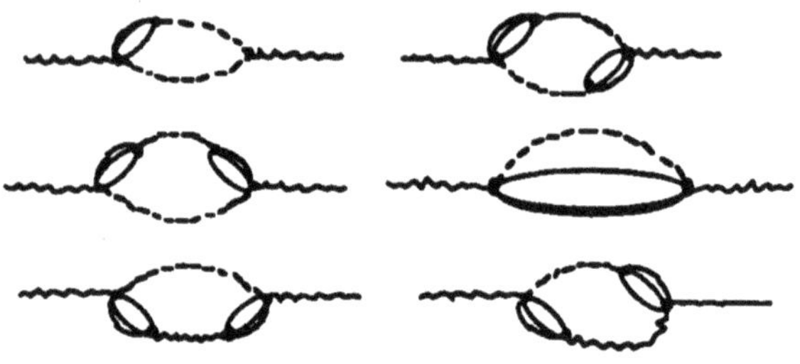

Fig. 26. Corrections for the rho-meson mass operator coming from gauge invariance.

It is of course understood that the Δ-h polarisation bubbles in Fig.26 are again corrected by short range screening (g' = 0.5) The explicit formula for the sum of all these diagrams are quite lengthy and will not be given here; they can be found in ref. [16].The occurrence of explicit breaking of Lorentz covariance due for instance to the treatment of short range correlations (g' parameter) certainly forbids to maintain gauge invariance, in general. However, we can show that even with g' included in using the same substraction procedure for the mass operator as in free space, that is $\Sigma^{\mu}(q) = \bar{\Sigma}^{\mu}(q) - \bar{\Sigma}^{\mu}(0)$, one can maintain gauge invariance [16] in the limit of small $|\vec{q}|$. Since $|\vec{q}| = 0$ actually corresponds to the experimental situation of dilepton detection one can be satisfied with this minimal compliance with gauge invariance. The result of the calculation for the structure function

$$S(q) = -\frac{\operatorname{Im}\Sigma(q)}{\left|q^2 - m_\rho^2 - \operatorname{Re}\Sigma(q)\right|^2 + \left|\operatorname{Im}\Sigma(q)\right|^2} \tag{27}$$

is shown in Fig. 27.

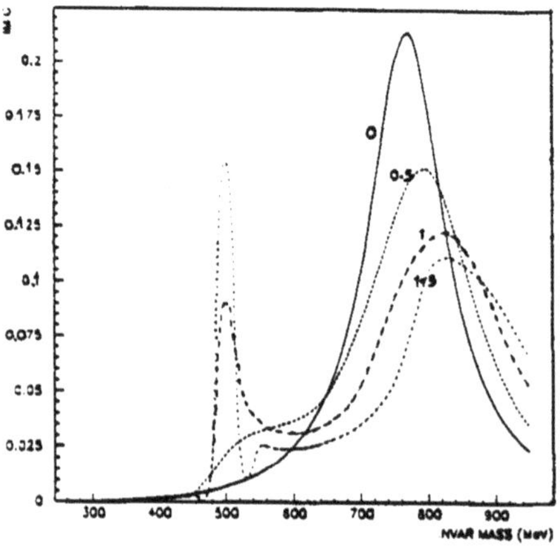

Fig. 27. In medium structure function of the rho meson with gauge invariance corrections included, at various densities ρ/ρ_0 =0, 0.5, 1, 1.5.

We see that the rho-meson itself as a function of density stays practically the same as without corrections from gauge invariance (Fig.23) : at normal nuclear matter density it is shifted upward by ≈ 50 MeV, quenched and broadened. The structure around threshold has completely disappeared, an effect which has already been noted by Korpa and Pratt [17]. On the other hand a spectacular narrow and high resonance now appears at roughly 3 m_π [18]. The analysis shows that this structure is due to the coupling of the rho-meson to ρ–Δh intermediate (doorway) states as shown in one of the graphs of Fig.26. At this point it is extremely important to estimate whether this new resonance will survive when the so far neglected width of the Δ-resonance is taken into account. Inclusion of this width (plus 2p-2h) on the quasi-pion and quasi-rho level yields the following modification of the rho meson structure function shown in Fig. 28.
As we can see from Fig.28 the effect of the width is to strongly wash out the narrow structure seen in Fig.27. However around normal nuclear matter density a shoulder survives and at $\rho = 2\,\rho_0$ even a small resonance can clearly be seen. On the other hand the quenching of the rho-meson is considerably reduced as well as its upward shift.

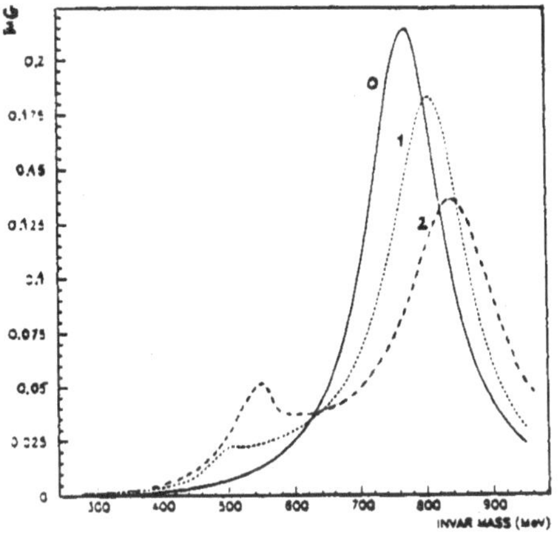

Fig. 28. Same as Fig. 27 but with Δ-width included.

With the structure function we can now calculate the dilepton production rate and compare with the experimental results of Roche [19] et al. obtained with Ca+Ca reaction at 1 and 2 GeV/nucleon. The dilepton production rate for not too high temperatures is related to the structure function by

$$\left[\frac{d^4R}{d^4q\,d^4x}\right]_{\vec{q}=0} \propto \frac{e^{-q_0/T}}{q_0^2}\,S(q) \qquad (28)$$

Our results are shown on Figs. 29a, 29b.

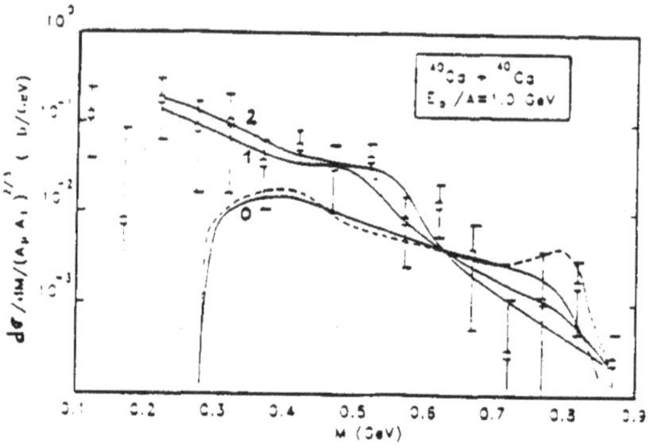

Fig. 29a. Dilepton production rate for ^{40}Ca + ^{40}Ca at 1 GeV per nucleon calculated for ρ/ρ_0=1 and ρ/ρ_0 =2 at T=65 MeV. The dashed curve corresponds to the π^+–π^- annihilation contribution obtained by Wolf [20] et al. used for the normalisation of our cross section (full line). The data are from [19] .

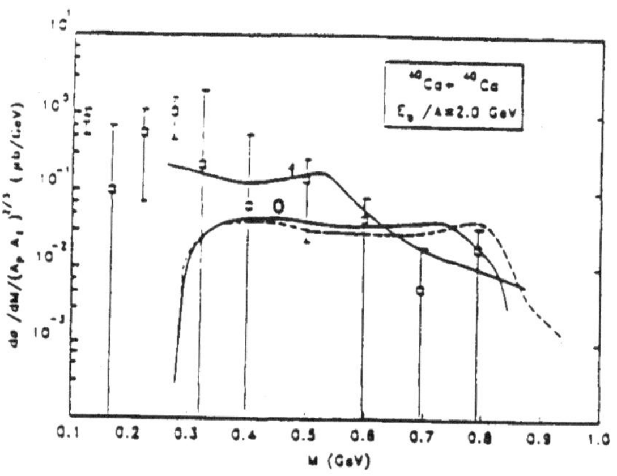

Fig. 29b. Same as Fig. 29 a but at 2 GeV per nucleon and for ρ/ρ_0 = 2 only (T = 95 MeV)

Since on the one hand the theoretical results presented here are very novel and on the other hand in future experiments the rho-meson structure function will certainly be measured with much higher precision it is important to contrast our results with those from other authors which recently appeared. Herrmann [21] et al used a more microscopic model and maintained gauge invariance for all q's. This was done, however, at the price to neglect short range screening, i.e. g' (the effect of g' was nevertheless simulated in reducing the pion-Δh coupling) and the width of the delta.

The results, as shown in Fig.30 are qualitatively similar to ours, for instance they also find the new structure at $\approx 3\ m_\pi$. However, the rho-meson in matter is much more broadened and quenched than in our case; also it remains almost unshifted. It would be interesting to check whether the latter feature is possibly due to their neglect of g'. On the other hand we have seen in our case that the introduction of the delta width considerably reduces the quenching of the rho meson peak. A third publication on the subject by Asakawa et al has recently appeared [22] which also confirms the structure at $3\ m_\pi$ as shown on Fig. 31.

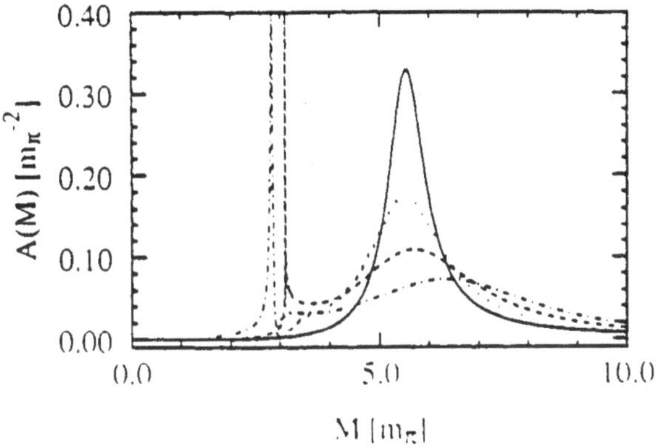

Fig. 30. Spectral function of the ρ-meson in vacuum (solid line) and in nuclear matter at saturation density $\rho=\rho_0$ (dotted) and at $\rho=2\rho_0$ (dashed) and $3\rho_0$ (dash-dotted). From [21].

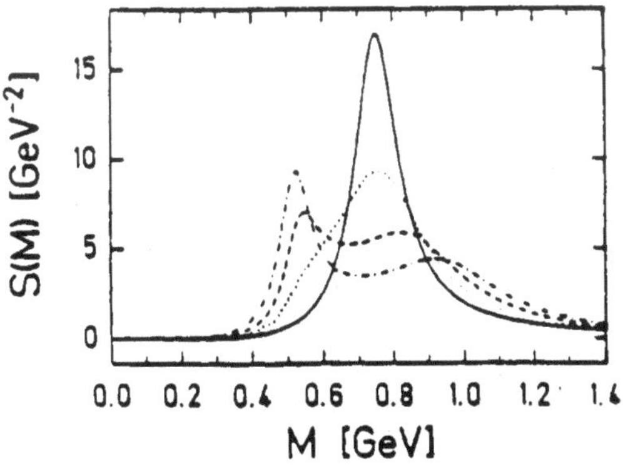

Fig. 31 The spectral function of the rho-meson. Solid curve is for a rho-meson in free space. For a rho-meson in the medium, the dotted, dashed, and dash-dotted curves correspond, respectively, to nuclear densities of ρ_0, $2\rho_0$, and $3\rho_0$, where ρ_0 is the normal nuclear matter density. From [22]

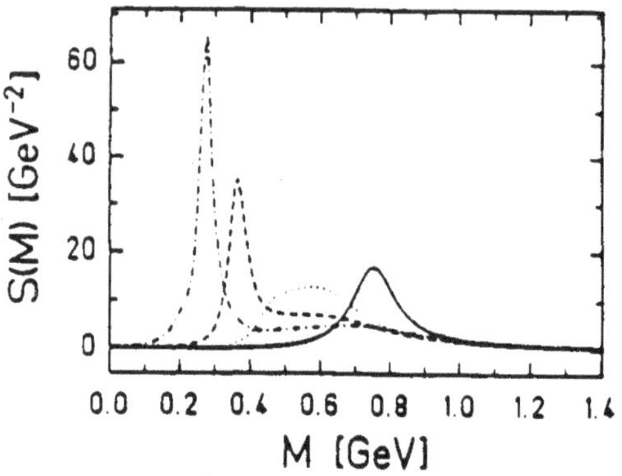

Fig.32 Same as fig.31 with the density-dependent hadron masses in the effective Lagrangian. From [22]

Delta width and g' are also included in this calculation. The rho-meson also broadens very strongly similar to the findings of Herrmann, however, at the same time it is shifted upwards similar to our result. The differences in the three calculations certainly come from the differing approximations and ingredients and it will be important in the future to confront these approaches more closely. On the other hand, Asakawa et al. also played with a reduction of the bare rho meson mass m_ρ as it was recently proposed [23] based on QCD sum rule arguments $[m_\rho{}^*/m_\rho = (1+0.25\ \rho/\rho_0)^{-1}]$. As shown in Fig.32 this has a dramatic effect on the spectrum shifting the distribution down to the threshold region. However this density dependence of m_ρ is certainly not a settled question and for instance vertex corrections could have an equally strong effect in the opposite direction. Furthermore in the rho-meson channel the coupling to p-h states could be quite important too (see the results in the σ-meson channel, Fig.16). However, maintaining gauge invariance will very much complicate these considerations. Again we would like to stress that the presence of a finite number of pions may deeply affect pion-pion rescattering in all cases through the presence of the boson occupation factor coth $(\omega_\mu-\mu)/2T$ reinforcing the scattering matrix elements.

Conclusion

In conclusion we here have given a review of recent theoretical progress of pion-pion correlations and of "sigma"-and rho-mesons in dense matter. We have seen that both the σ- and ρ - meson mass distribution can profoundly be restructured through in medium effects. We here have discussed the influence of a couple of them (the list of which is by far not exhaustive !). Coupling of pions to Δh and ph states shifts σ-meson strength to very low energy ($< 2\ m_\pi$). However vertex renormalization could slow down this feature. On the experimental side narrow structures in the σ-meson channel could possibly be detected from γ–γ coincidence measurements. In the rho-meson channel low energy structures are strongly suppressed by gauge invariance, a feature introduced from the VDM-hypothesis. On the other hand in medium effects influence the rho-meson proper, possibly shifting it upwards (some ten MeV), quenching and broadening it (the results of various authors still disagree on to what extent this happens); on the other hand, the density dependence of the "bare" rho-meson mass, as it was recently postulated on grounds of QCD sum rules, still can strongly modify the scenario in shifting the rho-meson in matter downwards. The very interesting feature of the rho-meson is the fact that it can be detected experimentally through dilepton production and refined measurements should come up in the near future. So we see that there are a lot interesting and novel features to be expected in both the σ - and ρ- meson channel as well on the theoretical as on the experimental side. This kind of nuclear physics will certainly play a very important role in the years ahead.

REFERENCES

[1] J.A.Johnstne and T-S.H.Lee, Phys. Rev. C34, 243 (1986).
[2] S.D.Protopopescu et al., Phys. Rev. D7 (1973), 1279.
[3] T. Hatsuda and T. Kunihiro, Prog. Theo. Phys. Supp. 91 (1987) 284.
[4] V. Mull, J. Wambach and J. Speth, Phys. Lett. B286 (1992) 13.
[5] H. A. Weldon, Phys. Lett. B274 (1992) 133.
[6] D.Vitterli et al., Nucl. Phys. A548 (1992) 541.
V. Sossi et al., Nucl. Phys. A548 (1992) 562.
[7] P.Camerini, N.Grion, R.Rui and D.Vetterli, Nucl. Phys. A552 (1993) 437.
[8] N. Grion et al., to be published.
[9] A.L. Fetter and J.D. Walecka. ED. Mcgraw-Hill, NY (1971).
[10] Z. Aouissat, G. Chanfray, P. Schuck, Mod.Phys. Lett. A. Vol 8, No.15 (1993) 1379.
[11] Z. Aouissat, Thesis report, I.S.N. Grenoble.
[12] J. Pluta et al., to appear in Nucl. Phys. A.
[13] W. Bauer, P. Danielewicz, D. Klakow, P. Schuck, Proceedings of the XXXI international Winter Meeting on Nuclear Physics, Bormio, Italy, page 358 (1993).
[14] Z. Aouissat, G. Chanfray, P. Schuck, G. Welke, Z.Phys. A340 (1992) 347.
[15] G. Chanfray, Z. Aouissat, P. Schuck, W. Nörenberg, Procedings of the International Workshop on Particle Correlations and Interferometry in Nuclear Collisions, Nantes. France 1990. Editor: D.Ardouin, World Scientific, p329.
G. Chanfray, Z. Aouissat, P. Schuck, W. Nörenberg, Phys. Lett. B256 (1991) 325.
[16] G. Chanfray, P. Schuck, Nucl. Phys. A555 (1993) 329.
[17] C. Korpa, S. Pratt, Phys. Rev, Lett. 69 (1990) 1502.
[18] G. Chanfray, P. Schuck, Proceedings of the international Workshop on Pions in Nuclei, Peniscola, Spain, June 1991.
G. Chanfray, P. Schuck, Nucl. Phys. A545 (1992) 271c.
[19] G. Roche et al., Phys.Rev. Lett. 61 (1988) 1069, Phys. Lett. B 226 (1989) 228.
[20] G. Wolf et al., Nucl. Phys. A517 (1990) 615.
[21] M. Herrmann, B.L. Friman, W. Nörenberg, Nucl.Phys. A 545 (1992) 267c.
[22] M. Asakawa, C.M.Ko, P. Lévai and X. J. Qiu, Phys. Rev, C46 (1992) 1159.
[23] G.E. Brown and M. Rho, Phys. Rev. Lett. 66 (1991) 2720.

Slopes and Source Velocities of Exclusive Hard Photons from ^{86}Kr + ^{nat}Ni Collisions at 60 $A\cdot$ MeV *

G. Martínez, J. Díaz, F. Ballester, A. Marín, M. Marqués
IFIC (Centro Mixto CSIC - Universidad de Valencia)
and Departamento de Física Atómica, Molecular y Nuclear
46100 Burjasot, Spain
and the TAPS collaboration:
GANIL-GIESSEN-GSI-KVI-VALENCIA

ABSTRACT

Hard photon production has been studied in a large range of impact parameters in ^{86}Kr+^{nat}Ni collisions at 60 $A\cdot$MeV. Different reaction channels were selected by using the multiplicity of light particles and mass of projectile-like fragments (PLF). The gamma ray multiplicity has been used to determine average impact parameters for selected reaction channels. A dependence of the slope of hard photon spectra with impact parameter is found.

1. Introduction

It is well established that hard photons ($E_\gamma \geq 30$ MeV) produced in heavy ion collisions between 15 and 120 $A\cdot$MeV incident energy are originated by neutron-proton incoherent bremsstrahlung in the earlier stage of nucleus-nucleus collisions[1]. This assumption is based on *i)* measurements of angular

*Experiment performed at the GANIL facility, Caen, France.

distributions which are almost isotropic in the nucleon-nucleon center-of-mass system, *ii)* measurements of the hard photon source velocities, which are compatible with the velocity of the N-N center-of-mass system and *iii)* the fact hard photon multiplicity is proportional to the average number of participant nucleons in the collision.[1] Only some experiments have been devoted to the question of how the γ-ray yield changes with the impact parameter of the collision[2-7]. This requires exclusive measurements in which information about the impact parameter can be obtained. In all these studies a strong variation of the γ-ray multiplicity with magnitudes related to the impact parameter is found, which confirms again p-n collisions as the main mechanism of γ-ray production; it is found that the photon multiplicity $M_\gamma(b)$ is determined by the number of individual p-n collisions during the reaction [3,5]. Turning the argument around, the photon multiplicity allows us to find the average impact parameter corresponding to a determined reaction channel[3]. It is also found that hard photon spectra become softer for peripheral reactions[2-5]. However, the data available neither are sufficiently accurate nor extend in a range of impact parameters large enough to allow the comparison with models.

In this paper we study the hard photon production as a function of the impact parameter for the system ^{86}Kr + ^{nat}Ni at 60 A·MeV . Different reaction channels have been selected by requiring conditions on the multiplicity of light particles and on the mass and charge of PLF. The average impact parameter of selected reaction channel were calculated from the photon multiplicity using a geometrical model[7]. The conditions imposed to the different channels allow to cover a large range of impact parameters.

2. Experimental Setup

The ^{86}Kr beam was delivered by the GANIL accelerator at an average intensity of 10 nAe. The nickel target was 11.89 mg/cm^2 thick and rotated 10° with respect to the beam axis. The total accumulated beam was $8.1 \cdot 10^{14}$ particles. Hard photons were detected by the 320 hexagonal BaF_2 scintillation detectors of TAPS[9], each one equipped with a charged particle plastic detector. To optimize the shower collection we selected a geometry in which the detectors were assembled in 5 square blocks of 64 detectors each, and they were positioned around the target at an average distance of 60 cm. The solid angle covered was $\Omega \approx 0.17 \times 4\pi\ sr$, spanning a nearly continuous angular range of θ between 35° and 165°. In addition to TAPS, the KVI hodoscope for light charged particle detection consisting of 60 phoswich modules was placed into the reaction chamber, covering angles between 3.7° and 24.5°. In the data analysis the charged particles detected by TAPS, mainly protons, electrons and positrons, were rejected with an efficiency of 80% by using the information from the plastic detector as a veto signal. In addition, the other conditions of having the correct time of flight and fast to slow component ratio of the BaF_2 signal were required which allow us to obtain an excellent rejection of charged particles. Before and after the experiment, the BaF_2 detectors were calibrated using γ-ray sources with energies up to 4.4 MeV. A high energy calibration point at 38.5 MeV was taken from cosmic-ray muons whose deposited energy was calculated with a GEANT Monte-Carlo simulation[10] . The time of flight and energy of all detector modules in which an energy of at least 1.5 MeV was deposited were recorded for events in which one or more detectors surpassed a 20 MeV energy threshold. The energy and emission angle of the photons were obtained reconstructing the electromagnetic shower[12]. For 50 MeV photons, the final energy resolution $\Delta E/E$ was 5% and the angular resolution $\Delta\theta$ was 1°.

Projectile-like fragments were detected by the magnetic spectrograph SPEG[8]. Four drift chambers, a Bragg chamber and a plastic scintillator detector were placed at the focal plane of the spectrograph. The spectrograph was located at 0°, covering an horizontal angular range from -0.5° to 0.5°. The magnetic spectrometer was set to 85% and 90% of the beam magnetic rigidity during the experiment.

3. Data analysis

Single photons with $E_\gamma \geq 30$MeV were assumed to be emitted by a moving source, with an exponential energy spectrum. Assuming the angular distribution to be a superposition of an isotropic and a dipolar term, the angular distribution in the source system would be[13]:

$$d\sigma_\gamma/d\Omega dE_\gamma = k \cdot (1 - \alpha + \alpha \cdot sin^2\theta) \cdot exp(E_\gamma/E_0) \qquad (1)$$

Fitting the angular distribution of hard photons in laboratory system, shown in Fig. 1, we obtain that $\beta = (0.172 \pm 0.007)$, $\alpha = (0.15 \pm 0.06)$ and $k = (71 \pm 4)$ μb/str/MeV, which is in good agreement with the velocity of the N-N center-of-mass frame, $\beta_{NN} = 0.177$. The hard photon energy spectrum between 40 to 100 MeV in the NN center of mass frame for inclusive data shows an inverse slope of $E_0 = (19.4 \pm 0.4)$MeV, corrected for detector response.

In a widely accepted geometrical model[7] the γ-ray multiplicity is parametrized as, $M_\gamma = P_\gamma^{incl} \cdot \langle N_{pn} \rangle$, where M_γ is defined as σ_γ/σ_R (σ_R being the geometrical cross section), P_γ^{incl} is the probability of producing a hard photon in a proton-neutron collision and $\langle N_{pn} \rangle$ is the number of p-n collisions calculated from the number of participants in the region where the projectile and target overlap. With the calculated average value of $\langle N_{pn} \rangle = 6.9$ the inclusive data yield $P_\gamma^{incl} = (1.20 \pm 0.03) \cdot 10^{-4}$. The cross section σ_γ for

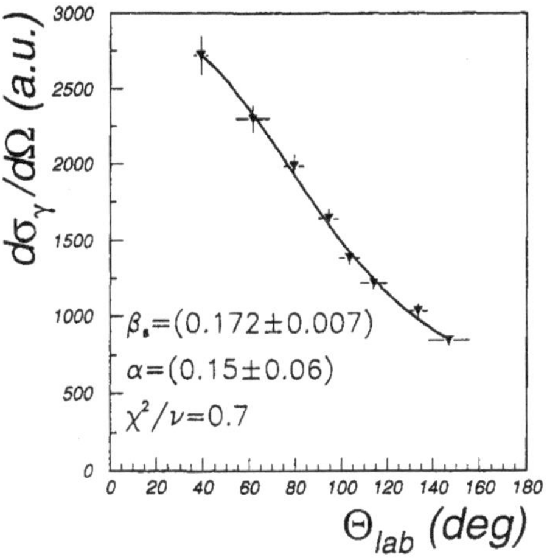

Figure 1: Angular distribution of hard photons in the laboratory system. This angular distribution has been fitted by the function given in Eq.1.

photons with energy greater than 30 MeV was found to be 3.6 ± 0.2 mb, in good agreement with the existing systematic[1].

For the exclusive events, where the hard photons were recorded together with the multiplicity of light particles and the mass of PLF, the analysis could be more detailed. The variation of the photon multiplicity with the multiplicity of light particles and the mass of PLF is shown in Fig.2. One observes an increase in the total photon yield with decreasing the mass of the PLF or increasing the multiplicity of light particles. This is demonstrated in in Fig.1b where we show gamma multiplicities versus mass of PLF fragments for SPEG set to 85% and 90% of the beam magnetic rigidity respectively. It is seen that the multiplicity is lower for the second configuration, and that for a given SPEG set-up the multiplicity decreases when the PLF mass increases.

We have selected a number of reaction channels in order to cover a large range of impact parameters. As it has been proposed in other works[3,4], the average impact parameter of each reaction channel has been calculated from the photon multiplicity using a geometrical model[7]. The conditions imposed to the reaction channels selected are described in Table 1, with their corresponding photon multiplicity, slope of hard photon spectra and calculated average impact parameter.

Angular distributions of hard photons associated to each reaction channel have been measured and fitted to the expression given in Eq. 1 above. In Fig. 3, angular distribution of hard photons corresponding to reaction channels of different centrality are shown. In all the cases, the source velocity of hard photons is compatible with the velocity of the N-N center-of-mass

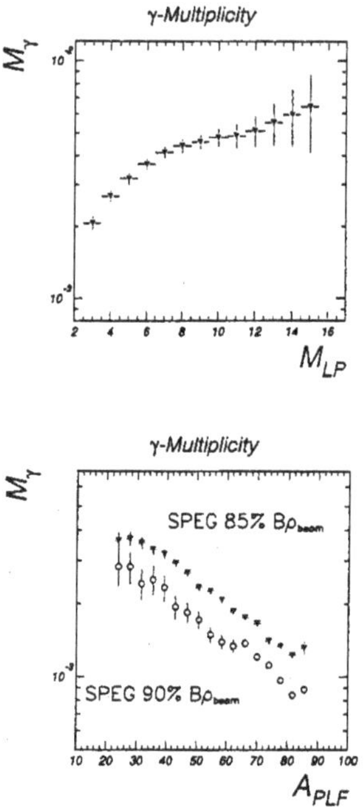

Figure 2: Photon multiplicity as a function of a) the multiplicity of light particle detected by the KVI hodoscope detector and b) the mass of PLF detected by the SPEG spectrograph set to 85% and 90% of the beam magnetic rigidity respectively.

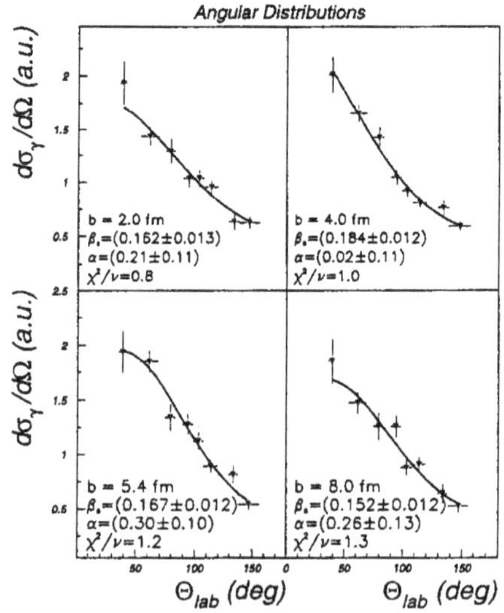

Figure 3: Angular distribution of hard photons for different values of impact parameter.

Table 1: Parameters describing exclusive photon spectra for different reaction channels. S1 and S2 stand for SPEG set to 85% and 90% of the beam magnetic rigidity respectively and M_H for the charged particle multiplicity in the hodoscope. The inequality denotes the PLF mass bin selected.

Condition	$M_\gamma \cdot 10^{-3}$	$E_0(MeV)$	$< b >_g$(fm)
$M_H = 10$	(3.18 ± 0.12)	(17.5 ± 0.7)	(1.0 ± 0.7)
$M_H = 9$	(3.03 ± 0.10)	(18.4 ± 0.6)	(2.0 ± 0.4)
$M_H = 8$	(2.91 ± 0.10)	(18.2 ± 0.7)	(2.1 ± 0.3)
$M_H = 7$	(2.73 ± 0.09)	(18.7 ± 0.8)	(2.7 ± 0.3)
$M_H = 6$	(2.44 ± 0.08)	(18.7 ± 0.5)	(3.3 ± 0.2)
$M_H = 5$	(2.12 ± 0.07)	(18.2 ± 0.4)	(3.8 ± 0.2)
$M_H = 4$	(1.79 ± 0.06)	(17.3 ± 0.4)	(4.1 ± 0.2)
$M_H = 3$	(1.37 ± 0.04)	(17.6 ± 0.5)	(5.3 ± 0.2)
S1, $20 < A < 40$	(2.24 ± 0.06)	(17.6 ± 0.8)	(3.3 ± 0.2)
S1, $40 < A < 60$	(1.57 ± 0.03)	(17.2 ± 0.5)	(4.6 ± 0.2)
S1, $60 < A < 70$	(1.19 ± 0.03)	(16.6 ± 0.7)	(5.4 ± 0.2)
S1, $70 < A < 80$	(0.94 ± 0.02)	(16.9 ± 0.4)	(6.2 ± 0.2)
S1, $60 < A < 90$, $M_H = 0$	(0.53 ± 0.02)	(15.6 ± 0.7)	(7.3 ± 0.2)
S1, $60 < A < 90$, $M_H = 1$	(0.67 ± 0.02)	(15.5 ± 0.6)	(6.7 ± 0.2)
S1, $60 < A < 90$, $M_H = 2$	(0.86 ± 0.02)	(15.4 ± 0.5)	(6.0 ± 0.2)
S1, $60 < A < 90$, $M_H = 3$	(1.06 ± 0.03)	(16.7 ± 0.7)	(5.8 ± 0.2)
S2, $60 < A < 70$	(0.87 ± 0.02)	(16.5 ± 0.4)	(6.3 ± 0.2)
S2, $70 < A < 80$	(0.69 ± 0.02)	(16.0 ± 0.4)	(6.8 ± 0.2)
S2, $80 < A < 90$	(0.58 ± 0.02)	(16.0 ± 0.7)	(7.2 ± 0.2)
S2, $60 < A < 90$, $M_H = 0$	(0.28 ± 0.02)	(13.5 ± 0.7)	(8.0 ± 0.2)
S2, $60 < A < 90$, $M_H = 1$	(0.44 ± 0.02)	(14.2 ± 0.7)	(7.3 ± 0.3)

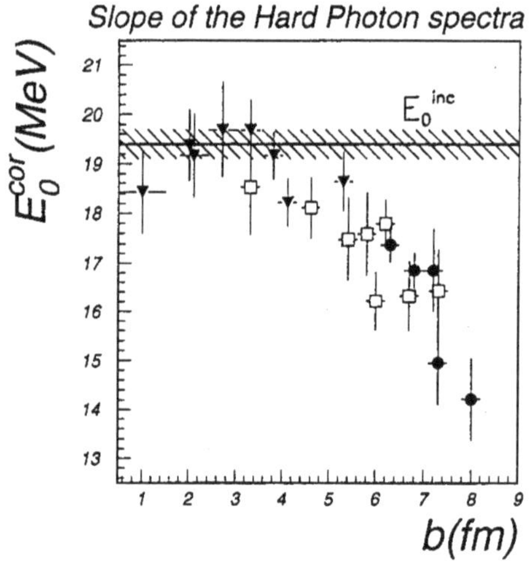

Figure 4: Variation of the slope of the hard photon spectra with the impact parameter. The closed triangle corresponds to the exclusive data obtained using the light particle multiplicity. The open squares and closed circles corresponds to exclusive data using the mass of the PLF for SPEG set to 85% and 90% of $B\rho$, respectively.

system. Therefore, p-n incoherent bremsstrahlung is the main mechanism of hard photon production for impact parameters from 2.0 fm to 8.0 fm. In Fig. 4 the slope parameter E_0 is plotted versus the average impact parameter for the different reaction channels. We observe that the slope of hard photon spectra decreases when the impact parameter increases, according with other works[2,3,4,5]. In addition, the correlation between E_0 and the impact parameter becomes stronger for large impact parameters.

4. Conclusions

We have measured the hard photon production in the reaction ^{86}Kr+^{nat}Ni at 60 A·MeV. Angular distributions of hard photons have been measured for different reaction channels, finding that hard photons are produced in individual NN collisions for a large range of impact parameters. We have done an exhaustive measurement of the variation of the slope parameter E_0 with impact parameter. We have confirmed that the more peripheral reactions lead to softer photon spectra, and that a significant variation of the slope parameter takes place at large impact parameters when the participants in the collision come mainly from the nuclear surface.

5. Acknowledgments

We wish to thank our colleagues at Oak Ridge National Laboratory, J.R. Beene, P. Mueller and R. Varner for their help in setting up the experiment. We thank the members of the technical staff of the Grand Accélérateur National d'Ions Lourds (GANIL) for their help and the Bordeaux target laboratory for producing the Ni targets. M. Marqués and G. Martínez acknowledge their Generalitat Valenciana research grant. The Spanish group was supported by the CICYT Research Project PB90-091.

6. References

[1] W. Cassing *et al.*, Phys. Rep. **188**, 365 (1990).
H. Nifenecker and J.A. Pinston, Prog. Part. Nucl. Phys. **23**, 271 (1989).
[2] R. Hingmann *et al.*, Phys. Rev. Lett. **58**, 759 (1987).
[3] S.Riess *et al.*, Phys. Rev. Lett. **69**, 1504 (1992).
[3] G. Martínez *et al.*, Nouvelles du GANIL **44**, 43 (1993).
[4] M. Kwato *et al.*, Nucl. Phys. **A489**, 368 (1988).
[5] L.G. Sobotka *et al.*, Phys. Rev. **C44**, R2257 (1991), **C46**, 819(E) (1992).
[6] E. Migneco *et al.*, Phys. Lett. **B298**, 46 (1993).
[7] H. Nifenecker and J.P.Bondorf, Nucl. Phys. **A442**, 478 (1985) .
[8] L. Bianchi *et al.*, Nucl. Instr. and Meth. **A276**, 509 (1989).
[9] R. Novotny, IEEE Trans. Nucl. Sc. **38-2**, 379 (1991).
[10] R. Brun *et al.*, CERN Report DD/EE/84-1 (1986).
[11] T. Matulewicz *et al.*, NIM **A289**, 194 (1990).
[12] M. Marqués *et al.*, Nouvelles du GANIL **45**, 45 (1993).
[13] R. Bertholet *et al.*, Nucl. Phys. **A474**, 541 (1987).

FORMATION OF RESONANCE MATTER IN NUCLEUS-NUCLEUS COLLISIONS

W. EHEHALT, W. CASSING, U. MOSEL, GY. WOLF
Institut für Theoretische Physik, Universität Giessen
35392 Giessen, Germany

ABSTRACT

We analyze the formation and decay properties of nucleon resonances formed in heavy-ion collisions at 1 – 2 GeV/u within a microscopic transport approach. In case of Au+Au reactions the density of Δ resonances reaches 0.15 fm^{-3} in the central cell for a time period of the order of 10 fm/c such that one can legitimately speak about *resonance matter*. The lifetime of the Δ's is found to be shortened at high density by only 20% due to the in-medium channel Δ+N$\rightarrow$N+N.

In heavy-ion reactions at bombarding energies of a few GeV/u pions are copiously produced. It is nowadays well known that they are created in individual nucleon-nucleon collisions [1, 2, 3]. Detailed analyses have shown that at these energies the probability for producing a pion in a nucleon-nucleon collision is about 30% [4]. The production of pions, except for the very low-energetic ones, proceeds dominantly via the Δ resonance, at least for pion kinetic energies up to about 300 MeV. This process between free nucleons has been investigated for a long time and is well understood.

However, if pion production happens inside the nuclear medium, as in the case of a heavy-ion collision, then two basically new phenomena appear. First, the final state of the nucleon – that is created in the decay $\Delta \rightarrow N\pi$– may be Pauli blocked and, second, a new decay channel opens for the Δ, i.e. the process $\Delta N \rightarrow NN$. Both are typical "in-medium" processes, which are well known from studies of pion interactions with nuclei [5]. The first effect leads to a decrease of the in-medium width of the Δ, the second to an increase.

At normal nuclear matter density $\rho_0 \approx 0.16$ fm^{-3} the two processes largely counterbalance each other as shown by photo absorption experiments [5]. The latter experimental studies also indicate that the resonance position is not significantly changed and that the potential, with which the Δ is bound, is nearly the same as that of a nucleon, possibly somewhat less attractive.

It would, therefore, be very interesting to see if this cancellation also works at higher densities as produced in heavy-ion collisions above 1 GeV/u. Furthermore, according to predictions by Brown *et al.* [6], precursor phenomena of the chiral symmetry restoration (at very high densities) should show up already at densities of 2-3 ρ_0. These densities actually are expected to be accessible in heavy-ion collisions, for example at the SIS accelerator ($\leq$ 2 GeV/u).

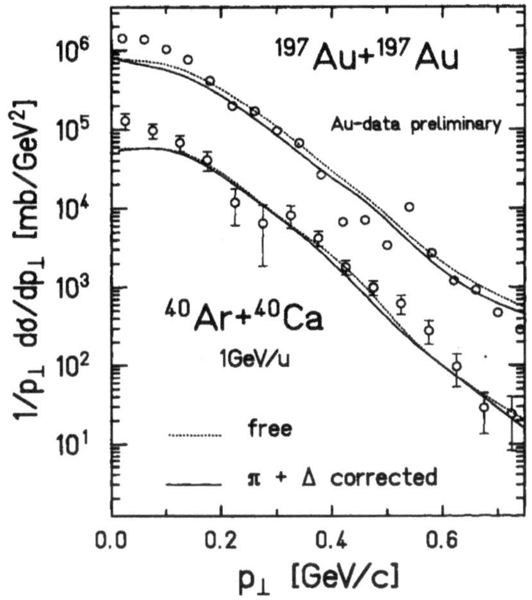

Figure1: Inclusive π^o transverse momentum spectra for Ar+Ca and Au+Au at 1 GeV/u employing free pions (dotted lines) and explicit pion and Δ selfenergies (solid lines). The experimental data are taken from Refs. [9, 10].

We have, therefore, performed calculations of the resonance population during such a collision within the transport approach described in Refs. [7, 8]; in this BUU calculation nucleons, Δ's, N(1440)'s and N(1535)'s are propagated explicitly with their isospin degrees of freedom as well as pions and η's.

That this approach works reasonably well can be seen from Fig. 1 where we compare the calculated inclusive transverse π^o momentum spectra for Ar+Ca and Au+Au at 1 GeV/u with the experimental data from [9, 10]. The dotted lines represent the calculated results assuming 'free' pions in the medium whereas the solid lines are obtained when explicitely including pion- and Δ- selfenergies as extracted from the Δ-hole model (cf. [16]). Except for very low transverse π^o momenta the spectra are well reproduced for both systems such that we can assure a realistic description of the Δ and pion abundancy throughout the reaction. Furthermore, when comparing the dotted and solid lines, we find no remarkable effect of Δ- and pion- selfenergies on the inclusive pion spectra.

In Fig. 2 we show the baryonic decomposition for the highest density region (maximal resonance/nucleon ratio) in the central cell ($\approx$ 33 fm^3) of the reaction volume for central Ne+Ne and Au+Au reactions as a function of the bombarding energy per nucleon. It is clearly seen that about 30% of all baryons are Δ's for both

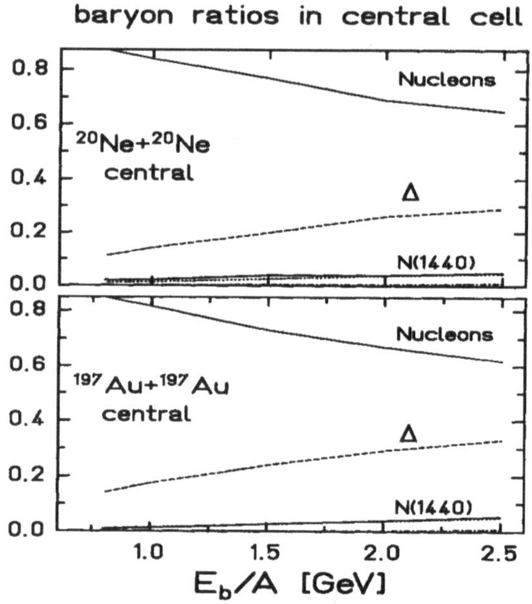

Figure 2: Baryon decomposition at the highest densities in a central cell of volume 33 fm^3 for head-on collisions of Ne+Ne and Au+Au as a function of the bombarding energy per nucleon.

systems around 2 GeV/u, whereas the population of the higher-lying resonances is small. Since the maximum density in this cell is about 3 times that of gound state nuclear matter (ρ_0), this result indicates a Δ density of about 0.8 - 1 ρ_0. Figure 3, which shows the time-development of the density ρ/ρ_0 (thick solid line) and the various hadronic components for Ne+Ne and Au+Au at 1.5 GeV/u in the central cell, indicates that this Δ density (dashed line) is maintained for about 8-10 fm/c in case of Au+Au, i.e., over a relatively long time compared to the typical nucleon-nucleon collision time ($\approx$ 1 fm/c) at this energy. We are thus looking at a strongly interacting system of relatively dense *resonance matter*, a new form of hadronic matter. On the other hand, the abundance of the higher resonances N(1440) (dotted line) and N(1535) (dash- dotted line) is down by roughly one or two orders of magnitude, respectively, in comparison to the nucleons (solid line).

In fact, our transport simulations indicate a strong correlation of the Δ abundance with the total baryon density (see also Ref. [11, 12]). This is demonstrated more closely in Fig. 4, where the Δ and N(1440) ratio to the total baryon density are plotted as a function of ρ/ρ_0 for central collisions of Ne+Ne and Au+Au at 1.5 GeV/u. In order to provide an integral information the latter correlation has been obtained by integrating over space and time of the whole reaction, thus including the

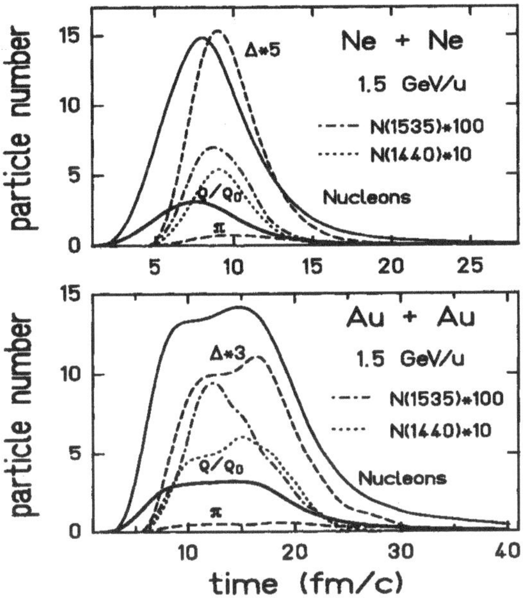

Figure 3: Time evolution of the density ρ/ρ_0 (thick solid line) and the number of baryons in the central cell for head-on collisions of Ne+Ne and Au+Au at 1.5 GeV/u. [Nucleons: solid line; Δ's: dashed line; N(1440): dotted line; N(1535): dash-dotted line.]

approach phase of the collision with average density $\rho/\rho_0 \approx 1$, however, $N_\Delta/N = 0$. As before the N(1440) (hatched histogram) and N(1535) (full histograms) appear to be of minor importance except when triggering on specific decay channels, e.g., N(1535) $\rightarrow \eta +$ N. We note in passing that the resulting η spectra [8] are also well in line with the experimental data.

An important consequence of the high Δ abundance (especially at high baryon density) is an amplification of particle production far below the free nucleon-nucleon threshold by two-body collisions in which one of the partners is a Δ. This has been illustrated for K^+ production in Ref. [13] where it was shown that at energies around 1 GeV/u about 90% of the kaons produced are created in nucleon-Δ collisions. Furthermore, in case of antiproton production around 2 GeV/u, almost all the yield stems from $N\Delta$ or even $\Delta\Delta$ channels [14]. This amplification is essentially due to the fact that a considerable amount of energy can be stored in internal nucleon excitations, which, in a subsequent reaction, can be used for the production of heavy mesons or a $p\bar{p}$ pair.

We have also analyzed the lifetime of the Δ resonance in matter, by looking at the questions raised above, namely the Pauli quenching and the collisional broadening. In the calculational method used the width Γ_Δ for populating a resonance is usually

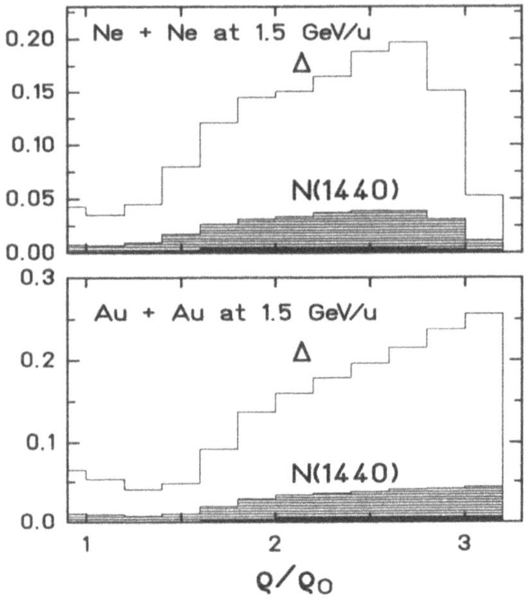

Figure 4: The ratios of Δ's, N(1440) (hatched histograms) and N(1535) (full histograms) to the total baryon number as a function of density ρ/ρ_0 for central collisions of Ne+Ne and Au+Au at 1.5 GeV/u integrated over space and time of the whole collision.

put in from the outside, whereas the effective decay width is determined dynamically, thus neglecting the effects of possible changes on the resonance population. We have, therefore, for the first time, performed these calculations of the widths self-consistently, i.e., they have been iterated by sequential BUU simulations until the prescribed width $\Gamma_{\rm in}(M_\Delta, \rho)$ for the population is the same as the dynamically evaluated decay width $\Gamma_{\rm out}(M_\Delta, \rho)$ of the resonances [12].

Figure 5 shows the results of such a study for various bins in the baryon density ρ/ρ_0 for a central collision of ^{40}Ca + ^{40}Ca at 1.5 GeV/u. It is seen that the iterated width of the Δ's (dashed lines) indeed gets somewhat larger at higher densities than the free width (solid line), but that this increase never amounts to more than 20%. Thus even at densities between 2 and 3 times ρ_0 Pauli quenching and collisional broadening cancel each other to a large extent. We also find that, contrary to earlier expectations, the self-consistency in the treatment of the resonance widths has only a very small effect.

Whereas the width or inverse lifetime of the Δ's varies only moderately with baryon density, we find a more significant shift of the Δ mass distribution as a function of the Δ generation. This is demonstrated in Fig. 6 (by the dotted lines) for central collisions of Ne+Ne and Au+Au at 1.5 GeV/u. Whereas the first generation of Δ

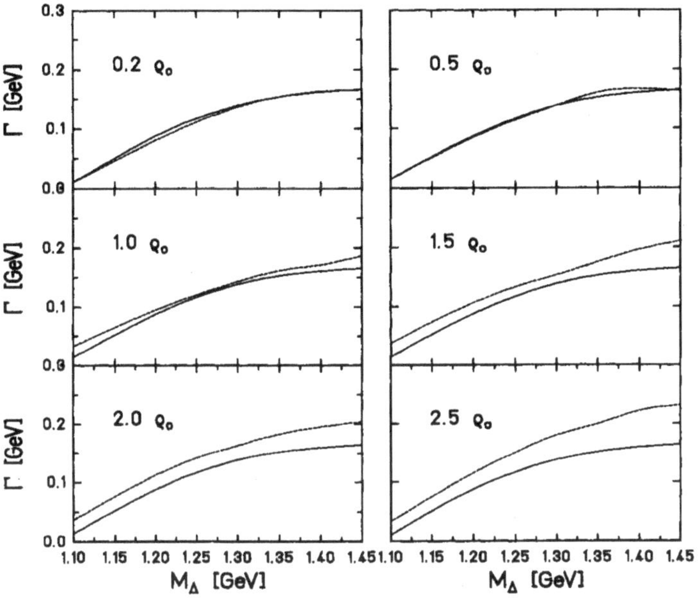

Figure 5: The iterated Δ width $\Gamma(M_\Delta, \rho)$ (dashed lines) as a function of mass and density in comparison to the free Δ width (solid lines) for a central collision of Ca+Ca at 1.5 GeV/u.

resonances – that are created in primary collisions – shows a distribution centered slightly below $M_\Delta = 1.232$ GeV (dashed line labeled 1), the sequential Δ generations are shifted downwards in mass due to Δ decay and pion reabsorption; this is the same effect as that discussed by Li and Bauer in Ref. [15]. The Δ mass distribution, which finally generates the asymptotically free pions, is shown in Fig. 6 in terms of the thick solid lines. The experimentally seen Δ mass distribution is only very moderately downward shifted for light Ne+Ne systems; for the heavier Au+Au the shift is more significant and amounts to about 50MeV. These results are relatively insensitive to the selfenergies of pions and Δ's if both are treated consistently [16].

In summary, our calculations indicate that Δ resonance matter does indeed exist in the central zone of the reaction volume in heavy-ion collision at energies around 2 GeV/u. Observable consequences are a strong enhancement of the far-subthreshold particle production cross sections [13, 14]; also the observed high pion production probabilities per nucleon-nucleon collision, extracted from heavy-ion reaction data in Ref. [4], point into that direction. Furthermore, the transverse pion spectra reflect the average Δ mass distribution and thus the average number of pion emission and reabsorption events.

We conclude that heavy-ion reactions in the energy regime about 1–2 GeV/u

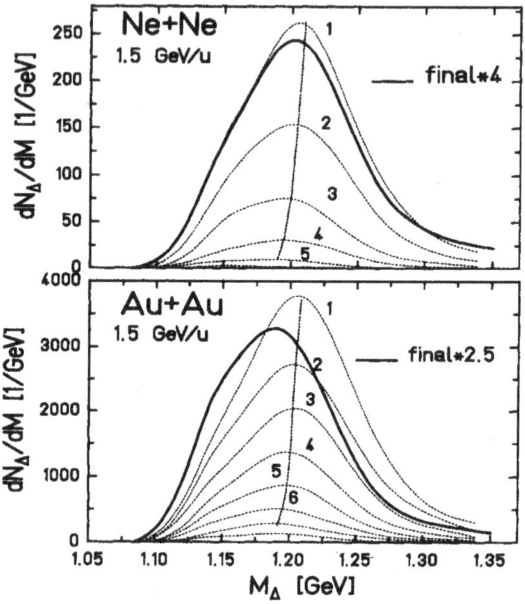

Figure 6: The Δ mass distribution for different Δ generations (labels 1-5). The dotted lines indicate the shift of the individual peak positions with the generation number. The thick solid line reflects the Δ mass distribution that decays into asymptotically observed pions.

should indeed be a good testing ground for in-medium modifications of hadronic properties.

We are grateful to V. Metag for many helpful and stimulating discussions on this subject. This work was supported by BMFT and GSI Darmstadt.

References

[1] G.F. Bertsch and S. Das Gupta, *Phys. Rep.* **160** (1988) 189.

[2] W. Cassing, V. Metag, U. Mosel, and K. Niita, *Phys. Rep.* **188** (1990) 363.

[3] U. Mosel, *Annu. Rev. Nucl. Part. Sci.* **41** (1991) 29.

[4] V. Metag, *Prog. Part. Nucl. Phys.* **30** (1993) 75.

[5] T. Ericson and W. Weise, *Pions and Nuclei* (Oxford University, Oxford, 1988).

[6] G. E. Brown and M. Rho, *Phys. Rev. Lett.* **66** (1991) 2720.

[7] Gy. Wolf, G. Batko, W. Cassing, U. Mosel, K. Niita, and M. Schäfer, *Nucl. Phys.* **A517** (1990) 615.

[8] Gy. Wolf, W. Cassing and U. Mosel, *Nucl. Phys.* **A545**, 139c (1992); **A552** (1993) 549.

[9] F.D. Berg et al, *Z. Phys.* **A340** (1991) 297.

[10] W. Kühn, *Proceedings of 8th winter workshop on nuclear dynamics*, Jan. 1992, Jackson Hole, Wyoming, USA.

[11] L. Xiong, Z.G. Wu, C.M. Ko and J.Q. Wu, *Nucl. Phys.* **A512** (1990) 772.

[12] W. Ehehalt, W. Cassing, A. Engel, U. Mosel, Gy. Wolf, *Phys. Rev.* **C47** (1993) 2467.

[13] A. Lang, W. Cassing, U. Mosel and K. Weber, *Nucl. Phys.* **A541** (1992) 507.

[14] G. Batko, W. Cassing, U. Mosel, K. Niita and Gy. Wolf, *Phys. Lett.* **B256** (1991) 331.

[15] B.A. Li and W. Bauer, *Phys. Rev.* **C44** (1991) 450.

[16] W. Ehehalt, W. Cassing, A. Engel, U. Mosel, and Gy. Wolf, *Phys. Lett.* **B298** (1993) 31.

PROPERTIES OF RHO AND ETA MESONS IN NUCLEAR MATTER

M. HERRMANN[a,*], CHR. SAUERMANN[a], B. L. FRIMAN[a,b] and W. NÖRENBERG[a,b]

[a] *Gesellschaft für Schwerionenforschung (GSI)*
Postfach 110552, D-64220 Darmstadt, Germany

[b] *Institut für Kernphysik, TH Darmstadt*
Schloßgartenstraße 9, D-64289 Darmstadt, Germany

ABSTRACT

The properties of ρ- and η-mesons in nuclear matter are studied within the scope of hadronic models. Unknown model parameters are obtained from fits to scattering data. – The treatment of the ρ-meson includes the coupling to two pions which, in matter, are strongly mixed with delta-particle–nucleon-hole states. The ρ-meson self-energy is evaluated in a current conserving approximation with in-medium pion propagators and vertex corrections. While the position of the original peak in the spectral function remains almost unchanged, its width grows rapidly with increasing density. Consequently, the ρ-meson strength function is strongly dispersed at high densities. Due to vertex corrections a new peak at a mass around $3m_\pi$ emerges with increasing density, while the spectral function around the two-pion threshold is found to be smooth at all densities. – The η-meson is strongly mixed with $N^*(1535)$-particle–nucleon-hole states in nuclear matter. The corresponding dispersion relations with an upper and a lower branch look similar to those of the $(\pi, \Delta N^{-1})$-modes. However, since the N^* is an S-wave resonance in the ηN-channel, the repulsion of the two branches survives at zero momentum.

1. Introduction

A main objective of heavy-ion collision experiments at intermediate energies is to determine the properties of nuclear matter at high densities and temperatures. Signals from hot and compressed matter, temporarily produced during relativistic heavy-ion collisions, are conveyed by hadronic spectra and abundances. Due to the large strength of nuclear interactions, the properties of hadrons are in general strongly modified in nuclear matter. Correspondingly, the in-medium values of cross sections differ from their vacuum values. Therefore, good knowledge of hadron properties in nuclear matter is essential for a reasonable description of heavy-ion reactions and a correct interpretation of experimental data. In this connection the ρ-meson and the η-meson are of particular interest.

*Present address: National Institute for Nuclear Theory, University of Washington, Seattle, WA 98195, USA

At energies below 1 GeV the neutral vector mesons $\rho^{(0)}$, ω and ϕ play an important role in the electromagnetic interactions of hadrons. In fact, in the *Vector Meson Dominance* (VMD) model, the coupling of a photon to a hadron always involves a neutral vector meson [1]. Although, in some cases there are additional contributions, the VMD mechanism remains a basic ingredient in any attempt to understand the electromagnetic properties of hadrons. The pion electromagnetic form factor $F(q^2)$ is a particularly striking example of VMD [2]. Since the pion current is a pure isovector, only the $\rho^{(0)}$ contributes when the small corrections due to G-parity violating processes are neglected. Such a model gives a good description of the pion electromagnetic form factor, and implies, e.g., that the annihilation of a $\pi^+\pi^-$ pair into an e^+e^- pair proceeds exclusively via the neutral ρ-meson.

Once produced in a relativistic heavy-ion reaction, photons and leptons escape virtually without subsequent collisions with the surrounding matter. Thus, such probes may provide undistorted information on the hot and dense matter created in the compression phase of the collision. Recently, dilepton production in nuclear collisions has attracted a lot of interest. Indeed, the first generation of experiments with p+Be collisions at 1.0, 2.1 and 4.9 GeV, Ca+Ca collisions at 1.0 and 2.0 GeV/u and Nb+Nb at 1.05 MeV show that e^+e^- production is a promising tool for probing the conditions in such collisions [3]. Simulations [4] of these heavy-ion reactions show that in the ρ-meson region the measured invariant-mass spectra of e^+e^- pairs should be sensitive to the $\pi^+\pi^-$ annihilation. From planned experiments with HADES at GSI one expects detailed information on this process [5]. We have studied the ρ-meson for various densities of nuclear matter in a hadronic model which satisfies unitarity and current conservation [6, 7]. The approach, the modifications of the ρ-meson in nuclear medium and the consequences for the dilepton spectra are discussed in section 2.

In contrast to photons and leptons, hadrons are modified after production by subsequent collisions with the surrounding matter at later stages of the reaction process. However, the production rates of certain particles are sensitive to the densities and temperatures which are reached in the hot compressed zone. Especially, particles like the η- and K-mesons, which are produced in the subthreshold energy regime, are very promising probes to study the properties of hot and dense nuclear matter [8]. First experiments at GSI show that the multiplicities and spectra of these mesons can be observed in heavy-ion collisions around 1 GeV/u bombarding energy [9, 10]. Motivated by these experiments the in-medium properties of the η-meson at various nuclear densities have been calculated [11]. The model and first results are discussed in section 3.

2. The ρ-Meson in Nuclear Matter and e^+e^- Production

In the first calculations [12, 13] of dilepton production in nuclear matter, medium corrections to the ρ-meson propagator were not taken into account. However, since the ρ-meson is essentially a two-pion resonance, large modifications of its properties (and thus, through VMD, also of the pion electromagnetic form factor) are expected from the strong pion-nucleon-delta ($\pi N\Delta$) coupling. For $\pi\pi$-scattering in the σ-channel, such interactions indeed lead to large medium effects [14]. Furthermore, unitarity is violated in models for dilepton production, where the properties of the annihilating pions are modified in medium, while the properties of the ρ-meson, and thus the form factor, are left unchanged. In the following parts of this section we report on calculations of the in-medium properties of ρ-mesons, which comply with unitarity, causality and current conservation [6, 7].

2.1. The ρ-Meson in Vacuum

Our Lagrange density for the photons, the neutral ρ-mesons and the charged π-mesons reads

$$\mathcal{L} = [D_\mu\pi]^* [D^\mu\pi] - m_\pi^2\pi^*\pi - \frac{1}{4}G_{\mu\nu}G^{\mu\nu} + \frac{(m_\rho^0)^2}{2}\rho_\mu\rho^\mu - \frac{1}{4}F_{\mu\nu}F^{\mu\nu} - \frac{e}{2g}F_{\mu\nu}G^{\mu\nu}, \quad (1)$$

where $G^{\mu\nu} = \partial^\mu\rho^\nu - \partial^\nu\rho^\mu$ and $F^{\mu\nu} = \partial^\mu A^\nu - \partial^\nu A^\mu$ are the field-strength tensors of the ρ-meson and the photon fields, respectively, and $D^\mu = \partial^\mu + ieA^\mu + ig\rho^\mu$ is a covariant derivative. This lagrangian is explicitly gauge invariant and leads to ρ-meson dominance [15]. The minimal coupling scheme implies a 4-point contact vertex which is essential for current conservation and gives rise to so-called "seagull" diagrams.

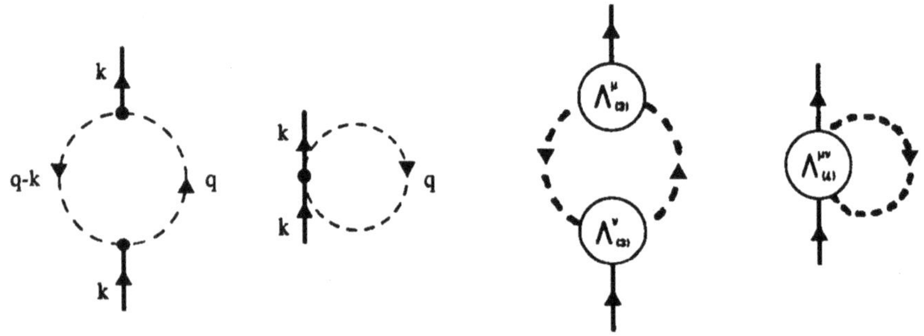

Fig. 1. Self-energy diagrams of the ρ-meson in vacuum (left part) and in nuclear matter (right part). The dashed lines represent pions. In nuclear matter the pion propagators and the vertices are dressed.

In eq.(1) the ρ-field corresponds to a stable "bare" ρ-meson with mass m_ρ^0. A model for the physical ρ-meson in vacuum is obtained by dressing the bare ρ with pions. We calculate the self-energy of the ρ-meson to order g^2 (see Fig. 1) and employ the Pauli-Villars scheme for regularization. The cut-off mass Λ, which enters in the regularization procedure, sets a scale beyond which one expects quarks and gluons to be the relevant degrees of freedom rather than hadrons. Our results, in vacuum as well as in nuclear matter, are almost independent of the cut-off mass for $\Lambda > m_\rho$. For a fixed cut-off mass of $\Lambda = 1$ GeV we fit the bare $\rho\pi\pi$ coupling constant ($g = 6.05$) and the bare ρ-mass ($m_\rho^0 = 875$ MeV) to the pion electromagnetic form factor as well as to the isovector P-wave phase shifts for $\pi\pi$ scattering (see Fig 2).

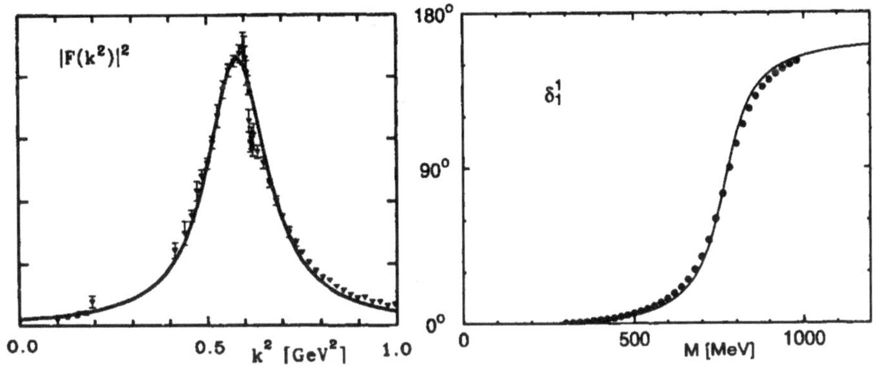

Fig. 2. Fits of the pion electromagnetic form factor as a function of k^2 in the time-like region and of the phase shift δ_1^1 for $\pi^+\pi^-$ scattering in the vector-isovector channel as a function of the invariant mass M.

2.2. The ρ-Meson in Nuclear Matter

The ρ-meson properties are modified in nuclear matter mainly due to the large $\pi N\Delta$ coupling which strongly mixes the pions in nuclear matter with delta-particle–nucleon-hole (ΔN^{-1}) excitations. The excitation of the Δ-resonance in pion-nucleon scattering at moderate energies is well described by the (non-relativistic) interaction lagrangian (cf. ref. [16])

$$\mathcal{L}_{\pi N\Delta} = -\frac{f_\Delta}{m_\pi}\psi_N^\dagger \left(\boldsymbol{S}\cdot\boldsymbol{\nabla}\right)\left(\vec{T}\cdot\vec{\pi}\right)\psi_\Delta + \text{h.c.}\,, \tag{2}$$

where $\boldsymbol{S}$ and $\vec{T}$ are the transition spin and isospin operators, respectively. The gradient of the isovector pion field $\vec{\pi}$ reflects the P-wave nature of the interaction in the spin-$\frac{3}{2}$ channel. The scattering data in the spin-$\frac{3}{2}$ isospin-$\frac{3}{2}$ channel and the width of the Δ-resonance are consistently described with the coupling constant $f_\Delta = 2$.

To keep the treatment as transparent as possible, we neglect the width of the Δ-resonance and approximately account for the short-range correlations by reducing the $\pi N\Delta$ coupling by 25% to reproduce the pionic dispersion relations for a realistic value of the Landau-Migdal parameter $g'_\Delta = 0.5$. In this fashion, we maintain current conservation without introducing additional vertex corrections of the kind considered in ref. [17]. Finally, the pion self-energy in nuclear matter (cf. Fig. 3) is calculated to lowest order in the baryon density (quasi-static approximation) including a monopole form factor with a cut-off mass of 1 GeV. The resulting dispersion relations for the coupled ($\pi, \Delta N^{-1}$)-modes are shown in Fig. 3.

The ρ-meson couples to the isospin current and, therefore, it couples to all particles which carry isospin in nuclear matter: to the pions, to the nucleons, to the deltas and consequently also to the ΔN^{-1} component of the pion-like modes. This implies corrections to the $\rho\pi\pi$ and $\rho\rho\pi\pi$ vertices, as indicated in Fig. 1. Since the ρ-meson couples to a conserved current also in nuclear matter, the resulting vertex functions satisfy the Ward-Takahashi identities. The vertex corrections give rise to a large number of additional ρ-meson self-energy diagrams. All of these have to be included in order that the coupling of the electromagnetic field to the pion remains gauge invariant in medium. Furthermore, since there are rather delicate cancellations between different diagrams, an approximation, where some diagrams are neglected or treated inconsistently, may result in sizable errors.

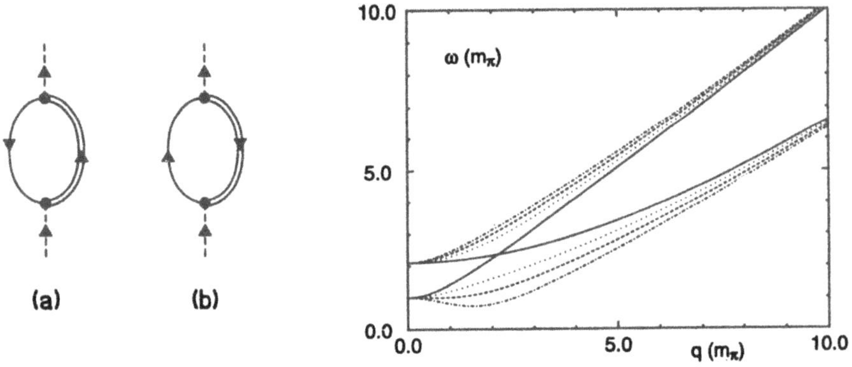

Fig. 3. Second-order self-energy diagrams for the pion and corresponding dispersion relations for the normal modes of pions and pionic ΔN^{-1} excitations in nuclear matter at zero temperature. The solid, dotted, dashed and dash-dotted lines, respectively, correspond to the baryon densities $\rho_B/\rho_0 = 0$, 1, 2 and 3.

2.3. Spectral Function and e^+e^- Spectra

The properties of the ρ-meson in nuclear matter are characterized by the spectral function which, for a ρ-meson at rest with energy $k_0 = M$, is given by

$$\mathcal{A}_\rho(M) = \frac{-2\mathrm{Im}\Sigma^+(M)}{\left(M^2 - \left((m_\rho^0)^2 + \mathrm{Re}\Sigma^+(M)\right)\right)^2 + \left(\mathrm{Im}\Sigma^+(M)\right)^2}. \tag{3}$$

We have calculated the spectral function of a ρ-meson at rest in zero-temperature nuclear matter (see Fig. 4) and the corresponding electromagnetic form factor of the pion as a function of the ρ-meson energy M for baryon densities 1, 2 and 3 times normal nuclear density. While the peak position remains almost unchanged, the width grows rapidly with increasing density, reaching about 500 MeV FWHM already at twice normal nuclear density. At threshold, around $M = 2m_\pi$, there is no structure which could be a signature for changes in the pion dispersion relation. The enhancement predicted by Gale and Kapusta[12] is cancelled by the vertex corrections and the medium corrections to the pion form factor. With increasing density a new structure develops around $M \approx 3m_\pi$, due to processes, where the ρ-meson decays into a pionic mode on the lower branch and a ΔN^{-1}-like mode. The latter is either an uncorrelated ΔN^{-1} state or a pionic mode on the upper branch. The enhancement develops when $d(\omega_1 + \omega_{\Delta N^{-1}})/dq$ becomes small. In this case, which involves transition matrix elements, the singularity at $\rho_B/\rho_0 = 3$ is not suppressed by vertex corrections.

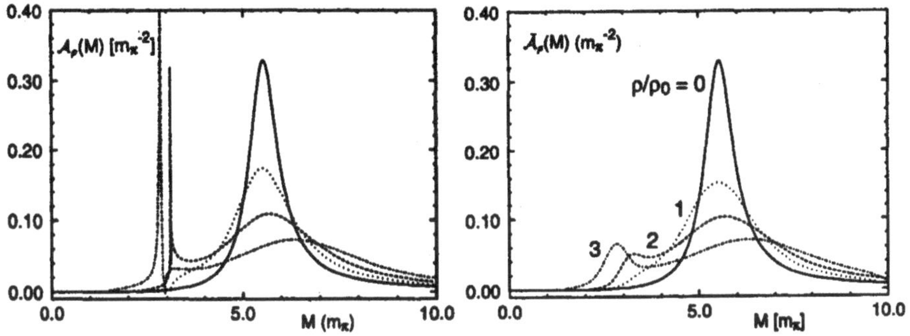

Fig. 4. Spectral functions of the ρ-meson at rest in nuclear matter for temperature $T = 0$ and densities $\rho_B/\rho_0 = 0$, 1, 2 and 3. For obtaining the smoothed curves of the right-hand figure, a gaussian with the energy-dependent standard deviation $\sigma(M) = (\Gamma_\Delta/\pi)\arctan[((M - M_{th})/m_\pi)^{3/2}]$, where M_{th} denotes the threshold energy, has been used for averaging the curves of the left figure. The smoothing procedure preserves the energy-weighted sum rule.

In order to simplify the treatment of the ρ-meson in nuclear matter we have introduced two major approximations: we have neglected the width of the Δ-resonance and treat the short-range correlations in the ΔN^{-1} channel by reducing the $\pi N \Delta$ coupling constant. In spite of these approximations, our model still contains the main characteristics of pion propagation in nuclear matter. Within this approach we are able to systematically implement current conservation without making further approximations. Our study demonstrates the importance of treating current conservation accurately. On the one hand, there are delicate cancellations between different diagrams. If current conservation is treated approximately, these cancellations are no longer exact, which may lead to sizable errors in the final results. On the other hand, the vertex corrections give rise to qualitative changes in the ρ-meson properties. For example, the narrow resonance structure around $M = 3m_\pi$ in the spectral function of Fig. 4 (l.h.s.) results from vertex corrections. Calculations without these vertex corrections[18] yield a smooth behaviour in this energy region.

We illustrate the effect of a finite Δ-width by folding the spectral functions with a gaussian. The right part of Fig. 4 shows the resulting strength functions; the main effect of the smoothing is to give an appreciable width to the narrow peak around $M = 3m_\pi$. Furthermore, it becomes clear that this peak carries only a small fraction of the total strength.

Asakawa *et al.*[19] find spectral functions which are rather similar to the smoothed curves of Fig. 4. Quantitatively, the main ρ-peak broadens faster than in our model, while the $3m_\pi$ bump seems to carry a larger fraction of the strength. Furthermore, in their calculation both peaks move slightly upwards in energy with increasing density. A similar upward shift is also reported by Chanfray and Schuck[20]. However, they find almost no broadening of the main peak. In these calculations[19, 20] both the Δ-width and the short-range repulsion in the effective ΔN^{-1} interaction were included. Whereas the Δ-width broadens the structure of the spectral function in quite the same way as seen in Fig. 4, the short-range correlations in the ΔN^{-1} states apparently cause a shift to higher energies with increasing density. The discrepancy between the results of ref.[19] and ref.[20] may be due to the different approximations in the treatment of the vertex corrections. A consistent calculation, in the spirit of this paper, including the Δ-width and short-range correlations without sacrificing an accurate treatment of current conservation, is still lacking.

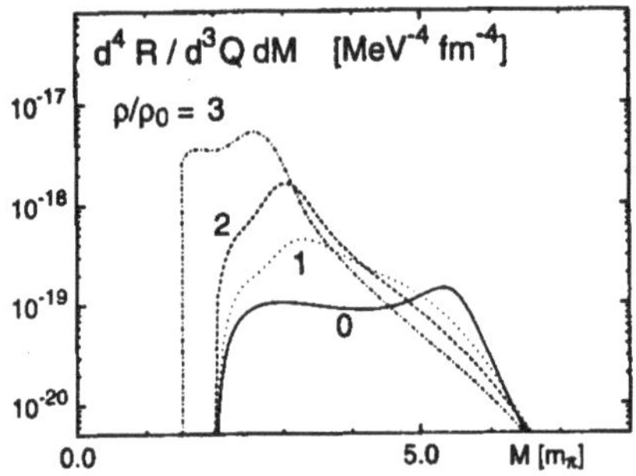

Fig. 5. Production rate for e^+e^- pairs of invariant mass M and total momentum $\boldsymbol{Q} = 0$ in nuclear matter at temperature $T = 100$ MeV and densities $\rho_B/\rho_0 = 0$, 1, 2 and 3. The curves are obtained with the smoothed spectral functions of fig. 4.

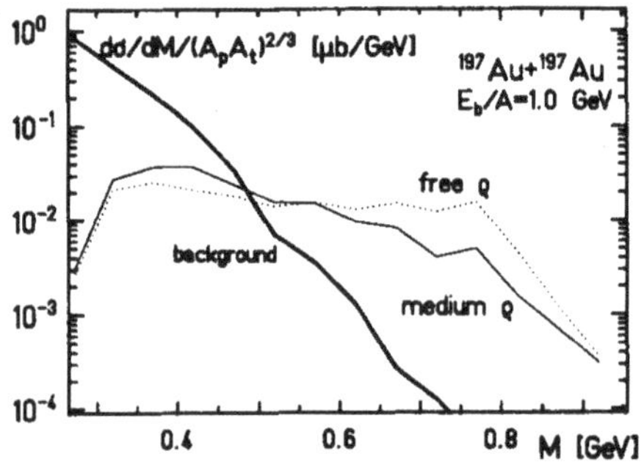

Fig. 6. Invariant-mass spectra of e^+e^- pairs with total momentum $\boldsymbol{Q} = 0$ produced by $\pi^+\pi^-$ annihilation in the collision Au(1GeV)+Au. The calculations have been performed by Gy. Wolf[28] with the free and the smoothed in-medium spectral functions of Fig. 4. The background contribution is due to all other processes, apart from the $\pi^+\pi^-$ annihilation, which produce e^+e^- pairs.

The spectral function of the ρ-meson is a fundamental quantity for all ρ-meson dominated processes. An interesting example, mentioned in the introduction, is the $\pi^+\pi^-$ annihilation into e^+e^- pairs. The corresponding production rate for e^+e^- pairs with invariant mass M, emitted back-to-back, i.e., with total momentum $\boldsymbol{Q} = 0$ is given by

$$\left.\frac{d^4R}{dMd^3Q}\right|_{Q=0} = \frac{4\alpha^2}{(2\pi)^3}\frac{(m_\rho^0)^4}{g^2}\frac{1}{M^2}\mathcal{A}_\rho(M)f_\rho(M) \tag{4}$$

for equilibrated nuclear matter (see also ref. [21]). Here, f_ρ denotes the ρ-meson equilibrium distribution function and $\alpha = e^2/4\pi$ is the fine-structure constant. The spectral function $\mathcal{A}_\rho$ is proportional to $\mathrm{Im}\Sigma^+|F(M)|^2$. Thus, using the vacuum form factor in (4) one violates unitarity, because only the numerator of (3) is modified in medium, while the self-energy in the denominator is left unchanged.

In Fig. 5 we show the thermal production rate for e^+e^- pairs with total momentum $Q = 0$ at $T = 100$ MeV and densities 1, 2 and 3 times normal nuclear density. In this calculation we have neglected the temperature dependence of the spectral function and used the smoothed ones of Fig. 4. The invariant-mass spectra are seen to depend strongly on the density. We note that, in spite of the relatively small spectral strength carried by the peak around $M = 3m_\pi$, it is quite prominent in the dilepton spectrum. However, as is seen from Fig. 6, this peak is masked by other processes in relativistic heavy-ion collisions. Only if shifted upwards by the short-range correlations in the ΔN^{-1} excitations into the mass region $M > 500$ MeV, where $\pi^+\pi^-$ annihilation dominates the dilepton spectra, also this new structure may become observable.

3. The η-Meson in Nuclear Matter

In order to understand the η-production in heavy-ion collisions it is necessary to know the properties of the η-meson in nuclear matter. Due to the interactions of the η-meson with the surrounding nucleons, N^*N^{-1}-pairs are excited, in analogy with the familiar $\pi N\Delta$-system (cf. subsect. 2.2). The lowest-lying resonance, which strongly couples to the ηN system, is the $N^*(1535)$. The lagrangian $L = g_{\eta NN^*}\bar{\Psi}_{N^*}\Psi_N\eta + h.c.$ describes the corresponding interaction.

3.1. Pion-Nucleon Scattering

We determine the ηNN^* coupling constant from fitting the πN scattering data in the S_{11}-channel with a model which describes both the elastic πN scattering data and the η-production.

The experimental data [22] for the T-matrix in the elastic S_{11}-channel are shown in Fig. 7. The lower-lying maxima in the real and imaginary parts are related to the $N^*(1535)$ resonance. In order to reproduce the double-hump structure we include the $N^*(1650)$ which has the same quantum numbers as the $N^*(1535)$. It couples only weakly to the η but quite strongly to the π and 2π channels. The coupling to two pions is simulated by including a fictitious pseudoscalar meson of mass 400 MeV.

We calculate the K-matrix in an approximation where the s-channel contributions from the two resonances N*(1535) and N*(1650) are taken into account considering the real parts of the self-energies as constant. An advantage of calculating the K-matrix is that it yields unitarity in any approximation. A further advantage of this approach is that one obtains coupling constants which do not depend on the cutoffs of the model. As shown in Fig. 7 the experimental data are quite well reproduced by fitting the model parameters including $g_{\eta NN^*} = 1.4$. The discrepancy in the real part at low energies is probably due to the non-resonant Born terms which are not yet included in the presented results. In the resonance region these terms are not expected to give an important contribution. Calculations including the Born terms are in progress.

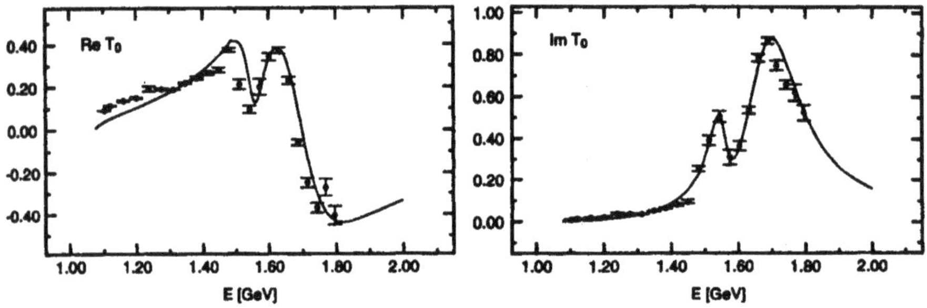

Fig. 7. Real and imaginary parts of the T-matrix T_0 for πN scattering in the elastic S_{11}-channel. The solid line is the fit of our K-matrix calculation to the experimental data of ref. [22].

3.2. The (η, N^*N^{-1})-Modes in Nuclear Matter

It is now straightforward to calculate the self-energy of the η-meson in the relativistic one-loop approximation (cf. Fig. 3). Figure 8 shows the location of the poles of the resulting η-propagator in the $(\omega, |\boldsymbol{q}|)$-plane. Also seen is how the spectral weight of the η is distributed in the region of the N^*N^{-1} continuum. The calculations have been performed with zero width for the N*.

The self-energy can be obtained analytically in the quasi-static approximation which is determined by the lowest-order terms of an expansion in k_F/m_N, where m_N is the nucleon mass (cf. subsect. 2.2). The dispersion relations of the free η-meson and the free N^*N^{-1}-pairs as well as the dispersion relations of the coupled (η, N^*N^{-1})-modes in nuclear matter are shown in Fig. 9. The result is obviously very similar to the $(\pi, \Delta N^{-1})$-system (cf. Fig. 3). An important difference, however, is that the N*(1535) is an S-wave resonance, which means that the free and the interacting modes do not coincide at $|\boldsymbol{q}| = 0$. As expected, the residues of the propagator reveal that at low energies the lower branch is mostly η and the upper branch is mostly N^*N^{-1}, while at higher energies the two branches exchange their roles.

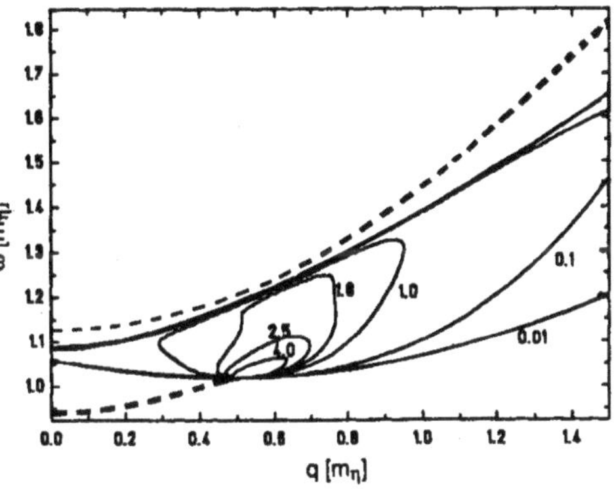

Fig. 8. Dispersion relation of the (η,N*N^{-1})-modes in normal nuclear matter outside the N*N^{-1} continuum (dashed lines) and spectral weight of the η-meson within the N*N^{-1} continuum.

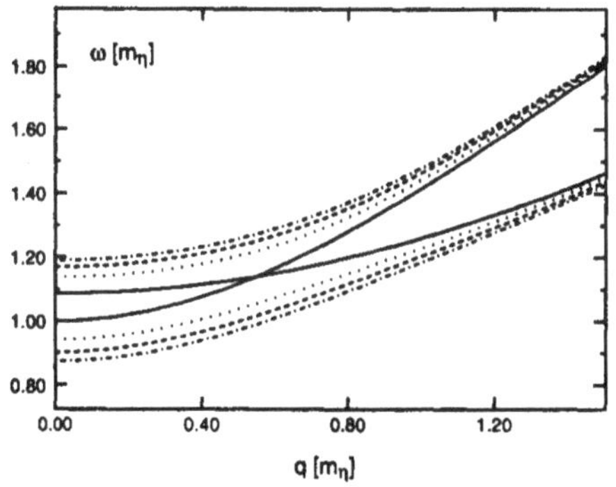

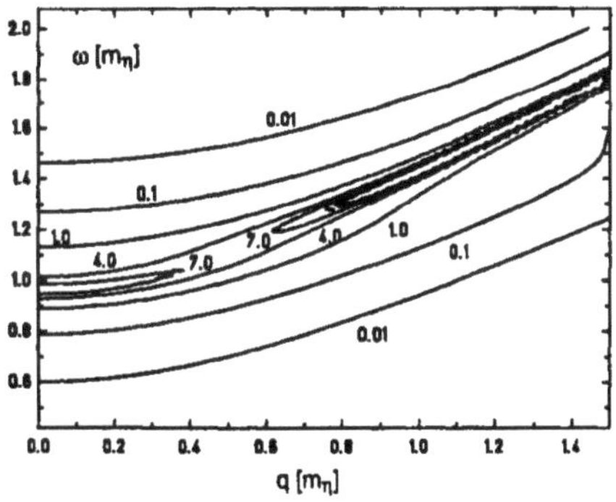

Fig. 9. Dispersion relations for the (η,N*N^{-1}) modes in nuclear matter at densities $\rho/\rho_0 = 0$ (solid lines), 1 (dotted lines), 2 (dashed lines) and 3 (dash-dotted lines).

Fig. 10. Spectral distribution of the η-meson in normal nuclear matter calculated for an N*-width of 150 MeV. Compare the ridge with the curves for $\rho/\rho_0 = 0$ and 1 in Fig. 9.

In order to obtain an idea about the effect of the width of the N*-resonance, we have performed a non-relativistic calculation in the quasi-static approximation with constant width Γ. Figure 10 shows the spectral distribution of the η for $\Gamma = 150$ MeV and normal nuclear density. We observe a ridge which is close to the dispersion relation of the η in vacuum, a result which is similar to the one obtained in ref. [23] for the spectral distribution of the pion in nuclear matter. For $|q|$ close to zero the same shift to lower energies is observed as for $\Gamma = 0$ (cf. Fig. 9).

So far our study suggests that the subthreshold production of η-mesons in relativistic heavy-ion collisions is enhanced due to medium effects, because the effective mass is reduced, and hence less energy is needed for the creation. Calculations of in-medium cross sections for η-meson production are in progress. We note, however, that in order to obtain results which can be compared with experimental data on η-production in heavy-ion collisions, it is not sufficient to calculate the excitation of the (η,N*N^{-1})-

modes. In addition, one has to solve the problem how these modes couple to the free η outside of the expanding fireball.

4. Conclusion

Our results show that ρ- and η-mesons are strongly modified in nuclear matter, with the changes increasing with increasing density. We have obtained these medium effects within hadronic models whose parameters are fitted to scattering data. There are good arguments [2] to believe that hadronic degrees of freedom are sufficient to describe hadronic reactions up to laboratory energies of 2 GeV/u. However, the possibility of effects from the quark and gluon degrees of freedom on meson properties in matter cannot be excluded. Based on the scale invariance of QCD, Brown and Rho [26] have suggested that the masses of π-, σ-, ρ- and ω-mesons decrease rapidly in nuclear matter with increasing density, a result which is in contrast to calculations within the (not scale-invariant) Nambu–Jona-Lasinio model [24, 25]. Similar rapid decreases of meson masses with increasing density have been predicted from QCD sum rules [27]. However, in our opinion, the situation concerning the density dependence of the $\bar{q}q$ components of the mesons is not yet settled. Consequently, we have restricted our studies of meson properties in matter to hadronic degrees of freedom and have considered no density dependence of the mass of the bare ρ-meson which is regarded as the $\bar{q}q$ component of the physical ρ-meson.

We thank Gy. Wolf for providing us with the results shown in Fig. 6.

References

1. J.J. Sakurai, *Currents and Mesons* (University of Chicago Press, Chicago, 1969) and Ann. Phys. (N.Y.) **11** (1960) 1
2. T. Ericson and W. Weise, *Pions and Nuclei* (Clarendon Press, Oxford, 1988)
3. G. Roche *et al.* (DLS collaboration), Phys. Rev. Lett. **61** (1988) 1069, Phys. Rev. Lett. **62** (1989) 2652, Phys. Lett. **B226** (1989) 228 and Phys. Rev. **C47** (1993) 2840
4. Gy. Wolf, W. Cassing and U. Mosel, Nucl. Phys. **A552** (1993) 549
5. HADES collaboration, Proc. Int. Workshop XXI on Gross Properties of Nuclei and Nuclear Excitations, Hirschegg, 1993, p. 144
6. M. Herrmann, *Eigenschaften des ρ-Mesons in dichter Kernmaterie*, PhD thesis (TH Darmstadt), Report 92-10 (GSI Darmstadt, ISSN 0171-4546)
7. M. Herrmann, B.L. Friman and W. Nörenberg, Nucl. Phys. **A560** (1993) 411
8. U. Mosel, Annu. Rev. Nucl. Part. Sci. **41** (1991) 29
9. V. Metag, Nucl. Phys. **A553** (1993) 283c

10. P. Senger *et al.* (KaoS collaboration), Nucl. Phys. **A553** (1993) 757c

11. Chr. Sauermann, *Resonanzen im S_{11}-Kanal der Pion-Nukleon-Streuung und die Propagation von η-Mesonen in Kernmaterie*, Diploma thesis (TH Darmstadt), GSI-Report 93-19; Chr. Sauermann, B.L. Frimann and W. Nörenberg, Annual Report 1992 (GSI-Report 93-1) p. 111

12. C. Gale and J. Kapusta, Phys. Rev. **C35** (1987) 2107

13. C.L. Korpa and S. Pratt, Phys. Rev. Lett. **64** (1990) 1502

14. P. Schuck, W. Nörenberg and G. Chanfray, Z. Phys. **A330** (1988) 119

15. N.M. Kroll, T.D. Lee and B. Zumino, Phys. Rev. **157** (1967) 1376

16. E. Oset and W. Weise, Nucl. Phys. **A319** (1979) 477

17. C.L. Korpa, L. Xiong, C.M. Ko and P.J. Siemens, Phys. Lett. **B246** (1990) 333

18. M. Herrmann, B.L. Friman and W. Nörenberg, Proc. Int. Workshop XIX on Gross Properties of Nuclei and Nuclear Excitations, Hirschegg, 1991, p. 134
G. Chanfray, A. Aouissat, P. Schuck and W. Nörenberg, Phys. Lett. **B256** (1991) 325
V. Mull, J. Wambach and J. Speth, Phys. Lett. **B286** (1992) 13

19. M. Asakawa, C.M. Ko, P. Lévai and X.J. Qiu, Phys. Rev. **C46** (1992) R1159

20. G. Chanfray and P. Schuck, Nucl. Phys. **A545** (1992) 271c

21. C. Gale and J. Kapusta, Nucl. Phys. **B357** (1991) 65

22. R.A. Arndt, J.M. Ford and L.D. Roper, Phys. Rev **D 32** (1985) 1085

23. L. Xia, P.J. Siemens and M. Soyeur, Proc. Int. Workshop XX on Gross Properties of Nuclei and Nuclear Excitations, Hirschegg, 1992, p. 183

24. V. Bernard and U.G. Meissner, Nucl. Phys. **A489** (1988) 647

25. M. Lutz, S. Klimt and W. Weise, Nucl. Phys. **A542** (1992) 521

26. G.E. Brown and M. Rho, Phys. Rev. Lett. **66** (1991) 2720

27. T. Hatsuda and S.H. Lee, Phys. Rev. **C46** (1992) R34

28. Gy. Wolf, private communication (Sept. 1993)

HEAVY-ION COLLISIONS WITHIN RELATIVISTIC MEAN-FIELD TWO-FLUID MODEL

YU.B. IVANOV[a,b,†], V.N. RUSSKIKH[a,b], YU.E. POKROVSKY[a,b],
P.A. HENNING[b,c], B.L. FRIMAN[b,c] and W. NÖRENBERG[b,c]

[a] *Kurchatov Institute, Moscow 123182, Russia*
[b] *Gesellschaft für Schwerionenforschung (GSI), 64220 Darmstadt, Germany*
[c] *Institut für Kernphysik, TH Darmstadt, 64289 Darmstadt, Germany*

Abstract

A three-dimensional realization of the relativistic mean-field 2-fluid model is described. The first results of analysizing the inclusive data on the yield of nuclear fragments and pions, as well as the Plastic-Ball rapidity distributions of nuclear fragments are presented. For comparison, the calculations within the conventional relativistic hydrodynamical model with the same mean fields are also performed. It is found that all the analyzed observables, except the pion spectra, appeared to be rather insensitive to the nuclear EOS. The sensitivity to the nuclear stopping power is slightly higher. The original sensitivity of the rapidity distributions is smeared out by the Plastic-Ball filter and selection criterion. The stopping power induced by the Cugnon cross-sections is not quite sufficient for a more adequate reproduction of the experimental data.

Applications of hydrodynamical models to the description of nuclear collisions already have a long history (cf. revs. [1]). The fluid dynamics was one of the first approaches applied to high-energy nuclear collisions, since it directly refers to an equation of state (EOS) of nuclear matter. The main disadvantage of the conventional hydrodynamical model, perceived from the very beginning, consists in the complete disregard of transport properties of nuclear matter, in particular, that of a finite stopping power of nuclei. The viscous hydrodynamics [2] somewhat improved this drawback but, apparently, it was not enough to properly describe the initial interpenetration stage of nuclear collisions. That is why already in 1978 a two-fluid hydrodynamical model [3] was proposed. This model explicitly treats the interpenetration stage. Later, such a two-fluid approach was further developed. However, these two-fluid models somewhat lost their consistency in introducing the EOS. At sufficiently high relative velocities they assumed that the fluids undergo only collisional interaction and do not feel each other through the EOS. Multi-fluid models with mean mesonic fields, originated in refs. [4,5], are primarily designed to overcome this problem. These models do not use any conventional EOS. They treat interacting hadronic fluids moving in self-consistent mean fields which correspond to some EOS only in the case of equilibrium.

In this contribution we present calculations within a 3-dimensional relativistic mean-field 2-fluid model (RMF-2FM) derived in ref. [5] especially for applications to intermediate-energy heavy-ion collisions. The input of this model consists of mean-field parameters and cross-sections, i.e. being just the same as that in the mean-field

† Alexander-von-Humboldt fellow

kinetic models. A comparison[5] with the corresponding results from the solution of the relativistic Landau-Vlasov equation within a simple model problem shows that this two-fluid approach reproduces gross properties of the collision dynamics reasonably well. In this sense the RMF-2FM can be treated as an approximate method for solving the relativistic kinetics with mean mesonic fields. Being an approximation to the full kinetics, the RMF-2FM can perform faster numerics. It opens up a possibility of taking into consideration a number of more sophisticated effects inherent in nuclear collisions, such as in-medium cross-sections, a momentum dependence of the mean-mesonic fields and stochastic fluctuations. However, in this paper we present calculations without such extensions.

The RMF-2FM is derived from the relativistic kinetic model with mean mesonic fields[6], assuming that the leading nonequilibrium in nuclear collisions occurs due to a finite stopping power of nuclear matter. We are looking for an approximate solution of the baryon kinetic equations in the form of two Fermi distributions:

$$f^{(a)}(x,p) = \frac{d_a}{(2\pi)^3} \sum_{\alpha} \left[\exp\{(u^{(\alpha)} \cdot p - \nu^{(\alpha)})/T^{(\alpha)}\} + 1\right]^{-1}, \tag{1}$$

where d_a is the degeneracy factor of the a-type baryons and $\alpha = p$ or t (projectile-like or target-like baryons). The $u^{(\alpha)}(x)$, $T^{(\alpha)}(x)$ and $\nu^{(\alpha)}(x)$ quantities denote the local 4-velocity, temperature and effective baryonic chemical potential of the α-fluid, respectively. In the spirit of this ansatz, pions are assumed to be in the thermal rather than in the chemical equilibrium within each baryonic fluid.

Applying the moment method[4] to the kinetic equations, one arrives at the equations of motion for the two-fluid dynamics[5]:

$$\partial^\mu J_\mu^{(\alpha)} = 0, \tag{2}$$

$$\partial^\mu T_{\mu\nu}^{(\alpha)} - g_\omega J_\mu^{(\alpha)} \omega_\nu^\mu + g_\sigma \rho^{(\alpha)} \partial_\nu \sigma = -D(J_\nu^{(\alpha)} \rho^{(\bar\alpha)} - J_\nu^{(\bar\alpha)} \rho^{(\alpha)}), \tag{3}$$

$$(\Box + m_\sigma^2)\sigma + dU/d\sigma = g_\sigma \sum_\alpha \rho^{(\alpha)}, \tag{4}$$

$$(\Box + m_\omega^2)\omega_\mu = g_\omega \sum_\alpha J_\mu^{(\alpha)} \tag{5}$$

with $\bar\alpha \neq \alpha$, where the baryon current and the energy-momentum tensor of the α-fluid are determined by

$$J_\mu^{(\alpha)} = n^{(\alpha)} u_\mu^{(\alpha)}, \tag{6}$$

$$T_{\mu\nu}^{(\alpha)} = (\varepsilon^{(\alpha)} + P^{(\alpha)}) u_\mu^{(\alpha)} u_\nu^{(\alpha)} - g_{\mu\nu} P^{(\alpha)}, \tag{7}$$

respectively. The pressure, $P^{(\alpha)}$, the internal energy density, $\varepsilon^{(\alpha)}$, the scalar density, $\rho^{(\alpha)}$, and the baryon density, $n^{(\alpha)}$, are defined by standard statistical expressions in terms of the corresponding temperature and chemical potentials, as well as in those of the effective baryon mass. The friction coefficient between the two fluids, D, is expressed in terms of Pauli-blocked transport cross-sections (cf. ref. [5]).

For the mean σ and ω fields the nonlinear parametrizations of ref. [7] were used. These parametrizations reproduce the ground state of nuclear matter but yield different compressibilities and effective nucleon masses in the ground state. According to the latter, they yield stiff, intermediate and soft EOS's. We solve field equations (4) and (5) in the local approximation, i.e. neglecting the derivative terms. Such an approximation is shown [8] to be fairly good at the incident energies lower than few GeV/u, since the retardation of the fields with respect to nuclear matter appears to be nonessential. The vacuum nucleon-nucleon cross-sections, according to the Cugnon parametrization [9], are used in the present calculation.

We solve these equations by a modified "particle-in-cell" method. The hydrodynamical calculation ends in a freeze-out which is performed locally, when the baryon density drops to a critical value, n_c. The dependence of the results on the n_c parameter in the $0.3 \leq n_c/n_0 \leq 0.5$ range is found to be weak enough. Not all the regions of the nuclear system are frozen out in this way. In the peripheral collisions there remain sizable pieces of spectator matter maintaining their density nearly constant during the collision. A slow evolution of this matter is stopped at the physical end of the hydrodynamical calculation. At the freeze-out density (and at higher ones for spectators) the mean fields are still nonzero. Thus, we still deal with quasinucleons rather than with real ones. To calculate the emission of real nucleons from the frozen-out system, we employ a decay-evaporation procedure. This procedure ensures the total energy-momentum conservation in transformation of quasinucleons into nucleons.

As a result of the above-described calculations, we obtain single-particle distribution functions for nucleons and pions at a fixed impact parameter. To simulate all the experimental selections of events, we apply an event generator which generates random events in accordance with the calculated single-particle distribution functions. To perform a meaningful comparison with the experimental data, light fragments are also generated by means of the coalescence model at the final stage of the calculation. All these aspects of the model are described in ref. [10] in detail.

First, we compare our results for inclusive spectra with data[11,12]. For the Ar + KCl collision two calculations, the RMF-2FM one and a conventional hydrodynamical one [13] (RMF-1FM) with the same "soft" mean fields, are presented in fig. 1. For a better reproduction of these data, one needs a higher stopping power as compared with that resulting from the Cugnon's cross-sections. By enhancing the friction coefficient, D, by a factor of $\approx$3, one can obtain a better reproduction of these spectra. Of course, this fact can be interpreted in terms of in-medium cross-sections. However, a more natural explanation of it apparently consists in multi-nucleon collisions [14] occurring during a nucleus-nucleus interaction. Introducing the multi-nucleon interactions as a by-product of their deutron dynamics, Danielewicz and Bertsch [14] obtained a fairly good description of these spectra. For a heavier system, like La + La, the lack of the stopping power is not so spectacular (cf. ref. [10]). In this case both the two- and one-fluid calculations are good at reproducing the proton spectra. Simulations with different EOS's (i.e. mean fields) exhibit a very week sensitivity of the proton spectra to variations of the EOS parameters in a wide range. The inclusive pion spectra are reproduced by the two- and one-fluid models approximately to the same extent (see fig. 2). Both the models yield too few soft pions, especially at small angles. Apparently, this is an effect of the Coulomb attraction by the target spectators [15], which is not taken into account in the present consideration. The "soft" RMF-2FM and "stiff" RMF-1FM are certainly better at reproducing the pion data. Thus, the stiffness of the EOS in the conventional hydrodynamics and the partial transparency in the two-fluid one affect the pion yield in the same way.

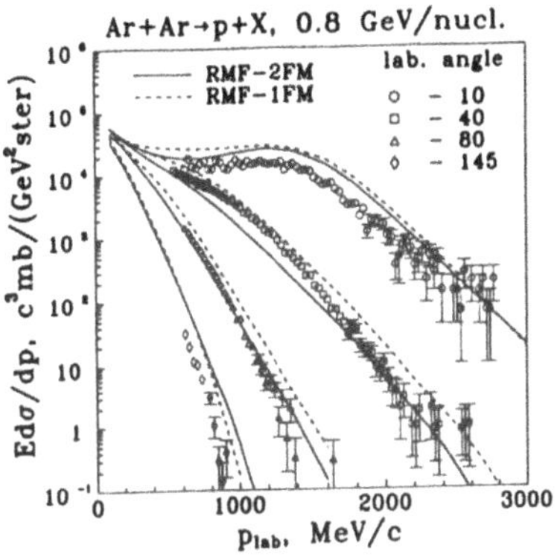

Fig. 1. Proton inclusive spectra from the $Ar(0.8\ GeV/nucl.) + KCl$ reaction. The two-fluid (solid lines) and conventional hydrodynamical (dashed lines) calculations are presented. The data are from ref. [12].

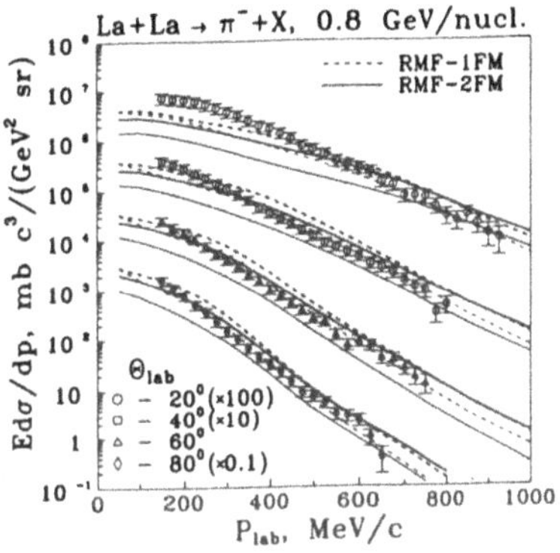

Fig. 2. Pion inclusive spectra from the $La(0.8\ GeV/nucl.) + La$ collisions. The two-fluid (full lines) and conventional hydrodynamical (dashed lines) calculations are presented. The solid lines correspond to the soft EOS, while the thin ones, to the hard EOS. The data are from ref. [11].

Let us turn to more exclusive data. In general, the Plastic-Ball data are subdivided into five multiplicity bins, MUL1-MUL5 (for an exact definition, see ref. [16]). These bins present a whole range of different collisions, from peripheral (MUL1) to the most central (MUL5) ones. The comparison with the Plastic-Ball data requires the application of the model to the full extent, including the event generator, the coalescence for the fragment production, as well as the experimental filter program. The latter simulates the acceptance of the Plastic-Ball detector. But before doing this, we would like to present a qualitative comparison with the data, i.e. a comparison with the calculations at fixed impact parameters without the event generator and filtering. In fig. 3 (the left column) we compare rapidity distributions in central collisions with the Plastic-Ball data [17]. The results of the calculations are renormalized to the total observed multiplicity, since in the experiment not all the particles were detected even in the central collisions due to the acceptance. From this comparison one could conclude that the data certainly indicate in favour of the partial transparency of the two-fluid model. However, such a conclusion is not true, since the experimental selection criterion and filtering wash out the differences between the model predictions. Fig. 3 (the right column) presents the results of calculations employing the real experimental event selection according to the total multiplicity of charged baryons after filtering[16] but without showing the result of this filtering in the rapidity distributions themselves. These results are also renormalized to the total observed multiplicity. One can see that such a selection itself significantly reduces the differences between the two models. When the selection criterion and filtering are applied to the full extent, the two- and one-fluid models reproduce the data approximately to the same extent (see fig. 4).

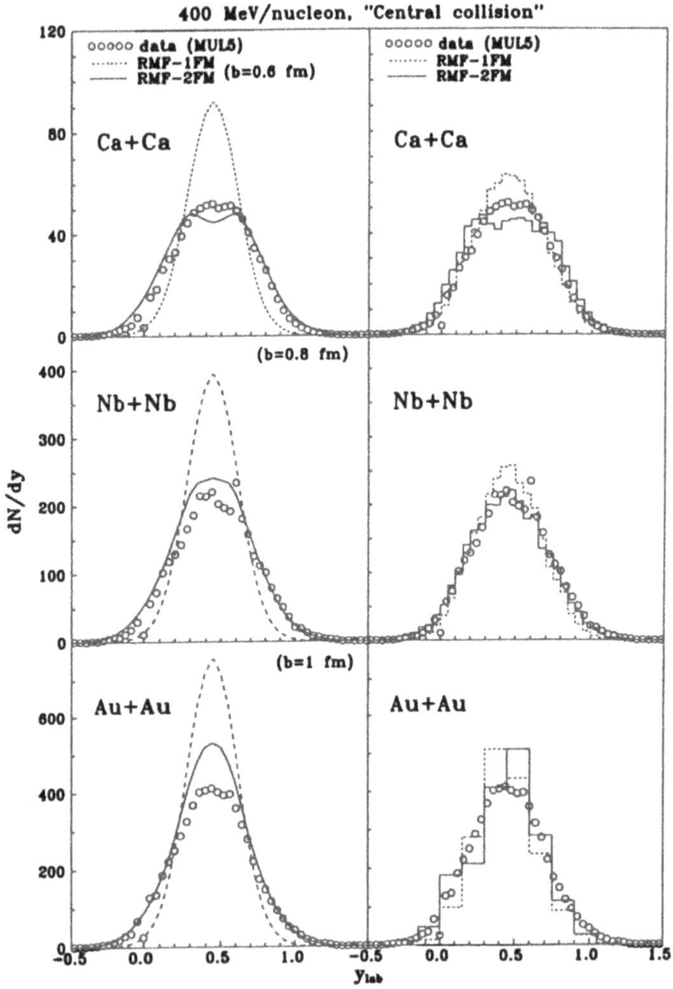

Fig. 3. Rapidity distributions of nucleons for the Au(0.4 GeV/nucl.) + Au and Nb(0.4 GeV/nucl.) + Nb collisions and those of protons for the Ca(0.4 GeV/nucl.) + Ca collision, calculated at "central" impact parameters (the left column) and employing the experimental event selection (the right column). The data (MUL5) are from ref.[17].

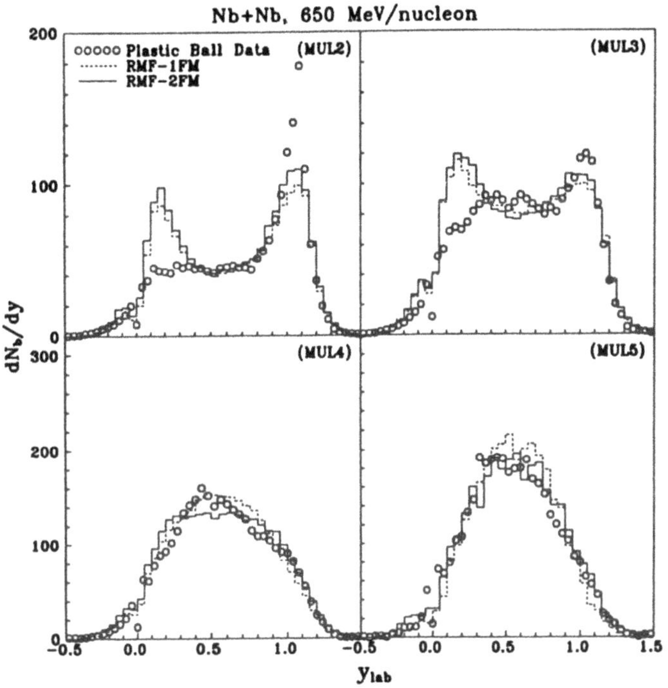

Fig. 4. Rapidity distributions of nucleons for the Nb(0.65 GeV/nucl.) + Nb collisions in different multiplicity bins (MUL2-MUL5) within the one- (RMF-1FM) and two-fluid (RMF-2FM) models. The Plastic-Ball data are from ref. [17].

In the case of lighter systems [Ca(0.4 GeV/nucl.) + Ca] and at a higher energies [Ne(2.1 GeV/nucl.) + Au] the conventional hydrodynamics appears even preferable (cf. ref. [10]). These rapidity distributions are fairly independent of the form of the mean-field parametrization already before filtering.

The main problem facing us here is that the calculated distributions are not asymmetrical enough in the noncentral multiplicity bins (MUL2 and MUL3 in fig. 4). Originally (before filtering), these distributions are symmetrical with respect to the c.m. rapidity, since they arise from the collisions of identical nuclei. The observed asymmetry is produced by the experimental acceptance. However, the target pick is not suppressed enough by the filter. This target pick prevents us from a better reproduction of the other parts of the rapidity distributions. The area under the distribution is fixed fearly well, due to the multiplicity selection in the bin. Since there are extra particles in the target pick, particles are lacking in the projectile and midrapidity regions.

We use the complete experimental filter program, SIMDAT_CLASTERS, which requires not only nucleons but also light fragments to be produced by the model. The quality of the fragment-yield reproduction is illustrated in fig. 5. Here we also have a satisfactory overall agreement with the data, yet the same problem with the target pick.

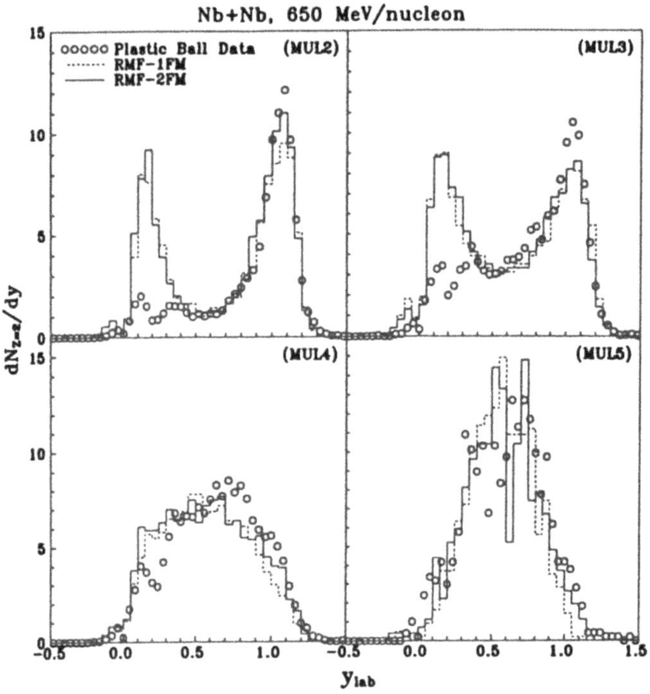

Fig. 5. Rapidity distributions of double-charged nuclear fragments, $N_{Z=2}$, for the $Nb(0.65\ GeV/nucl.) + Nb$ collisions in different multiplicity bins (MUL2-MUL5) within the one- (RMF-1FM) and two-fluid (RMF-2FM) models. The Plastic-Ball data are from ref. [17].

In view of the fact that even semi-exclusive data appear to be fairly insensitive to the EOS and stopping power, a paramount goal of the model calculations is to reveal global observables which are not only sensitive to the stopping power and EOS but allow us to disentangle their effects as well. Apparently, the mean transvese momentum versus rapidity [18] is one of such characteristics. Another important problem, closely connected with the former one, consists in revealing adequate criteria for the selection of impact parameters. As we have seen, the selection according to the total baryonic multiplicity is too rough and results in a loss of sensitivity to the dynamical characteristics of nucleus-nucleus collisions. With this preliminary theoretical work being done, there is hope that the new SIS experiments at the GSI will bring us a quantitative understanding of nuclear matter at high densities and temperatures.

Acknowledgments. The authors gratefully acknowledge fruitful discussions with the members of the theory group at the GSI. Three of us (V. R, Yu. I. and Yu. P.)

highly appreciate the hospitality and support rended to them at the GSI. We are sincerely grateful to T. Ivanova for thoroughly editing the English of the paper.

References

1. H. Stöcker and W. Greiner, Phys. Rep. **137** (1986) 277;
R. B. Clare and D. Strottman Phys. Rep. **141** (1986) 177;
J. A. Maruhn, in book "Relativistic Heavy-Ion Physics", eds. L. P. Csernai and D. Strottman, Int. Rev. of Nucl. Phys., vol. **5**, p. 81 (World Scientific, Singapore, 1991); I. N. Mishustin, V. N. Russkikh and L. M. Satarov, ibid. p. 180.

2. G. Buchwald et al., Phys. Rev. **C28** (1983) 1119; L. P. Csernai et al., Phys. Rev. **C28** (1983) 2001.

3. A. A. Amsden et al., Phys. Rev. **C17** (1978) 2080.

4. Yu. B. Ivanov, Nucl. Phys. **A474** (1987) 669; in *Relativistic Heavy-Ion Physics*, eds. L. P. Csernai and D. Strottman, Int. Rev. of Nucl. Phys., vol. **6**, p. 426 (World Scientific, Singapore, 1991).

5. Yu.B. Ivanov, V.N. Russkikh, M. Schönhofen, M. Cubero, B.L. Friman and W. Nörenberg, Z. Phys. **A340** (1991) 385.

6. J. D. Walecka, Ann. of Phys., **83** (1974), 491; B. D. Serot and J. D. Walecka, Adv. Nucl. Sci. **16** (1986) 1.

7. B. Waldhauser et al., Phys. Rev. **C38** (1988) 1003.

8. K. Weber et al., Nucl. Phys. **A515** (1990) 747.

9. J. Cugnon, Phys. Rev. **C22** (1980) 1885.

10. V.N. Russkikh, Yu.B. Ivanov, Yu.E. Pokrovsky and P.A.Henning, GSI-Preprint GSI-93-40 (1993), submitted to Nucl. Phys. **A**.

11. S. Hayashi et al., Phys. Rev. **C38** (1988) 1229.

12. S. Nagamiya et al., Phys. Rev. **C24** (1981) 971.

13. V. N. Russkikh, Sov. J. Nucl. Phys. **53** (1991) 1037; V. N. Russkikh and Yu. B.Ivanov, Nucl. Phys. **A543** (1992) 751.

14. P. Danielewicz and G. F. Bertsch, Nucl. Phys. **A533** (1991) 712;
A. Bonasera and F. Gulminelli, Phys. Lett. **259B** (1991) 399, **275B** (1992) 24;
G. Batko, J. Randrup and T. Vetter, Nucl. Phys. **A536** (1992) 786.

15. M. Gyulassy and S. K. Kauffmann, Nucl. Phys. **A362** (1981) 503.

16. K.-H. Kampert, J. Phys. G: Nucl. Phys. **15** (1989) 691.

17. H. H. Gutbrod et al., Preprint GSI-90-07 (1990).

18. P. Danielewicz and G. Odyniec, Phys. Lett. **157B** (1985) 146.

Recent Results of the Dilepton Spectrometer (DLS)

C. Naudet and the DLS Collaboration[1]
Lawrence Berkeley Laboratory
Berkeley, California, 94720, USA

Abstract

We report measurements of dielectron pair production in p+p and p+d interactions at 1.03, 1.26, 1.60, 1.85, 2.1 and 4.9 GeV kinetic beam energies and in α+Ca, d+Ca and C+C interactions at 1.0 GeV/nucleon. The invariant mass spectra for p+p and p+d collisions at 4.9 GeV is discussed. The ratio of dielectron production in p+p collisions to that of p+d collisions is found to be $\sim$ 2 for kinetic beams energies above 2 GeV increasing to $\sim$10, for masses above 150 MeV, at kinetic beam energies of 1 GeV. The invariant mass spectrum of dielectron production in α+Ca, d+CA and C+C interactions at 1 GeV/nucleon are presented and found to scale as A_pA_t.

1. Introduction

The Dilepton Spectrometer group has recently completed its six year program to measure dielectron production in nucleon-nucleon, nucleon-nucleus and nucleus-nucleus interactions from 1 to 5 GeV kinetic beam energy. The primary motivations of these measurements were to elucidate the dominate mechanisms of dielectron production in the low-mass region and to use dileptons as probes of the hot dense nuclear matter formed in heavy-ion collisions.

A series of experimental measurements[2–8] and theoretical calculations[20–26], for dielectron production in proton-nucleus collisions, showed that the low-mass dilepton ($m < 1$ GeV/c^2) and low-p_t single lepton yields were not fully understood. It has been believed that the low mass continuum could neither be explained by meson and vector meson decay nor by the Drell-Yan model (naive extrapolation to lower masses yields a underestimate by an order of magnitude). While many mechanisms have been suggested to explain the low mass region such as modified Drell-Yan, hadronic Bremsstralung and soft parton annihilations, this region is still not fully understood.

We present here the study of e^+e^- production in p+p and p+d interactions at 1.03, 1.26, 1.60, 1.85, 2.1, and 4.9 GeV kinetic beam energies and in α+Ca, d+CA and C+C at 1.0 GeV/nucleon kinetic beam energies.

2. Experimental Apparatus

The data were obtained with the Di-Lepton Spectrometer (DLS) at the Bevalac shown in Figure 1. Physics considerations guided us to employ a large acceptance spectrometer to compensate for the expected low production rate. Adequate detector segmentation were implemented to discriminate between the hadrons and electrons(e^+

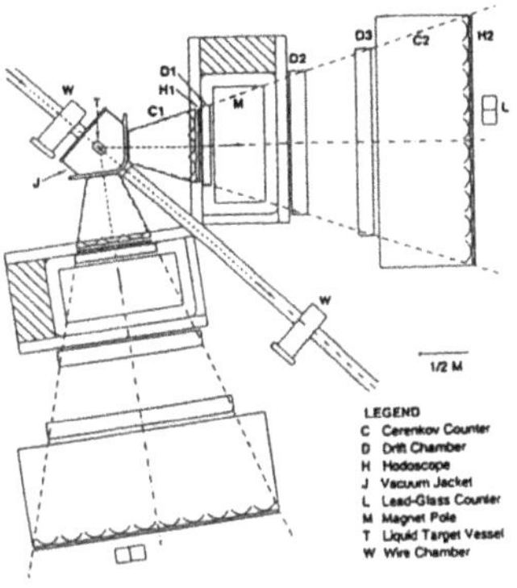

Figure 1: The DLS experimental setup

and e$^-$) in the heaviest system we planned to investigate (ie. Nb+Nb at 1.0 GeV).

The experimental apparatus consisted of segmented targets at the vertex of two symmetrical arms, each arm covering an angular range with respect to the beam axis of $17^0 < \Theta < 63^0$ and 14^0 in the transverse plane. This translates into a electron pair kinematic range of approximately .1 to 1 GeV/c^2 in mass, 0 to 800 MeV/c transverse momentum (p_t) and 0.5 to 1.9 units of laboratory rapidity (y). The electrons were identified by using large segmented gas Cerenkov counter's having approximately 95 percent electron detection efficiency and 10^{-5} hadron rejection per arm. The momentum reconstruction in each arm was obtained by use of one wire chamber before and two behind a large dipole magnet. The momentum resolution was balanced in favor of maximizing the acceptance of the low mass pairs by setting the magnetic field at 1.5 kilogauss. This yielded a momentum resolution of 12 percent. A sixteen segmented scintillator hodoscope in front and behind the dipole magnets were used in combination with the cerenkov counters to both trigger the apparatus and to give a first order tracking.

To an excellent approximation the dilepton signal (true pairs), which originates from a single elementary process is found by subtracting the same sign pairs from opposite sign pairs. This is because the electron pairs are independently produced thus creating charge symmetry of false pairs. Such electrons may be produced through Dalitz decay of a π^0 or gamma-ray conversion. This removes the main background (false pairs). The segmentation of the target and the designed minimal detector thicknesses reduces the rate of external conversions. In p+p collisions the false pairs are dominated by Dalitz produced electrons. One wide-angle Dalitz decay however may trigger our system at a very low rate and cannot be removed by this subtraction process. This contribution to the invariant mass cross sections is estimated by using the known meson production cross sections in pp interactions, branching ratios and Dalitz decay mass structure[9]. Additional information concerning the detailed

Table 1. DLS Dilectron Pair Statistics Summary

Interaction	Energy [GeV]	Dielectron Pairs	Ref
p + Be	4.90	531±31	ref 110,11
p + Be	2.10	419±27	ref 11
p + Be	1.04	146±16	ref 11
Ca + Ca	1.05	255±35	ref 12
Ca + Ca	2.10	131±30	ref 12
Nb + Nb	1.04	54±13	ref 16
p + p	4.90	7395±159	ref 15,17
p + d	4.90	9202±167	ref 15,17
p + p	2.10	2454±84	ref 18
p + d	2.10	3240±98	ref 18
p + p	1.80	911±46	ref 18
p + d	1.80	2109±66	ref 18
p + p	1.60	742±52	ref 18
p + d	1.60	850±51	ref 18
p + p	1.30	440±36	ref 18
p + d	1.30	762±41	ref 18
p + p	1.0	808±48	ref 18
p + d	1.0	915±45	ref 18
α + Ca	1.0	976±51	Here
d + Ca	1.0	810±41	Here
C + C	1.0	1938±94	Here
Ca + Ca	1.0	2000	In Progress

acceptance studies and the data analysis may be found in ref. 10-18.

3. Experimental Results

The DLS studied dielectron production in various beam and target combinations over a range of energies; a summary of the pair statistics is shown in Table 1. The existence of the dielectron signal above the combinatoric background is clearly established in all measured systems.

Figure 2 shows the dielectron yield for p+p and p+d at 4.95 GeV, without acceptance correction. The opposite sign backgrounds have been subtracted and the data have been corrected for inefficiencies in both the software and hardware. The dielectron yields for both p+p and p+d are seen to be very similar in shape for masses greater than 200 MeV. The ratio of p+d to p+p is found to be 2 for all masses above 200 MeV. This suggests a common production mechanism for dielectron production in pp and pd collisions at 4.95 GeV.

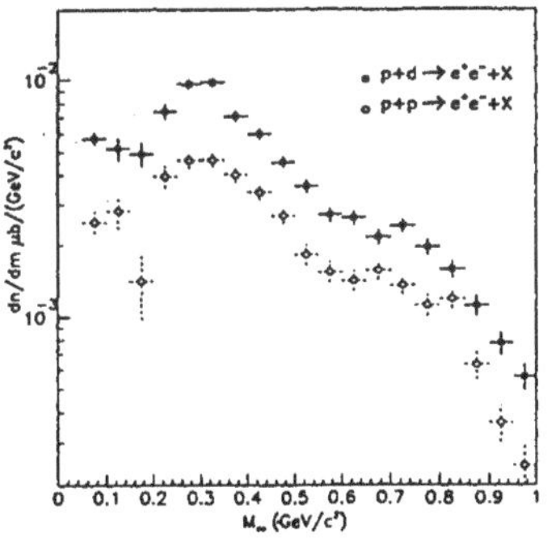

Figure 2: The True dielectron yields in (μb/(GeV/c)^2) from p+d and p+p collisions

The dielectron yield from p+p collisions at 4.95 GeV is shown in Figure 3 along with the expected contribution from decays of η, ρ, ω and δ. It's clear than additional sources are necessary to fully understand low mass dielectron production at these energies.

One hadronic source that has been suggested to fill this gap is p+n bremsstralung. However, our observed p+d to p+p dielectron ratio at 4.95 GeV is much smaller than the value of ten given by the soft-photon approximation, the one-boson exchange model or the external-line emission model. To understand the underlining mechanisms more fully the DLS understood a systematic study of the p+p and p+d dielectron production from 1 to 5 GeV . The ratio of p+d to p+p dielectron production as a function invariant mass is shown in Figure 4 for kinetic beams energies of 1.03, 1.26, 1.60, 1.85, 2.1 and 4.9 GeV. The measurements yield a surprising result, the ratio is constant (2) over the entire mass range for kinetic beam energies greater than 2.1 GeV then as the beam energy decreases the dielectron p+d to p+p ratio becomes mass dependent and increases in value at a fixed mass. This unexpected mass

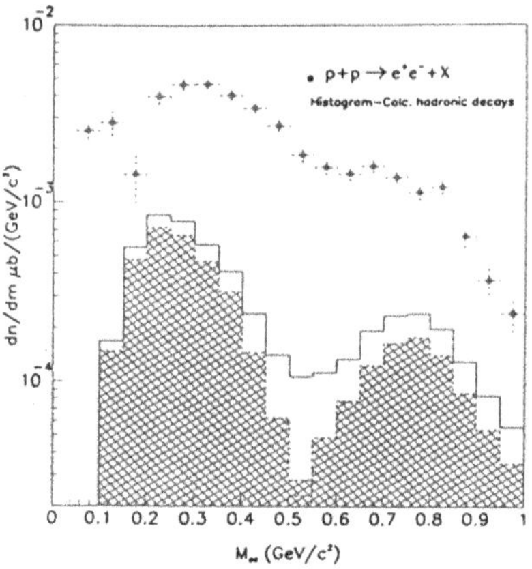

Figure 3: The dielectron yield from p+p collisions in comparison with the total probable contribution (solid histogram) in the DLS acceptance from decays of η, ρ, ω, and δ. The hatched areas are the individual contributions from the Dalitz decay of the η and two-body decay of the ω.

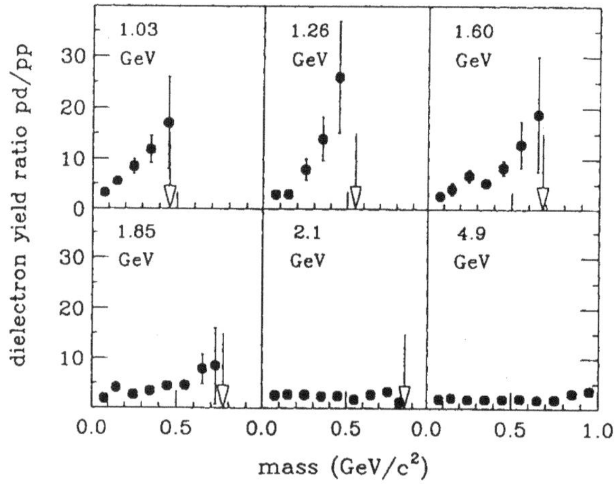

Figure 4: The p+d/p+p dielectron production ratio as a function of the invariant mass of the electron pair, for the kinetic beam energies of 1.03, 1.26, 1.60, 1.85, 2.1, and 4.9 GeV. The arrows indicate the maximum dielectron mass that is kinematically allowed in p+p collisions for each beam energy.

and beam energy dependence on the p+d to p+p dielectron ratio should provide tight constraints in the creation of dielectron production models.

The DLS heavy-ion dielectron production studies in Ca+Ca and Nb+Nb at 1.0 GeV/nucleon may be found in more detail elsewhere. Here the recent dielectron production in C+C, α+CA and d+CA at 1.0 GeV/nucleon is discussed. Figure 5 shows the invariant mass spectrum for dielectron production in C+C at 1.0 GeV. The data has been acceptance corrected and weighted by $1/A_tA_p$. When this spectra is compared with the e^+e^- production from p+p collisions at 1.0 GeV, also shown in Figure 5, it's clear that the C+C data produces a significantly higher cross-section at high mass. To assist in understanding this effect additional measurements were made on α+Ca and d+Ca collisions to help differeniate between fermi-motion and nuclear medium effects (rescattering, absorbtion or other collective effects).

Shown in Figure 6 are the mass spectra for both α+Ca and d+Ca at 1.0 GeV kinetic beam energy. The shapes for all three invariant mass spectra are similar when acceptance corrected and weighted with $1/A_tA_p$. Confidence level tests indicate that they are consistent with the same parent distribution. This suggests a common production mechanism for dielectron production in each system and that the production scales with the primary nucleon-nucleon interaction.

4. Summary

In summary, dielectron production at 1.03, 1.26, 1.60, 1.85, 2.1 and 4.9 GeV has been observed for both p+p and p+d. The production at 4.9 GeV cannot be explained by meson decays. The low ratio of p+d to p+p dielectron production seems to suggest a hadronic source and rules out simple p+n bremsstralung domination. The unexpected invariant mass and beam energy dependence of the p+d to p+p ratio will tightly constrain theoretical models. The C+C, α+Ca and D+CA invariant mass spectra were found to all have similar shape. The heavy-ion e^+e^- production cross-section scales as A_tA_p consisted with our earlier measurements[16]. Both the liquid target data and the heavy-ion data suggest that a common production mechanism for dielectron production in each system and that the production scales with the primary nucleon-nucleon interaction.

5. Acknowledgments

This work was supported by the Director, Office of Energy Research, Division of Nuclear Physics of the Office of High Energy and Nuclear Physics of the U.S. Department of Energy under the contracts DE-AC03-76SF00098, DE-FG03-88ER40424,DE-FG02-88ER40413 and DE-FG05-88ER40445. We would like to thank the staffs of the Bevalac and of the participating institutions whose contributions made this experiment possible.

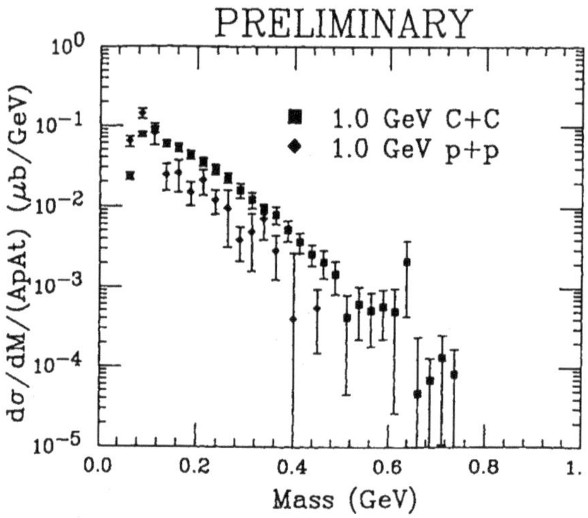

Figure 5: The PRELIMINARY invariant mass spectra for C+C at 1 GeV/nucleon kinetic beam energy. The cross-section is acceptance corrected and weight by $1/(A_pA_t)$, where A_p and A_t are the projective and target masses respectively.

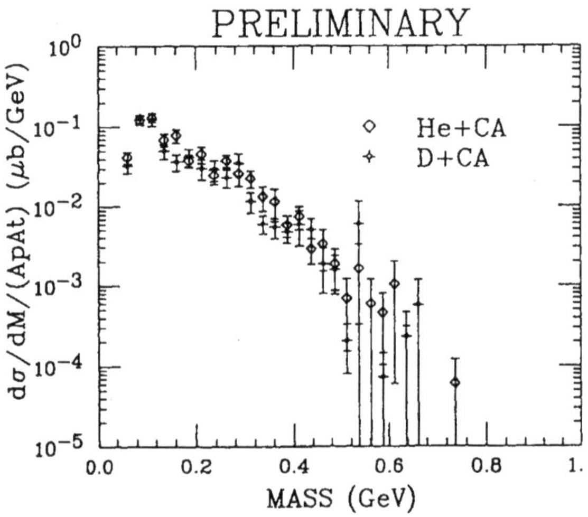

Figure 6: The PRELIMINARY invariant mass spectra for α+Ca and d+Ca at 1 GeV/nucleon kinetic beam energy. The cross-section is acceptance corrected and weight by $1/(A_pA_t)$, where A_p and A_t are the projective and target masses respectively.

6. References

[1] The DLS collaboration - Lawrence Berkeley Laboratory: S. Beedoe, L. Heilbronn, H.Z. Huang, G. Krebs, A. Letessier-Selvon, H. S. Matis, J. Miller, C. Naudet, R.J. Porter, L.S. Schroeder, P. Siedl, W.K. Wilson; University of California at Los Angles: J. Carroll, T. Hallman, G. Igo; The John Hopkins University: L. Madansky, R. Welsh; Louisiana State University: P. Kirk, Z.F. Wang; Northwestern University: D. Miller; Universite Blaise Pascal: M. Bouteg, F. Manso, M. Prunet, G. Roche; CEBAF: A. Yegneswaran.

[1] L. Lederman Physics Reports **C26**, 4, 149, (1976)

[2] A. Mikamo *et al* Phys. Lett. **106B**, 428, (1981)

[3] D. Blockus *et al* Nucl. Phys. **B201**, 205, (1982)

[4] M. R. Adams *et al* Phys. Rev. **D27**, 1977, (1983)

[5] K. Bunnel *et al* Phys. Rev. Lett. **40**, 136, (1978)

[6] J. Ballam *et al* Phys. Rev. Lett. **41**, 1207, (1978)

[7] B. Haber *et al* Phys. Rev. **D22**, 2107, (19780

[8] A. Browman *et al* Phys. Rev. Lett. **37**, 246, (1976)

[9] The particle data book supplied the branching ratios and inclusive meson cross sections may be obtained in F. Winkelmann LBL-3045. N.M. Knoll and W. Wada Phys. Rev. 98, 1325, (1955).

[10] G. Roche *et al* Phys. Rev. Lett. **61**, 1069 (1988)

[11] C. Naudet *et al* Phys. Rev. Lett. **62**, 2652 (1989)

[12] C. Roche *et al* Phys. Lett. **B226**, 228 (1989)

[13] A. Letessier-Selvon *et al* Phys. Rev. **C 40**, 1513 (1989)

[14] A. Yegneswaran *et al* Nucl. Instr. and Meth.. **A290**, 61 (1990)

[15] H.Z. Huang *et al* Phys. Lett. **B297**, 233 (1992)

[16] S. Beedoe *et al* Phys. Rev. **C 47**, 2840 (1993)

[17] H.Z. Huang *et al* to be published in Phys. Rev. C (1993)

[18] W.K. Wilson *et al* to be published in Phys. Lett. (1993)

[19] At 4.9 GeV the heavier meson cross sections are uncertain to within a factor of two.

[20] T. Goldman *et al* Phys. Rev. **D20**, 619, (1979)

[21] C. Gale and J. Kapusta Phys. Rev. **C35**, 2107, (1987)

[22] V. Cerny *et al* Phys. Rev. **D24**, 652, (1981)

[23] L.H. Xia *et al* Nucl. Phys. bf A485, 721, (1988)

[24] Gy Wolf *et al* Nucl. Phys. **A517**, 615, (1989)

[25] M. Schafer *et al* Phys. Lett. **B221**, (1989)

[26] K. Haglin *et al* Phys. Lett. **B224**, 433, (1989)

SHORT RANGE CORRELATIONS IN THE WEAK DECAY OF Λ HYPERNUCLEI

A. Parreño, A. Ramos
Departament d'Estructura i Constituents de la Matèria, Facultat de Física, Diagonal 647, E-08028 Barcelona, Spain

C. Bennhold
Center of Nuclear Studies, Dept. of Physics, The George Washington University, Washington, D.C. 20052, U.S.A.

D. Halderson
Dept. of Physics, Western Michigan University, Kalamazoo, MI49008, U.S.A.

ABSTRACT

The influence of short range correlations in the weak decay of Λ hypernuclei is studied using ΛN wave functions which are derived microscopically from the hyperon-nucleon Nijmegen interaction. It is observed that the origin of the discrepancies between the relativistic and non-relativistic models lies on the level at which the correlation function is included. The non mesonic width, the neutron to proton-induced ratio and the proton asymmetry are calculated for $^{12}_{\Lambda}C$ in an extension of our pion-kaon model which now includes the exchange of the vector mesons. It is found that the interferences between amplitudes are important and very sensitive to the form factor cut-off which determines the relative contribution of each meson.

1. Introduction

The weak decay of the Λ and other hyperons has been of interest for many years since it involves a weak interaction process without leptons. Hypernuclei in their ground states also decay via the weak interaction mechanism when they are stable against strong modes such as particle emission. However, the final nucleon produced in the mesonic decay of a Λ ($\Lambda \to \pi N$) has a very low momentum and it is Pauli blocked in the nuclear medium. Thus the dominant decay for all but the lightest hypernuclei is the non mesonic mode $\Lambda N \to NN$. The nucleons emerge with a large momentum and therefore the non mesonic process is an excellent tool to learn about the underlying decay mechanism since it is quite insensitive to the nuclear structure details.

Recent experimental data have obtained total and partial decay rates for several hypernuclei,[1] as well as the proton asymmetry from the decay of polarized hypernuclei.[2] New experiments are being proposed at the major laboratories,[3] which call for a theoretical effort to understand the mechanism of the $\Lambda N \to NN$ decay mode. Since the momentum carried by the exchanged meson is of the order of 400 MeV/c, this process is intrinsically of short range. For this reason, the aim of this work is to analyze in detail the influence of short range correlations (SRC) in the non mesonic decay of Λ hypernuclei, by considering different correlation functions for the initial ΛN system and comparing their effect in both the non-relativistic and relativistic models.

2. Transition potential for the non mesonic decay

As mentioned above, the weak decay of Λ-hypernuclei proceeds mainly through the non mesonic process $\Lambda N \to NN$. The weak vertex, for one pion exchange, is parametrized as

$$\mathcal{H}_{\Lambda N\pi} = iG_F\mu^2\bar{\psi}_N(A - B\gamma_5)\boldsymbol{\tau}\boldsymbol{\phi}_\pi\psi_\Lambda \ , \tag{1}$$

where $G_F\mu^2 = 2.21 \times 10^{-7}$. The empirical constants $A = 1.05$ and $B = 7.15$, adjusted to the free lambda decay, determine the strength of the parity violating (PV) and parity conserving (PC) rates and ψ_Λ is a $\tau_3 = -1/2$ field which is used to enforce the empirical $\Delta I = 1/2$ rule observed for the free lambda decay. Taking a pseudoscalar coupling for the strong vertex, the Feynman amplitude reads

$$\mathcal{M}_\pi = G_F\mu^2 g_{\pi NN}\int d^4x d^4y \bar{\psi}_{\mathbf{k}_1}(x)(A - B\gamma_5)\psi_\Lambda(x)\Delta_\pi(x-y)\bar{\psi}_{\mathbf{k}_2}(y)\gamma_5\psi_N(y) \ , \tag{2}$$

where spin and isospin indices have been supressed for simplicity. The non-relativistic reduction of Eq. (2) is

$$\mathcal{M}_\pi(\mathbf{k}) = G_F\mu^2\frac{g_{\pi NN}}{2M_N}\left(A - \frac{B}{2\bar{M}}\boldsymbol{\sigma}_1\mathbf{k}\right)\frac{F(k^2)}{k^2+\mu^2}\boldsymbol{\sigma}_2\mathbf{k} \ , \tag{3}$$

where $F(k^2)$ is a form factor which accounts for the finite size of the baryons and $\bar{M}$ is the average mass of the ΛN system. By Fourier transforming $\mathcal{M}_\pi(\mathbf{k})$ one obtains, for $F(k^2) = 1$, the following transition potential in r-space

$$V(\mathbf{r}) = V_{PC}(\mathbf{r}) + V_{PV}(\mathbf{r}) = V_C(\mathbf{r}) + V_T(\mathbf{r}) + V_{PV}(\mathbf{r}) \ , \tag{4}$$

where

$$V_C(\mathbf{r}) = G_F\mu^2\frac{g_{\pi NN}}{2M_N}\frac{B}{2\bar{M}}\frac{1}{3}\mu^2\frac{e^{-\mu r}}{4\pi r}\boldsymbol{\sigma}_1\boldsymbol{\sigma}_2 = V_C(r)\boldsymbol{\sigma}_1\boldsymbol{\sigma}_2$$

$$V_T(\mathbf{r}) = G_F\mu^2\frac{g_{\pi NN}}{2M_N}\frac{B}{2\bar{M}}\frac{1}{3}\mu^2\left(1+\frac{3}{\mu r}+\frac{3}{(\mu r)^2}\right)\frac{e^{-\mu r}}{4\pi r}S_{12}(\hat{\mathbf{r}}) = V_T(r)S_{12}(\hat{\mathbf{r}})$$

$$V_{PV}(\mathbf{r}) = -iG_F\mu^2\frac{g_{\pi NN}}{2M_N}A\mu\left(1+\frac{1}{\mu r}\right)\frac{e^{-\mu r}}{4\pi r}\boldsymbol{\sigma}_2\hat{\mathbf{r}} = V_{PV}(r)\boldsymbol{\sigma}_2\hat{\mathbf{r}}$$

with $S_{12}(\hat{\mathbf{r}}) = 3(\boldsymbol{\sigma}_1\hat{\mathbf{r}})(\boldsymbol{\sigma}_2\hat{\mathbf{r}}) - \boldsymbol{\sigma}_1\boldsymbol{\sigma}_2$. To obtain the decay rate we must evaluate matrix elements of the potential between the initial and final states characterized by the relative angular momentum L, spin S, total spin J and total spin projection J_z. The matrix elements of the spin operators can be straightforwardly evaluated using the standard rules of angular momentum algebra. For the spatial part one finds expressions of the type

$$\mathcal{M}_{L_0 S_0 J_0 \to (LL')SJ} \propto \int dr r^2 \Psi^{*J}_{LL'}(k_r, r) V(r) \Psi_{L_0 S_0 J_0}(r) \ , \tag{5}$$

where k_r is the relative momentum of the two outgoing nucleons. If the final state interactions induced by the strong force are neglected, the wave function $\Psi^{*J}_{LL'}(k_r, r)$ reduces to a spherical Bessel function $j_L(k_r r)$.

3. Short range correlations

In this section we explore the effects of different correlation functions on the rates and the results are compared with what is obtained with the relativistic model of Ref. 4.

For this study we will focus our attention on the four s-shell nucleons of $^{12}_{\Lambda}$C. In the ground state, the possible channels for the relative ΛN motion are $^1S_0$ and $^3S_1$ and the center-of-mass motion is also in a S wave. Correlated ΛN wave functions were obtained in r-space by solving the relative, $\Lambda\Sigma$ coupled channel Bethe-Goldstone equation as in Ref. 5. A kinetic energy propagator, assumed diagonal in the center-of-mass coordinate, was employed. The starting energy $\omega = \epsilon_\Lambda + \epsilon_N$ was set to -45 MeV, which is the sum of the experimental separation energies. Solutions were obtained with both the hard-core model D and the soft-core Nijmegen interactions.[6,7] The correlation function, defined as the ratio $f_{LSJ}(r) = \Psi_{LSJ}/\Phi_{LSJ}$ between the correlated and the uncorrelated wave functions, is plotted in Figs. 1a and 1b, for the hard-core and soft-core potentials respectively. The dotted line is a spin independent parametrization

$$f(r) = \left(1 - e^{-r^2/a^2}\right)^n + br^2 e^{-r^2/c^2} \tag{6}$$

with $a = 0.5$, $b = 0.25$, $c = 1.28$ and $n = 2$, which was used in Ref. 4. Note how the strong repulsion of the hyperon-nucleon force at short distances creates a "hole" in the wave function which prevents the lambda and nucleon from being too close.

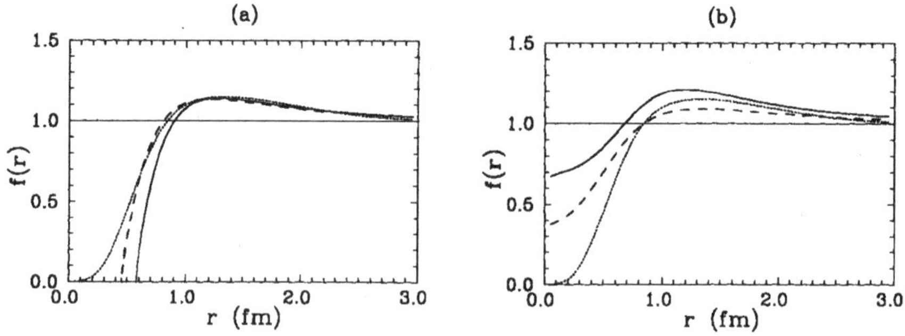

Figure 1. Correlation function for the ΛN channel $^1S_0$ (full line) and $^3S_1$ (dashed line) in the case of the hard-core D (1a) and the soft-core (1b) Nijmegen potential. The dotted line is the parametrization (6) used in Ref. 4.

The results in Table 1 show the contribution to the rate of the s-shell nucleons in $^{12}_{\Lambda}C$ for different correlation functions, in units of the free lambda decay rate Γ_Λ. It is observed that the SRC reduce the rate by a factor 1.3-1.5. This is consistent with what is obtained in other non-relativistic approaches in nuclear matter,[8,9] which obtain a reduction factor close to 2, where the slightly larger reduction is due to the use of a NN correlation function with a somewhat larger core.

Table 1. s-shell non mesonic rate $\Gamma_{nm}/\Gamma_\Lambda$ for $^{12}_{\Lambda}C$.

	FREE	HARD	SOFT	$f(r)$
P.C.	0.479	0.307	0.343	0.324
P.V.	0.208	0.146	0.192	0.162
TOTAL	0.687	0.453	0.535	0.486

Table 2. s-shell non mesonic rate $\Gamma_{nm}/\Gamma_\Lambda$ for $^{12}_{\Lambda}C$ for the relativistic model of Ref. 4

	FREE	$f(r)$
P.C.	0.456	0.130
P.V.	0.177	0.058
TOTAL	0.633	0.188

It is interesting to compare the results of Table 1 with what is obtained in a relativistic calculation using $f(r)$, as shown in Table 2. Surprisingly, the reduction factor introduced by the SRC is about 3, much larger than what is obtained in standard non-relativistic approaches. In the next section we will show that this is not a genuine relativistic effect, as it appears, but rather a consequence of the way correlations are considered in the transition amplitude.

4. Relativistic versus non-relativistic correlations

In a relativistic approach, one includes a correlation function $f(r)$ in Eq. (2) to account for the fact that the bound lambda and nucleon spinors feel the effect of the strong interaction and therefore the combined two-body spinor must show a "hole" at short distances. Taking free spinors and performing the usual non-relativistic reduction we obtain, for the parity conserving piece

$$\begin{aligned} \tilde{V}_{PC}(\mathbf{k}) &= -G_F\mu^2\frac{g_{\pi NN}}{2M_N}\frac{B}{2\bar{M}}\int d^3r e^{i\mathbf{kr}}\int\frac{d^3q}{(2\pi)^3}\frac{e^{-i\mathbf{qr}}}{q^2+\mu^2}f(r)(\boldsymbol{\sigma}_1\mathbf{k})(\boldsymbol{\sigma}_2\mathbf{k}) \\ &= -G_F\mu^2\frac{g_{\pi NN}}{2M_N}\frac{B}{2\bar{M}}\int d^3r e^{i\mathbf{kr}}\frac{e^{-\mu r}}{4\pi r}f(r)(\boldsymbol{\sigma}_1\mathbf{k})(\boldsymbol{\sigma}_2\mathbf{k})\ , \end{aligned} \tag{7}$$

which is the correlated parity conserving potential in momentum space. To compare later on with the standard non-relativistic procedure, it is useful to consider a correlation function of the type:

$$f(r) = 1 - e^{-\alpha r}\ , \tag{8}$$

with which Eq. (7) will give

$$\tilde{V}_{PC}(\mathbf{k}) = -G_F\mu^2\frac{g_{\pi NN}}{2M_N}\frac{B}{2\bar{M}}\left\{\frac{(\boldsymbol{\sigma}_1\mathbf{k})(\boldsymbol{\sigma}_2\mathbf{k})}{k^2+\mu^2} - \frac{(\boldsymbol{\sigma}_1\mathbf{k})(\boldsymbol{\sigma}_2\mathbf{k})}{k^2+(\mu+\alpha)^2}\right\}\ . \tag{9}$$

The corresponding correlated potential in r-space, $\tilde{V}(\mathbf{r})$, is the Fourier transform of Eq. (9) and the result will be the difference of two potentials of the type (4), one with parameter μ and the other with $(\mu+\alpha)$. The central channel reads

$$\tilde{V}_C(r) = G_F\mu^2 \frac{g_{\pi NN}}{2M_N} \frac{B}{2\bar{M}} \frac{1}{3} \left\{ \mu^2 \frac{e^{-\mu r}}{4\pi r} - (\mu+\alpha)^2 \frac{e^{-(\mu+\alpha)r}}{4\pi r} \right\} \boldsymbol{\sigma}_1 \boldsymbol{\sigma}_2 \,. \qquad (10)$$

In the standard non-relativistic calculations, one uses the potential (4) to calculate transition matrix elements between an initial correlated two-body wave function and a final state as in Eq. (5). Assuming $\Psi_{L_0S_0J_0} = f(r)\Phi_{L_0S_0J_0}$ one may define a correlated potential in r-space as the product $V(\mathbf{r})f(r)$ which, for the central channel and in the case of the correlation function (8), reads

$$\tilde{V}_C^{st}(r) = G_F\mu^2 \frac{g_{\pi NN}}{2M_N} \frac{B}{2\bar{M}} \frac{1}{3} \left\{ \mu^2 \frac{e^{-\mu r}}{4\pi r} - \mu^2 \frac{e^{-(\mu+\alpha)r}}{4\pi r} \right\} \boldsymbol{\sigma}_1 \boldsymbol{\sigma}_2 \,. \qquad (11)$$

From comparing expressions (10) and (11) it is clearly seen that both procedures are not equivalent and end up giving different correlated potentials. This is illustrated quantitatively in Table 3, where the non mesonic rates obtained with both procedures are shown for two values of α.

Table 3. s-shell non mesonic rate for the correlated potentials $\tilde{V}^{st}$ and $\tilde{V}$.

	$\tilde{V}^{st}$	$\tilde{V}$	$\tilde{V}^{st}$	$\tilde{V}$
	$\alpha = 3.33$		$\alpha = 1.67$	
P.C.	0.311	0.287	0.231	0.137
P.V.	0.168	0.129	0.114	0.064
TOTAL	0.480	0.417	0.346	0.201

We wish to emphasize here that the differences appear from having included $f(r)$ before or after the non-relativistic reduction is performed. In a relativistic approach, one is close to the results obtained by $\tilde{V}$ (Eq. (10)) since the correlation function is naturally included at the level of the Feynman amplitude. Being more fundamental, one may assume that this is the right procedure. Moreover, it can be shown[10] that a correlated potential of the type $V(\mathbf{r})f(r)$ is also obtained from Eq. (7) if a pseudovector coupling ($\gamma_5\gamma_\mu\partial^\mu \leftrightarrow (\boldsymbol{\sigma}\mathbf{q})/(2M)$) is assumed rather than the pseudoscalar one ($\gamma_5 \leftrightarrow (\boldsymbol{\sigma}\mathbf{k})/(2M)$). In other words, the standard way of including correlations can also be thought as coming from Eq. (7) if both vertices are taken as pseudovector, which is the preferred coupling for the strong NN interaction. In the case of the non mesonic decay studied here one of the vertices is weak. Thus, keeping the weak $(A - B\gamma_5)$ parametrization of the free lambda decay, one faces a hybrid problem (pseudoscalar×pseudovector) for the PC amplitude and the resulting effect of SRC would be something in between both approaches. However, for the PV amplitude, the assumption of a strong pseudovector coupling provides the standard correlated potential $(V_{PV}(\mathbf{r})f(r))$.

5. Results

We have calculated the non mesonic decay rate of ${}^{12}_{\Lambda}C$ using the relativistic model described in Ref. 4, which has been extended to include also the vector mesons (ρ, ω, K^*) where the weak couplings are obtained from a $SU(6)_w$ model.[11] For the one pion exchange mechanism the dynamics of the decay is contained in the transition amplitude (2). The bound lambda and nucleon spinors are obtained from a solution of the time-independent Dirac equation with scalar and time-like vector potentials, adjusted to reproduce the charge form factor of ${}^{12}C$ and the separation energies of ${}^{12}_{\Lambda}C$. To account for SRC, and in the absence of yet clear answers on how they should be included on the weak decay of hypernuclei, we have constructed an effective correlation function $f_{\text{eff}}(r)$ with $a = 0.3$, $b = 0.5$, $c = 0.75$ and $n = 2$, which yields, when used in the relativistic code, the same reduction factor as the non-relativistic calculations. In this way, the standard procedure of including correlations is simulated. A monopole form factor $F(q^2) = (\Lambda_c^2 - m^2)/(\Lambda_c^2 - q^2)$ is used at each vertex. The cut-off is $\Lambda_c = 1.3$ GeV for the lighter π, K and η mesons and $\Lambda_c = 2.0$ GeV for K^*, ρ and ω. We have also tried a cut-off $\Lambda_c = 1.5$ GeV for the ρ and ω mesons to be more consistent with other models of the strong hyperon-nucleon interaction.[12]

The results for the non mesonic rate are given in Table 4, where the effect of including the different mesons is explicitely shown. The third and fourth columns correspond to the original parametrization $f(r)$ and the effective one $f_{\text{eff}}(r)$, respectively, when no form factor is assumed. Since the calculations are time consuming, only results which include a form factor were obtained for the other mesons. It is interesting to note that there are important interferences between the different mesons. The cut-off Λ_c controls the relative strength of each meson and therefore may change the interferences, giving rise to different cancellations. It is observed that the relative contribution of the K^* meson is much larger in the sixth column, where the ρ and the ω are more supressed, such that in the end the rates are very similar and both in agreement with the experimental result.

Table 4. Non mesonic decay rate $\Gamma_{\text{nm}}/\Gamma_{\Lambda}$ for ${}^{12}_{\Lambda}C$

	FREE	$f(r)$	$f_{\text{eff}}(r)$		
			$\Lambda_c = \infty$	$\Lambda_c = 2$ GeV (ρ, ω)	$\Lambda_c = 1.5$ GeV (ρ, ω)
π	1.59	0.53	1.21	0.87	0.87
$+K$	1.13	0.39	0.86	0.65	0.65
$+\rho$	2.54	0.57	1.45	0.90	0.68
$+\omega$				1.33	0.67
$+K^*$				1.16	0.88
$+\eta$				1.22	1.12
Exp:		1.14 ± 0.20			[1]

The neutron to proton ratio Γ_n/Γ_p is shown in Table 5 . Again, it is observed that the contribution of K^* is substantially different when the cut-off for ρ and ω is changed from $\Lambda_c = 2$ GeV to $\Lambda_c = 1.5$ GeV and the final result, modified by a factor of two, is clearly closer to the experimental value.

Table 5. Ratio Γ_n/Γ_p for ${}^{12}_{\Lambda}$C

	FREE	$f_{\rm eff}(r)$	
		$\Lambda_c = 2$ GeV	$\Lambda_c = 1.5$ GeV
		(ρ,ω)	(ρ,ω)
π	0.19	0.19	0.19
$+K$	0.48	0.28	0.28
$+\rho$	0.13	0.19	0.25
$+\omega$		0.28	0.24
$+K^*$		0.17	0.37
$+\eta$		0.20	0.41
Exp:	$1.33^{+1.12}_{-0.81}$		[1]

We have also calculated the angular asymmetry in the distribution of the ejected nucleons from the decay of polarized hypernuclei. The proton asymmetry is proportional to both the polarization P_y of the hypernucleus, created in (π^+, K^+) reactions, and the asymmetry parameter A_y, characteristic of the weak process.

Our results for A_y are shown in Table 6. The relative contribution of the ω meson is now sensitively different for the two choices of the cut-off parameter, and the final results differ by a factor of two. To obtain the total asymmetry it is necessary to multiply by the polarization P_y, which, as shown in Ref. 4 is of the order of -0.25 for ${}^{12}_{\Lambda}$C. The resulting asymmetry of -0.11 will be reduced when distortions in the final state are considered[4] and will be closer to the experimental value[2] of -0.01 ± 0.11.

Table 6. Hypernuclear asymmetry A_y for ${}^{12}_{\Lambda}$C

	FREE	$f_{\rm eff}(r)$	
		$\Lambda_c = 2$ GeV	$\Lambda_c = 1.5$ GeV
		(ρ,ω)	(ρ,ω)
π	0.279	0.292	0.292
$+K$	0.273	0.334	0.334
$+\rho$	0.190	0.233	0.322
$+\omega$		0.026	0.303
$+K^*$		0.099	0.368
$+\eta$		0.211	0.441

6. Conclusions

The influence of SRC in the weak decay of Λ hypernuclei has been studied. Different correlation functions for the initial ΛN system have been considered in both the non-relativistic and the relativistic models. The larger reduction of the rates obtained in the relativistic approach has been understood as coming from including the correlation function at the level of the Feynman amplitude. It has been observed that the standard way of including correlations can be obtained if pseudovector couplings are assumed. Although this provides the right interpretation for the PV amplitudes, it is not correct for the PC contribution since it corresponds to a hybrid weak pseudoscalar-strong pseudovector case and further study is necessary.

An effective correlation function has been derived and used in the relativistic model of Ref. 4, which has been extended to include also the vector mesons. It is observed that the interferences between the exchanged mesons are substantially different for two choices of the cut-off Λ_c in the form factors of the ρ and ω mesons. The non mesonic rate turns out to be rather similar for both choices of Λ_c, whereas the neutron to proton ratio Γ_n/Γ_p and the hypernuclear asymmetry change by about a factor of 2.

7. Acknowledgments

We are grateful to E. Oset for helpful comments and enlightening discussions. This research was supported by DGICYT Grant No. PB92-0761, by US-DOE Grant No. DE-FG05-86-ER40270 and by NSF Grant No. PHYS-9202938. Some calculations of this work were performed at the Centre de Supercomputació de Catalunya (CESCA).

References

1. J.J. Szymanski *et al*, *Phys. Rev.* **C43** (1991) 849.
2. S. Ajimura *et al*, *Phys. Lett.* **B282**, (1992) 293.
3. Proceedings of the *Particles and Nuclei XIII International Conference*, Perugia (Italy), World Scientific, Singapore (in press).
4. A. Ramos, E. van Meijgaard, C. Bennhold and B.K. Jennings, *Nucl. Phys.* **A544**, (1992) 703.
5. D. Halderson, *Phys. Rev.* **C48** (1993) 581.
6. M.M. Nagels, T.A. Rijken and J.J. de Swart, *Phys. Rev.* **D15** (1977) 2547.
7. P.M.M. Maessen, T.A. Rijken, and J.J. de Swart, *Phys. Rev.* **C40** (1989) 2226.
8. E. Oset and L.L. Salcedo, *Nucl. Phys.* **A443** (1985) 704.
9. B.H.J. McKellar and B.F. Gibson, *Phys. Rev.* **C30** (1984) 322.
10. E. Oset, private communication
11. J.F. Dubach, G.B. Feldman, B.R. Holstein, L. de la Torre, University of Massachusetts preprint UMHEP-366 (1992).
12. A. Reuber, K. Holinde and J. Speth, *Nucl. Phys.* **A** (in press).

POSSIBLE SIGNATURE OF MEDIUM EFFECTS WITH PION INVARIANT-MASS CORRELATIONS

J. PLUTA[1], A. RAHMANI, D. ARDOUIN, R. LEDNICKY[2] , P. STEFANSKI[3] and H. DABROWSKI[4]
Laboratoire de Physique Nucléaire
IN2P3/CNRS et Université de Nantes
2, rue de la Houssinière - 44072 Nantes cedex 03- France

J.P. ALARD[5], J. AUGERAT[5], N. BASTID[5], C. CAVATA [6], P. CHARMENSAT [6], P. DUPIEUX [5], P. GORODETZKY [7], J. GOSSET [6], A. LE MERDY [6], D. L'HOTE [6] M.C. LEMAIRE [6], B. LUCAS [6], J. MARRONCLE [5], G. MONTAROU[5], M. J. PARIZET [5], J. POITOU [6], D. QASSOUD [5] and O. VALETTE [6]
DIOGENE COLLABORATION

H.N. AGAKISHIEV and V.G. GRISHIN †
Joint Institute for Nuclear Research, 141980 Dubna, Russia

W. BAUER, P. DANIELEWICZ and D. KLAKOW
National Superconducting Cyclotron Laboratory, Michigan State University
East Lansing, MI 48824-1321, USA

P. SCHUCK
Institut de Physique Nucléaire, Université de Grenoble, 53 avenue des Martyrs
38026 Grenoble cedex, France

† Deceased in 1991.
Permanent address :
[1] Institute of Physics, Warsaw University of Technology, ul. Koszykowa 75, 00-662 Warsaw, Poland.
[2] Joint Institute for Nuclear Research, 141980 Dubna, Russian Federation.
[3] Institute of Nuclear Physics, ul. Kawiory 26a, 30-055 Krakow, Poland.
[4] Institute of Nuclear Physics, ul. Radzikowskiego 152, Krakow, Poland
[5] Laboratoire de Physique Corpusculaire, Clermont Ferrand, Boîte Postale No. 45 63170 Aubière, France
[6] Département de Physique Nucléaire, Centre d'Etudes Nucléaires de Saclay, 91191 Gif-sur-Yvette, cedex, France.
[7] Centre de Recherches Nucléaires, Boîte Postale No 20 CR, 67037 Strasbourg, cedex, France.

ABSTRACT

The target mass dependence of invariant-mass distributions of two pions as well as azimuthal distributions are shown to contain signatures of medium effects.

1. Introduction

The goal of this work is to show that the target mass dependence of invariant-mass distributions of two pions bear the signatures of nuclear medium effects. Pions are known as a tool to probe hot and dense hadronic matter in nuclear collisions. Particularly, such observables like pion multiplicities, emission patterns or energy distributions may reveal the existence of collective effects. The scattering and absorption of the pions by the nuclear medium decisively influence the above observables mainly through the modification of the ΔN process. Such an observation was for instance made by the DIOGENE collaboration using $Ne + A$ collisions at $E = 800A.MeV$ where a positive in-plane transverse momentum was found[1] for pions, even at backward rapidities. It is expected that the corresponding pattern of emission can also generate geometrical correlations between coincident pions, thus, motivating a correlation study. The usual technique of the correlation function will then allow the study of the two-pion density in the phase-space.

For this purpose we have used existing data on proton and light nuclei-nucleus collisions from Dubna [2-3] and Diogène [4-5]) collaborations when non-inclusive interactions have been used to study target mass, energy and impact parameter dependences of π^+ and π^- multiplicity distributions and their comparison with Intranuclear Cascade Calculations. In order to get rid of quantum statistical effects in final-state interactions, we chose to analyse $\pi^+\pi^-$ pairs. Another reason of this choice stems from predicted modifications of hadron properties in nuclear matter ; as an example, the influence of nucleon-hole and Δ-hole excitations on the binding energy of a two-pion state has shown [6-7]) the possibility of a bound two-pion pair. This could affect the description of the intermediate-range NN interaction in conjunction to the σ meson in effective field theories. The possible link of an experimental enhancement at low $\pi^+\pi^-$ invariant mass with the hypothetical existence of the σ-strength or strong s-wave interaction has been only studied with elementary interactions like πN interactions (where medium effects are absent). From these studies [8], deviations from phase-space are found showing a clear indication of a resonance around $M_{\pi\pi} = 400$ MeV caused by a $I = J = 0$ s-wave $\pi\pi$ interaction. Searches for similar effects using $p + A$ collisions are currently under investigation at TRIUMF.

A subsequent interest for the study of $\pi^-\pi^+$ pairs density at high energy is the possibility to look for the signature of the decay of other mesonic resonances ($\rho\omega$) and their modifications in nuclear medium.

Some detailed aspects of the present analysis can be found in a recent published paper by J. Pluta et al [9].

2 - Presentation of the Data

We have analysed the $\pi^+\pi^-$ correlations resulting from the following data (1) $p+C$, $p+Nb$, $p+Pb$ reactions at 1600 MeV projectile kinetic energy (Diogène collaboration data [4-5]) where pions over 50 MeV/c are analyzed in the range $20° < \theta < 132°$.
(2) ^1H,^2H,^4He,^{12}C+ propane at momentum 4.2 GeV/c per nucleon, $p+Ta$ at 10 GeV/c (Dubna, propane bubble-chamber data [2-3]) with pion threshold 70 MeV/c.

2.1. Analysis of the pion correlation

Fig. 1 shows the Diogène data for C, Nb and Pb targets at $E = 1.6$A.GeV in terms of a two-pion invariant mass correlation-function : $\Delta N = N(M_{inv}) - BN_{ref}(M_{inv})$ were N, N_{ref} are the number of pairs for coincident and uncorrelated events and B is a normalization constant. Fig. 2 shows the same representation of the Dubna data for H, C and Ta targets. Both data show an average correlation (per event) close to zero for lighter targets while a clear enhancement is observed between the threshold and 450 MeV for heavy targets.

Negative correlations as appearing in the case of H target (Fig.2) result from momentum conservation. These phase space effects are destroyed in the N_{ref} term and also randomized in the case of secondary interactions in heavy targets. We have examined these effects as well as the contribution of other physical processes such as pion-nucleus coulomb interaction, final state $\pi\pi$ interaction and occurence of mesonic resonances. In ref.9 we have shown that they are not the origin of the observed trends and distributions for the heavy targets.

2.2 Discussion of medium effects.

The target mass evolution may contain nuclear medium effect like absorption. To look for an absorption effect we have analysed azimuthal correlations between emitted pions. An analysis of particle correlations in the plane perpendicular to the reaction axis has the advantage of being independent of a boost from one reference frame to another along the beam axis. Some azimuthal correlations appear naturally from a total-momentum conservation, leading particle effects etc...These types of correlations should be present mainly in elementary reactions. In nuclear interactions one can expect rather an isotropic azimuthal distribution due to rescattering effects and particle production in different secondary intranuclear collisions.

To study the azimuthal $\pi\pi$ correlations we analysed the distributions of the opening angle between the vectors of the transverse pion momenta, p_{T_i}, relative to the beam direction :

$$\Delta\phi = cos^{-1}\left(\frac{p_{T_i} \cdot p_{T_j}}{|p_{T_i}| \cdot |p_{T_j}|}\right). \tag{1}$$

The azimuthal anisotropy can be quantitatively characterized by the asymmetry coefficient defined as :

$$B = \frac{N(\Delta\phi < 90°) - N(\Delta\phi \geq 90°)}{N(\Delta\phi < 90°) + N(\Delta\phi \geq 90°)}, \quad (2)$$

where N denotes the number of pion pairs in the selected angular interval.

Fig. 3 shows the $\Delta\phi$ distributions for the pion pairs from the Diogène data. A small enhancement in the region of opposite pion transverse momenta for the carbon target is a trivial consequence of the phase-space restrictions and corresponds to the effect of negative correlations seen in the invariant-mass distribution.

A qualitatively different type of correlations appears for the interactions with heavy targets. The reverse correlation effect shows a collimation of pions in the region of small $\Delta\phi$. Its strength increases with an increase of the target mass. Similar results are observed for $\pi^+\pi^+$ correlations while a 50% reduction of the asymetry (B) is found for $(\pi^-\pi^-)$ correlations. There exists a clear relation between the effect observed in the azimuthal correlations and the enhancement in the region of small invariant masses. A collimation of pion pairs with small opening angles in the azimuthal plane shifts the values of their invariant masses towards the low invariant-mass region due the suppression of the number of pion pairs whose transversal momenta are opposite. An asymmetry in the azimuthal plane can be related to a shadowing effect which appears due to pion absorption inside nuclear matter.

In this domain of energy, the Δ production decay and interactions are responsible for the creation, rescattering and absorption of pions with subsequent reduction of the mean free path and influence on the emission patterns. The observation for an asymmetric system of a positive in-plane transverse momentum by the DIOGENE collaboration [1] was attributed to pion absorption by the heavier target spectator remnant opposite to the projectile side (Fig. 4). This was corroborated by transport calculations [10] showing a preferential emission toward the participant region and negligible correlation with the baryonic flow (N, Δ) itself. (Fig. 5). Recent measurements [11] of azimuthal pion correlations for protons + A reactions at CERN by the WA 80 collaboration completed by Bevalac Plastic Ball data (Fig. 6) provided evidence for a preferred side-by-side emission as opposed to a back-to-back pattern for protons. Again pion absorption by the target matter was considered and corroborated by relativistic QMD calculations (Fig. 7).

More recent VUU and QMD, calculations by Bass [12] examined the origin of the in-plane and out-of-plane pion transverse momentum spectra in $Ne+A$ collisions at 2 A GeV. By deactivating pion absorption channels $(\Delta N \rightarrow NN)$ or scattering channels $(\pi N \rightarrow \Delta)$, they show (Fig. 8) that pion re-scattering on target nucleons has the main influence on pion distributions : target nucleons act as spectator pieces reflecting the observed pions in the positive in-plane side ; whereas true absorption only influences the rapidity distribution dN/dY and the total multiplicity. Furthermore, the lack of spectator matter outside the reaction plane favors the perpendicular emission of high momentum pions as observed.

To verify one of these assumptions we examined from our Diogène data the dependence of the mean number of emitted pions on the proton-like multiplicity N_p which we are using for the impact parameter selection. Fig. 9 shows this dependence for the interaction of protons with different nuclear targets (Diogène data). No clear dependence is seen for the interaction with the carbon target. Conversely for the medium and heavy targets, the mean pion multiplicity decreases rapidly with the increase of N_p which may be attributed to the pion absorption inside nuclear medium.

2.3 Theoretical calculations

We have investigated the possibility that the enhancement of the correlation function close to the invariant mass of $2m_\pi$ could be caused by the effect of focussing of the two correlated pions into one hemisphere due to the presence of the target spectator matter, which causes some of the produced pions to be rescattered and absorbed. For this purpose, we performed transport calculations with the BUU code, which had been successfully used to calculate pion spectra in heavy-ion collisions [13] and to show that the observed pion collective flow was caused by nuclear shadowing effects [10]. The code was recently amended following the results of ref. [14].

In our transport code we propagate protons, neutrons, Δ and N^* resonances and pions. The process of greatest interest here is the creation and reabsorption of pions, which is dominated by the two step process.

$$N + N \leftrightarrow N + \Delta \leftrightarrow N + N + \pi \tag{3}$$

In obtaining the correlation function comparable to the data, we calculate the invariant masses of correlated pions using particles from simulated events corresponding to the same impact parameter and same fixed orientation of the reaction plane. The invariant masses of background pairs are calculated taking the pair members from events at different random impact parameters and with a random relative orientation of the reaction planes. The correlation at given invariant mass M_{inv} is then computed from :

$$R(M_{inv}) = \frac{dN_{corr}/dM_{inv}}{dN_{back}/dM_{inv}}, \tag{4}$$

where the subscript "corr" and "back" refer, respectively, to the correlated and background pairs. With the above method we approximate the experimental situation where in generating the background event one is forced to average over random relative orientations of the reaction planes and random impact parameters, while sampling the correlated pairs at one fixed orientation of the reaction plane and at one impact parameter, with the pair members focussed in the same way by spectator matter. For a more detailed description of the numerical procedure and the effect of pion absorption on two pion correlation functions see ref (15 and 9). In the fig. 10; we compare the results of our calculations with the experimental data for the C and Pb targets. In this case, the experimental background calculations were performed without the selections of "proton-like" or "participant" numbers. We find a satisfactory agreement with the data as well as the target mass dependence.

3. Conclusions

The analysis of $\pi^+\pi^-$ correlations in the interactions of light relativistic nuclei with light and heavy targets has been performed for the first time, using a correlation function technique. Clear positive correlations for invariant masses less than 450 MeV have been observed. The effect is strongly target dependent and increases with increasing target mass.

After examination of other non-nuclear effects, a shadowing effect due to pion absorption and scattering inside nuclear matter has been considered as a possible source of the correlation. Observed azimuthal correlations and transport calculations are in agreement with this assumption. Additional support comes from the analysis of the mean pion multiplicity which decreases with the decreasing impact parameter evaluated by the proton-like multiplicity. Although the same trends are observed for $(\pi^+\pi^+)$ and $(\pi^-\pi^-)$ correlations, a different pattern arises for $(\pi^-\pi^-)$ pairs which cannot be accounted for by simple coulomb static calculations.

References

1. J. Gosset and Diogene Collaboration *Phys. Rev. Lett.* 62 (1989) 1251.
2. I. Angelov et al *Yad. Fiz.* 30 (1979) [Sov. J. Nucl. Phys. 30 (1979) 824]
3. G.N. Agakishiev et al, *Yad. Fiz* 37 (1983) 938 [Sov. J. Nucl. Fiz. 37 (1983) 559]
4. M. Trzaska and Diogene Collaboration, *Z. Phys. A* 340 (1991) 325
5. M.C. Lemaire and Diogene Collaboration *Phys. Rev. C* 43 (1991) 2711.
6. P. Schuck, W. Norenberg and G. Chanfray, *Z. Phys. A* 330 (1988) 119.
7. G. Chanfray, Z. Aouissat, P. Schuck and W. Norenberg, *Phys. Lett. B* 256 (1991) 325.
8. G. Kernel et al, *Z. Phys. C* 51 (1991) 377 and references therein.
9. J. Pluta, A. Rahmani, D. Ardouin et al, *Nucl. Phys. A* 560 (1993) in press.
10. Bao-An Li, W. Bauer and G. Bertsch, *Phys. Rev. C* 44 (1991) 2095.
11. H.R. Schmidt et al, *Nucl. Phys. A* 544 (1992) 449 c.
12. S. Bass, R. Mattiello, H. Stöcker, W. Greiner, *Phys. Lett. B* 302 (1993) 381 and GSI Annual Report (1992) p. 121.
13. B.A. Li and W. Bauer, *Phys. Rev. C* 44 (1991) 450.
14. P. Danielewicz and G.F. Bertsch, *Nucl. Phys. A* 533 (1991) 712.
15. D. Klakow, W. Bauer, P. Danielewicz and P. Schuck, in preparation.

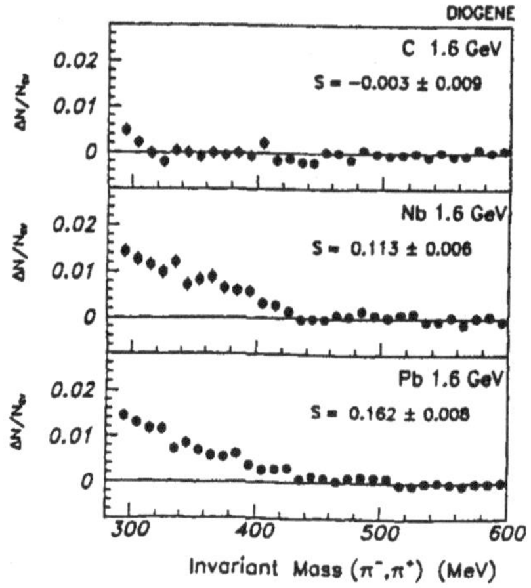

Fig. 1. The correlation function from the Diogene data divided by the total number of events. S denotes the integrated value of the function up to M_{inv} = 500 MeV.

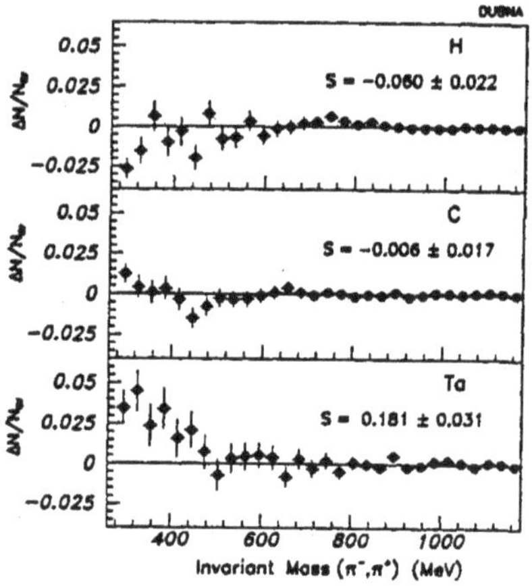

Fig. 2 Same as in fig. 1 from the Dubna data. The results from three targets : hydrogen (H), carbon (C) and tantalum (Ta) are shown.

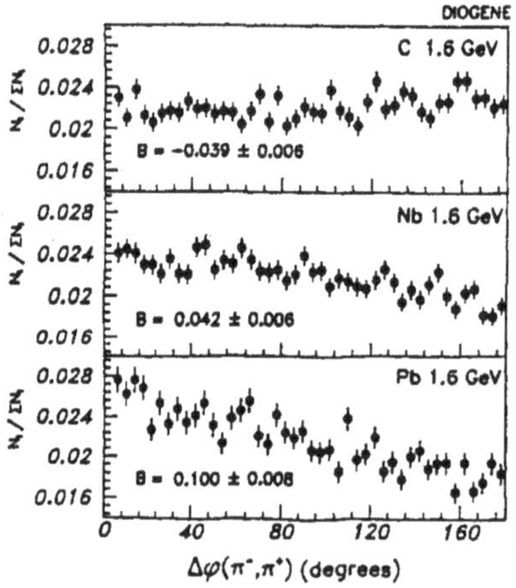

Fig. 3. Normalized distributions of the relative azimuthal angle between two pions from the Diogene data. B denotes a value of the asymmetry coefficient defined by the formula 2.

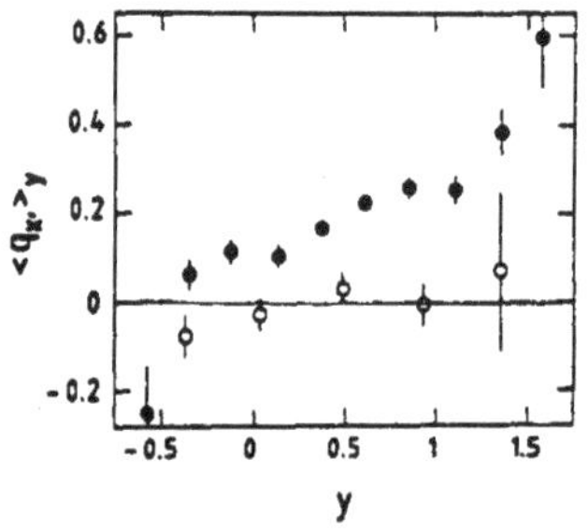

Fig. 4. $< q_{x'} >_y$ vs rapidity y for π^+ produced in Ne + Pb collisions at E/A = 800 MeV. $(q_{x'})$ is the transverse momentum normalized to the pion mass, in the estimated reaction plane. Experimental results (filled circles) are compared to the results of intranuclear-cascade (INC) calculations (open circles) for collisions with squared reduced impact parameter $b^2/(R_1 + R_2)^2 \sim 0.18$ (from ref.1).

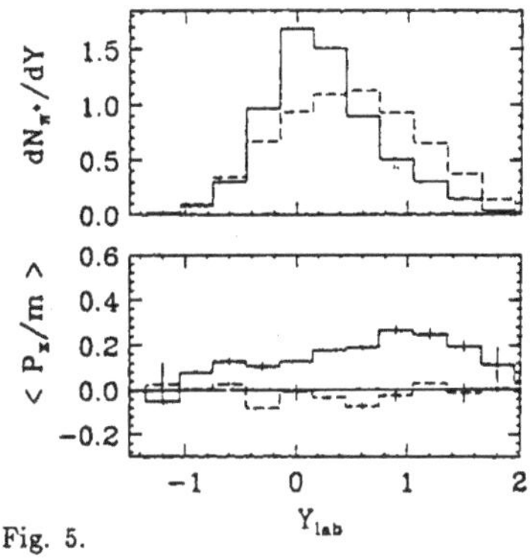

Fig. 5.

Upper figure : π^+ rapidity distribution calculated with (solid histogram) and without (dashed histogram) the pion reabsorption channels for the reaction of Ne + Pb at E/A = 800 MeV and the impact parameter of 3fm. For ease of comparison, the overall normalization of both curves was fixed to the same value. Lower figure : calculated π^+ transverse momentum distributions in the true reaction plane with (solid histogram) and without (dashed histogram) the pion reabsorption channels (from ref.10).

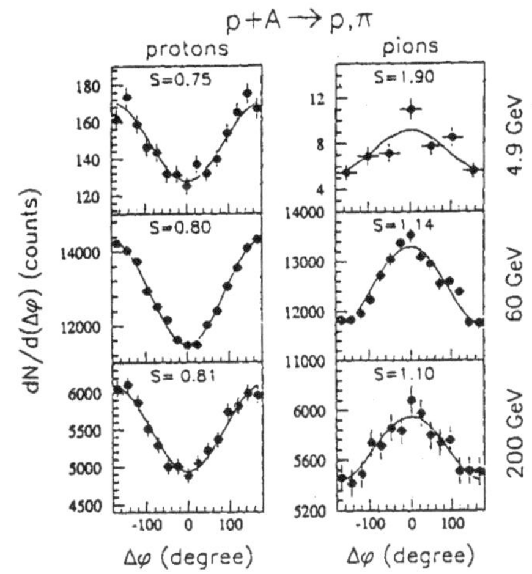

Fig. 6. The correlation function ($C\Delta\phi$) for p + Au collisions at 4.9 (top), 60 (middle) and 200 (bottom) GeV. The left and right hand plots contain $C(\Delta\phi)$ for protons and pions, respectively (from ref. 11.).

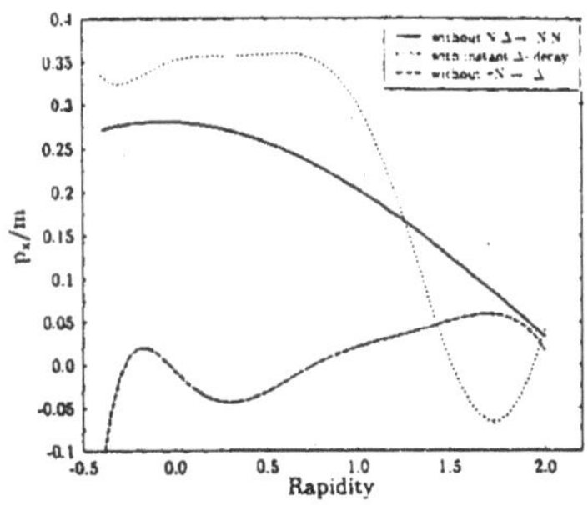

Fig. 7 Rapidity y versus $< p_x > /m$ for π^+ in a Ne (2A GeV) Pb collision at impact parameter b = 4.1fm. We compare modes of calculation with deactivated pion absorption, deactivated Δ absorption and with zero lifetime for the Δ resonance (instant Δ decay) from ref. 12).

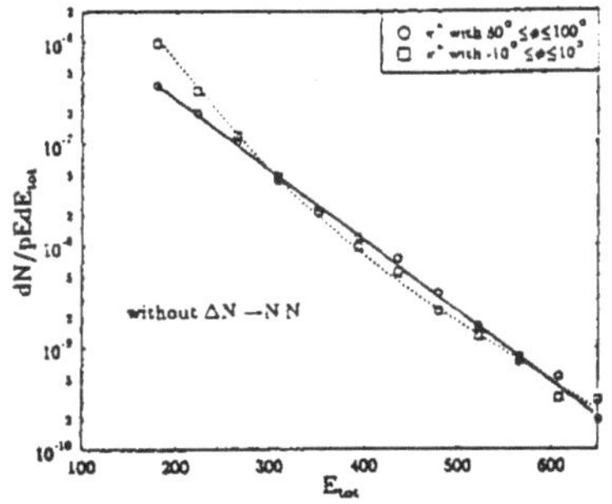

Fig. 8. Spectra of π^+ emitted in the reaction plane and perpendicular to the plane with deactivated $\Delta N \rightarrow NN$ channel. The shapes of the spectra resemble the default calculation but not pioniç sqeeze-out is detected (from ref.12).

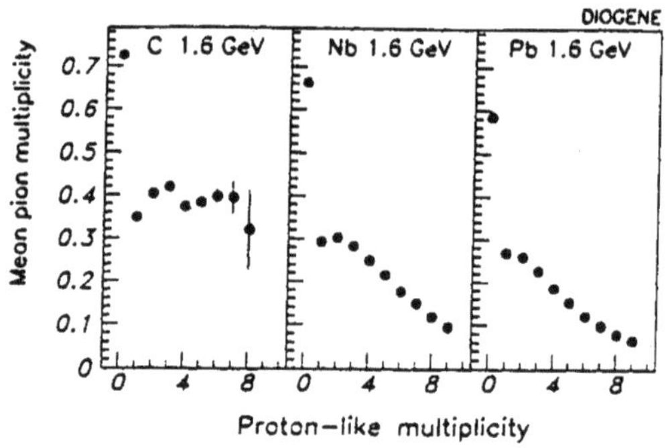

Fig. 9.The mean pion multiplicity as a function of the proton-like multiplicity from the Diogene data.

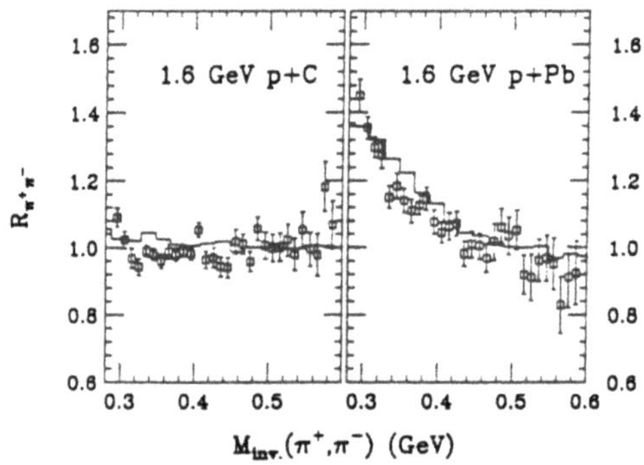

Fig. 10. A comparison of the experimental correlation function (full circles) with the theoretical calculations (histogram) for the Diogene data. See the text for details of calculations.

PROBING THE NUCLEUS-NUCLEUS REACTIONS WITH HARD PHOTON AND LIGHT CHARGED PARTICLE EMISSION *

P. Sapienza[1], R. Alba[1], R. Coniglione[1], A. Del Zoppo[1], E. Migneco[1,2],
C. Agodi[1], G. Bellia[1,2], P. Finocchiaro[1], C. Lukashin[1,+]
C. Maiolino[1], P. Piattelli[1], D. Santonocito[1], A. Peghaire[3]

(1) *INFN - Laboratorio Nazionale del Sud, Via S. Sofia 44, 95123 Catania (ITALY)*
(2) *Dipartimento di Fisica dell'Università di Catania (ITALY)*
(3) *GANIL, Caen (FRANCE)*

ABSTRACT

The hard photon and pre-equilibrium proton production has been studied by means of exclusive experiments in reaction induced by Ar and Xe beams at 44 MeV/u. In particular the dependence on the centrality of the reaction has been investigated. Preliminary results on the γ-proton correlation are also presented.

1. Introduction

Heavy ion collisions at intermediate energy ($20 \leq E \leq 200$ MeV/u) are a very useful tool to study the behaviour of nuclei in conditions far from the stability. In this regime, the evolution of the nucleus-nucleus collision from the initial stage, in which the highest temperatures and densities are reached, to the expansion and decay stage is very rapid. Most of the reaction products are mainly related to the final stages of the collision, therefore it is very difficult to find probes sensitive to the initial phase of the reaction. Subthreshold pions, hard photons and very energetic nucleons have been indicated by theoretical calculations as the best probes to investigate the first stages of the nucleus-nucleus collisions. It is therefore necessary to perform *exclusive* experiments which could give detailed information on the production mechanism in order to clarify if and to what extent photons, pions and very energetic nucleons can be considered sensitive probes of the initial stages of the collision. In the following we will present some of the experimental results obtained with the MEDEA detector at GANIL with regards to the production of hard photons and pre-equilibrium protons.

---

* Experiment performed at GANIL (France)
+ On leave from S. Peterburg Nuclear Physics Institute, Russia

2. Hard photon production

Due to their electromagnetic nature, hard photons are a very attractive probe that can provide unperturbed information on the nuclear dynamics. Consequently their emission in heavy ion collisions at intermediate energy has been intensively studied in the last years[1,2]. This has provided a wealth of inclusive data that span a broad range of incident energies and projectile-target systems. From the analysis of the inclusive γ spectra one can extract the γ emitting source velocity, that has been found to be close to the half-beam velocity independently of the projectile-target mass asymmetry[3]. Since this velocity coincides with the average nucleon-nucleon center of mass velocity, this result strongly supports the commonly accepted interpretation of these hard photons as bremsstrahlung radiation from first n-p collisions. Some *exclusive* experiments, aimed mainly at the study of the hard photon emission dependence on the centrality of the reaction, have also been performed[4-8].

In our experiments we have measured, on an event-by-event basis, hard photons ($E_\gamma \geq 25$ MeV) in coincidence with light charged particles (Z=1,2) (LCP) in the reactions $^{40}Ar+^{51}V$ and $^{132}Xe+^{197}Au$ at 44 MeV/u. The LCP multiplicity can be assumed as an indicator of the impact parameter: the larger the multiplicities are the more violent the collisions are (small impact parameters). The experiment was performed with the MEDEA detector[9]. 150 BaF_2 detectors, covering the angular range $42^o \leq \theta \leq 180^o$, and 80 plastic phoswhich detectors covering the polar angles $10^o \leq \theta \leq 30^o$ were operating with a total geometrical efficiency of 80%. Hard photons were measured in the BaF_2 ball while the LCP were measured and identified both in the BaF_2 ball and in the phoswich wall. In the analysis the high energy γ-rays were identified by combining the time of flight and pulse shape information. To get rid of the cosmic ray contamination in the γ energy spectra the events triggered by cosmic rays were rejected off-line requiring a coincidence with at least one LCP detected in the phoswich wall. From a simultaneous fit of the unfolded inclusive γ spectra at different angles in the laboratory we have extracted the γ emitting source velocity, the anisotropy and the γ cross section which are in good agreement with the existing systematics for both the reactions.

However such result cannot give a conclusive answer on the production mechanism. A way to get more information is to study the hard photon emission dependence on the impact parameter which allows a more detailed comparison with the models.

The hard photon cross section is reported in the lower part of fig.1 as a function of the LCP multiplicity[10]. This distribution exhibits a pronounced maximum at intermediate multiplicities. This behaviour, different from that of the inclusive LCP cross section shown in the upper part of the figure, confirms the fact that hard photons are more likely emitted in central collisions, as indicated by previous data[2].

The conversion of the LCP multiplicity into impact parameter b is an essential

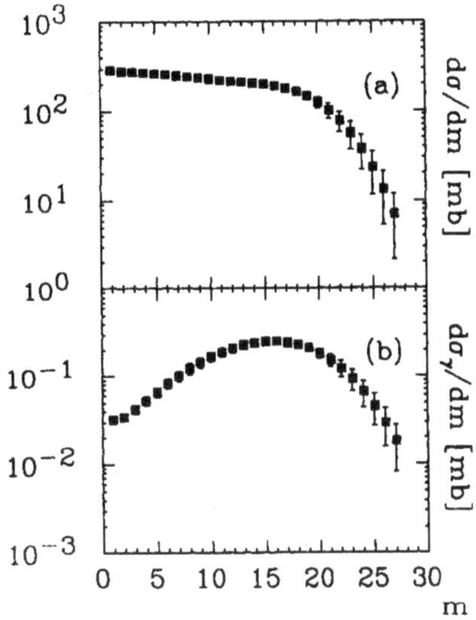

Fig.1 - (a) LCP production cross section as a function of the LCP multiplicity; (b) hard photon production cross section ($E_\gamma \geq 40$ MeV) as a function of the LCP multiplicity in the reaction ^{129}Xe + ^{197}Au at 44 MeV/u.

point in order to compare the data with the theoretical calculations. We have adopted the method of Cavata et al.[11] which is based on the assumption that the impact parameter and the multiplicity are connected by a monotone correlation. Moreover, since this relation is expected to hold between the impact parameter and the corresponding average emitted multiplicity, we have performed an unfolding of the experimental multiplicity spectra before applying the conversion into b to remove the statistical fluctuactions on the observed multiplicities.

More detailed information can be obtained by looking at the hard photon multiplicity which is defined as $M_\gamma(b) = \frac{d\sigma_\gamma(b)}{d\sigma_R(b)}$. The experimental $M_\gamma(b)$, reported in fig.2, increases steeply with increasing centrality from peripheral (large b) to mid-central collisions and then saturates for the most central collisions. This behaviour is in very good agreement with the BNV[12] calculations also reported in fig.2 (diamonds). On the other hand, it is worth to be noticed that also the geometrical model based on the first n-p collision bremsstrahlung hypothesis[1] (solid line of fig.2) reproduces the data quite well, at least up to the mid-central collisions. In this model M_γ is parametrized as $M_\gamma(b) = P_\gamma \cdot N_{np}(b)$ where P_γ is the probability of emitting a nucleon in a n-p collision and $N_{np}(b)$ is the number of n-p collisions at a given b, evaluated on the basis of the surface dependence proposed by Bauer et al.[13] which follows the data better than standard volume prescription proposed by Nifenecker et al.[1]. In conclusions the good agreement with the geometrical model indicates that $M_\gamma(b)$ is mainly affected by the size of the participant region formed during the collision, while the flattening observed for the most central collisions needs further investigations.

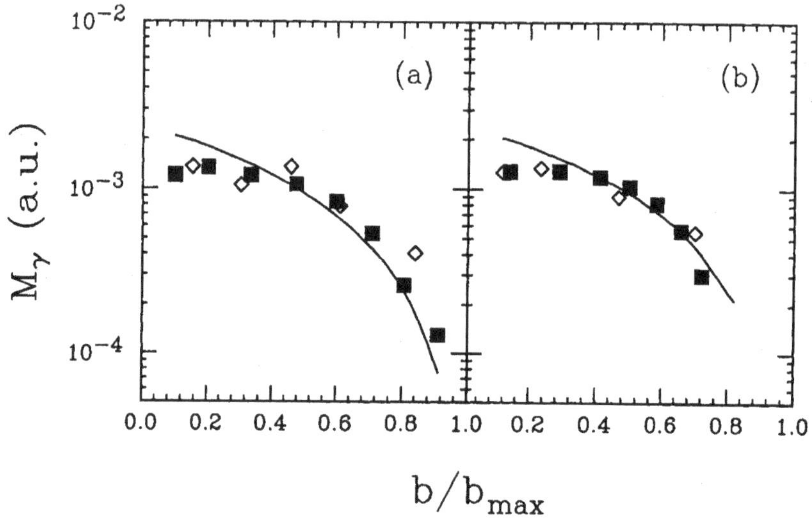

Fig.2 - Experimental (squares) and BNV (diamonds) photon production multiplicity M_γ ($E_\gamma \geq 40$ MeV) as a function of the impact parameter for the reactions (a) ^{132}Xe + ^{197}Au and (b) ^{40}Ar + ^{51}V at 44 MeV/u. The solid line is the result of a geometrical model calculation (see text).

Another piece of information concerns the dependence of the inverse slope parameter E_o on the centrality of the reaction. In fig.3 the value of E_o in four bins of impact parameter is reported for the reaction ^{40}Ar+^{51}V at 44 MeV/u. E_o is almost constant for the central reactions while decreases for the most peripheral ones. This behaviour, observed also in other experiments[6,7], is usually interpreted, still in the frame of the n-p bremsstrahlung hypothesis, as due to the softening of the nucleon momenta at the nuclear surface. In fact, since for peripheral reactions the n-p collisions take place between nucleons at the surface of the nuclei, the softer nucleon momentum distribution reflects into a softer bremsstrahlung γ spectrum emitted. Also this result confirms the fact that a careful analysis of the high energy γ spectra could give very interesting information on the nucleon momentum distributions.

3. Pre-equilibrium proton emission

Several experimental works, aimed at the study of the proton emission, have evidenced the presence of a half-beam velocity source with characteristics which have been interpreted in terms of pre-equilibrium emission from the partecipant

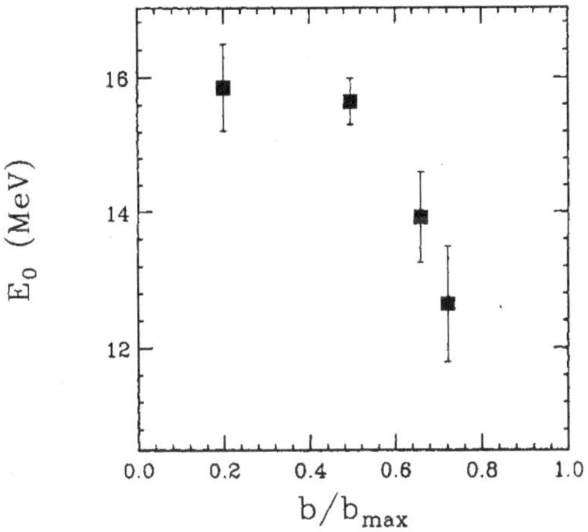

Fig.3 - Experimental inverse slope parameter E_o for four impact parameter bins for the reaction ^{40}Ar + ^{51}V at 44 MeV/u.

zone[14]. In particular, very energetic protons have been measured also at rather large polar angles with the energy spectrum slopes which imply very high *temperatures*.

We have undertaken a study of this intermediate velocity source by measuring the energy spectra, angular distributions and multiplicities of the protons emitted in the reactions induced by ^{40}Ar on ^{27}Al, ^{51}V and ^{197}Au and by ^{132}Xe on ^{51}V, ^{158}Gd and ^{197}Au at 44 MeV/u. We have performed a standard analysis by fitting the inclusive proton spectra with three maxwellian moving sources which were found to correspond to a projectile-like, a target-like and an intermediate source respectively. In table 1 the parameters related to the intermediate source are reported.

	Ar + Al	Ar + V	Ar+ Au	Xe + V	Xe + Gd	Xe + Au
T (MeV)	12.3	12.5	12.2	13.1	14.1	14.2
v_s/v_{beam}	0.51	0.50	0.45	0.56	0.51	0.49

Table 1

From this data some characteristics can be extracted. The intermediate source velocity is close to the half-beam velocity, independently of the projectile-target mass asymmetry. The high inverse slopes parameters can be explained[15] in terms of composition of the beam energy with the average nucleon fermi energy. At this incident energy the intermediate source accounts for a relevant part of the fast proton ($v_p \geq v_{beam}$) multiplicity.

To study in more details the pre-equilibrium proton emission the analysis was restricted to protons with energy higher than 40 MeV at angles larger than 42° to strongly reduce the contribution of the other sources. From the inclusive analysis we have found[16] that the average pre-equilibrium proton multiplicity increases linearly with respect to the average number of partecipant nucleons, calculated as in ref.13, indicating a dependence on the size of the partecipant region of the colliding system.

An important piece of information concerns the proton emission dependence on the impact parameter. In fig.4 the average experimental multiplicities $< m_{p40} >$ are reported as a function of the impact parameter for the reactions $^{40}Ar + {}^{51}V$ and $^{132}Xe + {}^{197}Au$. It is worth to be noticed that $< m_{p40} >$ exhibits the same impact parameter dependence as the hard photon multiplicity m_γ. The solid line represents the impact parameter dependence of the number of first nucleon-nucleon collisions occurring in the overlap area of the two colliding nuclei and it gives a reasonable description of the experimental data.

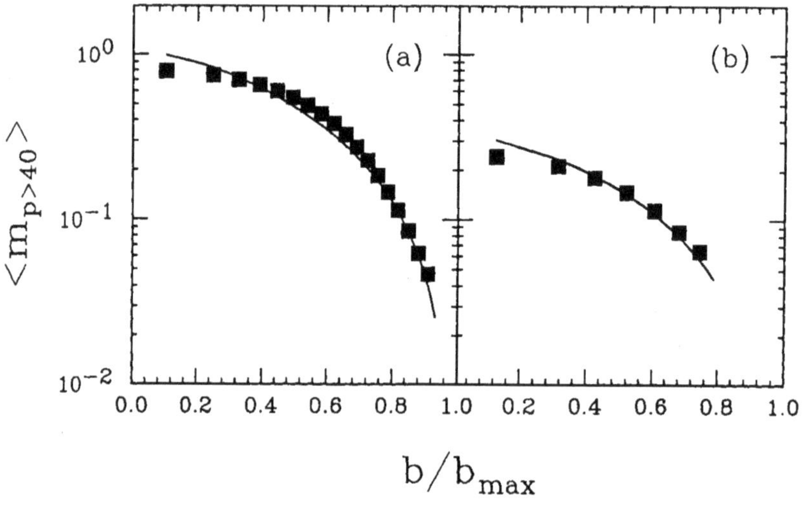

Fig.4 - Impact parameter dependence of the average pre-equilibrium proton multiplicity for the reactions $^{40}Ar + {}^{51}V$ and $^{132}Xe + {}^{197}Au$ at 44 MeV/u.

This result is consistent with the interpretation that also the pre-equilibrium protons are emitted from the nucleon-nucleon collisions which take place at the first stages of the nuclear reaction.

The data can be also explained assuming a pre-equilibrium emission from a fireball with an equal numbers of partecipants from the target and the projectile[17,18].

4. γ-proton correlations: preliminary results

The study of the correlation between very energetic protons and hard γ is very interesting since it could provide a clear signature of the hard photon and pre-equilibrium proton production mechanism.

In the frame of the first chance n-p bremsstrahlung, the probability of observing a coincidence between a hard photon and a very energetic proton should decrease with increasing proton energy. In fact, if a very energetic γ is emitted, the energy of the bremsstrahlung-generating proton is constrained by the limited energy available in the n-p collision and a reduced cross section for the γ-proton coincidence is expected. On the other hand, in the case of the emission from higher order n-p collisions or from a more thermalized system in a later stage of the reaction the γ and proton emission are almost independent and therefore no correlation is expected.

In our analysis the correlation factor $R(E_p) = \frac{\sigma_{\gamma p}(E_p)}{\sigma_\gamma \cdot \sigma_p(E_p)}\sigma_R$, which represents the ratio between the probability of emitting a γ and a proton in coincidence and the product of the independent probabilities of emitting a γ and a proton, has been studied as a function of the proton energy E_p for the reaction $^{40}Ar+^{51}V$ at 44 MeV/u in very central collisions. The results are shown in fig.5. For $E_\gamma \geq 25$ MeV $R(E_p)$ is close to one and it doesn't show any significant dependence on the proton energy as expected for uncorrelated γ-proton emission. This result is consistent with the data by Lampis et al[19] for $E_\gamma \geq 20$ MeV/u in the reaction ^{14}N + Zn at 40 MeV/u. On the other hand, if we rise the γ threshold up to 70 MeV, in spite of the very large error bars, a different trend is observed. In fact in this case the correlation factor is close to one for the lower proton energies but decreases with increasing energy indicating an anticorrelation between the most energetic γ and protons.

These are preliminary data and a more detailed analysis with higher statistics is needed to extract more quantitative information which should be compared with the theoretical calculations.

5. Conclusions

In summary we have studied hard photon and pre-equilibrium proton production and its dependence on the impact parameter. A description in terms of nucleon-nucleon collisions occurring at the first stage of the reaction is consistent with both the hard photon and pre-equilibrium proton data. The preliminary results on the γ-proton correlations strongly indicate that at least the most energetic γs and pre-equilibrium protons are produced by first chance n-p collisions.

We would like to thank Dr. A. Bonasera who provided us with the BNV calculations.

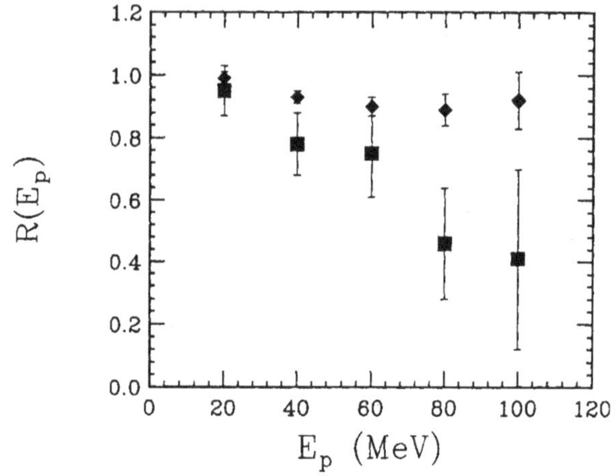

Fig.5 - Experimental correlation factor $R(E_p)$ as a function of the proton energy for ($E_\gamma \geq 25$ MeV) (diamonds) and ($E_\gamma \geq 70$ MeV) (squares) in very central collisions.

References

1. H. Nifenecker and J.A. Pinston, *Annual Rev. Nucl. Part. Science* **40**, (1990) 113.
2. W. Cassing et al., *Phys. Rep.* **188**, (1990) 365.
3. V. Metag, *Nucl. Phys.* **A488**, (1988) 483c.
4. R. Hingmann et al., *Phys. Rev. Lett.* **58**, (1987) 759.
5. H. Herrmann et al., *Phys. Rev. Lett.* **60**, (1988) 1630.
6. M. Kwato Njock et al., *Nucl. Phys.* **A489**, (1988) 368.
7. T. Reposeur et al., *Phys. Lett. B* **276** (1992) 418.
8. S. Riess et al., *Phys. Rev. Lett.* **69**, (1992) 1504.
9. E. Migneco et al., *Nucl. Instr. and Meth.* **A314** (1992) 31.
10. E. Migneco et al., *Phys. Lett. B* **298** (1993) 46.
11. C. Cavata et al., *Phys. Rev. C* **42** (1990) 1760.
12. A. Bonasera et al., *Phys. Lett. B* **246** (1990) 337 and *Il Nuovo Cimento A* **105** (1992) 301.
13. W. Bauer et al., *Phys. Rev. C* **34** (1986) 2127.
14. B.H. Jacak et al., *Phys. Rev. C* **35** (1987) 1751.
15. H. Fuchs and H. Homeyer, *International workshop on particle correlation and interferometry in nuclear collisions*, Nantes (1990) 305.
16. R. Alba et al., *XXXI International Winter meeting on nuclear physics* , Bormio(1993)
17. H. Nifenecker and J.P. Bondorf, *Nucl. Phys.* **A442**, (1985) 478.
18. A. Bonasera et al., *Nucl. Phys.* **A483**, (1988) 738.
19. A.R. Lampis et al., *Phys. Rev. C* **38** (1988) 1961.

Clusters

PLASMON DISPERSION IN ALKALI METALS

E. LIPPARINI
Dipartimento di Fisica, Universitá di Trento,
38050 Povo, Italy and INFN Trento

and

K. TAKAYANAGI
Tokyo Denki University, Department of Mathematical Science,
Hatoyama, Saitama, 350-03 Japan

ABSTRACT

We study the dynamical response of the valence electron gas in nearly-free-electron metals. We show that the measured dispersion relations of the plasma resonance in the bulk, surfaces and clusters of alkali atoms can be explained by linear response theory with a non local effective interaction. The electron-electron interaction is derived in the framework of the g-matrix approach that describes the scattering of two electrons in the presence of a many-electron medium. The electron-ion interaction is derived from non-local ionic pseudo-Hamiltonians, which replace core electrons in atomic calculations. The non-locality affects both the electron-electron and electron-ion components of the effective interaction.

1. Introduction

Alkali metals are simple systems. Each atom has one valence electron outside a ionic core which is a nearly-free-electron so that one can regard to alkali metals as nature's closest realization of the Fermi-electron gas. As a matter of fact they offer a unique possibility to study correlation effects of the nearly-free electron gas system at various densities. Furthermore a large experimental information on alkali metals is available now not only for the metal interior but also for metal surfaces and clusters so that it is possible to test many-body theories for the correlated electron gas in different geometries.

One of the more important sources of information on the structure of the correlated electron gas arises from the study of the plasma oscillation in alkali metals. In fact the existence of a plasmon is a direct consequence of the Coulomb interaction between the valence electrons. Recent electron-energy-loss experiments have allowed an accurate measurement of both the volume plasmon dispersion[1] and the surface plasmon dispersion[2] in alkali metals. The surface plasmon mode has been also observed in small metal particles of Na, K and Li by means of fragmentation experiments[3,4,5] and we know now to a good accuracy the location in energy of the plasma resonance in alkali clusters as a function of the number of atoms in the cluster.

This paper deals with the problem to reproduce the dispersion of the plasmon mode in the bulk, surface and clusters within a single microscopic many-body approach. In section 2 we review the mean field approach to this problem within the jellium model (JM). In section 3 we compare the results of the mean field calculations with experiments. Section 4 is dedicated to the building of an effective electron-electron interaction by means of the g-matrix theory[6]. In section 5 we investigate the electron-ion interaction. Finally in section 6 we draw our conclusions and we discuss some future perspectives.

2. Mean field theories within the JM

Static and dynamic properties of alkali metals are determined by the N valence electrons for which one assumes the following Hamiltonian (atomic units are used throughout this paper):

$$H = \sum_{i=1}^{N}(\frac{-\nabla^2}{2} + v_{ext})_i + \sum_{i<j}^{N} \frac{1}{\mid \mathbf{r}_i - \mathbf{r}_j \mid},$$

$$v_{ext}(i) = \sum_{I=1}^{N_I} h_{i,I}. \tag{1}$$

The first term in (1) is the kinetic energy term, the second one is the external field generated by the ions of the system interacting with the valence electrons, finally the last one is the Coulomb interaction between the valence electrons. N_I is the number of ions. In the jellium model (JM) one ignores the structure of the ions and takes

$$h_{i,I} = \frac{1}{\mid \mathbf{r}_i - \mathbf{r}_I \mid}. \tag{2}$$

Assumption (2) amounts to a point-charge approximation of the ionic potential, which cannot resolve any chemical difference between, say, two ions belonging to the same group of the periodic table. Moreover in the JM one assumes that the positive ions have a uniform distribution with density $n^+(\mathbf{r}_I)$. In the bulk $n^+(\mathbf{r}_I) = n_0 = 3/4\pi r_s^3$ is constant and defines the Wigner-Seitz radius r_s. In metal surfaces the positive background occupies the half space $x \leq 0$ and $n^+(\mathbf{r}_I) = n_0\Theta(-x)$, finally in spherical clusters the ionic density is constant inside a sphere of radius $R = r_s N_I^{\frac{1}{3}}$: $n^+(\mathbf{r}_I) = n_0\Theta(R - r_I)$. The electron-ion interaction is then simply described by the electrostatic potential due to such a model ionic density

$$v_{ext}(\mathbf{r}_i) = -\int \frac{n^+(\mathbf{r}_I)}{|\mathbf{r}_i - \mathbf{r}_I|} d\mathbf{r}_I. \tag{3}$$

Mean field theories with the JM have been developed by many people. In the Hartree-Fock (HF) approach[7] one takes the expectation value of the electron Hamiltonian (1) on a Slater determinant (SD)

$$E =< SD|H|SD >, \tag{4}$$

and then minimizes the energy with the constraint that the single particle wave functions φ_i entering the SD are normalized:

$$\delta(E - \sum_i \epsilon_i \int |\varphi_i|^2 d\mathbf{r}) = 0 \ . \tag{5}$$

This procedure generates HF equations which can be solved by iteration, yielding a complete set of single particle wave functions and energies ϵ_i. The HF energy functional writes:

$$E = \frac{1}{2}\int \tau d\mathbf{r} - \frac{1}{2}\int |\rho(\mathbf{r},\mathbf{r}')|^2 \frac{1}{|\mathbf{r}-\mathbf{r}'|} d\mathbf{r}d\mathbf{r}' + \frac{1}{2}\int \frac{(\rho - n^+)_{\mathbf{r}}(\rho - n^+)_{\mathbf{r}'}}{|\mathbf{r}-\mathbf{r}'|} d\mathbf{r}d\mathbf{r}', \tag{6}$$

where

$$\tau = \sum_{i,\sigma} |\nabla\varphi_i(\mathbf{r},\sigma)|^2,$$

$$\rho(\mathbf{r},\mathbf{r}') = \sum_{i,\sigma} \varphi_i^*(\mathbf{r},\sigma)\varphi_i(\mathbf{r}',\sigma),$$

$$\rho(\mathbf{r}) = \sum_{i,\sigma} |\varphi_i(\mathbf{r},\sigma)|^2, \tag{7}$$

are the electron kinetic energy , one-body non-diagonal and diagonal densities, respectively. In eq. (6) the three terms are, respectively, the kinetic, exchange and electrostatic parts of the total energy of the system. The electrostatic term containing the electron-electron direct Coulomb, electron-ion interaction and ion-ion interaction, is the classic potential energy of the system. In the bulk, due to the translational invariance of the system, the electron density ρ is equal to the ionic density n^+ and the electrostatic energy exactly vanishes. In surfaces and clusters its contribution to the total energy is very small so that the binding is still due to the Coulomb exchange electron-electron term. It is a quantum effect.

Starting from the HF single particle basis one has performed RPA calculations of the excited states of the electron gas. In the bulk the RPA equations for the response function $\chi(q,\omega)$, where q is the momentum transfer and ω the excitation energy, are written:

$$\chi^{RPA}(q,\omega) = \frac{\chi^0(q,\omega)}{1 - v(q)\chi^0(q,\omega)} , \tag{8}$$

where $\chi^0(q,\omega)$ is the Lindhard free response function and $v(q) = \frac{4\pi}{q^2}$ is the Fourier transform of the Coulomb potential (direct term). As a solution of eq.(8) one gets a collective mode, the bulk plasmon, which corresponds to the collective oscillation of the valence electrons against the rigid positive ionic background. The energy of the bulk plasmon is obtained as solution of the equation

$$\chi^0(q,\omega) = \frac{1}{v(q)} , \tag{9}$$

which gives the poles of the RPA equations (8). In the $q \to 0$ limit one gets[8] the following RPA dispersion for the bulk plasmon

$$\omega_b(q) = \omega_p + \frac{3\epsilon_F}{5\omega_p}q^2 , \tag{10}$$

where $\omega_p = \sqrt{\frac{3}{r_s^3}}$ is the plasma frequency and $\alpha^{RPA} = \frac{3\epsilon_F}{5\omega_p}$, where ϵ_F is the Fermi energy, is the so-called RPA volume plasmon dispersion coefficient.

In the case of metal surfaces the solution of the RPA equations yields for the surface plasmon the dispersion[9]

$$\omega_s(q) = \frac{\omega_p}{\sqrt{2}}(1 - \frac{1}{2}qd) , \tag{11}$$

where d is a coefficient related to the centroid of the induced surface charge. For alkali metals d is positive so that the surface plasmon first shifts linearly to lower frequencies at small q.

Finally for clusters, in the limit $\frac{1}{R} \to 0$, the RPA equations give a plasmon located at the energy[10]

$$\omega_R = \frac{\omega_p}{\sqrt{3}}(1 - \frac{1}{2}d_R),$$

$$d_R = \frac{1}{N}\int_{r>R} \rho(\mathbf{r})d\mathbf{r}, \tag{12}$$

where d_R gives the percentage of electrons outside the jellium radius R.

HF is not able to reproduce experimental data, for example, on ionization potentials and cohesive energies since correlation effects in nearly-free-electron metals are important. One has therefore developed a local density approximation (LDA) approach to simple metals. In this approach one takes the bulk expression of the exchange Coulomb energy per particle and makes the LDA getting the density dependent exchange term:

$$\epsilon_x(\rho) = -\frac{3}{4\pi}(3\pi^2)^{\frac{1}{3}}\rho^{\frac{1}{3}}. \tag{13}$$

Then one adds to the energy functional a correlation term $\epsilon_c(\rho)$ obtained by fitting the correlation energy per particle of a quantum Montecarlo calculation in the bulk[11]. The LDA energy functional is then given by

$$E = \frac{1}{2}\int \tau d\mathbf{r} + \int(\epsilon_x(\rho) + \epsilon_c(\rho))\rho(\mathbf{r})d\mathbf{r} + \frac{1}{2}\int \frac{(\rho - n^+)_{\mathbf{r}}(\rho - n^+)_{\mathbf{r}'}}{|\mathbf{r} - \mathbf{r}'|}d\mathbf{r}d\mathbf{r}'. \tag{14}$$

Starting from eq.(14) one has performed self-consistent mean field calculations for free surfaces[12] and clusters[13] both static and dynamic. Linear response theory in the framework of the energy functional (14) is called time dependent local density approximation (TDLDA). In the bulk the TDLDA equations look like the RPA equations (8)

$$\chi^{TDLDA}(q,\omega) = \frac{\chi^0(q,\omega)}{1 - v(q)\chi^0(q,\omega)} , \tag{15}$$

with the only difference that now the interaction $v(q)$ in eq.(15) takes contribution also from the exchange and correlation terms in eq.(14):

$$v(q) = \frac{4\pi}{q^2}(1 - G(q)). \tag{16}$$

The function $G(q)$ in (16) is related to the second functional derivative with respect to the density ρ of $E_{x,c}(\rho) = \int(\epsilon_x(\rho) + \epsilon_c(\rho))\rho(\mathbf{r})d\mathbf{r}$ in eq.(14) , and in the $q \to 0$ limit tends to the incompressibility of the system[14].

3. Comparison of mean field theories with experiments

In figure 1 we report the electron energy loss experiment on Na films by Fink et al.[1].

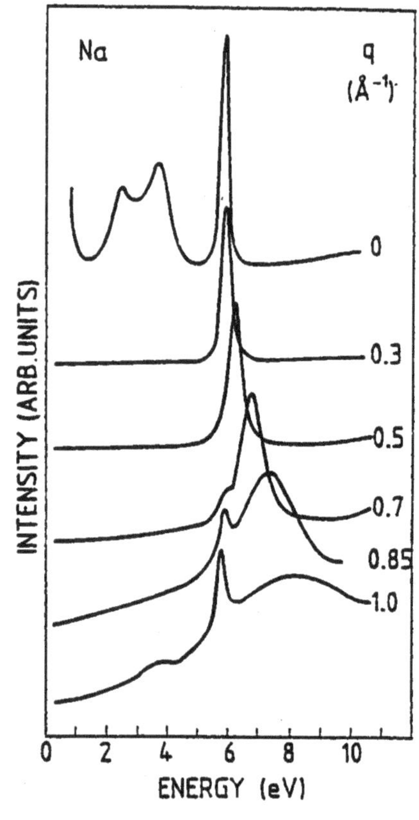

Figure 1

At q=0 one observes a very narrow resonance at $\sim$ 5.9 eV, which is exactly the prediction of the RPA (eq.(10)) for the bulk plasmon in the case of sodium (r_s=4). Increasing q, the resonance disperds towards high energy and its width rapidly increases due to the Landau damping. Figures 2 are plots of the bulk plasmon energies in Na, K, Rb and Cs versus the square of the tranferred momentum in the excitation process. The dashed and full lines are theoretical calculations. The dashed line is obtained[15] starting from the TDLDA (eqs.(15,16)). The full line is the same calculation after additional inclusion of some structure effect of core electrons of the ions[1]. The inclusion of the core polarizability lowers the plasma energy especially at q=0 where brings the theoretical result closer to the experimental value. Comparison between the measured volume plasmon dispersion and TDLDA calculations shows the inadequacy of the theory to explain experimental data at $q \neq 0$.

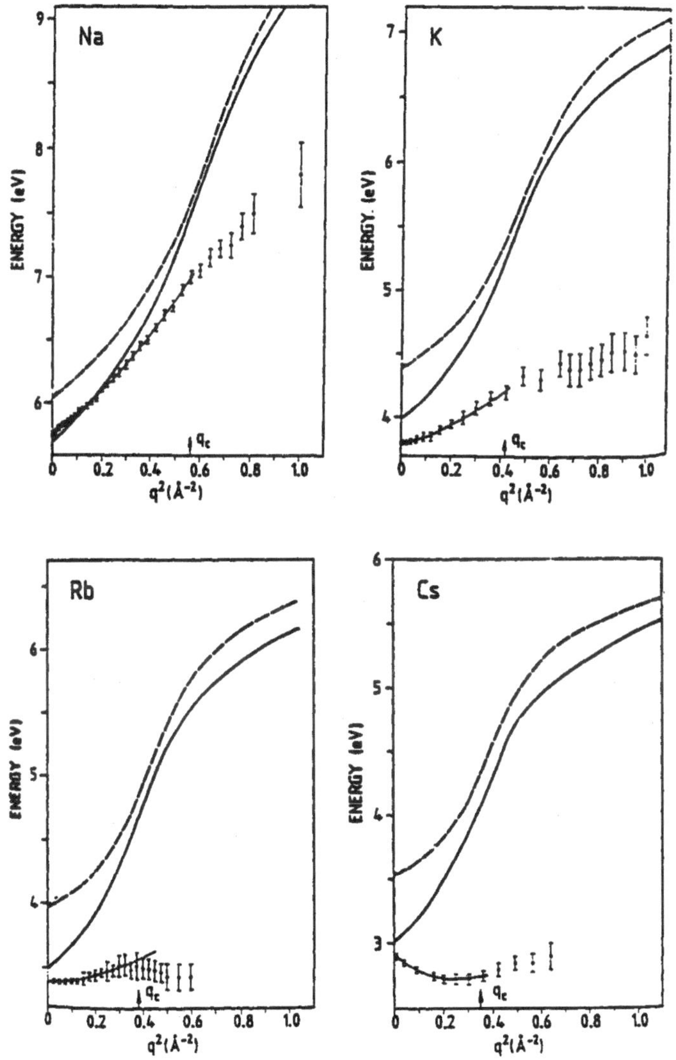

Figure 2

This is particularly evident in fig. 3 where we plot the experimental volume plasmon dispersion coefficient α ($\omega_b(q) = \omega_p + \alpha q^2$), normalized to the dispersion coefficient in RPA (see eq.(10)) versus the Wigner-Seitz radius r_s, together with the RPA (dashed line) and TDLDA (dotted line) predictions.

Figure 3

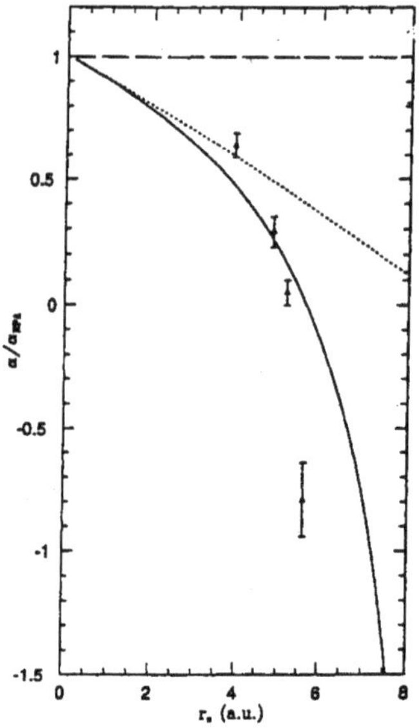

In fig.4 we report the experimental[2] momentum dispersion of the surface plasmon normalized to the plasma frequency for K and Na, together with the RPA and TDLDA response calculations[2,12]. The dashed lines are the calculated linear slopes at small q (see eq.(11)). From the figure one sees that the predicted RPA linear slope is closest to the experimental data than the TDLDA one. This result is surprising and not yet understood. Finally in fig.5 we report the measurement[5] of the photoabsorption cross sections of two lithium clusters together with the predictions of the TDLDA response (dashed line). The striking feature is the large shift between experimental and calculated curves. This shift is systematically found in all the measured lithium and K clusters (see fig. 6 where the cubes are the TDLDA peak energies and the black circles with bars are the experimental peak energies) and is very similar to the shift observed in the bulk plasmon dispersion at q=0 (see fig. 1). Like in the case of the bulk plasmon we believe that this is due to the neglecting in the JM of the ionic structure. The problem to take into account the effect of core electrons on the location of the plasma resonance in alkali clusters is treated in section 5. The ionic structure is not however able to explain the large discrepancy between experimental data and calculations in the bulk plasmon dispersion at $q \neq 0$ (see figs. 1,2). In fact inclusion of core polarization in the calculation of the plasmon dispersion coefficient α of fig. 2 makes worse this comparison between theory and experiment[1]. We believe that the observed discrepancy is due to the presence of non local components in the effective electron-electron interaction which in the LDA approach are washed out. In the following section we then develop the g-matrix approach[6] to the electron gas. The g-matrix effective interaction describes the scattering of two electrons in the presence of a many-electron medium. This scattering process induces a considerable change in the short range behaviour of two electrons compared to the one in free space. In particular it introduces important non local components in the effective interaction.

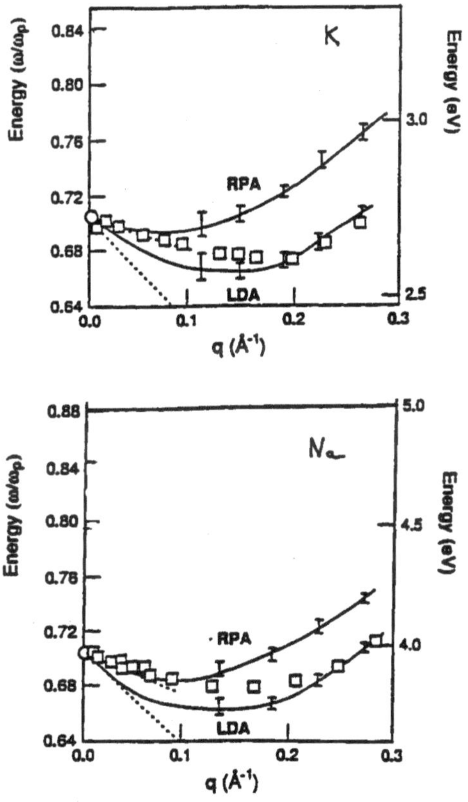

Figure 4

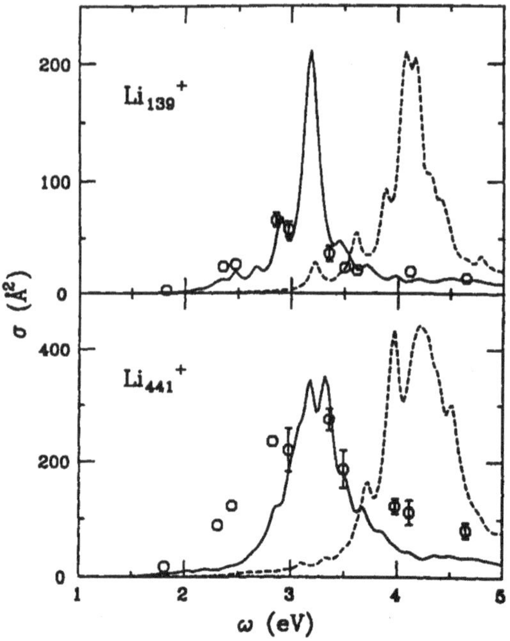

Figure 5

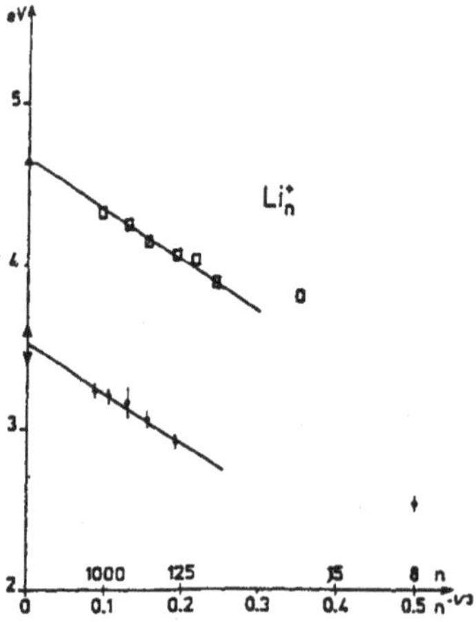

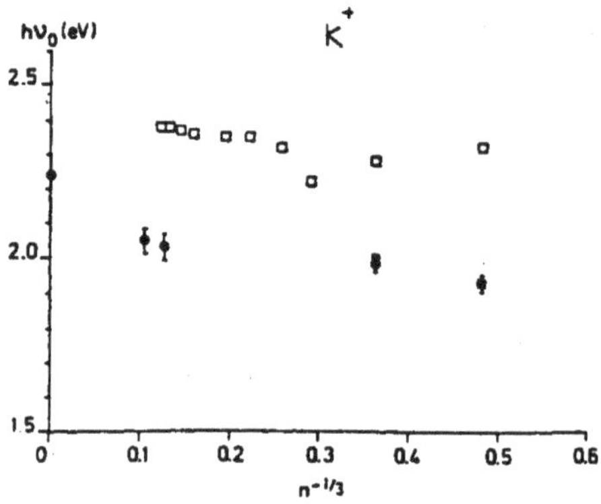

Figure 6

4. The effective electron-electron interaction

In the LDA approach one takes into account the effect of dynamical electron-electron correlations by introducing a local density dependent term in the energy functional: $\epsilon_c(\rho)$. As was pointed out many years ago[16,17], an alternative way to take into account these correlations is to use effective interactions obtained by solving, in analogy with nuclear matter calculations, the Bethe-Goldstone integral equation which describes the scattering of two electrons in the presence of an electron medium. In this section we adopt this approach to study in a microscopic way the effect of electron correlations on the bulk plasmon dispersion. The Brueckner g-matrix in an electron gas is the analogue of the scattering matrix for two electrons in the free space. The starting point of the theory is then the Lippman-Schwinger equation for the scattering matrix t of two electrons:

$$< \mathbf{k_1 k_2}|t|\mathbf{k'_1 k'_2} >=< \mathbf{k_1 k_2}|v|\mathbf{k'_1 k'_2} > \\ + \frac{1}{2}\sum_{\mathbf{p_1 p_2}} < \mathbf{k_1 k_2}|v|\mathbf{p_1 p_2} > \frac{1}{E - p_1^2/2 - p_2^2/2} < \mathbf{p_1 p_2}|t|\mathbf{k'_1 k'_2} > , \tag{17}$$

where $\mathbf{k_1}, \mathbf{k_2}$ and $\mathbf{k'_1}, \mathbf{k'_2}$ are the momenta of the incoming and outgoing electrons, respectively, and $E = k_1^2/2 + k_2^2/2$ is the incident energy. Since the center of mass momentum $\mathbf{K} = \mathbf{k_1} + \mathbf{k_2}$ is conserved, the scattering matrix t is a function of the three momenta $\mathbf{K}$, $\mathbf{q} = \frac{1}{2}(\mathbf{k_1} - \mathbf{k_2})$ and $\mathbf{q}' = \frac{1}{2}(\mathbf{k'_1} - \mathbf{k'_2})$. The matrix element of the bare Coulomb interaction is given by

$$< \mathbf{k_1 k_2}|v|\mathbf{k'_1 k'_2} >= lim_{\eta \to 0} \frac{4\pi}{|\mathbf{q} - \mathbf{q}'|^2 + \eta^2} \quad . \tag{18}$$

In the presence of an electron medium, the interaction of the two electrons could be modified. If we are interested only in the short-range piece of our interaction, we can neglect all the terms other than two-body terms and approximate the many-body medium as a filled Fermi sea of non-interacting electrons. The role of the Fermi sea is to restrict the summation over the intermediate states so that it does not include states below the Fermi surface. This idea is realized in terms of the Bethe-Goldstone integral equation for the g-matrix:

$$< \mathbf{qK}|g|\mathbf{q'K} >=< \mathbf{q}|v|\mathbf{q}' > \\ + \int \frac{d\mathbf{p}}{(2\pi)^3} < \mathbf{q}|v|\mathbf{p} > \frac{Q(\mathbf{p}, \mathbf{K}, k_F)}{E - \epsilon_{|\mathbf{K}/2+\mathbf{p}|} - \epsilon_{|\mathbf{K}/2-\mathbf{p}|}} < \mathbf{pK}|g|\mathbf{q'K} > , \tag{19}$$

where in the same way as eq.(18), we have introduced relative and center of mass momentum coordinates. The Pauli projection operator $Q(\mathbf{p}, \mathbf{K}, k_F)$, which excludes all the occupied states, is given by

$$Q(\mathbf{p}, \mathbf{K}, k_F) = \Theta(|\mathbf{K}/2 + \mathbf{p}| - k_F)\Theta(|\mathbf{K}/2 - \mathbf{p}| - k_F), \tag{20}$$

where $\Theta(x)$ is the usual step function.

The solution of eq.(19), which is referred to as the Brueckner g-matrix, expresses the sum of all ladder diagrams as shown in fig. 7.

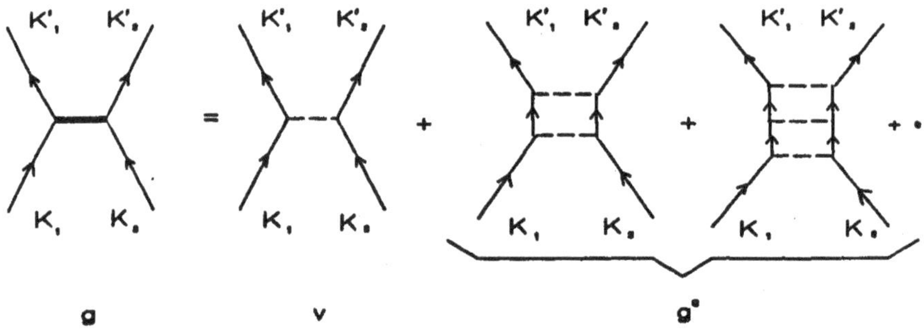

Figure 7

We have solved[18] eq.(19) using standard techniques. The obtained g-matrix has been written in the form $g = v + g^c$, where v is the bare Coulomb force and g^c is the correction due to the multiple scattering processes at short distances. Since g^c is a slowly varying function of the center of mass momentum K, we have averaged over K to obtain the following expression for g^c

$$< \mathbf{q}|g^c|\mathbf{q}' >= 4\pi \sum_{\ell}(2\ell + 1)P_\ell(\hat{q} \cdot \hat{q}')g^c_\ell(q, q'). \tag{21}$$

In order to illustrate the features of the g-operator (21), let us assume the following separability:

$$< \mathbf{q}|g^c|\mathbf{q}' >= C(\frac{|\mathbf{q} + \mathbf{q}'|}{2})V(|\mathbf{q} - \mathbf{q}'|), \tag{22}$$

where $C(\frac{|\mathbf{q}+\mathbf{q}'|}{2})$ and $V(|\mathbf{q} - \mathbf{q}'|)$ represent the nonlocal and local parts of the g-operator, respectively. Approximation (22) has turned out to be a reasonable one.

The behaviour of C and V is shown in fig.8 in the case of $r_s = 4$, from which one can see that the range of V in the configuration space is of the oredr of the average electron-electron distance and C is a peaked function around 1.3 $\AA^{-1}$.

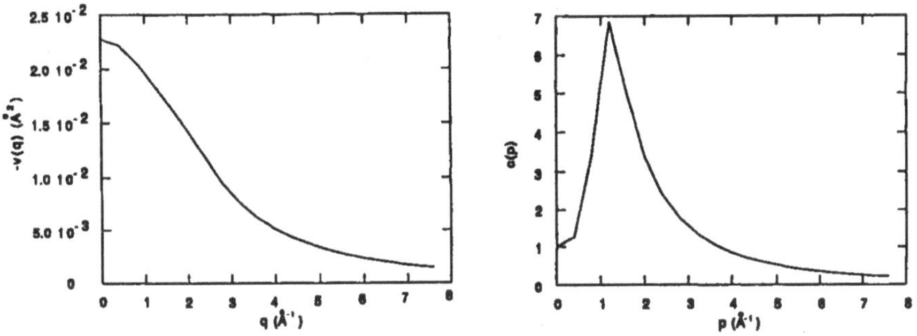

Figure 8

The consenquences of this behaviour of C on observable quantities, as for example the plasmon dispersion, have never been explored previously. Previous authors[16,17] have considered mainly the effect of the local part $V(|\mathbf{q} - \mathbf{q}'|)$ on the electron pair correlation function of the infinite system. This local part is not however sufficient to explain the experimental data of figs. 1 and 2. More precisely it gives the same results of the TDLDA theory (see the dotted curve of fig. 3).

Starting from eq.(22) we have calculated the ground state Brueckner correlation energy $E^B = \frac{1}{N} < SD| \sum_{i<j} g^c(\mathbf{r}_i, \mathbf{r}_j)|SD >$, which takes into account the effect of two-body rescattering processes on the ground state energy, and the plasmon dispersion coefficient α. Numerical values for the correlation energy per particle E_B are given in table 1 for some values of r_s. In column two of the table are given the values obtained by Ceperley[11] in his Montecarlo calculation. These values should be considered as benchmarks against which other values should be compared. As one can see our values compare very favorably with the Montecarlo ones.

Table 1	r_s	$E_B(eV)$	$E_{QM}(eV)$	$E_{DM}(eV)$
	2	-1.22	-1.23	-1.13
	3	-1.15	-1.10	-0.92
	4	-1.03	-0.92	-0.80
	5	-0.92	-0.80	-0.71

In fig. 3 we report our predictions (solid line) for the plasmon dispersion coefficient as obtained in a sum rule approach[19] which makes explicit use of our non local effective interaction. As one can see from the figure, the effect of non locality is crucial to reproduce the correct r_s dependence of α. In order to exploit better the non locality of the g-matrix correction g_c we have performed a density matrix expansion[18] of g_c. This density matrix expansion corresponds to keep in the multipole expansion (21) only s (ℓ=0) and p (ℓ=1) waves. One gets in the configuration space:

$$g_c(\mathbf{r}) = I_2\delta(\mathbf{r}) + \frac{1}{24}(I_2C_4 - 2I_4)(\nabla_{\mathbf{r}}^2\delta(\mathbf{r}) + \delta(\mathbf{r})\nabla_{\mathbf{r}}^2 - 2\nabla_{\mathbf{r}}\cdot\delta(\mathbf{r})\nabla_{\mathbf{r}}), \qquad (23)$$

where $\mathbf{r} = \mathbf{r}_1 - \mathbf{r}_2, \nabla_{\mathbf{r}} = \frac{1}{2}(\nabla_1 - \nabla_2)$ and the r_s-depenedent coefficients I and C are obtained as integrals of the Fourier transform of V and C (see eq. (22) and fig.(8)):

$$I_k = 4\pi\int r^k V(r)dr \ ; \ C_k = 4\pi\int r^k C(r)dr. \qquad (24)$$

In table 1 in column three we report the correlation energy obtained starting from the density matrix expansion (23). As one can see from this table the density matrix expansion result is close to the calculation with the full g-matrix correction which includes all the multipoles. This is very encouraging to using the simple non local effective interaction (23) in calculations for free surfaces and clusters. The relevant HF energy functional would be:

$$E = \frac{1}{2}\int (1-2t_1\rho)\tau d\mathbf{r} + \int \rho(\mathbf{r})v^{ext}d\mathbf{r} + \frac{1}{4}t_0\int \rho^2 d\mathbf{r} + \frac{1}{2}\int \frac{\rho(\mathbf{r})\rho(\mathbf{r}') - |\rho^{(1)}(\mathbf{r},\mathbf{r}')|^2}{|\mathbf{r}-\mathbf{r}'|}d\mathbf{r}d\mathbf{r}' \tag{25}$$

where

$$t_0 = I_2, \qquad t_1 = \frac{1}{24}(I_2C_4 - 2I_4) \tag{26}$$

are the density dependent coefficients of the g-matrix correction (23). These values (in atomic units) are reported in table 2 for different values of r_s.

	r_s	t_0	t_1
	2	– 2.54	1.35
Table 2	3	– 6.99	8.40
	4	– 14.4	30.7
	5	– 25.2	83.7

5. The electron-ion interaction

In this section we study the electron-ion interaction with the aim to explain the shift between measured and calculated peak energies of the plasma resonance observed in electron energy loss experiments at q=0 in the bulk (see fig.2) and in photoabsorption experiments in clusters (see figs. 5,6). We illustrate the results of a recent work[20] on Li clusters where the JM has been generalized by substituting the assumption (3) for the electron-ion interaction with the following one:

$$\hat{v}_{ext} = -\int n^+(\mathbf{r}_I)h(|\mathbf{r}_i - \mathbf{r}_I|)d\mathbf{r}_I, \tag{27}$$

where $h(|\mathbf{r}_i - \mathbf{r}_I|)$ is the pseudohamiltonian[21]

$$h(|\mathbf{r}_i-\mathbf{r}_I|) = -\frac{1}{2}\nabla_{\mathbf{r}_i-\mathbf{r}_I}a(|\mathbf{r}_i-\mathbf{r}_I|)\cdot\nabla_{\mathbf{r}_i-\mathbf{r}_I} + \frac{1}{2}\mathbf{L}_{\mathbf{r}_i-\mathbf{r}_I}\frac{b(|\mathbf{r}_i-\mathbf{r}_I|)}{|\mathbf{r}_i-\mathbf{r}_I|^2}\cdot\mathbf{L}_{\mathbf{r}_i-\mathbf{r}_I} + v(|\mathbf{r}_i-\mathbf{r}_I|), \tag{28}$$

which describes the relative motion of an electron with coordinate $\mathbf{r}_i$ and an ion at point $\mathbf{r}_I$. We shall refer to the model that makes use of this pseudohamiltonian as the pseudo-jellium model (PJM). In Eq.(28), a, b, v are radial functions and

$\mathbf{L}_{\mathbf{r}_i-\mathbf{r}_I}$ is the angular momentum operator with respect to the position $\mathbf{r}_I$. The pseudohamiltonian (28) takes into account the ionic structure (core electrons) in the interaction between the valence electron and the ion. Outside the core of an atom, a and b vanish and $v = -1/|\mathbf{r}_i - \mathbf{r}_I|$. Inside the core the radial functions a, b and v are determined to optimally replace a full core atom with a pseudo- (valence only) atom[21]. They are shown in Fig.9 for the case of the Li atom.

The potential $\hat{v}_{ext}$ of eqs.(27,28) is non local and gives rise to mean field equations

$$\{-\frac{1}{2}\nabla\,[1+\alpha(r)]\cdot\nabla + \mathbf{L}\beta(r)\cdot\mathbf{L} + u(r) + \int \frac{\rho(\mathbf{r}')}{|\mathbf{r}-\mathbf{r}'|}d\mathbf{r}' + \frac{\delta E_{xc}(\rho)}{\delta\rho}\}\varphi_i(\mathbf{r}) = \epsilon_i\varphi_i(\mathbf{r})\,, \tag{29}$$

with a modified kinetic energy operator with respect to the one obtained in the LDA within the JM. The functions $\alpha(r), \beta(r)$ and $u(r)$ in eq. (29) are related[20] to the functions a, b and v entering the pseudohamiltonian (28). The striking feature of those modified Kohn-Sham equations is the appearence of a radial effective mass

$$m^*(r) = 1/[1+\alpha(r)] \tag{30}$$

which, for example, has the effect of changing the integrated oscillator strength $m_1 \equiv \sum_n \omega_n| < O|D|n > |^2$ of the dipole operator $D = \sum_{i=1}^N z_i$:

$$m_1^{PJM} = \frac{1}{2}\int \left[1+\alpha(r)+\frac{4}{3}r^2\beta(r)\right]\rho(r)dr \tag{31}$$

as compared to the JM result $m_1^{JM} = \frac{N}{2}$. In Li the effective mass (30) is larger than the free electron mass, but for the other alkali metals it is generally smaller than the free electron mass.

Eqs.(29) have been solved self-consistently in the case of single ionized Li clusters, where experiments have shown the deficiency of the JM. Using the obtained single-particle spectrum, one has calculated the dipole response function in the TDLDA. The result of the calculation is shown in fig.5 where the solid and dashed curves correspond to the PJM and JM calculations respectively. Fig. 5 shows a good agreement of the experimental peaks with PJM results whereas the JM predictions are off by about 1 eV. In Fig. 10 we compare the JM (crosses) and PJM (stars) calculated peak energies for Li cluster ions ranging from Li_9^+ (N=8) to Li_{441}^+ (N=440), together with measured (open circles with bars) peak energies taken from ref. 5 . On the same figure are shown calculated values for the neutral cluster Li_8 which has been also measured[22]. The PJM energies are systematically lower than the JM ones by about 1 eV, resulting in a remarkable agreement with experiment.

Figure 9

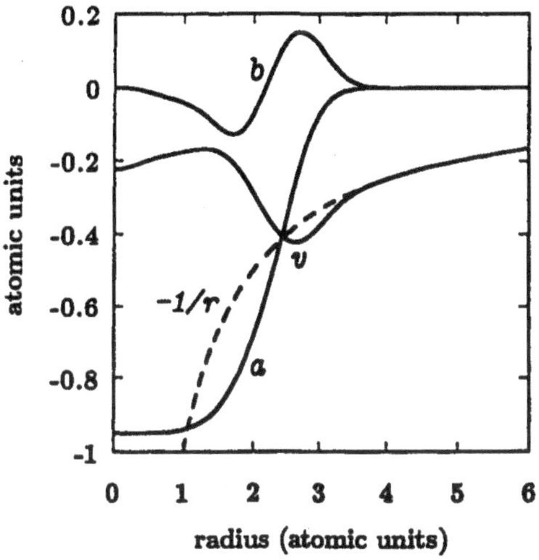

Figure 10

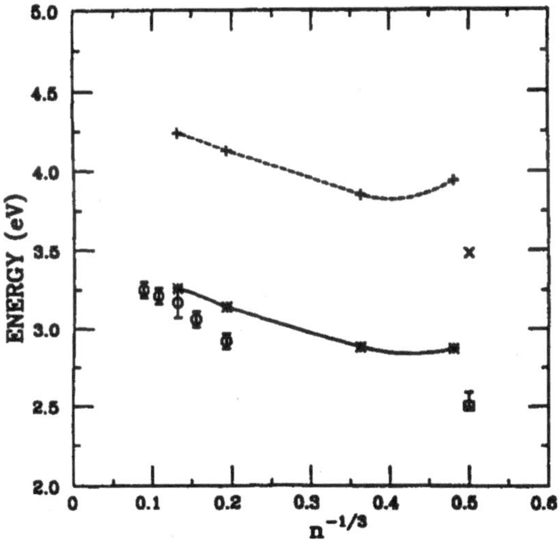

6. Conclusions

We have proved that non local components of the g-matrix effective electron-electron interaction are crucial in explaining the q-dependence of the bulk plasmon dispersion in alkali metals. A density matrix expansion of the g-matrix which includes only s and p waves is able to reproduce closely the results for the correlation energy of quantum Montecarlo calculations. This expansion gives rise to a very simple energy functional with an effective mass and a gradient term in the density.

An electron-ion interaction which takes into account the structure of the ions by means of a pseudohamiltonian (pseudojellium model) greately improves the agreement between theory and experiments on photoabsorption cross sections in Li clusters.

Some problems related to the use of the effective electron-electron and electron-ion interactions remain to be explored and deserve further theoreticel effort. In particular we mention

i) HF-RPA calculations in surfaces anf clusters with the effective hamiltonian

$$H = \sum_i (\frac{p_i^2}{2} - \int n^+(\mathbf{r}_I) h(|\mathbf{r}_i - \mathbf{r}_I|) d\mathbf{r}_I) + \sum_{i<j} (\frac{1}{|\mathbf{r}_i - \mathbf{r}_j|}$$

$$+ t_0 \delta(\mathbf{r}_{i,j}) + t_1 (\nabla_\mathbf{r}^2 \delta(\mathbf{r}) + \delta(\mathbf{r}) \nabla_\mathbf{r}^2 - 2\nabla_\mathbf{r} \cdot \delta(\mathbf{r}) \nabla_\mathbf{r})_{i,j}) \quad ,$$

ii) HF-RPA calculations with effective forces in low-dimensional systems.

REFERENCES

1. A. vom Felde, J. Sprosser-Prou and J. Fink, Phys. Rev. **B 40** (1989) 10181.
2. K.D. Tsuei, E.W. Plummer, P.J. Feibelman, Phys. Rev. Lett. **63** (1989) 2256; K.D. Tsuei, E.W. Plummer,, A. Liebsch, K. Kempa and P. Bakshi, Phys. Rev. Lett. **64** (1990) 44.
3. K. Selby, K. Kresin, V. Masui, J. Vollmer, M. Scheidemann and W.D. Knight, Z. Phys. **D 19** (1991) 43 and references therein.
4. C. Bréchignac, Ph. Cahuzac, N. Kebaili, J. Leygnier and A. Sarfati, Phys. Rev. Lett. **68** (1992) 3916 and references therein.
5. C. Bréchignac, Ph. Cahuzac, F. Carlier, J. Leygnier, J.Ph. Roux and A. Sarfati, Z. Phys. **D** to be published; C. Bréchignac, Ph. Cahuzac, J. Leygnier and A. Sarfati, Phys. Rev. Lett. **70** (1993) 2036.

6. see for example, A.L. Fetter and J.D. Walecka, Quantum Theory of Many Particle Systems, McGraw-Hill, New York, 1971.
7. C. Guet and W.R. Johnson, Phys. Rev. **B 45** (1992) 11283 .
8. D. Pines and P. Nozières, The Theory of Quantum Liquids, Benjamin, New York, 1966, Vol. I.
9. P.J. Feibelman, Progr. Surf. Sci., **12** (1982) 287.
10. M. Brack, Phys. Rev. **39** (1989) 3533; E. Lipparini and S. Stringari, Z. Phys. **D 18** (1991) 193.
11. D. Ceperley, Phys. Rev. **B 18** (1978) 3126 .
12. A. Liebsch, Phys. Rev. **B 36** (1987) 7378.
13. W. Ekardt, Phys. Rev. **B 31** (1985) 6360.
14. N. Iwamoto, E. Krotscheck and D. Pines, Phys. Rev. **B 29** (1984) 3936.
15. P. Vashista and K.S. Singwi, Phys. Rev. **6** (1972) 875.
16. D.N. Lowy and G.E. Brown, Phys. Rev. **B 12** (1975) 2138 .
17. K. Bedell and G.E. Brown, Phys. Rev. **B 17** (1978) 4512.
18. K. Takayanagi and E. Lipparini, Phys. Rev. **B** submitted.
19. E. Lipparini, S. Stringari and K. Takayanagi, Phys. Rev. Lett. , submitted.
20. Ll, Serra, G.B. Bachelet, N. Van Giai and E. Lipparini, Phys. Rev. **B** in press.
21. G.B. Bachelet, D.M. Ceperley and M.G.B. Chiocchetti, Phys. Rev. Lett. **62** (1989) 2088 .
22. J. Blanc, M. Broyer, J. Chevaleyre, Ph. Dugourd, H. Kuhling, P. Labastie, M. Ulbricht, J.P. Wolf and L. Woste, Z. Phys. **D 19** (1991) 7.

SHELL STRUCTURE AND FISSION OF METAL CLUSTERS

T.P. MARTIN, S. FRANK, N. MALINOWSKI[1], U. NÄHER
AND U. ZIMMERMANN
Max-Planck-Institut fuer Festkoerperforschung
Heisenbergstr. 1, D-70569 Stuttgart (Germany)

ABSTRACT

Cluster science has greatly benefited from the work carried out by nuclear physicists decades ago. The shell model and the liquid droplet model provide a ready and appropriate framework for understanding the properties of metal clusters.

1. Introduction

It is not obvious that metal clusters should behave like atomic nuclei - but they do. Of course, the energy and distance scales are quite different. But aside from this, the properties of these two forms of condensed matter are amazingly similar. The shell model developed by nuclear physicists describes very nicely the electronic properties of alkali metal clusters. The giant dipole resonances in the excitation spectra of nuclei have their analogue in the plasmon resonances of metal clusters. Finally, the droplet model describing the fission of unstable nuclei can be successively applied to the fragmentation of highly charged metal clusters. The similarity between clusters and nuclei is not accidental. Both systems consist of fermions moving nearly freely in a confined space.

Many years ago it was noticed that atomic nuclei containing either 8, 20, 50, 82 or 126 protons or neutrons have very long lifetimes. It was a challenge for the nuclear physicists back in the forties to explain these so-called magic numbers. Since physicists tend to see most objects as perfectly round, it should come as no surprise that they assumed atomic nuclei are spherically symmetric. Under this assumption they had to solve only a radial Schrödinger equation,

$$\left[-\frac{d^2}{dr^2}+\frac{l(l+1)}{r^2}+V(r)\right]P_{nl}(r)=E_{nl}P_{nl}(r) \tag{1}$$

where l is the angular momentum quantum number and V(r) is the radial dependence of the potential in which the nucleons move. They assumed further that the potential could be described by a simple potential well. Some confusion can arise because nuclear physicists and atomic physicists use slightly different definitions for the principal quantum number n,

[1]Permanent address: Central Lab. of Photoprocesses, Bulgarian Academy of Science, 1040 Sofia

$$n(\text{atomic}) = n(\text{nuclear}) + l. \tag{2}$$

Throughout this paper we will use the principal quantum number from nuclear physics, i.e. n denotes the number of extrema in the radial wavefunction.

Eigenstates of the radial Schrödinger equation are often called subshells. The subshells of the infinite spherical potential well are shown ordered according to momentum in Fig. 1. The lowest energy state is 1s, then comes 1p, 1d, 2s, 1f, 2p,... etc. That is, with 2, 8, 18, 20, 34, 40, 58, 90,... nucleons, subshells are completely filled and the corresponding nuclei could be expected to be exceptionally stable. However, these are not the observed magic numbers.

In 1949 Maria Goeppert-Meyer [1] and Haxel, Jensen and Suess [2] came up with a modified model which yielded the observed magic numbers. Their idea was that the spin-orbit interaction is unusually strong for nucleons. Subshells with high angular momentum split and the states rearrange themselves into different groups. As we shall see the original shell model, which the nuclear physicist had to discard, describes very nicely the electronic states of metal clusters [3–18].

2. Subshells, Shells and Supershells

If it can be assumed that the electrons in metal clusters move in a spherically symmetric potential, the problem is greatly simplified. Subshells for large values of angular momentum can contain hundreds of electrons having the same energy. The highest possible degeneracy assuming cubic symmetry is only 6. So under spherical symmetry the multitude of electronic states condenses down into a few degenerate subshells. Each subshell is characterized by a pair of quantum numbers n and l. Under certain circumstances the subshells themselves condense into a smaller number of highly degenerate shells. The reason for the formation of shells out of subshells requires more explanation.

The concept of shells can be associated with a characteristic length. Every time the radius of a growing cluster increases by one unit of this characteristic length, a new shell is said to be added. The characteristic length of shells of atoms is approximately equal to the interatomic distance. The characteristic length for shells of electrons is related to the wavelength of an electron in the highest occupied energy level (Fermi energy). For the alkali metals these lengths differ by a factor of about 2. This concept is useful only because the characteristic lengths are, to a first approximation, independent of cluster size.

The concept of shells can also be described in a different manner. An expansion of N, the total number of electrons, in terms of the shell index K will always have a leading term proportional to K^3. One power of K arises because we must sum over all shells up to K in order to obtain the total number of particles. One power of K arises because the number of subshells in a shell increases approximately linearly with shell index. Finally, the third power of K arises because the number of particles in the largest subshell also increases with shell index. Expressing this slightly more quantitatively, the total number of particles needed to fill all shells, k, up to and including K is

$$N_K = \Sigma_{k=1}^{K} \Sigma_{l=0}^{L(k)} 2(2l+1)\ K^3 \tag{3}$$

where L(k) is the highest angular momentum subshell in shell k.

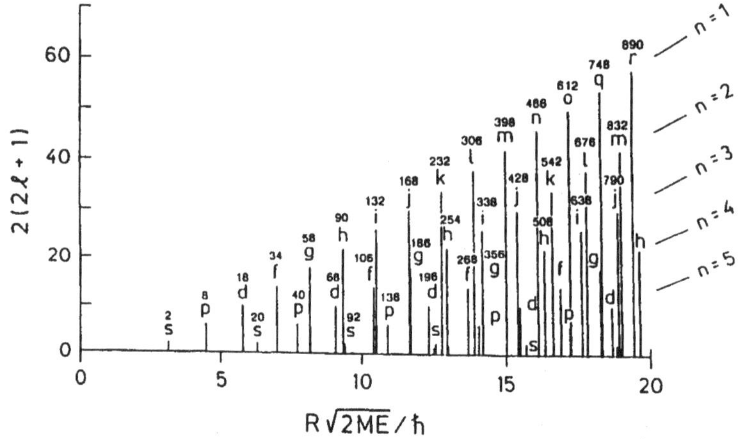

Fig. 1 The degeneracy of states of the infinitely deep spherical well on a momentum scale. The total number of fermions needed to fill all states up to including a given subshell is indicated above each bar.

Shell structure is not necessarily an approximate and infrequent bunching of states as in the example of the spherical well, Fig. 1. Clearly, almost none of the subshells occur exactly at the same energy for this potential. Shell structure can be the result of exactly overlapping states. Such degeneracies signal the presence of a symmetry higher than spherical symmetry. Subshells of hydrogen for which $n+l$ have the same value have exactly the same energy. This additional degeneracy in the states of hydrogen is a result of the form of its potential, 1/r, which bestows on hydrogen 0(4) symmetry. Subshells of the spherical harmonic oscillator for which $2n+l$ have the same value also have exactly the same energy due to the form of the potential, r^2, and the resulting symmetry, SU(3). For this reason it is said that these systems, hydrogen and oscillator, have quantum numbers $n+l$ and $2n+l$ that determine the energy. We have shown that $3n+l$ is an approximate energy quantum number for alkali metal clusters [16]. As the cluster increases in size, electron motion quantized in this way would finally be described as a closed triangular trajectory [19]. The grouping of large subshells into shells is illustrated in Fig. 2 for the spherical potential well. Here, it can again be seen that in certain energy or momentum regions the subshells bunch together. However, the states are so densely packed in this figure that the effect is perceived as an alternating light-dark pattern. That is, for the infinite potential well, bunching of states occurs periodically on the momentum scale. The periodic appearance of shell structure is referred to as supershell structure [20,21]. Although supershell structure was predicted by nuclear physicists more than 15 years ago, it has never been observed in nuclei.

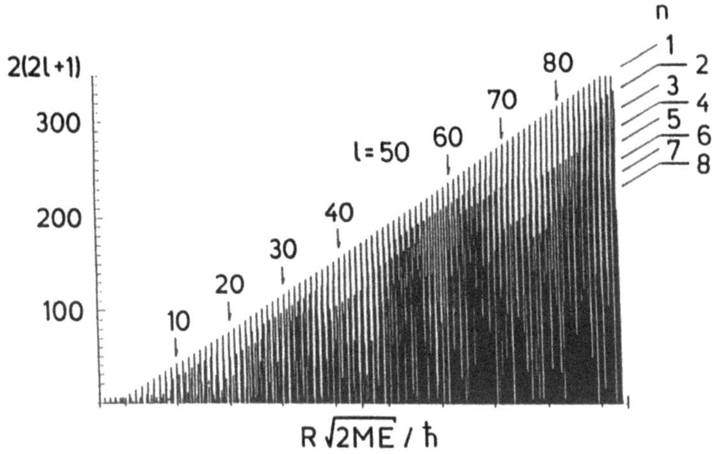

Fig. 2 The states of the infinitely deep spherical well for very large values of l. Notice the periodic bunching of states into shells. This periodic pattern is referred to as supershell structure.

The reason for this is very simple. The first supershell beat or interference occurs for a system containing 800 fermions. There exist, of course, no nuclei containing so many protons and neutrons. It is possible, however, to produce metal clusters containing such large numbers of electrons.

3. The Experiment

The technique we have used to study shell structure in metal clusters is photoionization time-of-flight (TOF) mass spectroscopy. The mass spectrometer [22–24] has a mass range of 600 000 amu and a mass resolution of up to 20 000. The cluster source is a low pressure, rare gas, condensation cell. Sodium vapor was quenched in cold He gas having a pressure of about 1 mbar. Clusters condensed out of the quenched vapor were transported by the gas stream through a nozzle and through two chambers of intermediate pressure into a high vacuum chamber. The size distribution of the clusters could be controlled by varying the oven-to-nozzle distance, the He gas pressure, and the oven temperature. The clusters were photoionized with a dye laser pulse.

4. Observation of Electronic Shell Structure

Knight, Clemenger, de Heer, Saunders, Chou and Cohen [3] first reported electronic shell structure in sodium clusters in 1984. Electronic shell structure can be demonstrated experimentally in several ways: as an abrupt decrease in the ionization energy with increasing cluster size, as an abrupt increase or an abrupt decrease in the intensity of peaks in mass spectra. The first type of experiment can be easily understood. Electrons in newly opened shells are less tightly bound, i.e. have lower ionization energies. However, considerable experimental effort is required to measure the ionization energy of even a single cluster. A complete photoionization spectrum must be obtained and very often an appropriate source of tunable light is simply not available. It is much easier to observe shell closings in photoionization, TOF mass spectra. However, depending upon the intensity and wavelength of the ionizing laser pulse, the new shell is announced by either an increase or a decrease in mass peak height.

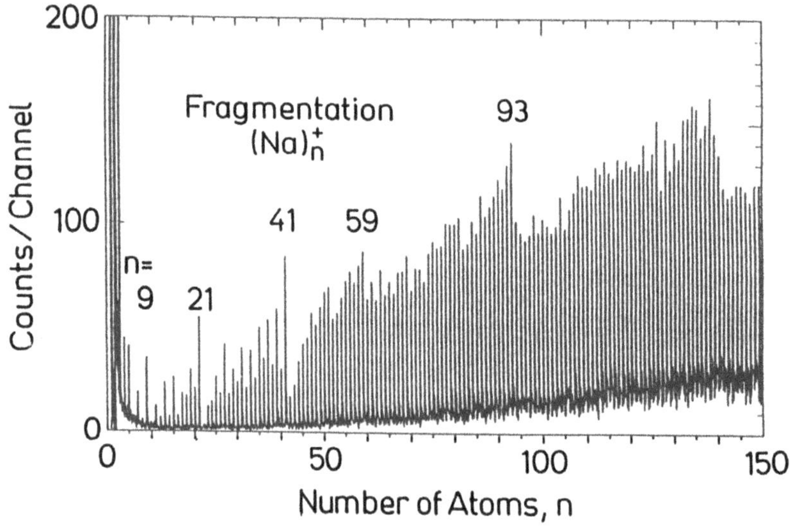

Fig. 3 Mass spectrum of $(Na)_n^+$ clusters ionized with high intensity, 2.53 eV light. The clusters are fragmented by the ionizing laser. Fragments having closed shell electronic configurations are particularly stable.

For high laser intensities, multiple-photon processes cause the mass spectra to be less wavelength sensitive and also cause considerable fragmentation of large clusters. The resulting mass spectrum reflects the stability of cluster ion fragments. Clusters with newly opened shells are less stable and are weakly represented in the mass spectra.

Notice in Fig. 3 that as each new shell is opened there is a sharp step downward in the mass spectrum. Remember that cluster ions containing 9, 21, 41, 59,... sodium atoms contain the magic number (8, 20, 40, 58,...) of electrons. For low laser fluence and wavelengths near the ionization threshold, the mass spectra have a completely different character. As each new shell is opened there is a sharp step upward in the mass spectra, Fig. 4 (top).

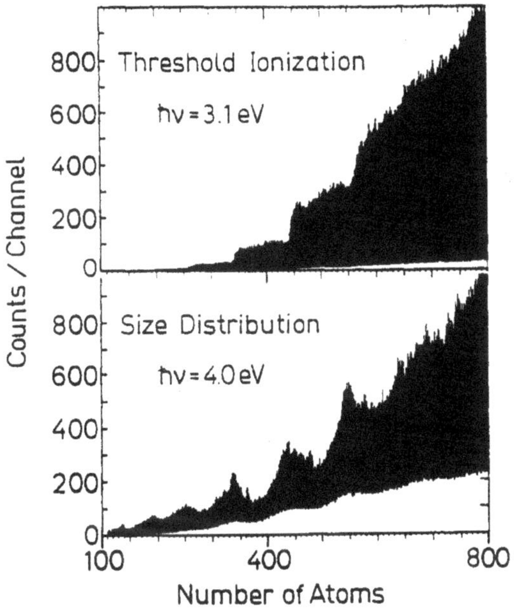

Fig. 4 Mass spectra of $(Na)_n$ clusters obtained using ionizing light near the ionization threshold (top) and well above the ionization threshold (bottom). In both cases the neutral cluster beam was heated with 2.54 eV and 2.41 eV laser light.

Open shell clusters have low ionization thresholds which fall below the energy of the incident photons, while closed shell clusters remain unionized.

Finally, for low laser fluence and wavelengths well above the ionization threshold, it is possible to observe the neutral distribution of clusters sizes. If the source conditions are appropriately chosen, this distribution can peak at sizes corresponding to closed electronic shells, Fig. 4 (bottom).

Cluster intensities can sometimes be increased by a factor of ten by using a seed to nucleate the cluster growth. For example, by adding less than 0.02 % SO_2 to the He cooling gas, Cs_2SO_2 molecules form which apparently promote further cluster growth. Mass spectra of $Cs_{n+2}(SO_2)$ clusters obtained [15] using four different dye-laser photon energies are shown in Fig. 5. Although it is not possible to distinguish the individual mass peaks in this condensed plot, it is evident that the spectra are characterized by steps. For example, a sharp increase in the mass-peak intensity occurs between n = 92 and 93. In addition to the steps for n = 58 and 92 in Fig. 5, there are broad minima in the 2.53 eV spectrum at about 140 and 200 Cs masses. These broad features become sharp steps if the ionizing photon energy is decreased to 2.43 eV. By successively decreasing the photon energy, steps can be observed for the magic numbers n = 58, 92, 138, 198±2, 263±5, 341±5, 443±5, and 557±5 [15,17]. However, the steps become less well defined with increasing mass. We have studied the mass spectra of not only $Cs_{n+2}(SO_2)$ but also $Cs_{n+4}(SO_2)_2$, $Cs_{n+2}O$, and $Cs_{n+4}O_2$. They all show steplike features for the same values of n.

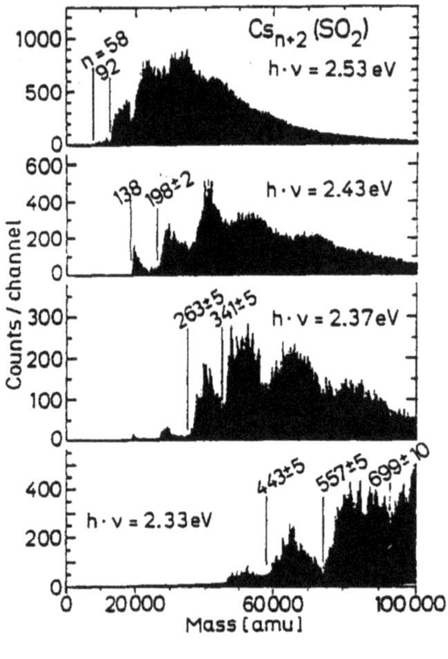

Fig. 5 Mass spectra of $Cs_{n+2}(SO_2)$ clusters with decreasing photon energy of the ionizing laser from 2.53 eV (top) to 2.33 eV (bottom). The values of n at the steps in the mass spectra have been indicated, Ref. 15.

First, we would like to offer a qualitative explanation for these results and then support this explanation with detailed calculation. Each cesium atom contributes one delocalized electron which can move freely within the cluster. Each oxygen atom, and each SO_2 molecule, bonds with two of these electrons. Therefore, a cluster with composition $Cs_{n+2}(SO_2)$, for example, can be said to have n delocalized electrons. The potential in which the electrons move is nearly spherically symmetric, so that the states are characterized by a well-defined angular momentum. Therefore, the delocalized electrons occupy subshells of constant angular momentum which in turn condense into shells. When one of these shells is fully populated with electrons, the ionization energy is high and the clusters will not appear in mass spectra obtained using sufficiently low ionizing photon energy.

In other experiments [9] the closing of small subshells of angular momentum was shown to be accompanied by a sharp step in the ionization energy for Cs-O clusters having certain sizes, namely, for $Cs_{n+2z}O_z$ with n = 8, 18, 20, 34, 58, and 92. The closing at n = 40 seen in all other alkali-metal clusters could not be observed, neither in the experiments nor in the calculations. The steps were observed for clusters containing from one to seven oxygen atoms.

5. Observation of Supershells

Although nuclear physicists speculated on the possible existence of supershells several decades ago, the phenomenon has never been observed in atomic nuclei for a very simple reason. No nucleus contains enough fermions to allow supershell formation. However, there is almost no limit to the number of electrons that can be contained in metal clusters.

Supershells are the periodic appearance and disappearance of shell structure in the energy density of states of a fermion system. In order to make clear the physical origin of supershells, it is necessary to go back one step to the semiclassical description of shells. Shells are associated with a characteristic length. Each time an integral number of fermi wavelengths fits into this length, a new shell has formed. The systems that we are studying are so large that the classical picture of an electron bouncing back and forth inside a metal cluster is not completely without meaning. The characteristic length associated with a set of shells is just the length of a closed electron trajectory within the clusters. For spheres, two closed trajectories with almost the same length turn out to be the most important - a triangular path and a square path. This leads to two sets of shells with nearly the same energy spacing. These two contributions interfere with one another to produce a beat pattern known as quantal supershells.

The first attempts to observe supershell structure in our laboratory were hindered by the unexpected appearance of a second set of shells in clusters containing more than 1500 atoms. These proved to be geometric shells of atoms that masked the weaker electronic shell structure. In the new experiments the geometric shell structure was surpressed by "melting" the clusters through heating with a continuous laser beam tuned to the plasmon frequency of the electron system.

The clusters were warmed prior to ionization with a continuous Ar-ion laser beam running parallel to the neutral cluster beam. The laser light entered the ionization chamber through a heated window, passed through the ionization volume, through 2.2 mm ø and 3.0 mm ø skimmer apertures, through a 3.0 mm nozzle, through the oven chamber and finally exited through a second window where the laser intensity was recorded. Short wavelength light was found to warm much more efficiently. Using the 458 nm (2.71 eV) laser line, 10 mW proved sufficient to appreciably alter the neutral size distribution.

The mass spectra obtained with ionizing photons having energy well above threshold are quite different from the spectra discussed in section 4. Without the warming laser the mass spectra are without structure, i.e. the size distribution of the cold clusters emerging from our source is smooth. If the warming laser is turned on we obtain not steps but peaks. We believe these peaks reflect the neutral size distribution of the laser-warmed clusters. It appears that it is usually possible to correlate a falling edge of the size distribution with a

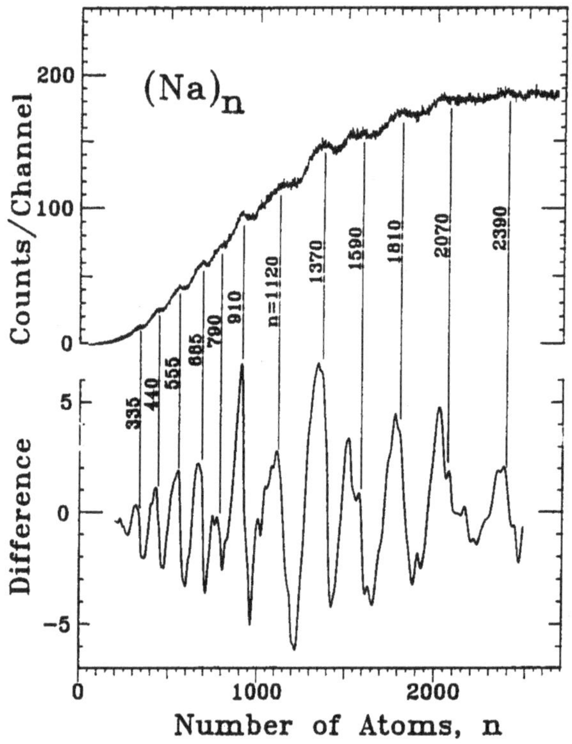

Fig. 6 Mass spectrum of $(Na)_n$ clusters using 4.0 eV ionizing light and (458 nm) 2.71 eV continuous axial warming light having an intensity of 500 mW/cm^{-2}. The spectrum has been smoothed over one hundred 16 ns time channels (top). In order to emphasize the shell structure an envelope function (obtained by smoothing over 20 000 time channels) is subtracted from a structured mass spectrum (smoothed over 1500 time channels). The difference is shown in the bottom spectrum.

step in the threshold ionization spectrum. Because of this correlation, we will characterize mass spectra obtained using excimer light by the number of atoms at steep negative slopes. A more extended mass spectrum of laser-warmed sodium clusters obtained with 4.0 eV ionizing photons is shown in the top of Fig. 6. The photon energy

(4.0 eV) of the ionizing laser has been chosen so that it is well above the ionization threshold (3.0 eV) of the sodium clusters investigated. The excess energy (1 eV) is sufficient to cause only one atom to evaporate. This is a negligible loss on the mass scale we will be considering. For this reason, we believe that the magic numbers obtained reflect variations in the size distribution of the neutral clusters induced by the warming laser.

It has been suggested [19–28] that shell structure might periodically appear and disappear with increasing cluster size. Such a supershell structure can be understood as a beating pattern created by the interference of two nearly equal periodic contributions. Quantum mechanically the contributions can be described as arising from competing energy quantum numbers. Classically, the contributions can be described as arising from two closed electron trajectories within a spherical cavity. One trajectory is triangular, the other square.

6. Fission

The fission of clusters was one of the first subjects [29–43] to be investigated in the newly developing field of cluster research. It is often referred to as Coulomb explosion, since the fission is caused by the Coulomb repulsion of like charges concentrated in a cluster smaller than a critical size. The kinetic energy that the charged fragments acquire can be as high as several eV. Most of these studies have delt with the fission of doubly or triply charged clusters. Recently, we have shown that it is possible to induce charges as high as + 14 or large Na clusters by photoionization [44]. In this section we will discuss fission in these highly charged clusters.

The technique we have used to study fission in sodium clusters is photoionization time-of-flight mass spectroscopy. The cluster source is a low pressure, inert gas, condensation cell. The clusters we photoionized with a 50 mJ, 15ns, 193nm (6.4 eV) excimer laser pulse focussed onto the neutral cluster beam with a 150 cm focal length quartz lens. The ionized clusters were heated 30ns later with a second 5mJ/mm^2, 470nm (2.6 eV) laser pulse.

The energy (I) required to remove an additional electron from a cluster that already has charge +z can be written

$$I(z,R) = W + (\alpha + z)e^2/r \tag{4}$$

where W is the bulk work function, e is the electronic charge, and R is the radius of the cluster. Clearly, we have assumed that the cluster can be modelled as a conducting sphere. Various values of α have been used in the literature. We will assume α is 0.5 and point out that for large values of z the value of α used becomes unimportant. Since the radius of the cluster can be related to the number of electrons (or in our case atoms) through the Wigner-Seitz radius, $R^3 = r_s^3 n$, Eq. 4 can be rewritten as

$$I(z,n) = W + (\alpha + z)e^2/r_s n^{1/3}. \tag{5}$$

It can be seen from this expression that a large amount of energy is required to remove electrons from small, highly charged clusters. If the amount of energy available is limited to that in one photon, then the maximum charge attainable for a cluster of a given size is,

$$z_{max} = 1 - \alpha + (h\nu - W)r_s n^{1/3}/e^2. \tag{6}$$

This means that a two-dimensional cluster space (n,z) can be divided by a line into clusters that can be formed with, for example, an ArF excimer laser and those that cannot, Fig. 7.

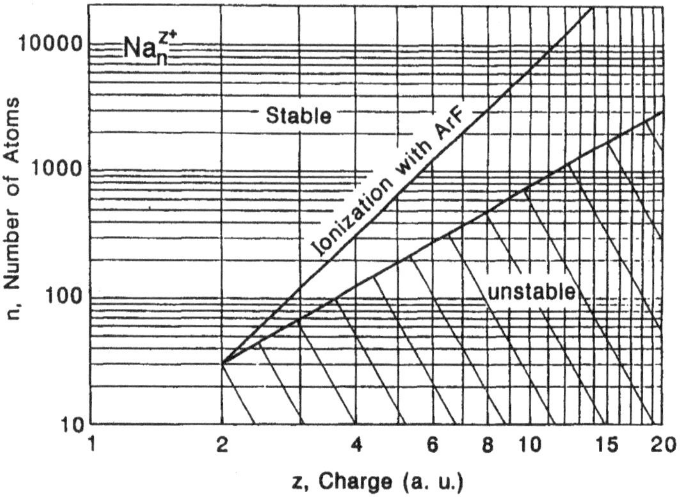

Fig. 7 Two-dimensional cluster space (n,z) can be divided by straight lines into Na_n^z clusters which can (cannot) be produced with an ArF laser and into cluster which are (are not) stable against Coulomb explosion. Notice that all clusters (except for z=2) that can be ionized with 6.4 eV photons are stable.

Also indicated in this figure is a line dividing the space into stable and unstable clusters. Notice that all of the values of n and z accessible with the ArF photons characterize stable clusters. The unstable clusters which we would like to investigate cannot be produced by direct multi-step ionization with this laser. There is, however, a way out of this dilemma.

The ArF laser can be used to prepare a stable, highly charged, large cluster and then this large cluster can be reduced in size by heating and subsequent evaporation. A second laser pulse, containing photons with energies near the plasmon resonance of the sodium clusters, is used for heating. The clusters shrink down in size without losing charge until they reach a critical size at which they undergo fission. A mass spectrum, or better said, an n/z spectrum for Na_n^z clusters produced in this way is shown in Fig.s 8 and 9. The log scale emphasizes, perhaps overemphasizes, the effect we wish to show. The highest set of mass peaks belongs to singly-charged sodium clusters. The peaks

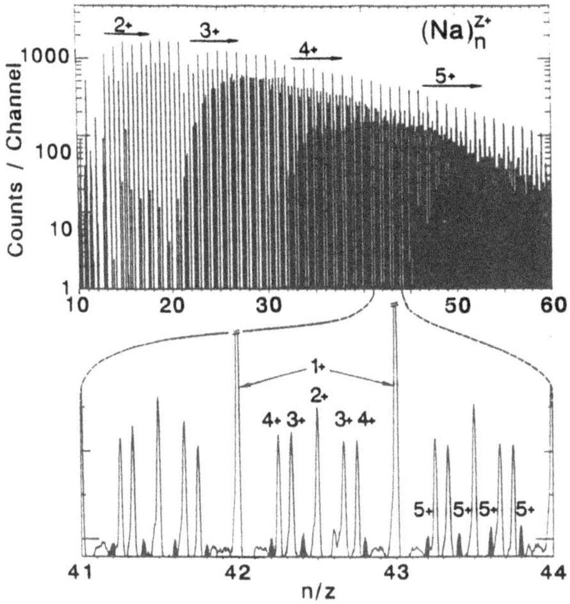

Fig. 8 An n/z spectrum of Na_n^{Z+} clusters. Large clusters were first charged by multi-step ionization using a high fluence, ArF laser. The clusters were then heated to reduce their size by evaporation. A portion of the spectrum is expanded to show the appearance of Na_n^{5+} clusters (black filled).

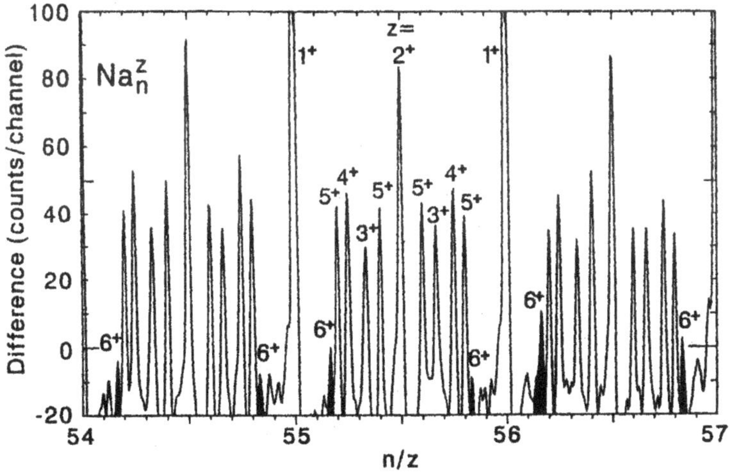

Fig. 9 An expanded portion of Fig. 8 showing the first appearance of Na_n^{6+} clusters. Smaller clusters in this charge-state are not stable.

which occur exactly half-way between the Na_n^+ peaks are due to Na_n^{2+} clusters. Notice that new sets of peaks appear in the spectrum at various critical values of n/z. This is perceived as a step-wise darkening of the mass spectrum. In the lower part of Fig. 8 we see the threshold region for the appearance of Na_n^{5+} on an expanded scale.

In this way, by careful examination of the fine structure in the mass spectra, it is possible to determine that the critical sizes for z = 1,2,3,4,5,6 and 7 are 27±1, 64±1, 123±2, 208±5, 321±5 and 448±10 atoms, respectively. These values are plotted on a double log scale in Fig. 10.

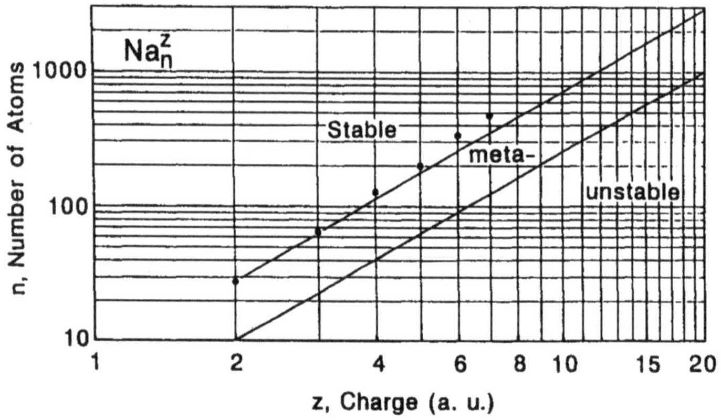

Fig. 10 The number of atoms in the smallest experimentally observed Na_n^{z+} clusters (filled circles). Cluster space can be divided into stable, metastable, and unstable regions, according to Ref. 46, using the symmetric liquid drop model.

They lie on a straight line with slope 2. This means that the critical condition for stability is

$$z^2/n \leq 0.125. \tag{7}$$

z^2/n is proportional to the so-called fissility parameter used in nuclear physics as a measure of stability. It has been shown the past that this parameter is also useful for clusters with small total charge [40,41]. Here we see that it continues to be applicable for values of z up to 7.

The results of an extensive theoretical investigation of fission in Na clusters have recently been published [45,46]. An important assumption made in this work was that the fission is symmetric, i.e.,
the mass and charge of the original cluster are divided nearly equally between the fission products. Unfortunately, we have no evidence at this time to either support or to challenge this assumption. Still, it is useful to compare the results of this calculation with our experiment, Fig. 10. Here, cluster space (n,z) has been divided into stable, metastable and unstable regions according to the tunneling criteria appropriate in nuclear physics. The fission process for nuclei can be described qualitatively in terms of three energies; the initial energy (E_i) of the charged, nondeformed clusters, the final

energy (E_f) which is the sum of the energies of the noninteracting fission products. The third energy necessary to characterize fission is the energy (E_b) of the lowest barrier separating the initial and final states. If $E_b < E_i$, the nucleus is unstable. If $E_f < E_i < E_b$ the nucleus is metastable to fission by tunneling through the barrier. Finally, if $E_f > E_i$ then the nucleus is stable. Since tunneling for clusters has negligible probability, it is more accurate to say at zero temperature clusters are either stable or unstable depending on whether there is a barrier or not. At finite temperature clusters can be classified as either unstable ($E_B < E_i$) or as metastable ($E_B > E_i$) to thermal hopping over the barrier. In practice it is useful to further subdivide the set of metastable clusters [36–38]. At finite temperatures clusters can lose mass and thermal energy by the evaporation of neutral atoms. Evaporation will always compete with fission and will, in fact, dominate if E_B is greater than E_v, the energy needed to evaporate on atom. For this reason we have the conditions,

a) unstable to fission, $E_B < E_i$

b) metastable to fission, $E_i < E_B < E_v$

c) metastable to evaporation, $E_i < E_v < E_B$

Since previous experiments [36–38] on doubly-charged Na and K clusters indicate that fission is strongly asymmetric, it would be appropriate now to consider this alternative. Even using the droplet model it is not easy to calculate the height of the barrier if the mass and charge can be distributed arbitrarily between the fission products. For this reason we will consider an energy which does not exactly characterize a real system, but it is trivial to estimate and therefore useful. It is the energy at the instant of scission, E_s. That is, starting from the final state we merely bring the fission products together until they just touch. Of course, the energy increases monotonically from E_f to E_s according to Coulomb's law. We assume that E_s is nearly equal to the barrier height for the fission process. There is no unique value of E_s for a cluster in initial state n and z. Rather, a whole set of values exist corresponding to the various ways of distributing charge and mass between the fission products. However, one value of E_s has special significance and that is the minimum value. If this minimum value of $E_s < E_i$ the cluster is unstable and will spontaneously fission, even at zero temperature. If $E_s > E_i$ the cluster is stable against fission. Strictly speaking, one should say metastable because at finite temperatures the final state can be reached by jumping over the barrier, no matter how high.

The results of these calculations are summarized in Fig. 11.

The initial and final energies are determined using the spherical droplet model assuming only two fission fragments with arbitrary size and charge and assuming a surface tension parameter $\sigma = 200$ dyne/cm appropriate for sodium. One might expect that the experimental points would fall on the line corresponding to $E_B = E_v$. Clearly, this is not the case, since E_v is known [36–38] to have a value of about 1eV. That is, this rough model overestimates the barrier height by a factor of two. Various refinements are clearly needed; proper treatment of the Coulomb energy allowing for electron redistribution as the fragments move away from one another [48], a description of the

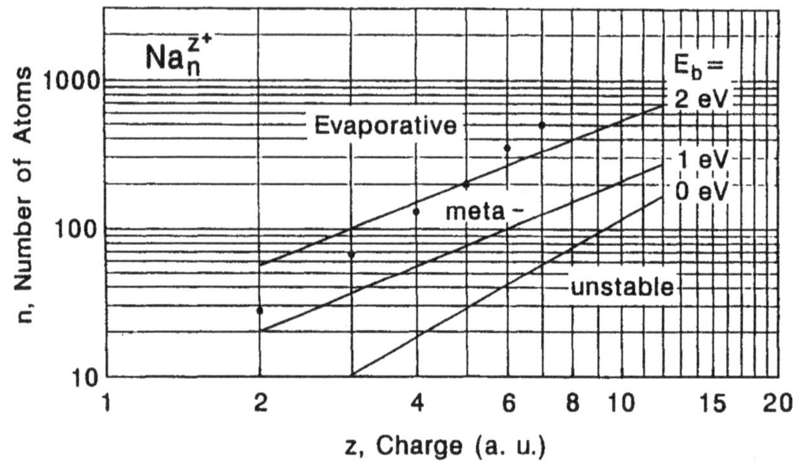

Fig. 11 Cluster space divided into evaporative, metastable and unstable regions. The barrier height has been obtained using an oversimplified model (see text). The number of atoms in the smallest experimentally observed Na_n^{z+} clusters are shown by the filled circles.

asymmetric fission before scission, shell effects [49,50] and entropy effects. Also needed are experiments demonstrating how mass and charge are distributed between the fission fragments.

7. Acknowledgement

One of us (N.M.) would like to thank the Deutsche Forschungsgemeinschaft for financial support.

References

[1] M. G. Mayer, *Phys. Rev.* **75**, (1949) 1969L.

[2] O. Haxel, J. H. D. Jensen, H. E. Suess, *Phys. Rev.* **75** (1949) 1766L.

[3] W. D. Knight, K. Clemenger, W. A. de Heer, W. A. Saunders, M. Y. Chou, M. L. Cohen, *Phys. Rev. Lett.* **52** (1984) 2141.

[4] M. M. Kappes, R. W. Kunz, E. Schumacher, *Chem. Phys. Lett* **91** (1982) 413.

[5] I. Katakuse, I. Ichihara, Y. Fujita, T. Matsuo, T. Sakurai, T. Matsuda, *Int. J. Mass Spectrum. Ion Processes* **67** (1985) 229.

[6] C. Bréchignac, Ph. Cahuzac, J.-Ph. Roux, *Chem. Phys. Lett.* **127** (1986) 445.

[7] W. Begemann, S. Dreihofer, K. H. Meiwes-Broer, H. O. Lutz, *Z. Phys.* **D3** (1986) 183.

[8] W. A. Saunders, K. Clemenger, W. A. de Heer, W. D. Knight, *Phys. Rev.* **B32** (1986) 1366.

[9] T. Bergmann, H. Limberger, T. P. Martin, *Phys. Rev. Lett.* **60** (1988) 1767.

[10] J. L. Martins, R. Car, J. Buttet, *Surf. Sci.* **106** (1981) 265.

[11] W. Ekardt, *Ber. Bunsenges. Phys. Chem.* **88** (1984) 289.

[12] K. Clemenger, *Phys. Rev.* **B32** (1985) 1359.

[13] Y. Ishii, S. Ohnishi, S. Sugano, *Phys. Rev.* **B33** (1986) 5271.

[14] T. Bergmann, T. P. Martin, *J. Chem. Phys.* **90** (1989) 2848.

[15] H. Göhlich, T. Lange, T. Bergmann, T. P. Martin, *Phys. Rev. Lett.* **65** (1990) 748.

[16] T. P. Martin, T. Bergmann, H. Göhlich, T. Lange, *Chem. Phys. Lett.***172** (1990) 209; *Z. Phys. D* **19** (1991) 25.

[17] S. Bjørnholm, J. Borggreen, O. Echt, K. Hansen, J. Pederson, H. D. Rasmussen, *Phys. Rev. Lett.* **65** (1990) 1627.

[18] J. L. Persson, R. L. Whetten, Hai-Ping Cheng, R. S. Berry, *Chem. Phys. Lett.* **186** (1991) 215; E. C. Honea, M. L. Homer, J. L. Persson, and R. L. Whetten, *Chem. Phys. Lett.* **171** (1990) 147; J. Lerme, M. Pellarin, J. L. Vialle, B. Baguenard and B. Broyer, *Phys. Rev. Lett.***68** (1992) 2818.

[19] R. Balian, C. Bloch, *Ann. Phys.***69** (1971) 76.

[20] A. Bohr, B. R. Mottelson, *Nuclear Structure (Benjamin,* London) (1975).

[21] H. Nishioka, K. Hansen, B. R. Mottelson, *Phys. Rev. B* **42** (1990) 9377.

[22] T. P. Masrtin, S. Bjørnholm, J. Borggreen, C. Bréchignac, Ph. Cahuzac, K. Hansen, and J. Pedersen, *Chem. Phys. Lett.***186** (1991) 53.

[23] J. Pedersen, S. Bjørnholm, J. Borggreen, K. Hansen, T. P. Martin, and H. D. Rasmussen, *Nature* **353** (1991) 733.

[24] C. Bréchignac, Ph. Cahuzac, F. Carlier, M. de Frutos and J. Ph. Roux, *Phys. Rev. B* **47** (1993) 2271.

[25] O. Genzken and M. Brack, *Phys. Rev. Lett.***67** (1991) 3286.

[26] J. Mansikka-aho, E. Hammarén and M. Manninen, *Phys. Rev. B* **46** (1992) 12649.

[27] P. Stamfli and K. H. Bennemann, *Phys. Rev. Lett.***69** (1992) 3471.

[28] C. Bréchignac et al., to be published.

[29] N. Parloff and S. C. Creagh, submitted to *Phys. Rev. B*

[30] O. Echt, in "Physics and Chemistry of Small Clusters", eds. P. Jena, B. K. Rao and S. N. Khanna, *Plenum*, New York (1987).

[31] O. Echt, D. Kreisle, E. Recknagel, J. J. Saenz, R. Casero and J. M Soler, *Phys. Rev. A* **38** (1988) 3236.

[32] T. D. Märk, P. Scheier, M. Lezius, G. Walder and A. Stamatovic, *Z. Phys. D* **12** (1989) 279.

[33] N. G. Gotts, P. G. Lethbridge and A. J. Stace, *J. Chem. Phys.* **96** (1992) 408.

[34] O. Kandler, K. Athanassenas, O. Echt, D. Kreisle, T. Leisner and E. Recknagel, *Z. Phys. D* **19** (1991) 151.

[35] K. Sattler, J. Mühlbach, O. Echt, P. Pfau and E. Recknagel, *Phys. Rev. Lett.* **47** (1981) 160.

[36] C. Bréchignac, Ph. Cahuzac, F. Carlier and M. de Frutos, *Phys. Rev. Lett.* **64** (1990) 2893.

[37] C. Bréchignac, Ph. Cahuzac, F. Carlier, M. de Frutos, and J. Leygneir, *Z. Phys. D*, **19** (1991) 1.

[38] C. Bréchignac, Ph. Cahuzac, F. Carlier, J. Leygnier, and A. Sarfati, *Phys. Rev. B* **44** (1991) 11386.

[39] I. Katakuse, H. Itoh, T. Ichihara, *Int. J. Mass Spectrum. Ion Proc.* **97** (1990) 47.

[40] W. A. Saunders, *Phys. Rev. Lett.* **64** (1990) 3046.

[41] W. D. Saunders and N. Dam, *Z. Phys. D* **20** (1991) 111.

[42] W. Schulze, B. Winter and I. Goldenfeld, *J. Chem. Phys.* **87** (1987) 2402.

[43] I. Rabin, C. Jackschath and W. Schulze, *Z. Phys. D* **19** (1991) 153.

[44] U. Näher, H. Göhlich, T. Lange and T. P. Martin, *Phys. Rev. Lett.* **68** (1992) 3416.

[45] S. Sugano, *Microcuster Physics*, Springer, Berlin, Heidelberg (1991)

[46] M. Nakamura, Y. Ishii, A. Tamura and S. Sugano, *Z. Phys. D* **19** (1991) 145.

[47] E. Lipparini and A. Vittori, *Z. Phys. D* **17** (1990) 57.

[48] F. Garcias, J. A. Alonso, J. M Lopez, and M. Barranco, *Phys. Rev. B* **43** (1991) 9459.

[49] B. K. Rao, P. Jena, M. Manninen, and R. M Nieminen, *Phys. Rev. Lett.* **58** (1987) 1188.

[50] R. N. Barnett, U. Landmann and G. Rajagopal, *Phys. Rev. Lett.* **67** (1991) 3058.

FRAGMENTATION OF CHARGED SODIUM CLUSTERS

F. GARCIAS
Dep. Física, Universitat de les Illes Balears, E-07071 Palma de Mallorca, Spain

J. A. ALONSO and J. M. LÓPEZ
Departamento de Física Teórica, Universidad de Valladolid, E-47011 Valladolid, Spain

and

M. BARRANCO
Departament ECM, Facultat de Física, Diagonal-647, E-08028 Barcelona, Spain

Abstract

The competition between asymmetric fission and neutral monomer evaporation as dissociation channels of excited charged sodium clusters has been investigated within the Density Functional Theory and the jellium model. Results from an axially symmetric, fully selfconsistent Kohn-Sham calculation (including ellipsoidal deformations) are compared with those previously obtained using an Extended Thomas-Fermi approximation. A simple formula for the fission barrier height of multiply charged clusters is presented, which leads to good agreement with experimental critical numbers.

1. Introduction

Since the experimental discovery of Coulomb explosion phenomenon, the study of stability of metal clusters in charged states has motivated an important experimental and theoretical research. Basically, the fundamental open questions are referred to the number of atoms for which a multiply charged cluster is stable, the possible dissociation channels for unstable and metastable clusters and the charge and mass distribution of fragments.

It is well established that multiply charged metal clusters M_N^{q+} can be observed in mass spectra only if the number of atoms N exceeds a critical size N_c that depends on the metal species M and on the charge state q. For instance, the critical sizes for doubly charged clusters are 25 for Li [1, 2], 27 for Na [3], 21 for K [1, 2] and 19 for Rb [2] and Cs [4]. Very recently, Martin *et al* [2] have determined the critical sizes for multiply charged alkali clusters up to $q = 7$.

These critical sizes are experimentally obtained studying charged clusters produced by multi-step ionization of hot larger clusters that lose a sizeable part of their excitation energy mainly evaporating neutral atoms. This causes the ionized clusters M_N^{q+} to shrink up to sizes $N \sim N_c$. For cluster sizes around N_c, monomer evaporation competes with another dissociation channel, asymmetric fission, in which two charged fragments of very different sizes are emitted. Electron shell effects are manifested, with preferential emission of magic fragments such as M_3^+ [5] or M_9^+ [1].

Old theoretical studies of cluster stability were based on pure energetic criteria which only involved the energies of the initial and final states [6]. This permited to study different dissociation channels easily and yielded the correct trends for the de-excitation of neutral and singly charged aggregates. However, stable multiply charged clusters have been detected below $N = N_c$. This occurs if they are created in a multi-step ionization process starting from cold neutral clusters [3], and indicates that cluster fragmentation is a barrier-controlled process. For neutral or singly charged clusters the dissociation channels are given by the energy balance because of the absence of barrier. For multiply charged clusters there exists a finite energy barrier against fragmentation, even if the total energy of the dissociation products is smaller than the energy of the initial configuration. These facts have prompted the interest of theoreticians to obtain these barriers.

In this contribution we review our results on the fragmentation of doubly charged sodium clusters, obtained within the Density Functional Theory and the jellium model. An extension of the model to the case of multiply charged clusters is also presented.

2. The model for cluster fragmentation

The model used to study the dissociation problem in the case of doubly charged alkali clusters [3, 7-9] can be straightforwardly applied to higher charge states. We consider as the main fragmentation mechanisms for excited charged clusters the evaporation of a neutral monomer

$$M_N^{q+} \longrightarrow M_{N-1}^{q+} + M \ , \tag{1}$$

and the fission into two charged fragments

$$M_N^{q+} \longrightarrow M_{N-p}^{(q-1)+} + M_p^+ \ . \tag{2}$$

Different fission channels can be regarded: monomer emission ($p = 1$), charged trimer emission ($p = 3$), symmetric fission ($p = N/2$), etc.

The heat of monomer evaporation ΔH_e corresponding to reaction (1) is a positive quantity defined as the difference between the total energy of fragments at infinite separation and the energy of parent cluster:

$$\Delta H_e = E(M_{N-1}^{q+}) + E(M) - E(M_N^{q+}) \ . \tag{3}$$

The heat of fission ΔH_f for the process (2) is calculated as:

$$\Delta H_f = E(M_{N-p}^{(q-1)+}) + E(M_p^+) - E(M_N^{q+}) \ . \tag{4}$$

The heat of fission is related to the *fission* F_m and *fusion* B_m maxima of the corresponding barriers by the relationship [7]:

$$F_m = B_m + \Delta H_f \ . \tag{5}$$

Fission is favored over monomer evaporation when F_m is smaller than ΔH_e, and the opposite if it is larger. If the fission barrier vanishes, the cluster is absolutely unstable against fission.

ΔH_e, ΔH_f and F_m (or equivalently B_m) are calculated within the Density Functional Theory using the jellium model. This model assumes a background of positive charge with constant density representing the ions, and a distribution of interacting valence electrons (one per atom in the case of alkali metal clusters) whose density $n(\vec{r})$ is selfconsistently calculated minimizing the total energy E of the cluster:

$$E[n] = T + U_{ee} + U_{Je} + U_{JJ} + E_{xc} \ , \tag{6}$$

where T is the electron kinetic energy, U_{ee} is the classical Coulomb energy of the valence electrons, U_{Je} gives the electron-jellium electrostatic interaction, U_{JJ} is the jellium self-energy, and E_{xc} is the sum of the electron exchange and correlation energies taken in the Local Density Approximation.

ΔH_e and ΔH_f are easily obtained if one is dealing with spherical clusters. These two quantities are rather sensitive to shell effects which cause some clusters to be tightly bound (spherical magic clusters), and to adopt deformed equilibrium configurations to many others. Even in the jellium model, the calculation of ΔH_e and ΔH_f is a formidable task when the parent clusters and/or final products are not magic. In this case, to obtain the energy of each cluster one has to carry out a deformed calculation in which the positive background adopts shapes whose equilibrium deformation has to be determined selfconsistently [10].

To obtain the fission (fusion) barrier height F_m (B_m), the cluster fragmentation path has been parametrized in terms of the distance between the centers of two jellium spheres representing the fragments, starting from the touching configuration (in the Kohn-Sham calculations to be presented below we only deal with fission reactions in which the fragments are magic, and thus spherical) [7-9]. A more detailed description of the barrier requires a sequence of jellium shapes continuously connecting the initial configuration of the parent cluster (spherical or deformed) with the final one corresponding to two separated fragments. Several shape parametrizations of this kind exist [11]. However, if the goal is the calculation of the barrier height and not the barrier itself, one can take advantage of the fact that the maximum of the

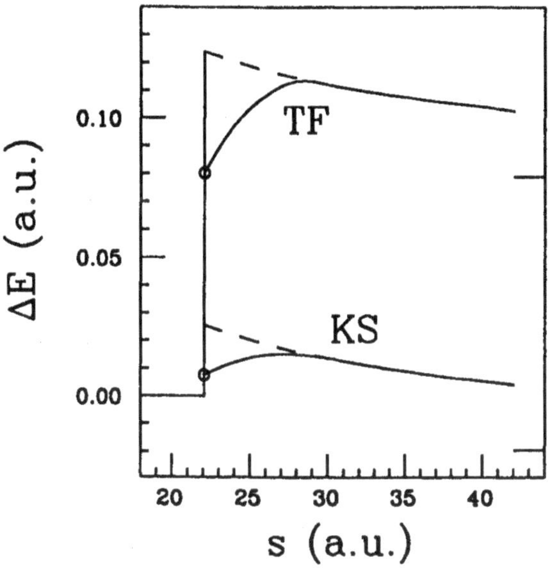

Figure 1: Fission barrier in atomic units (a.u.) for symmetric dissociation of Na_{42}^{2+} obtained from TF and KS approximations.

barrier (i.e., the saddle point) corresponds to a configuration in which the emerging fragments are already formed [12, 7-9] and slightly tied up by the electron cloud. The shapes of fission and fusion barriers differ at shorter distances, which is completely irrelevant for the problem at hand.

3. Results for doubly charged sodium clusters

The most asymmetric fission ($p = 1$) and its competition with monomer evaporation in Na_N^{2+} clusters has been investigated in some detail using in Eq. (6) an Extended Thomas-Fermi (TF) approximation to the kinetic energy [7, 8]. We have taken an effective value β=0.5 for the coefficient of the von Weizsäcker correction. The critical value for the competition between evaporation and fission obtained from the TF model was $N_c = 36$, in qualitative agreement with the experimental result (27 for Na). Inclusion of other fission channels, like symmetric fragmentation [9] or Na_3^+ emission [8], leads to a decrease of the critical number. As a general trend, the latter processes are hindered by a barrier higher than that of completely asymmetric fission, thus indicating that monomer emission is the preferred fission channel within

Table 1: Experimental critical numbers N_c for Na_N^{q+} clusters [2].

q	2	3	4	5	6	7
N_c	27±1	63±1	123±2	206±4	310±10	445±10

the TF model. The critical size for spontaneous fission was found to be $N_c^* \sim 10$.

The semiclassical TF method cannot account for shell effects, which have been included to study the dissociation channels of small alkali clusters. In this section we present some results obtained from an axially symmetric, fully selfconsistent Kohn-Sham (KS) calculation.

Figure 1 shows the fission barrier for $Na_{42}^{2+} \longrightarrow 2\ Na_{21}^{+}$ obtained in the KS approximation as compared with the TF result. The vertical line at $s \sim 22$ a.u. corresponds to the touching distance of the jellium spheres representing the fragments. Both barriers are referred to the corresponding initial spherical configuration of Na_{42}^{2+}. Horizontal lines at the right indicate the final state energy in each case. Shell structure confers the daughter clusters a special stability and ΔH_f becomes negative in the KS method. One can see that, when shell effects are taken into account, the fission barrier height reduces drastically, from $F_m(\text{TF}) = 3.1$ eV to $F_m(\text{KS}) = 0.4$ eV. However, the fusion barrier height B_m is rather insensitive to shell effects and has the same value in both approximations. The $1/s$ curve (dashed lines) is the classical Coulomb repulsion between the fragments as point charges. Contrarily to the results obtained in the TF approximation, within the KS approximation the symmetric fission of Na_{42}^{2+} is favored over monomer evaporation and completely asymmetric fission. The KS method gives $F_m(Na_{21}^{+}) = 0.4$ eV, $\Delta H_e = 0.9$ eV and $F_m(Na^{+}) = 1.2$ eV, respectively. Unfortunately it is not possible to detect symmetric fission using unimolecular dissociation techniques [3].

Figure 2 shows the KS fission barrier and heats corresponding to $Na_{44}^{2+} \longrightarrow Na_{41}^{+} + Na_{3}^{+}$. The horizontal dashed line at -81.1 eV represents the energy of the non-magic Na_{44}^{2+} cluster treated as spherical, whereas the horizontal solid line at ~ -81.6 eV indicates its energy when ellipsoidal deformations are considered. In this case ΔH_f is negative and very small. For Na_{44}^{2+} the deformed KS model yields $F_m(Na_3^{+}) = 0.7$ eV and $\Delta H_e = 1.1$ eV, thus predicting charged trimer emission as most favorable dissociation channel. The ΔH_e value has been obtained by treating Na_{43}^{2+} as a nonspherical cluster.

As a final example, let us comment that for the dissociation $Na_{24}^{2+} \longrightarrow Na_{21}^{+} + Na_{3}^{+}$, the ellipsoidal initial configuration of Na_{24}^{2+} has an energy 0.3 eV higher than the saddle point one. Consequently, Na_{24}^{2+} will spontaneously fission.

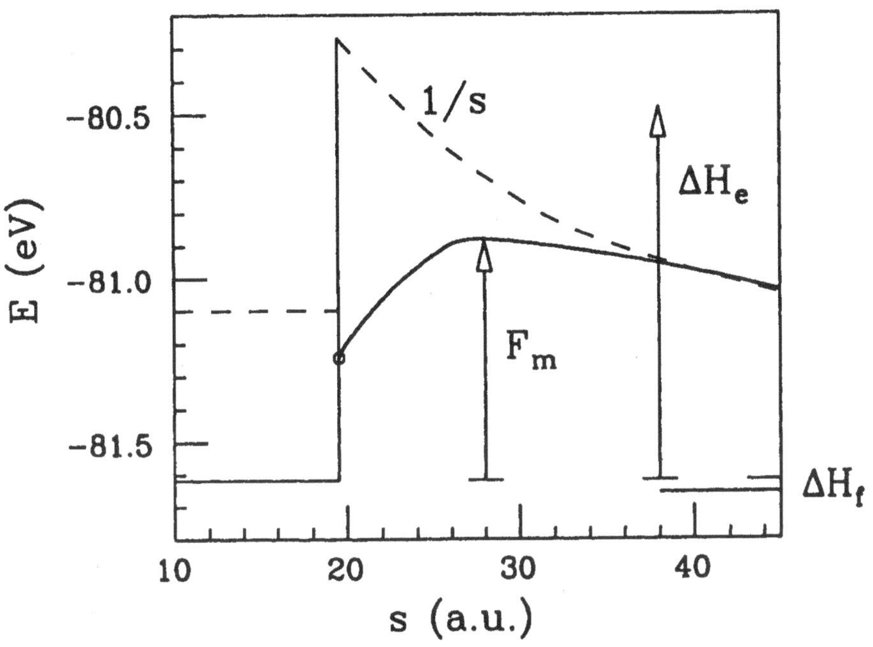

Figure 2: KS fission barrier for $Na_{44}^{2+} \longrightarrow Na_{41}^{+} + Na_{3}^{+}$.

4. Results for multiply charged clusters

Table 1 shows the experimental critical numbers $N_c(q)$ below which Na_N^{q+} clusters are not observed in mass spectra [2]. For large cluster sizes, shell effects are expected to play a minor role, so the semiclassical TF model can provide a first estimate of critical numbers and main trends of the fragmentation process of excited highly charged clusters. From the experience gained in the study of doubly charged clusters, it is reasonable to assume that charged trimer emission is the predominant fission channel, in competition with neutral monomer evaporation.

Figure 3 displays the TF fission barrier and heats corresponding to the process $Na_{312}^{6+} \longrightarrow Na_{309}^{5+} + Na_{3}^{+}$ considering now all clusters as spherical. In this case, monomer evaporation is favored over fission but the small difference between the fission barrier height $F_m = 1.5$ eV and the heat of evaporation $\Delta H_e = 1.2$ eV indicates that this cluster is not very far from the critical one.

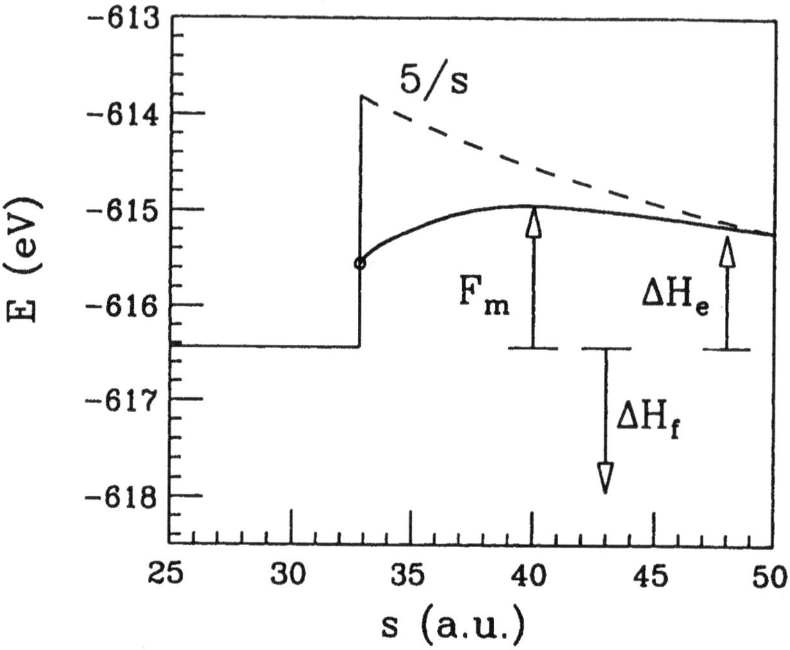

Figure 3: TF fission barrier for $Na_{312}^{6+} \longrightarrow Na_{309}^{5+} + Na_3^+$.

The comparison between the TF result and the pure Coulomb barrier (dashed line) suggests a simple approximation for the fission barrier height of the reaction $Na_N^{q+} \longrightarrow Na_{N-3}^{(q-1)+} + Na_3^+$:

$$F_m \text{ (a.u.)} = \frac{q-1}{r_s[(N-3)^{1/3} + 3^{1/3}] + D_3} + \Delta H_f \ , \tag{7}$$

where r_s = 4 a.u., that is, the Wigner-Seitz radius for metallic sodium. The D_3-parameter serves to take care of the large electron spill-out, and a judicious choice of its value may allow for a fair and very quick estimate of F_m [9, 5]. We have verified that Eq. (7) with an effective value $D_3 \sim 18$ a.u. for the spill-out parameter is able to reproduce the critical sizes for $q > 2$ within the experimental error bars, although figure 3 seems to point to the value $D_3 \sim 13$ a.u. in the case of Na_{312}^{6+}.

5. Acknowledgements

We would like to thank Judith Németh and Angel Mañanes for their contribution to some parts of the work here presented, which has been supported by the DGICYT (Spain), grants PB89-0332, PB89-0352, PB92-0645 and PB92-0021-C02-02. One of us (F.G.) acknowledges the financial support from the Vice-Rectorat d'Investigació de la Universitat de les Illes Balears.

6. References

1. C. Bréchignac, Ph. Cahuzac, F. Carlier and M. de Frutos, submitted to *Phys. Rev. B* (1993).
2. T.P. Martin, U. Näher, H. Göhlich and T. Lange, *Chem. Phys. Lett.* **196** (1992) 113; U. Näher, S. Frank, N. Malinowski, U. Zimmermann and T.P. Martin, submitted to *Z. Phys. D* (1993).
3. C. Bréchignac, Ph. Cahuzac, F. Carlier and M. de Frutos, *Phys. Rev. Lett.* **64** (1990) 2893.
4. T.P. Martin, *J. Chem. Phys.* **81** (1984) 4426.
5. C. Bréchignac, Ph. Cahuzac, F. Carlier, J. Leygnier and A. Sarfati, *Phys. Rev. B* **44** (1991) 11386.
6. M.P. Iñiguez, J.A. Alonso, M.A. Aller and L.C. Balbás, *Phys. Rev. B* **34** (1986) 2152.
7. F. Garcias, J.A. Alonso, J.M. López and M. Barranco, *Phys. Rev. B* **43** (1991) 9459.
8. J.M. López, J.A. Alonso, F. Garcias and M. Barranco, *Ann. Physik* 1 (1992) 270.
9. M. Barranco, J. A. Alonso, F. Garcias and J. M. López, in *Clustering Phenomena in Atoms and Nuclei*, eds. M. Brenner, T. Lönnroth and F.B. Malik (Springer, Berlin, 1992) p. 305.
10. W. Ekardt and Z. Penzar, *Phys. Rev. B* **38** (1988) 4273.
11. R.W. Hasse and W.D. Myers, *Geometrical Relationships of macroscopic nuclear physics* (Springer, Berlin, 1988).
12. R.N. Barnett, U. Landman and G. Rajagopal, *Phys. Rev. Lett.* **67** (1991) 3058.

Giant Resonances; Other Topics

NUCLEAR GIANT RESONANCES: A NEVER ENDING STORY

Cai Yanhuang[1,2], *M.Di Toro*[1], *M.Papa*[1], *A.Smerzi*[1] *and Zhong Jiquan*[1,3] *

[1])*Laboratorio Nazionale del Sud, INFN*
44, Via S.Sofia, 95125 Catania, Italy
and
Dipartimento di Fisica dell' Universitá di Catania,

[2])*Institute of Nuclear Science, Shangai, China*

[3])*Institute of Modern Physics, Lanzhou, China*

ABSTRACT

Some new features of nuclear giant resonances are discussed, which will certainly lead to interesting developments in future years: 1) The disappearance of nuclear collective motions in hot compound nuclei; 2) The observation of pre-equilibrium giant dipole photons in relatively fast fusion reactions induced by charge asymmetric entrance channels; 3) The use of giant dipole γ-decays from long-living intermediate nuclear systems to get a direct insight into the dynamics of fusion and deep inelastic collisions; 4) The use of giant resonance properties to develop models of neutron excess nuclei. An attempt to make predictions useful for planning future experiments is always present.

1. Introduction

Nuclear Giant Resonances, and in particular the Giant Dipole Resonance (GDR), represent universal collective excitation modes involving all the nucleons, very well established in all nuclei. They have always been precious sources of information for nuclear structure as well as for the dynamics. Aim of our report is to show how this field of nuclear physics is still extremely alive. We will discuss some new features, of course not exhaustive, that represent quite challenging problems for the next years.

We will first address the question of the possibility of using giant resonances as a signal of the presence of a nuclear system at high excitation energy. The problem is to find the limiting temperature for disappearance of collective motions in hot nuclei. Is the value comparable to the limiting temperature for nuclear disassembling? Which is the T-dependence of the giant resonance damping? Which experiments do we still need in order to clarify these points? Sections 2,3 are devoted to the discussion of hot GDR's, that are well observed through the photon decay. We show that an increase of the spreading width leads to a sharp quenching of the giant photon emission due to the competition with particle, mostly neutron,

---

* *Report presented by M.Di Toro*

evaporation. This seems to take place at temperatures of the order of $4MeV$, well below the limits for nuclear stability.

Since in order to get such highly excited compound nuclei we need fusion reactions at relatively high energy (between 15 and $30MeV/u$ beam energy), we have looked at the possibility of a pre-equilibrium collective dipole γ- emission due to the fact that the charge can be not fully equilibrated when a fused nucleus is formed. This is discussed in Sect.4. The results are quite appealing. The effect seems to be noticeable for N/Z asymmetric entrance channels.

At lower beam energy the times involved in the dynamics of fusion and most dissipative deep inelastic collisions are long enough to strongly reduce such charge asymmetry entrance channel effects. However the dynamics seems now to allow in some cases the formation of long living intermediate very deformed compound nuclei, close to di-nuclear systems. We show this effect in Sect.5, where we also evaluate the very characteristic features of a GDR γ-decay from such exotic systems.

Section 6 is devoted to a fast growing new field of nuclear physics: the study of nuclei far from the stability valley. In particular we introduce a simple phase-space model for neutron-halo nuclei, tuned on the properties of their giant resonances. The idea is then to use this picture for the dynamics of collisions with radioactive beams.

Finally some conclusions and perspectives are presented in section 7.

2. Disappearance of collective motions in hot nuclei

There is considerable interest in the disappearance of nuclear collectivity in compound nuclei at high excitation energy : it could provide information on new damping mechanisms and on the limits of the existence of hot nuclear systems. However quite controversial conclusions have been extracted from data on GDR photon emission in the high E_x-region [1-6]. The main point is that the shape of the experimental γ-ray spectrum can be equally well reproduced by using a GDR spreading width which either saturates or increases continuously with E_x [5,6].

On the other hand there is some clear experimental evidence which deserves a thorough explanation and which can shed some light on the problem. It is a decrease of the dipole strength in the giant resonance region at high excitation energies, with a consequent saturation of the number of emitted photons [2-6]. This observation has been recently associated with a pre-equilibrium mechanism [7,8]. Indeed , as remarked by D.M.Brink [7], if at high temperature the GDR states do not reach a statistical equilibrium with the compound nucleus before it cools down by neutron emission, the GDR states will not be populated and a smaller gamma emission strength will appear in the GDR energy region.

However a strong increase of the GDR width, with a related smoothing of the resonant form factor of the γ-absorption cross section, also leads to a saturation in the γ-multiplicity. The increase in the GDR width would be due to nucleon-nucleon collisions that become more and more important in the damping of the collective

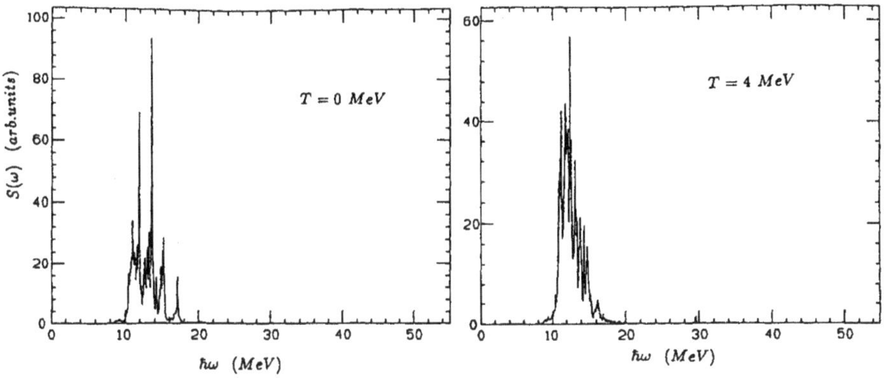

Fig. 1. *GDR* strength distribution for Sn without collision term, at $T = 0$ and $T = 4MeV$.

motion because of the Pauli-blocking suppression with increasing temperature.

In order to get the GDR spreading width we use a semiclassical approach based on exact solutions of the linearized Vlasov equation [9,10]. A way to go beyond this semiclassical RPA is to follow the relaxation time method [10,11], based on the equilibration time due to two-body collisions. From this analysis a very interesting interplay between collisional dissipation and one-body (Landau-) damping has been revealed to build up the observed widths in cold nuclei [10,11]. The procedure can be easily extended to finite temperatures [11,12].

Here we will concentrate on isovector dipole modes in Sn isotopes. In fig. 1 we show the strength fragmentation (Landau Damping) at $T = 0$ and $4MeV$ respectively, obtained solving the linearized Vlasov equation with a Woods-Saxon mean field and a dipole-dipole residual interaction, with parameters not depending on temperature. The relaxation time due to two-body collisions is microscopically calculated solving the Landau kinetic equation for an initially heated and distorted momentum distribution [11,12]. For the GDR we consider an out-of-phase neutron-proton oscillation in momentum space [13]. We use the test-particle propagation method simulating the Uehling-Uhlenbeck integral with a random test-particle scattering based on the concept of nucleon mean free path [14]. Since we have small amplitude distortions the Pauli blocking is playing a fundamental role. In particular it is essential to consider the momentum distribution rearrangement at each time step. This is achieved using the notion of a time-dependent temperature, given by the collective energy loss at each time step, in a distorted and heated Fermi distribution [11,12].

Fig.2 shows the clean exponential decays of the collective dipole energy at $T = 0, 2, 4MeV$ for the Sn case. We extract collisional widths $\Gamma = \hbar/\tau$ rang-

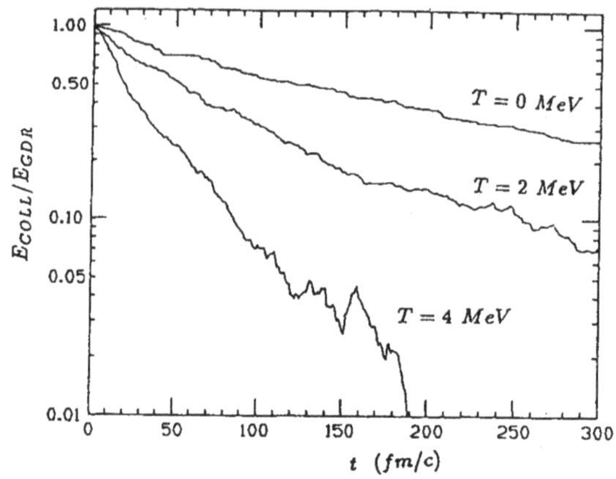

Fig. 2. Collisional decay of the collective GDR energy.

ing from $0.8MeV$ at zero temperature to $5MeV$ at $T = 4MeV$. We remark that the temperature dependence, although always clearly present, is weaker than in a previous calculation [see fig.4 of ref. 12]. Indeed we are now using a new collision simulation procedure which has been accurately tested in exactly solvable cases of cold quadrupole distortions and finite temperature classical and Fermi systems [15,16].

The collisional widths lead to imaginary parts, given by i/τ, in the unperturbed frequencies of the response function. The interplay with the Landau Damping is then evident. From fig.1 with a $\Gamma = 0.8MeV$ we get a final GDR spreading width at zero temperature of about $5MeV$. With increasing temperature, larger values of Γ imply also a frequency shift with a further spreading of the strength distribution. The excitation energy dependence of the final GDR widths, for Sn isotopes, computed with this new procedure is reported as full line in fig.3 and seems to lie fairly well on the higher E_x extrapolation of lower excitation energy data, given by the parametrization $\Gamma_{GDR}(E_x) = \Gamma_0 + 0.0026\ E_x{}^{1.6}\ (MeV)$ [1,3]. We must remark that this result likely represents an upper limit to the increase in the width with excitation energy due to collisional damping. Two main corrections should both act in the direction of reducing it: i) Surface effects from a smoothing density will imply larger mean free paths and so less collisions, particularly at higher temperatures, when we expect a further smearing of the nuclear surface; ii) A free nucleon-nucleon cross section has been used: a decrease due to medium effects would also lead to larger relaxation times.

At $E_x \simeq 230MeV$ we get a spreading width of the order of the resonance energy and we expect the GDR γ-emission to disappear, as discussed later. In the same figure we report the saturating behaviour suggested in ref. [8] (dashed line). All the curves are deduced with a level density parameter $a = A/10$. The rotational energy

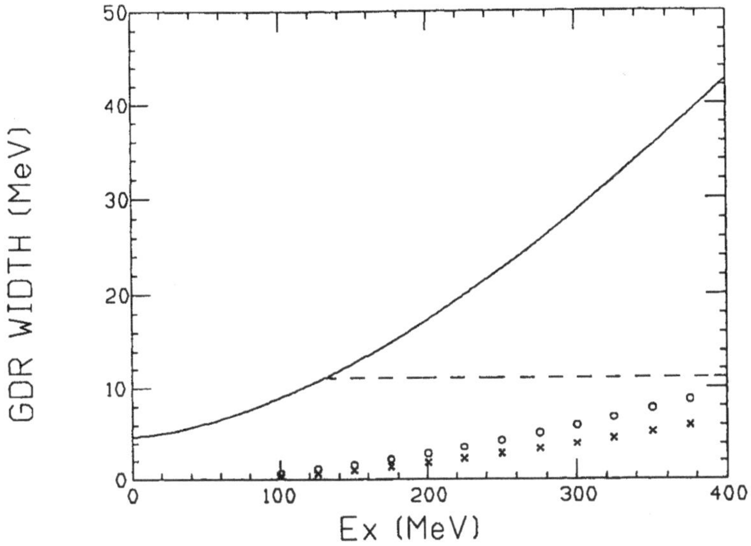

Fig. 3. Excitation energy dependence of the total *GDR* widths in *Sn* isotopes. The evaporation widths used in our Cascade analyses are also reported. The various curves are explained in the text.

has been accounted for using the prescription of ref.[6]. The particle evaporation widths, computed as in [12] and used in the statistical decay analysis discussed below, are also shown: open circles for $a = A/12$, crosses for $a = A/10$.

3. Giant photon emission rate

We make a quantitative analysis by considering the maximum possible reduction of the *GDR* strength due to the pre-equilibrium effect. This is achieved just assuming that at the time of compound nucleus formation (defined as $t = 0$ in the following) no *GDR* components are present in the system. As it will be discussed later this assumption is actually not fully justified in the medium beam energy region of interest here [17,18].

If we call $P_D(t)$ and $P_C(t)$ respectively the probabilities for the compound nuclei to exist with a *GDR* excitation and without, choosing as initial condition $P_D(t = 0) = 0$ and $P_C(t = 0) = 1$ we obtain [8,17]:

$$P_D(t) = \frac{\lambda}{\lambda + \mu} \, exp(-\gamma_{ev} t) \, [1 - exp[-(\lambda + \mu)t]$$

$$P_C(t) = \frac{\lambda}{\lambda + \mu} \, exp(-\gamma_{ev} t) \, [\frac{\mu}{\lambda} + exp[-(\lambda + \mu)t] \qquad (1)$$

where μ is the transition rate between compound states with the GDR built in and without, λ is the inverse transition rate and γ_{ev} is the transition rate of particle evaporation.

We have a strong competition between the value of γ_{ev} and μ in order to reach the equilibrium. The probability for γ-decay is given by:

$$P_\gamma = \int_0^\infty dt \; \gamma_\gamma \; P_D(t) \qquad (2)$$

where γ_γ is the transition rate of photon emission. Finally the probability of photon emission integrated over the resonance region is:

$$P_\gamma \simeq \int_{12MeV}^{20MeV} dE \; \frac{\Gamma_\gamma(E)}{\Gamma_{ev}} \left[\frac{\Gamma^\downarrow}{\Gamma^\downarrow + \Gamma_{ev}} \right] \qquad (3)$$

where

$$\Gamma_\gamma(E) = \frac{E^2}{(\pi \hbar c)^2} \; \sigma_{\gamma-abs}(E) \; e^{-E/T} \qquad (4)$$

with

$$\sigma_{\gamma-abs}(E) = (60 \; MeV \; mb) \frac{2NZ}{\pi A} \; \frac{S}{\Gamma_{GDR}} \times$$

$$\frac{(\Gamma_{GDR} E)^2}{(E^2 - E_{GDR}^2)^2 + (\Gamma_{GDR} E)^2} \qquad (5)$$

and S, E_{GDR}, $\Gamma^\downarrow = \hbar\mu$, Γ_{GDR} are respectively strength (in percent of the EWS-R), energy , spreading width and total width of the giant resonance (Full Width Half Maximum); $\Gamma_{ev} = \hbar\gamma_{ev}$ is the energy integrated evaporation width. In our calculations we have used an accurate statistical evaluation of Γ_{ev} [12], in good agreement with the emission widths obtained in [19,20]. The corresponding values are shown in fig.3. Once we have the particle and photon decay widths we can compute, using a Montecarlo technique, the actual number of neutrons, protons and photons emitted as a function of the initial excitation energy .

We have performed the analysis for a Sn compound nucleus, where many data are presently available [1-6,21] using the different E_x-behaviours of the total GDR widths shown in fig.3.

In fig.4 we report the excitation energy dependence of photon multiplicities. Fig.4a shows photon yields integrated in the GDR energy window ($12 \leq E_\gamma \leq$

$20MeV$) in the Sn case. The calculation is performed with a simplified CASCADE code where only photon, neutron and proton emissions are considered. All the theoretical curves are normalized at the same point at low excitation energy. $5 * 10^5$ events are considered in the Montecarlo procedure. Data are from ref.s [5,6], where the neutron emission has been also measured at each excitation energy. Data below $E_x \simeq 150MeV$ are from a different entrance channel [JAERI exp. in ref.6] and so the emitting compound nuclei are not exactly the same for the whole E_x range. Nevertheless we can have an idea of the general trend of the excitation function of the GDR γ-decay for $A \simeq 120$.

We see that an increasing E_x-behaviour of the GDR width can reproduce the experimentally observed saturation of the GDR photon emission at $E_x \simeq 200MeV$ [3,5,6,21] . We can get a corresponding limiting temperature for the GDR in Sn isotopes $T \simeq 4.0MeV$.

The use of a constant spreading width (dashed line of fig.4a) leads to a too large γ-yield at high E_x. Moreover the photon quenching is obtained only at excitation energies of about $400MeV$, corresponding to limiting temperatures $T \simeq 7-8MeV$, in disagreement with data. A larger Γ_{ev}, which could be obtained with a level density parameter $a = A/12$(see fig.3), is not sensitively affecting this result, while the corresponding estimate for neutron emission will overshoot the data, as shown later. However since large uncertainties are present in statistical emission calculations at high E_x (level densities, binding energies, opening of new channels) we stress that the important region of careful comparison with data should be in the range $200 - 300MeV$ excitation energy.

It has been recently observed that an increase of Γ_{GDR} should affect the high energy part of the γ-spectrum [22]. In fig.4b we plot photon yields integrated in a higher energy window ($25 \leq E_\gamma \leq 40MeV$), above the GDR region. Now there is no saturation: the γ-multiplicity is expected to increase with excitation energy, almost independently on the behaviour of the GDR width. The temperature effect, due to level densities, on the slope of the spectrum seems to dominate although the curve with increasing Γ_{GDR} (solid) appears systematically higher than the one for a constant width (dashed). This γ-energy region is difficult to analyse experimentally because of the competition with the direct bremsstrahlung component [6,21]. Evaluation of the latter just from extrapolation of higher beam energy data is not trivial since the Pauli blocking for nucleon-nucleon collisions is playing a quite different role.

To show that the statistical particle emission is correctly reproduced within our Montecarlo approach, in fig. 5 we present the corresponding neutron and proton multiplicities, computed using the E_x-increasing parametrization of Γ_{GDR}. The normalization is the same as for the γ- decay. In this case we do not see any saturation effect. The neutron increasing rate with excitation energy is in a nice quantitative agreement with data [5,6].

Since in our picture the GDR limiting temperature derives from a competition with particle evaporation in hot compound nuclei, we would expect larger values

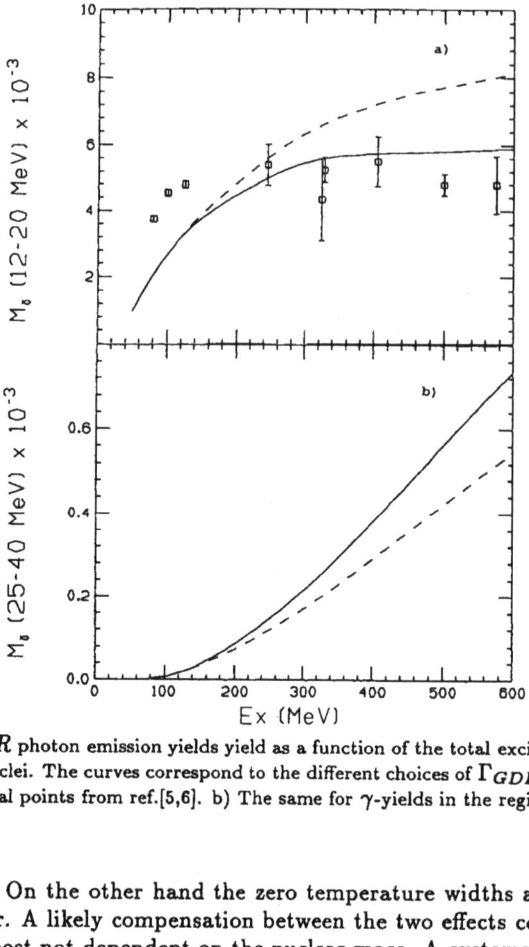

Fig. 4. a) *GDR* photon emission yields yield as a function of the total excitation energy in *Sn* compound nuclei. The curves correspond to the different choices of $\Gamma_{GDR}(E_x)$ shown in fig.1. Experimental points from ref.[5,6]. b) The same for γ-yields in the region $25 \le E_\gamma \le 40 MeV$.

for lighter nuclei. On the other hand the zero temperature widths are decreasing with mass number. A likely compensation between the two effects could lead to a $T_{lim} \simeq 4MeV$ almost not dependent on the nuclear mass. A systematics in theory and experiments would be quite interesting.

Our conclusion is that pre-equilibrium effects computed with realistic particle emission probabilities are not large enough to reproduce the quenching of giant dipole photons observed at high excitation energy. In order to obtain the correct limiting temperatures for the *GDR* it is necessary to use a resonance spreading width which increases with E_x (and in this case the pre-equilibrium effects are less important). A reduction of the resonant form factor eq.(5) will be sufficient to enhance the cooling by particle emission , in agreement with experiments.

This conclusion can be tested performing new experiments in a larger range of masses and in an energy region just above the one of saturation of the angular

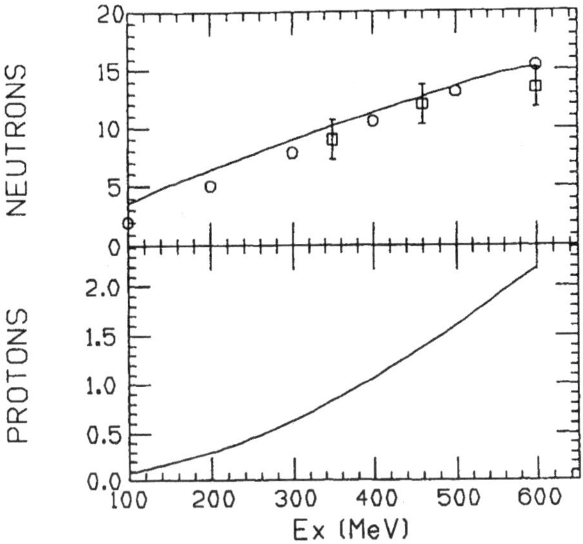

Fig. 5. Neutron and proton emission yields as a function of the total excitation energy in Sn compound nuclei. The solid lines are the results of our Montecarlo analysis using the E_x-increasing behaviour for the GDR width. Neutron data are from ref.s [5,6] and circles show the result of an evaporation calculation performed by the authors within a full Cascade approach with their sharp E_x- increasing parametrization of the GDR width.

momentum of the compound nucleus.

Finally we would like to stress that the initial conditions $P_D = 0, P_C = 1$ which lead to the correction in eq.(3) are certainly overestimating such pre-equilibrium effects. Actually some dipole oscillations in the fused system are likely present due to the long equilibration time of the charge degree of freedom [17,18]. This is quite evident for N/Z-asymmetric entrance channels but it should be seen also for N/Z-symmetric partners particularly at the high energies of interest here. Indeed N/Z-fluctuations due to pre-equilibrium emissions in incomplete fusion events will bring some isovector dipole strength in the fused system .

4. Pre-equilibrium *GDR* strength.

A thorough analysis has been performed in ref.s [17,18] where a GDR phonon gas coupled to a compound nucleus without any phonon has been considered. The γ-emission probability, integrated over the Giant Dipole region, will be much dependent on the GDR strength present in the fused system before a complete equilibration:

$$P_\gamma = P_\gamma^{eq} \frac{\mu}{\mu + \gamma_{ev}} [1 + n_0 \frac{\gamma_{ev}}{\lambda}] = P_\gamma^{eq} + (n_0 - \lambda/\mu) \frac{\gamma_\gamma}{\mu + \gamma_{ev}} \tag{6}$$

where μ and λ are the phonon decay and excitation rates introduced before, γ_{ev} is the particle evaporation rate and n_0 is the mean number of GDR phonons present at the time of compound nucleus formation ($n = P_D/(P_C + P_D)$ using the Brink's notations of the previous section).

$$P_\gamma^{eq} = \frac{\gamma_\gamma}{\gamma_{ev}} \frac{\lambda}{\mu} \tag{7}$$

is the statistical equilibrium prediction, γ_γ being the partial width for photon emission.

For n_0=0 we get back the reduction factor used in eq.(3), but in general eq.(6) says that if $n_0 > \lambda/\mu$ (equilibrium value of the number of GDR phonons) we should expect to actually see an enhancement of Giant Dipole photon emissions. This seems indeed to be a quite likely case since, from level density considerations,

$$\frac{\lambda}{\mu} = \frac{\rho(E_x - E_{GDR})}{\rho(E_x)} << 1. \tag{8}$$

and particularly for N/Z asymmetric entrance channels [17]. However if the spreading width becomes large this entrance channel effect will disappear since the mean number of GDR phonons is quickly reaching the equilibrium value. This can be seen from fig. 6 where the GDR-integrated γ-emission probability is plotted as a function of the excitation energy for various values of n_0 $(0, 1/2\lambda/\mu, \lambda/\mu, 3/2\lambda/\mu, 3\lambda/\mu,$ bottom to top): fig.6a for an increasing spreading width (solid curve of fig.3), fig.6b for a saturating one (dashed curve of fig.3).

Our conclusion is that these entrance channel effects could give some independent experimental information on the behaviour of hot GDR's. Moreover there is a good chance of a direct observation of pre-equilibrium collective dipole photons. More experimental work is certainly needed in this medium energy region, possibly also with radioactive beams in order to enhance the range of N/Z asymmetries.

In order to check this point we have performed a fully microscopic dynamical calculation [14,15] for a central collision, leading to a fused system in the Sn region, at 26 MeV/n beam energy in different entrance channels ^{36}Ar $(N/Z = 1.0)$ + ^{96}Zr $(N/Z = 1.4)$, asymmetric, and ^{40}Ar $(N/Z = 1.22)$ + ^{90}Zr $(N/Z = 1.25)$, symmetric.

In fig.7 we report the time evolution of the isovector dipole moment in momentum space for the composite system

$$D(t) = \frac{N\,Z}{A} \left[\frac{< P_z >_{proton}}{N} - \frac{< P_z >_{neutron}}{Z} \right] \tag{9}$$

in the N/Z asymmetric fig(7a) and N/Z symmetric (fig.7b) case. The difference seems quite evident. In the same figure we report the corresponding isoscalar

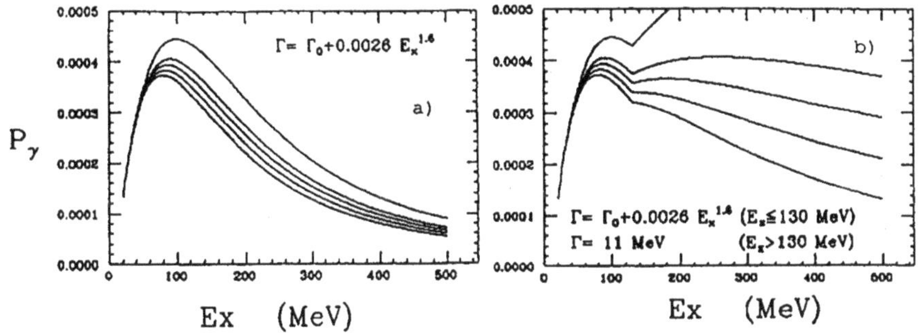

Fig. 6. E_x-dependence of the GDR-integrated γ-emission probability for Sn isotopes: a) increasing spreading width; b) saturating spreading width.

quadrupole moment, which gives a measure of the overall equilibration of the fused nucleus. We see that at $t \simeq 140 fm/c$ the fused system is equilibrated for both entrance channels. At the same time in the N/Z asymmetric entrance channel we still see quite clear dipole oscillations.

This effect is particularly evident from the Fourier transforms, shown in Fig.8, defined as

$$S(\omega) = |F(\omega)|^2$$

with

$$F(\omega) = \int_0^{t_{max}} dt\, e^{i\omega t} D(t) \tag{10}$$

We have chosen $t_{max} = 300 fm/c$ in order to avoid the contribution from the compound nucleus. For the asymmetric case (fig.8a) we clearly see a nice peak around 16 MeV, which roughly corresponds to the GDR energy for the compound system. It is interesting to notice a peak in the same position, but with a much reduced strength (the scale is different) also for the symmetric entrance channel (fig.8b): this could be interpreted as due to N/Z fluctuations present in incomplete fusion events.

We can conclude that studying different reactions with or without asymmetry in the N/Z ratio one should be able to observe the presence of a pre-equilibrium collective dipole γ-emission [23].

Inspired by the quite exciting results obtained at intermediate energies we have also looked at the entrance channel N/Z asymmetry effect at lower energies, where actually more data are available. When we decrease the beam energy the initial amplitude of the dipole oscillation in the asymmetric case will decrease due to a

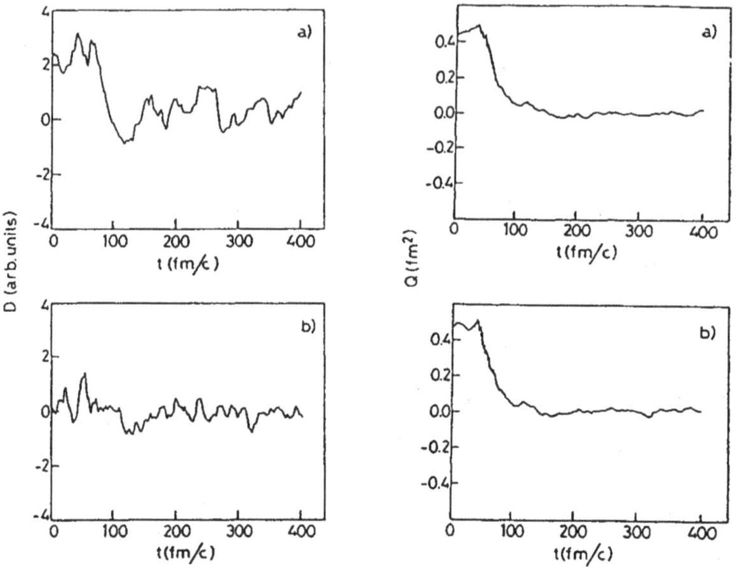

Fig. 7. Time evolution of isovector dipole (left) and isoscalar quadrupole (right) moments for central collisions of charge asymmetric (a) and symmetric (b) $Ar + Zr$ entrance channels.

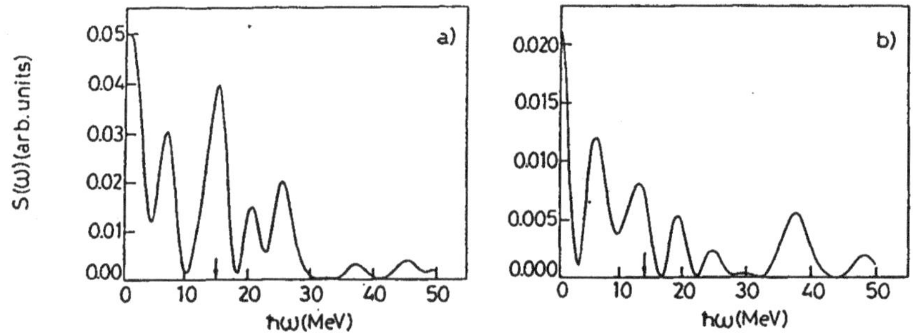

Fig. 8. Fourier transforms of the dipole oscillations of fig.7 for a) the asymmetric case and b) symmetric.

smaller shift of the neutron and proton Fermi spheres. In fact what we finally observe is a clear disappearance of the asymmetry effect with decreasing beam energy [18].

Now the time involved to form a fused system is longer and therefore entrance channel effects cannot be seen. However the fusion dynamics seems to allow the possibility of a direct observation of pre-fusion GDR's in highly deformed intermediate nuclear system. But this is another chapter of the never ending Nuclear Giant Resonance Story.

Indeed in the low energy region only some mass-asymmetry effects have been observed so far, which are related to different time scales for the fusion process [24]. Of course this effect would be enhanced from the presence of some angular momentum, i.e. for more peripheral collisions. We have started a systematic study in order to check if it would be possible to detect pre-equilibrium giant photon decays from these intermediate states in fusion and deep inelastic collisions at low energies.

5. Collective Dipole emission from a very deformed di-nuclear system.

It is well know that properties of the GDR give direct information on the structure of the nuclear system on top of which the resonance is built [25]. So the possibility of observing some GDR photon emission from intermediate dinuclear states in fusion, fusion-fission and deep inelastic collisions looks extremely stimulating in order to gain a direct insight into the dynamics of the process. As discussed at the end of the previous section we are addressing now our attention to a γ-emission in relatively slow processes, where a long living di-nuclear system can be formed. The charge degree of freedom will equilibrate but not the shape [26] and there is a chance of observing a statistical emission of GDR-photons from a very deformed *compound* nucleus. Now the dynamics of the reaction is playing a major role and important entrance channel effects will be related to the mass asymmetry, rather than to charge asymmetry, at variance with the fast fusion processes at higher energies analysed in the previous section.

The most evident feature of a Giant Dipole γ-emission from a very deformed system will be a clear splitting of the resonance: i.e. for a quadrupole leading shape we expect to see two main peaks in the γ-yield with a separation proportional to the amount of deformation [25]. There are some first experimental evidences of this effect in fusion reactions [24,27]. In ref.[24] such two-bump γ-emission is needed to account for the measured spectrum when the fused system, ^{164}Yb at $E_x = 49MeV$ is formed in a more mass-symmetric entrance channel, where one expects longer times for the fusion dynamics [28]. In ref. [27] coincidence α-particles are used as a clock for the emission times. When the γ-spectrum is measured, in the $^{nat}Sn+^{20}Ne$ reaction at $164MeV$ beam energy, in coincidence with forward emitted α's (short living source), the two bumps are neatly seen. They disappear in spectra with α-backward coincidences.

Evidences of GDR-photon emission from very deformed systems have been observed also in exit fission channels, either in fusion-fission reactions [29] or in spon-

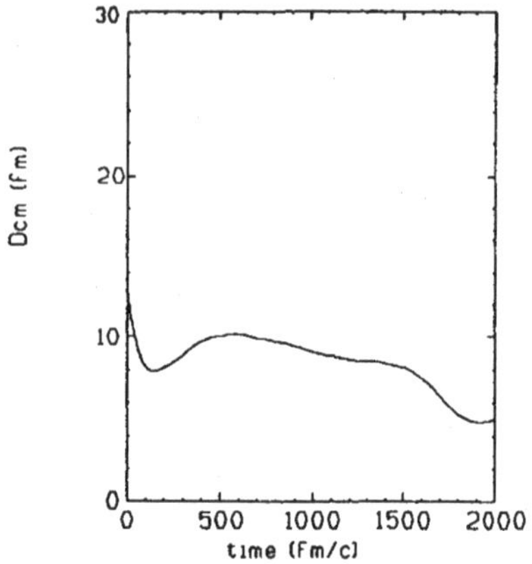

Fig. 9. Time evolution of the $PLF - TLF$ relative distance for the collision $^{35}Cl + ^{64}Ni$ at $271 MeV$ beam energy, $b = 6fm$.

taneous fission decays [30].

Recently some discrepancies between measured γ-spectra and calculated statistical fragment decays in deep inelastic collisions [31]. The presence of an intermediate very deformed emitting source seems needed in order to account for a quite large GDR strength around $8MeV$ excitation energy. For the reaction studied in that experiment, $^{35}Cl + ^{64}Ni$ at $271MeV$ beam energy, one can perform a microscopic BNV [14, 45] calculation. We will show some interesting results obtained for quite dissipative collisions around an impact parameter $b = 6fm$. Using an appropriate clusterization procedure we can follow the time evolution of the relative distance $D_{cm}(t)$ between projectile-like and target-like fragments, which is shown in fig.9.

For this impact parameter the system is going to fuse but the relative distance is staying almost constant, around $8fm$, for a quite long time interval, from 200 to $1500fm/c$. Meanwhile the system is rotating.

In figs. 10a), 11a) we report dipole oscillations, of the vector $\vec{D}(t) = \Sigma_i z_i \vec{r}_{i(cm)}(t)$, respectively along the rotating elongation axis and orthogonal to it. The difference in frequencies is quite evident, as also clearly shown in the corresponding Fourier transforms plotted in fig.s 10b) and 11b). In particular we remark the large strength around $8MeV$ for the component along the deformation axis (fig.10b)).

Our conclusion is that there is a chance to observe directly the GDR γ-emission from very deformed intermediate di-nuclear systems in low energy fusion and deep inelastic collisions. We are now performing a systematical study in order to evaluate the corresponding cross sections and to select the best systems to look at. This method could give also valuable information on the best entrance channel choices for the population of very deformed bands in highly rotating compound nuclei.

It would be also interesting to check if for this type of GDR emission the en-

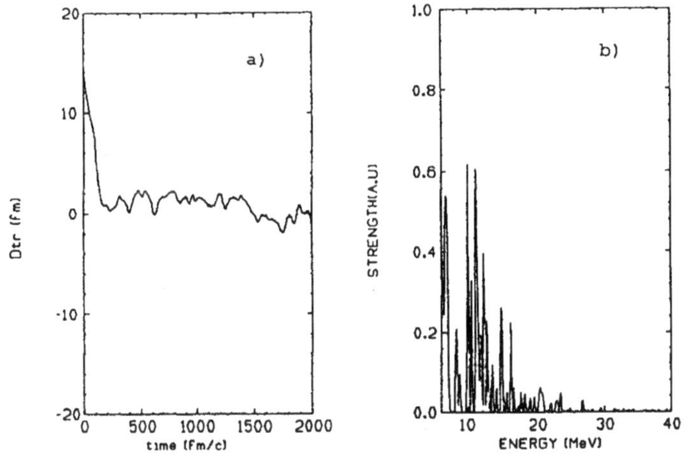

Fig. 10. a) Dipole oscillations along the axis which connects projectile-like and target-like fragments for the collision of fig.9; b) Corresponding Fourier transform defined as in eq.(10).

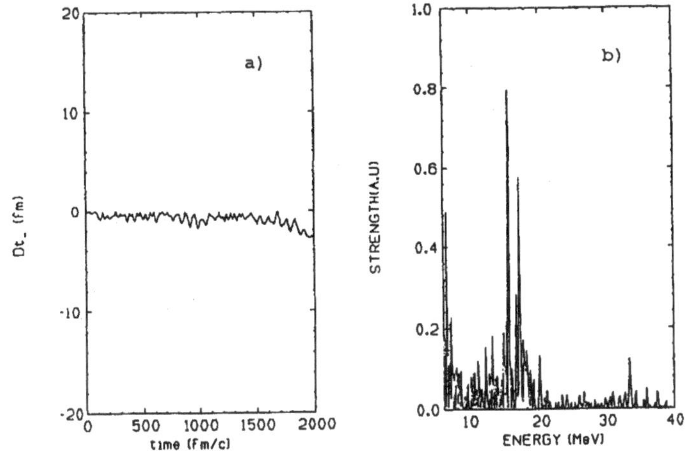

Fig. 11. a) Dipole oscillations along the direction orthogonal to the axis which connects projectile-like and target-like fragments for the collision of fig.9; b) Corresponding Fourier transform defined as in eq.(10).

trance channel charge asymmetry is not so important as in the fast fusion processes discussed in the previous section.

6. Halo-nuclei in phase space: Soft Giant Resonances

This final section is devoted to some considerations involving giant resonances in the opening new field of exotic nuclei, quite far from the stability line. This is to show how important tools can be the giant modes to study static and dynamical properties of such new nuclides.

Light neutron-rich nuclei have shown many new interesting features [32,33]: large nuclear matter radii, narrow momentum distributions of fragments after removing valence neutrons and large e.m. dissociation cross sections, coupled to very small neutron separation energies. The qualitative understanding is through an extended neutron distribution, the neutron "halo", which seems to be outside any fully consistent mean field picture [34,35,36]. A consequence of this exotic nuclear structure should be seen also in the presence of a different strength distribution for collective excitations [37,38,39]. In particular a soft isovector dipole mode (SDR) is expected as due to low frequency oscillations of valence neutrons against the core [34, 37-41]. This could explain the observed large e.m. dissociation cross section [42].

Important dynamical effects from the neutron halo are expected also in nuclear collisions, particularly in fusion reactions [43]. Therefore it would be nice to construct a phase space picture of halo nuclei to use in dynamical analyses. In order to do that we will use the giant resonance properties as guidelines.

A semiclassical microscopic model for the SDR in ^{11}Li was already introduced in ref.[44]. The linearized Vlasov equation for a coupled three-fluid dynamics (core neutrons and protons, halo neutrons) was solved using simplified interactions in order to get the isovector dipole response. A noticeable $E1$ strength at low excitation energy was found using large r.m.s. radii for halo neutrons. Indeed this strength was obtained just from the uncorrelated response, i.e. almost without any enhancement due to the residual interaction. In this sense we cannot speak of collective behaviour for the soft mode.

The model seems to be then a good starting point for the construction of a phase-space halo nucleus. The ^{11}Li system is formed considering two Fermi-type density distributions. The momentum space is occupied by spheres with Fermi momentum given by the local density. The core density distribution has a central value $\rho_0 \simeq .16 fm^{-3}$, r.m.s. radius $R_c = 2.6 fm$ and diffuseness $a = .65 fm$. The halo distribution contains two neutrons in a radius $R_h = 4.5 fm$ with diffuseness $a = 1.3 fm$. Switching on the Skyrme effective forces we get a mean binding energy of $5 MeV$ for the core nucleons and $.3 MeV$ for the valence neutrons.

The dynamical response of this system to an external perturbation can be analysed. We have studied the isoscalar quadrupole response function just imposing an initial quadrupole "kick" in momentum space:

$$p_z \rightarrow p_z/(1+2\alpha)^2; p_x \rightarrow p_x(1+2\alpha); p_y \rightarrow p_y(1+2\alpha) \tag{11}$$

An initial amplitude $\alpha = .125$ was chosen, which corresponds to a deformation collective energy of about $20 MeV$ [11]. The time evolution of the quadrupole moment in momentum space is shown in fig. 12. We have followed a Vlasov dynamics with 200 test particles per nucleon. The corresponding strength distribution obtained from the Fourier transform is plotted in fig.13.

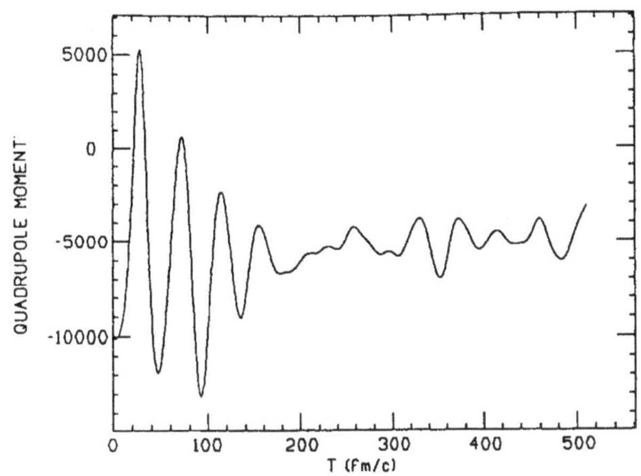

Fig. 12. Time evolution of the quadrupole moment for the ^{11}Li phase space model

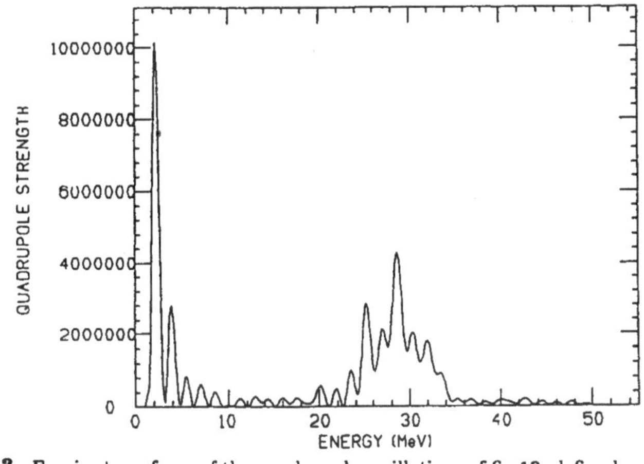

Fig. 13. Fourier transform of the quadrupole oscillations of fig.12, defined as in eq.(10) with $t_{max} = 800 fm/c$.

We see a clear quadrupole strength at excitation energies below 10 MeV. The related fraction of the EWSR is 5.6 percent, in good agreement with the values obtained in RPA-type calculations by S.Fayans [37] and H.Sagawa [38].

We have finally used the analytical model of ref.[44] to check the results. The linearized Vlasov equation was solved for coupled core nucleons and valence neutrons, respectively sitting in Woods- Saxon potentials with the parameters used for the density distributions. A quadrupole-quadrupole separable residual interaction has been considered with standard coupling constant K_c for the core part and a

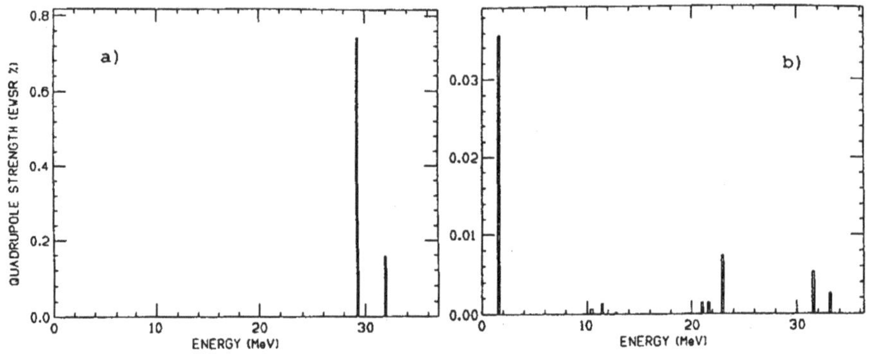

Fig. 14. Core (a) and Halo (b) quadrupole strength distributions from the solution of the linearized Valsov equation for ^{11}Li

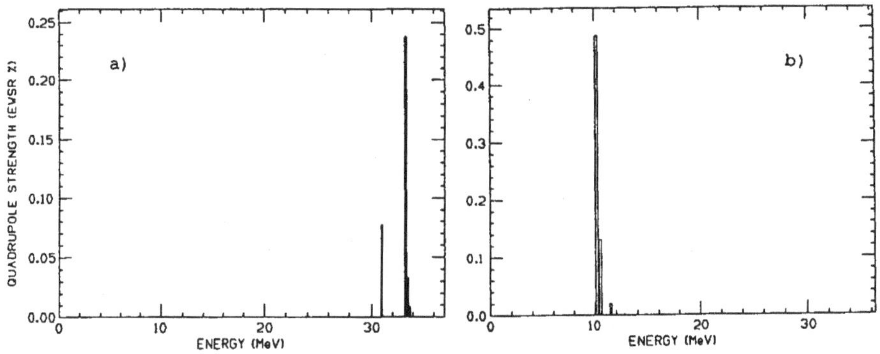

Fig. 15. Like fig.14, but for the uncorrelated response: a) Core, b) Halo.

suitably rescaled one for the halo

$$K_h = (R_c/R_h)^2 K_c \tag{12}$$

The core and halo strength distributions are shown in fig.14, and are in good agreement with the microscopic phase-space results of fig.13. In fig.15 we report the corresponding uncorrelated strength distributions. It seems that the effect of the residual interaction is quite noticeable also for the halo part, at variance with the isovector dipole case [44].

7. Conclusions

We have shown that the story of nuclear giant resonances is still quite fascinating and continuously reveals new stimulating features which seem to deserve a lot of theoretical and experimental work for the next years. Some points are extensively discussed in this report.

We have first suggested a mechanism to understand the disappearing of giant dipole resonances in very excited nuclei. More data are certainly needed. In particular it would be of interest to study the mass dependence of the limiting temperature. Moreover some independent informations could be brought from the γ-yield above the GDR region.

Pre-equilibrium emission is a well known mechanism in nuclear collisions. Pre-equilibrium light particles and incoherent photons have been well detected. We stress here the possibility of observing pre-equilibrium emissions of collective dipole photons in *fast* fusion processes. Charge asymmetry in the entrance channel is playing an essential role and consequently these experiments are very appropriate for the new available radioactive beams.

At lower beam energies this charge effect should disappear, but we have shown the possibility then of detecting GDR γ-decays from long living intermediate very deformed nuclear systems. This would represent a direct picture of the fusion and DIC dynamics, particularly interesting for the study of fusion paths leading to hyperdeformed bands.

Finally new features are certainly expected for giant resonances in nuclides near the drip lines. We have discussed the use of giant modes to tune dynamical models of neutron-halo nuclei.

To conclude we would like to stress the more general validity of the topics related to giant resonances in nuclei. We expect the giant modes to be of great importance for the study of properties of other finite Fermi systems.

We warmly thank A.Bonasera, D.M.Brink, Ph.Chomaz and F.Gulminelli for very helpful discussions.

8. References

1) D.R. Chakrabarty et al., Phys.Rev. C36 (1987) 1886
2) A. Bracco et al., Phys.Rev.Lett. 62 (1989) 2080
3) J.J.Gaardhoje et al., Phys.Rev.Lett. 59 (1987) 1409 and Nucl.Phys. A538 (1992) 573c
4) G.Enders et al., Phys.Rev.Lett. 69(1992)249
5) K. Yoshida et al., Phys.Lett. B245 (1990) 7
6) J.Kasagi et al., Nucl.Phys. A538 (1992) 585c
7) D.M.Brink, Nucl.Phys. A519(1990)3c
8) P.F.Bortignon, A.Bracco, D.M.Brink and R.A.Broglia, Phys.Rev.Lett. 67(1991)3360
9) D.M.Brink, A.Dellafiore and M.Di Toro, Nucl.Phys. A456 (1986) 205

10) G.F.Burgio and M.Di Toro, Nucl.Phys. A476(1988)189
11) A.Bonasera, M.Di Toro and F.Gulminelli, Phys.Rev. C42(1990)966
12) A.Smerzi, A.Bonasera and M.Di Toro, Phys.Rev. C44 (1991)1713;
13) G.F. Bertsch, Z.f.Physik A289 (1978) 103
14) A.Bonasera et al. Phys.Lett. B221 (1989) 233; B259 (1991) 399
15) A.Smerzi, Ph.D.Thesis, Catania 1993
16) A.Bonasera, M.Di Toro, A.Smerzi and D.M. Brink "Pre- equilibrium and temperature effects on the disappearing of Giant Dipole Resonances in highly excited nuclei" Int.Conf. on Giant Resonances, Gull Lake 1993, Nucl.Phys. A in press
17) Ph. Chomaz, M. Di Toro and A. Smerzi, Nucl.Phys. A563 (1993) 509
18) Ph.Chomaz, Zhong Jiquan, A.Smerzi and M. Di Toro "New pieces into the Hot Giant Dipole Resonance puzzle" Int.Winter Meeting on Nuclear Physics, Bormio 1993, Ed.I.Iori, p.427-450
19) V.Weisskopf "Statistics and Nuclear Reactions", Phys.Rev. 52 (1937) 295; A.Friedman, G.Lynch "Statistical formalism for particle emission" Phys.Rev. C28 (1983) 16
20) F.Pulhofer, Nucl.Phys. A280(1977)267
21) T.Suomijarvi et al. "Study of hot Giant Dipole Resonance with the MEDEA detector" Preprint Orsay IPNO DRE 92-35, Proc. INFN-RIKEN meeting "Perspectives in Heavy Ion Physics", Ed.s M.DiToro and E.Migneco, SIF Bologna 1993, pp.189-197
22) T.Suomijarvi private communication and J.H.Le Faou et al. "Towards limiting temperatures in nuclei: the behavior of collective motion" IPN-Orsay preprint Oct.93
23) R.Alba et al., OUVERTURE Proposal with the first test Ni beams of the LNS Superconducting Cyclotron, Catania 1993.
24) M.Thoennessen et al., Phys.Rev.Lett. 70 (1993) 4055
25) K.A.Snover, Ann.Rev.Nucl.Part.Sci. 36 (1986) 545
26) Ch.Gregoire, "Winter College on Fundamental Nuclear Physics" , Ed.s K.Dietrich, M.Di Toro and H.J.Mang, World Sci. 1984, Vol I p.501
27) V.V.Kamanin et al., Z.f.Physik A337 (1990) 111
28) H.Feldmeier, Rep.Progr.Phys. 50 (1987) 915
29) J.J.Gaardhoje, Ann.Rev.Nucl.Part.Sci. 42 (1992) 483 and ref.s therein
30) H.Van der Ploeg et al., Phys.Rev.Lett. 68(1992) 3145 and 69 (1992) 1148 (errata)
31) TRASMA Exp., Collab. LNS Catania-Napoli-LNL Legnaro-Saclay 1993 and L.Campajola et al. "Dynamic excitation of giant dipole modes in dissipative heavy ion collisions", subm. Phys.Rev.Lett.
32) I.Tanihata et al., Phys.Lett. B160 (1985) 380, Phys.Rev.Lett. 55 (1985) 2676, Phys.Lett. B206 (1988) 592
33) T.Kobayashi et al., Phys.Rev.Lett. 60 (1988) 2591, Phys.Lett. B232 (1989) 51
34) P.G.Hansen and B.Jonson, Europhys.Lett. 4 (1987) 409
35) G.F.Bertsch and J.Foxwell, Phys.Rev. C42 (1990) 1159
36) H.Sagawa, Phys.Lett. B286 (1992) 7

37) S.A.Fayans Phys.Lett. B267 (1991) 443
38) H.Sagawa, Nucl.Phys. A538 (1992) 619c
39) H.Sagawa, N.Van Giai, N.Takigawa and M.Ishihara, "Soft Multipole Resonances", to be published
40) C.Bertulani and G.Baur, Nucl.Phys. A538 (1988) 615
41) Y.Suzuki, K.Ikeda and H.Sato, Pr.Theo.Phys. 83 (1990) 180
42) T.Kobayashi, Nucl.Phys. A538 (1992) 343c
43) N.Takigawa, M.Kuratani and H.Sagawa, Phys.Rev. C47 (1993) R2470
44) M.Di Toro and Cai Yanhuang, Phys.Rev. C42 (1990) 2748
45) A.Bonasera, F.Gulminelli and J.Molitoris, "The Boltzmann Equation at the borderline: a decade of Monte Carlo simulations of a quantum kinetic equation" Phys. Reports (1994) to appear

Dynamic Response of Deformed Fermi Systems

A. DELLAFIORE

Istituto Nazionale di Fisica Nucleare, Sezione di Firenze,
Largo E. Fermi 2 Firenze, I-50125 Italy

and

F. MATERA

Dipartimento di Fisica, Università di Firenze,
Istituto Nazionale di Fisica Nucleare, Sezione di Firenze,
Largo E. Fermi 2 Firenze, I-50125 ,Italy

ABSTRACT

A solution of the linearized Vlasov equation, suitable for the study of collective vibrations in deformed Fermi systems, is discussed in terms of the polarization propagator in momentum space.

The linearized Vlasov equation can be solved for a system of interacting fermions moving in a static cavity of spheroidal shape (ellipsoids with one axis of symmetry).This model can be used to interpret the features of collective excitations in deformed nuclei or in deformed microclusters.

While for a spherical system the RPA-type integral equation obtained for the polarization propagator separates into different, simpler, equations for each partial-wave component L of the propagator [1,2],

$$D_L(q',q,\omega) = D_L^0(q',q,\omega) + \frac{1}{(2\pi)^3}\int_0^\infty dk k^2 D_L^0(q',k,\omega)\tilde{u}(k)D_L(k,q,\omega)\ , \quad (1)$$

for the axially symmetric systems studied here we get a somewhat more complicated set of equations in which the responses of different multipolarity are coupled by the two-body interaction between the constituents:

$$\begin{aligned} D_{L'LM}(q',q,\omega) &= D^0_{L'LM}(q',q,\omega) \\ &+ \frac{1}{(2\pi)^3}\sum_{\ell=|M|}^{\infty}\int_0^\infty dk k^2 D^0_{L'\ell M}(q',k,\omega)\tilde{u}(k)D_{\ell LM}(k,q,\omega)\ . \end{aligned} \quad (2)$$

The full propagator in momentum space is given by:

$$D(\mathbf{q}',\mathbf{q},\omega) = \sum_{L=0}^{\infty}\sum_{M=-L}^{L}\sum_{L'=|M|}^{\infty} D_{L'LM}(q',q,\omega)Y_{L'M}(\hat{\mathbf{q}}')Y^*_{LM}(\hat{\mathbf{q}})\ . \quad (3)$$

For separable mean-field Hamiltonians, the "free" propagator $D^0(\mathbf{q}\,',\mathbf{q},\omega)$ can be expressed as:

$$D^0(\mathbf{q}',\mathbf{q},\omega) = \frac{-2}{(2\pi\hbar)^3}\sum_{\mathbf{n}}\int d\mathbf{J}\ \delta(E_F - E)\ \frac{\mathbf{n}\cdot\boldsymbol{\omega}(\mathbf{J})}{\mathbf{n}\cdot\boldsymbol{\omega}(\mathbf{J}) - (\omega + i\eta)}\ Q^*_{\mathbf{n}}(\mathbf{J},\mathbf{q}')\ Q_{\mathbf{n}}(\mathbf{J},\mathbf{q})\ , \tag{4}$$

with the combination of frequencies $\mathbf{n}\cdot\boldsymbol{\omega}$ taking the place of the particle-hole excitation energies, and the Fourier coefficients

$$Q_{\mathbf{n}} = \frac{1}{(2\pi)^3}\int d\boldsymbol{\Phi}\ e^{-i\mathbf{n}\cdot\boldsymbol{\Phi}}\ e^{i\mathbf{q}\cdot\mathbf{r}} \tag{5}$$

replacing the quantum matrix elements. The variables $(\boldsymbol{\Phi},\mathbf{J})$ are angle-action variables; in Eq. (4) we have assumed a zero-temperature Thomas-Fermi distribution for the system of particles at equilibrium[1], E is the particle energy, and E_F is the Fermi energy. In order to evaluate the coefficients (5), the explicit relation between the angle coordinates $\boldsymbol{\Phi}$ and the particle coordinate $\mathbf{r}$ are needed. For this purpose it is useful to introduce the prolate or oblate spheroidal coordinates u, v, φ[4,5]. The relation between u, v, φ and Φ_α can then be obtained either from the definition of the angle variables

$$\Phi_\alpha = 2\pi\frac{\partial S}{\partial J_\alpha}, \tag{6}$$

with S Hamilton's characteristic function[6], or from the solution of the Vlasov equation[3]. Rather than reporting the explicit expressions of the Φ_α's, here we simply note that, for given values of the action variables $\mathbf{J}$,

$$\Phi_u = \Phi_u(u,v), \tag{7a}$$

$$\Phi_v = \Phi_v(u,v), \tag{7b}$$

$$\Phi_\varphi = \varphi + \tilde{\Phi}_\varphi(u,v). \tag{7c}$$

The functions $\Phi_u(u,v)$,$\Phi_v(u,v)$, and $\tilde{\Phi}_\varphi(u,v)$ involve elliptic integrals, and can be evaluated numerically.

Because of the trivial dependence of Φ_φ on the cyclic variable φ, the Fourier coefficients (5) can be simplified. Expanding $e^{i\mathbf{q}\cdot\mathbf{r}}$ in multipoles, gives

$$Q_{\mathbf{n}}(\mathbf{J},\mathbf{q}) = \sum_{LM} Q^{(LM)}_{\mathbf{n}}(\mathbf{J},q)Y^*_{LM}(\hat{\mathbf{q}})\ , \tag{8a}$$

with

$$Q^{(LM)}_{n_u,n_v,n_\varphi} = \frac{1}{(2\pi)^{5/2}}\delta_{M,n_\varphi}\oint du \oint dv\ \left|\frac{\partial(\Phi_u,\Phi_v)}{\partial(u,v)}\right|\ e^{-i(n_u\Phi_u + n_v\Phi_v + n_\varphi\tilde{\Phi}_\varphi)} \times (4\pi)i^L Y_{LM}(\theta(u,v),0)j_L(qr(u,v))\ , \tag{8b}$$

thus, because of the rotation symmetry about the z-axis, the integer n_φ must be equal to M, and the coefficients vanish if $|n_\varphi| > L$. The other important ingredients necessary to evaluate the propagator (4) are the components of the frequency vector $\boldsymbol{\omega}$. They are defined as

$$\omega_\alpha \equiv 2\pi \frac{\partial}{\partial J_\alpha} h_0(\mathbf{J}) , \tag{9}$$

where h_0 is the equilibrium hamiltonian (describing a spheroidal cavity with perfectly reflecting wall in our case). For a prolate cavity [4,5] the fundamental frequencies are given explicitly by

$$\omega_u = \frac{2\pi}{m\xi^2} \frac{\oint \frac{dv}{p_v}}{\left(\oint \frac{dv}{p_v} Ch^2 v \oint \frac{du}{p_u} - \oint \frac{dv}{p_v} \oint \frac{du}{p_u} sin^2 u\right)} , \tag{10a}$$

$$\omega_v = \frac{2\pi}{m\xi^2} \frac{\oint \frac{du}{p_u}}{\left(\oint \frac{dv}{p_v} Ch^2 v \oint \frac{du}{p_u} - \oint \frac{dv}{p_v} \oint \frac{du}{p_u} sin^2 u\right)} , \tag{10b}$$

$$\omega_\varphi = \frac{\lambda_z}{2\pi} \left(\omega_v \oint \frac{dv}{p_v Sh^2 v} + \omega_u \oint \frac{du}{p_u cos^2 u}\right) . \tag{10c}$$

The momenta p_u and p_v are, for a prolate cavity,

$$p_u = \sqrt{2m\xi^2 \left(\epsilon - E\, sin^2 u - \frac{\lambda_z^2}{2m\xi^2 cos^2 u}\right)} , \tag{11a}$$

$$p_v = \sqrt{2m\xi^2 \left(E\, Ch^2 v - \epsilon - \frac{\lambda_z^2}{2m\xi^2 Sh^2 v}\right)} , \tag{11b}$$

m is the mass of the constituent particles, ξ is half the focal distance, and ϵ is the separation variable, which, for prolate deformation, is related to the scalar product of the particle angular momenta with respect to the two foci by:

$$2m\xi^2(\epsilon - E) = \boldsymbol{\lambda}_1 \cdot \boldsymbol{\lambda}_2 . \tag{12}$$

For oblate cavities Eqs.(11) are replaced by

$$p_u = \sqrt{2m\xi^2 \left(\epsilon - E\, cos^2 u - \frac{\lambda_z^2}{2m\xi^2 cos^2 u}\right)} , \tag{13a}$$

$$p_v = \sqrt{2m\xi^2 \left(E\, Ch^2 v - \epsilon + \frac{\lambda_z^2}{2m\xi^2 Ch^2 v}\right)} . \tag{13b}$$

The motion in the u and v variables can be viewed as two separate one-dimensional motions in effective potentials of the kind (oblate cavity)

$$U_{eff}(u,\lambda_z) = 2m\xi^2 E(cos^2 u + \frac{\lambda_z^2}{2mE\xi^2 cos^2 u}) \tag{14a}$$

and

$$V_{eff}(v,\lambda_z) = 2m\xi^2 E(Ch^2 v + \frac{\lambda_z^2}{2mE\xi^2 Ch^2 v}) \,. \tag{14b}$$

It has been pointed out in [5] that, contrary to the analogous expressions for the prolate case, the effective potentials (14) are non-monotonic. This results in a separation of the oblate phase space into two distinct regions, and makes the description of three-dimensional orbits in an oblate cavity slightly more involved than for a prolate cavity. As a consequence, for oblate cavities, the propagator (4) can be split into two terms, each containing only contributions from orbits in one of the two distinct regions of phase space (the integration limits in the u and v variables are different in the two regions of phase space, see[3] for details). Finally, changing variables from $\mathbf{J}$ to the more convenient set of constants of the motion E, ϵ, λ_z, for prolate cavities the propagator (4) can be written as

$$\begin{aligned} D^0_{L'LM}(q',q,\omega) = & -\frac{2}{(2\pi\hbar)^3} \sum_{n_u,n_v,n_\varphi} \delta_{M,n_\varphi}\, 2\pi \int d\lambda_z \int d\epsilon \, \left| \frac{\partial(J_v,J_u)}{\partial(E_F,\epsilon)} \right| \\ & \times \frac{\mathbf{n}\cdot\boldsymbol{\omega}(E_F,\epsilon,\lambda_z)}{\mathbf{n}\cdot\boldsymbol{\omega}(E_F,\epsilon,\lambda_z) - (\omega + i\eta)} \\ & \times Q^{(LM)*}_{\mathbf{n}}(E_F,\epsilon,\lambda_z;q') Q^{(LM)}_{\mathbf{n}}(E_F,\epsilon,\lambda_z;q) \,, \end{aligned} \tag{15}$$

with $Q^{(LM)}_{\mathbf{n}}$ given by Eq. (8b). For oblate cavities, because of the phase–space separation mentioned before, we can write

$$D^0_{LL'M}(q',q,\omega) = D^{(1)}_{L'LM}(q',q,\omega) + D^{(2)}_{L'LM}(q',q,\omega) \,, \tag{16}$$

with $D^{(1,2)}_{L'LM}$ each having an expression similar to Eq. (15).

In order to describe collective effects, we must evaluate D, rather than D^0, so, once the $D^0_{L'LM}$ have been evaluated, we must still solve the system of coupled integral equations (2). The strength function for a given multipolarity can then be obtained from

$$S_{LM}(q,\omega) = -\frac{1}{\pi}\, Im\, D_{LLM}(q,q,\omega) \,. \tag{17}$$

The expression (15) has to be evaluated numerically, it looks rather involved, but it is still simpler than its quantum counterpart, consequently we believe that the Vlasov theory can be useful in the study of collective vibrations in deformed Fermi systems.

References

1. D.M.Brink, A.Dellafiore, and M.Di Toro, Nucl.Phys. **A456** (1986) 205.
2. A.Dellafiore and F.Matera, Phys. Rev. **B41** (1990) 3488.
3. F.Matera and A. Dellafiore, *Solution of the Vlasov Equation for deformed systems*, to be published.
4. V.M.Strutinsky, A.G.Magner, S.R.Ofengenden and T.Dossing, Z.Phys. **A283** (1977) 269.
5. R.Arvieu, F.Brut, J.Carbonell and J.Touchard, Phys.Rev. **A35** (1987) 2389.
6. H.Goldstein, *Classical Mechanics* (Addison-Wesley Reading, Ma. 1980).

COLLECTIVE EFFECTS IN THE CONTINUUM OF LIGHT NUCLEI

F. ARICKX, J. BROECKHOVE and P. VAN LEUVEN
Department of Mathematics and Informatics
University of Antwerp (RUCA)
Groenenborgerlaan 171, B2020 Antwerpen, Belgium
and
V. VASILEVSKY and G. FILIPPOV
Institute for Theoretical Physics, Kiev, 143, 252143
Republic of Ukraine

Abstract

In light nuclei, the interesting dynamics occurs in the continuum along channels corresponding to the different decay modes. For the purpose of this contribution, we distinguish between cluster channels and collective channels. In the case of (binary) cluster channels, the system decays into two fragments and the active dynamical variable is the inter-cluster distance. As collective modes, we consider the monopole channel defined as the total break up of the nucleus with all nucleons flying apart isotropically and the quadrupole mode responsible for deformation effects of axial stretching.

The method that we employ for theoretical description and numerical calculation is based on two key elements. The first is the microscopic realisation of the cluster, monopole and quadrupole state spaces with their appropriate bases. The calculation of the corresponding matrix elements is achieved using the Generating Function Method that was developed in the context of the symplectic model. The second key element is the Algebraic Method for calculating the spectrum, including phase shifts. It is based on a matrix formulation of scattering theory.

We consider the system ^6Li and study the interplay of two cluster channels and two collective channels. More specifically we investigate the transfer reaction $^3He(t,d)\alpha$ at low energies. This reaction is of great interest to astrophysics and thermonuclear physics. We compare calculations involving only two cluster channels with the full four coupled channels results. In particular, we determine the "astrophysical factor" and we find that its value is strongly affected by the inclusion of the collective modes.

1 Introduction

The s- and p-shell nuclei have often served as a testing ground for theoretical models and approximations. Extended shell-model calculations [1] have revealed that the shell model converges slowly. Enlarging the shell-model basis to include all $10\hbar\omega$ states of ^4He, one finds that the low energy spetrum can still not be consistently described. The cluster model has also been tested extensively in the region of light nuclei and it is well-known that many bound states contain an appreciable amount of cluster configuration [2]. Although the distinction between single-particle and collective properties stems originally from the theory of heavy nuclei, it has been shown that also in the p-shell collective quadrupole effects [3] play an important role. The breathing mode of ^4He is another example of collective motion occuring in light nuclei.

In spite of their relative simplicity, light nuclei display a rich variety of physical phenomena, ranging from single-particle shell effects and clustering phenomena to full collective motion. Most of the interesting dynamics is contained in the continuum because many light nuclei have only very few bound states, sometimes none, as in ^8Be, or only one, as in ^6Li. The major theoretical problem in the description of light nuclei therefore is the application of the various models to the description of the continuous spectrum or, in other words, to the calculation of scattering and reaction parameters.

The different models mentioned above are, of course, not contradictory. They all serve to define a subspace structure of the entire hilbert space and hence, in principle, any state of the system may be represented in either model. Cluster wave functions can e.g. be expanded in shell-model states and collective states can e.g. be considered as excited cluster configurations. The common feature is that for light nuclei it is possible to write the wave functions in terms of the nucleonic coordinates and thus provide a microscopic version of each of the models.

Although each model, in its most generalized form, could provide an exact description, which model to choose is a matter of which kind of dynamics one is interested in. It is the contention of this paper that a rapid convergence and a good physical insight are obtained if stucture and dynamics of light nuclei are described by a combination of cluster- and collective modes. This assertion is based on the fact that these modes have a direct intuitive physical basis and that collective and cluster channels contain very high lying shell-model states.

In the past the cluster model and the collective model have been treated as quite different descriptions, as indeed they are. The cluster model, which is usually implemented through the "resonating group method", contains as the active degree of freedom the cluster separation(s) which is a translational variable. The collective model on the other hand, customarily formulated by the symplectic model, is described with scaling degrees of freedom : a radial stretching as in the monopole mode or an axial stretching as in the quadrupole mode.

There are strong arguments, however, for a joint treatment of cluster and collective modes, based on considerations concerning elastic and inelastic scattering and reactions. Usually the thresholds [1] for cluster reactions lie much lower in energy than the thresholds for collective channels. Nevertheless at the culmination of a reaction, i.e. when, in a time dependent picture, the reacting clusters are interpenetrating, there is a coupling between all channels both open and closed. The interaction effects might be quite strong and may be estimated by comparing reaction data calculated with and without coupling of cluster and collective channels.

The description of the continuum of light nuclei in terms of combined cluster and collective effects thus requires a well-adapted method. In this paper we wish to describe such a method and an application of it to a particular process in the ^6Li system. The method we propose is designed with the purpose of treating the essentially different nature of the cluster and collective degrees of freedom on an equal footing. The fundamental idea, which we shall develop more extensively below, is to transform from a representation of the

[1]In this paper we shall intend by "threshold" of a channel the excitation energy of the compound system for zero incident energy of the fragments

wave function in coordinate space, or more precisely, in generator coordinate space, to a representation in terms of the basic oscillator states of the underlying shell-model potential. The number of oscillator quanta is then considered as a discrete variable which takes over the role of the continuous coordinates.

As we are mainly interested in the continuous spectrum, we are led, in this way, to consider a theory of scattering in terms of square integrable wave functions. Lately such "algebraic methods" of scattering have been discussed also in atomic physics [4]. The main feature of such methods is the formulation of asymptotic boundary conditions. This is done, not by way of asymptotic wave functions but in terms of the Fourier coefficients of the discrete basis. As will be explained below, the phase shift, or more generally, the S-matrix elements can be expressed in these coefficients quite easily and the algorithm for calculating them is straightforward.

As an application we shall investigate here the simplest system with a clear binary clusture structure i.e. the compound nucleus ^6Li. Here the ground state is believed to be mainly alpha+deuteron with small corrections. The threshold for free alpha-deuteron scattering is about $1.5MeV$ above the ground state and at about $16MeV$ opens the tritium-^3He channel. Still higher up in the energy spectrum is the threshold for the collective monopole channel, which, by definition, is at the binding energy of ^6Li, because it corresponds to all six particles separating with zero kinetic energy. The threshold for the collective quadrupole channel would have a still higher energy, because here the particles would decay along the axial direction, while they remain confined in the transverse direction, a restriction that costs kinetic energy.

The continuum of the ^6Li system can be divided in three regions : the low-energy region where experimental data are available and are consistent with the cluster model, the intermediate energy region where experimental results are scarce and where we would expect to see effects of the closed collective channels, and the high energy region above the monopole threshold where no experimental analysis has been made. In our application we shall be interested mainly in the intermediate energy range.

2 Method

2.1 Generating Functions

Let us begin by defining the four channels we will employ to describe ^6Li. We shall designate the various clusters by the following abreviations $^2H = d$, $^3H = t$, $^3He = h$, $^4He = \alpha$. We consider two cluster modes : the (α, d) and the (t, h) (respectively referred to as C and C') and two collective modes : the monopole and the quadrupole (referred to ass M and Q). These modes or channels, as we shall call them, can be envisaged as subspaces of the six-particle hilbert space. These subspaces are most easily defined by means of generating functions.

For the cluster channels the generating function is the one usually used in the resonating group method and has the form

$$\Phi^C = \mathcal{A}\mathcal{N}\phi^c\phi^{c'}f(R) \tag{1}$$

where ϕ^c and $\phi^{c'}$ stand for the simplest shell-model ground state of each cluster and f is the wave function for the relative motion of two clusters. As we have in mind applications to $L = 0$ states, only the relative distance R will be involved. $\mathcal{A}$ is the total antisymmetrization operator and $\mathcal{N}$ is a normalizing factor. We have to consider two such generating functions, one for the cluster pair (α, d) and one for (t, h).

For the collective channels we start from the translationally invariant shell-model ground state $\Phi(\mathbf{r})$ of ^6Li, depending on the position vectors (collectively denoted by $\mathbf{r}$) relative to the centre-of-mass. For the monopole mode we consider as generating function the radially scaled wave function with α as scaling parameter

$$\Phi^M = \mathcal{N}\Phi(\alpha\mathbf{r}) \tag{2}$$

For the quadrupole mode we use a scaling parameter β in the z-direction only

$$\Phi^Q = \mathcal{N}\Phi(x, y, \beta z) \tag{3}$$

Here $\mathcal{N}$ is a normalizing factor. No antisymmetrization is needed as Φ was already antisymmetric.

The above generating functions define, for varying R, α and β, a manifold which can be interpreted in the spirit of the generator coordinate method. These parameters can be regarded as generator coordinates which serve as a continuous label, while the states of the manifold form a continuous basis that subtend the subspace.

So far the subspaces have been defined on the basis of a physical picture. The forms of the generating functions clearly show the different dynamics of the modes. In order to obtain a more unified representation of the four channels we transform to a new representation. To that end we introduce an oscillator basis for each of the four spaces. For the cluster channels this can be done by choosing in Eq. (1) for f the one-dimensional radial oscillator functions

$$f_n(R) = exp(-R^2/2b^2)L_n^{\frac{1}{2}}(R^2/b^2) \tag{4}$$

wher b is the basic oscillator length parameter.

For the collective channels it is advantageous to take a slightly different definition of the scaling parameters

$$\alpha = \sqrt{\frac{1-\epsilon}{1+\epsilon}} \qquad \beta = \sqrt{\frac{1-\eta}{1+\eta}} \tag{5}$$

In terms of the raising operators

$$\begin{aligned} A_0^+ &= \sum_{i=1}^{A}\sum_{\alpha=1}^{3} a_{i,\alpha}^+ a_{i,\alpha}^+ \\ A_{zz}^+ &= \sum_{i=1}^{A} a_{i,z}^+ a_{i,z}^+ \end{aligned} \tag{6}$$

where $a_{i,\alpha}^+$ are the familiar oscillator creation operators, we can write the above generating functions as

$$\begin{aligned} \Phi^M(\epsilon) &= exp(\epsilon A_0^+)\Phi \\ \Phi^Q(\eta) &= exp(\eta A_{zz}^+)\Phi \end{aligned} \tag{7}$$

or also

$$\begin{aligned} \Phi^M(\epsilon) &= \sum_{n=0}^{\infty} a_n^{(M)}\epsilon^n \Phi_n^M \\ \Phi^Q(\eta) &= \sum_{n=0}^{\infty} a_n^{(Q)}\eta^n \Phi_n^Q \end{aligned} \tag{8}$$

The functions thus generated are the new basis states.

We have now at our disposal, for each channel, a discrete basis of six-particle wave functions labeled by an index n. The quantum number n corresponds to the number of oscillator quanta. It can easily be seen from the above constructions that they belong entirely to the shell-model configuration with total number of quanta equal to n. Hence the basis functions with different n will be orthogonal. It should be stressed, however, that functions with same n but belonging to different channels are not orthogonal. For $n = 0$ they all coincide with the shell-model ground state. The overlaps $< \Phi_n^i \mid \Phi_n^j >$ decrease rapidly with increasing n.

In the multichannel model we are considering, the states of the ^6Li system are represented as

$$\Psi = \sum_n (a_n^C \Phi_n^C + a_n^{C'} \Phi_n^{C'} + a_n^M \Phi_n^M + a_n^Q \Phi_n^Q) \tag{9}$$

The coefficients, which form the "wave function" in the discrete basis, contain the physical information and describe the different modes in a uniform way.

2.2 Algebraic Method for Scattering

Traditionally, in Quantum Mechanics, a distinction is made between methods for bound states and methods for scattering. This is clear from the mathematical point of view, owing to the distinction between the point spectrum and the continuum spectrum of the Hamiltonian operator. In scattering theory square integrable basis functions are often used to expand the continuum wave function in the interior region of configuration space. Recently some proposals have been made [5] to use an expansion in square integrable functions for the entire space, i.e. also for the asymptotic part of the wave function. This provides a novel conceptual framework in which asymptotic configuration space wave functions are not introduced. Instead only square integrable functions and matrix manipulations are required. Here we use such a method using the basis of harmonic oscillator states introduced above.

Let us illustrate the main ideas of the method by the example of a single particle (with mass m) moving in a short range potential $V(r)$. For the discrete spectrum, the most direct information is given by the energy eigenvalues of the bound states. As for the continuous spectrum, the most important information is hidden in the wave function, in particular in the phase shift, which, in principle, can be measured experimentally. To extract this information we have to consider the wave function $\Psi_E(\mathbf{r})$ with energy E at large distances r. We know that the asymptotic form (of the partial wave with angular momentum l) is given by

$$\Psi_E(r \to \infty) \to j_l(kr) + \tan(\delta_l(k))n_l(kr) \tag{10}$$

Here we have put $k = \sqrt{2mE/\hbar^2}$ and δ_l is the phase shift of the l-th partial wave. j_l and n_l are the usual spherical Bessel and Neumann functions.

Let us expand Ψ_E in a series of discrete states $|n>$

$$|\Psi_E >= \sum_n a_n(k)|n> \tag{11}$$

The coefficients $a_n(k)$ are the projections of the continuum wave function on the basis vectors and can be obtained from the set of linear equations

$$\sum_n < m|H - E|n > a_n(k) = 0 \tag{12}$$

where H is the Hamiltonian operator. The nature of the expansion (11) would require a thorough mathematical discussion in view of the fact that the wave function Ψ_E is neither square integrable nor has it the same boundary conditions as the basis functions. However, the precise mathematical analysis of the convergence properties of Eq. (11) is beyond the scope of this paper. It will be clear from the following that we require only the existence of the $a_n(k)$ for finite n.

The spectral information contained in the continuum wave function is now coded in the coefficients $a_n(k)$. The description of the scattering phenomena by means of the discrete representation is the subject of the Algebraic Method. In the traditional theory of scattering the asymptotic form of the wave function is obtained by considering the free particle motion. Likewise in the Algebraic Method we shall first study the coefficients $a_n(k)$ in the case where $V = 0$. If we disregard the potential energy contribution in (12) then the Hamiltonian reduces to the radial kinetic energy operator

$$T = -\frac{\hbar^2}{2m}\left(\frac{d^2}{dr^2} + \frac{2}{r}\frac{d}{dr} - \frac{l(l+1)}{r^2}\right) \tag{13}$$

We now introduce basis states $|n>$ consisting of the radial eigenfunctions of the three-dimensional harmonic oscillator

$$< r|n >= R_{n,l}(r) = (-1)^n N_{n,l}(r/b)^l \exp(-r^2/2b^2) L_n^{l+\frac{1}{2}}(r^2/b^2) \tag{14}$$

where b is the length parameter of the oscillator, $L_n^{l+\frac{1}{2}}$ are the Laguerre polynomials [6], and

$$N_{n,l} = \left(\frac{n!2}{\Gamma(n+l+3/2)}\right)^{\frac{1}{2}} b^{-\frac{3}{2}} \tag{15}$$

is the factor that normalizes the state to one. With those states the matrix $< m|T|n >$ becomes tri-diagonal, i.e. $< m|T|n >= 0$ unless $m = n, n \pm 1$, and Eq. (12) becomes the recursion relation

$$T_{m,m-1}a_{m-1,l}(k) + (T_{m,m} - E)a_{m,l}(k) + T_{m,m+1}a_{m+1,l}(k) = 0 \tag{16}$$

with

$$T_{n,n} = \frac{1}{2}(2n + l + 3/2)\hbar\omega \quad T_{n-1,n} = T_{n,n-1} = -\frac{1}{2}[n(n+l+1/2)]^{\frac{1}{2}}\hbar\omega \tag{17}$$

where $\omega = \hbar^2/mb^2$ is the basic oscillator frequency.

It can be shown [7] that the above recursion relation can be transformed into that for the confluent hypergeometric functions $M(a,b;z)$ of reference [6] and that two independent solutions can be obtained for $a_{n,l}(k)$, namely a "regular" solution $a_{n,l} = f_{n,l}$ and an "irregular" solution $a_{n,l} = g_{n,l}$. One finds that

$$\begin{aligned} f_{n,l}(k) &= A_{n,l}k^l \exp(-k^2/2)M(-n, l+3/2; k^2) && (18)\\ g_{n,l}(k) &= B_{n,l}k^{(-l-1)} \exp(-k^2/2)M(-n-l-1/2, -l+1; k^2) && (19) \end{aligned}$$

where

$$A_{n,l} = \left(\frac{2\Gamma(n+l+3/2)}{\Gamma(n+1)}\right)^{\frac{1}{2}} \frac{1}{\Gamma(l+3/2)} \tag{20}$$

$$B_{n,l} = \left(\frac{2\Gamma(n+1)}{\Gamma(n+l+3/2)}\right)^{\frac{1}{2}} \frac{(-1)^l}{\Gamma(-l+1/2)} \tag{21}$$

We shall also be interested in the asymptotic form (for large n) of the expansion coefficients. From the well-known properties of the Laguerre polynomials and of the functions $M(a,b;z)$ (see reference [6]) one finds that

$$f_{n\to\infty,l}(k) \propto (r_n)^{\frac{1}{2}} j_l(kr_n) \tag{22}$$

$$g_{n\to\infty,l}(k) \propto (r_n)^{\frac{1}{2}} n_l(kr_n) \tag{23}$$

where

$$r_n = b\sqrt{4n + l + 3} \tag{24}$$

So we see that the asymptotic form of the radial wave functions and of their expansion coefficients are formally the same. This can be understood by the following heuristic argument. The value of r_n corresponds to the classical turning point of the oscillator in the state $|n, l>$. For high values of n the oscillator wave function is extremely well localized in the neighbourhood of this turning point and therefore it acts as a delta function.

From the above discussion we can derive a computational strategy. We propose to exploit the correspondence between the configuration space of points with coordinates r and the function space of oscillator states with quantum numbers n. In both spaces we can consider the partitioning into an "interior" region, in which the potential is active, and an "exterior" or "asymptotic" region, in which the potential can be neglected. In configuration space the cut-off radius is defined as is done in most of the traditional theories of scattering (R-matrix theory e.g.) and the interior wave function is matched to the asymptotic form, thereby defining the phase shift. In function space we define a cut-off value of n, say N, and assume that the potential matrix elements $< m|V|n >$ can be neglected as soon as n or m is larger than N. As will be shown further, we can then also assume that the expansion coefficients $a_n(k)$ take on their asymptotic form (23) as soon as $n > N$. Matching the interior and exterior components of Ψ_E is then achieved by solving the Schrödinger equation (12).

Let us write again the Schrödinger equation

$$\sum_{n=0}^{N}(T_{m,n} + V_{m,n} - E\delta_{m,n})a_n + \sum_{\nu=N+1}^{\infty}(T_{m,\nu} + V_{m,\nu} - E\delta_{m,\nu})a_\nu = 0 \tag{25}$$

We have written explicitely the matrix elements of kinetic and potential energy separately and split up the summation in interior and exterior contributions.

We shall now define an approximation scheme henceforth referred to as the "Algebraic Method". It is based on two basic assumptions. The first is that the potential V is such that its matrix elements $V_{m,n}$ become negligibly small when m or n are larger than some critical N. The second basic assumption is that for these large values ν of n the expansion coefficients may be replaced by their asymptotic values (23). In coordinate space the asymptotic wave function is

$$\Psi_E^{as}(r) = j_l(kr) + t\, n_l(kr) \tag{26}$$

which defines the normalisation and the phase shift through $t = \tan\delta_l(k)$. Referring to the correspondence between r-space and n-space and Eq. (23), we shall therefore put

$$a_\nu = (r_\nu)^{\frac{1}{2}}[j_l(kr_\nu) + t\, n_l(kr_\nu)] \equiv f_\nu + t\, g_\nu \tag{27}$$

If we now introduce both approximations in the set of homogeneous equations (25), we can distinguish 4 types of equations :

1. for $0 \leq m \leq N$, both kinetic and potential energy contribute :

$$\sum_0^N (T_{m,n} + V_{m,n} - E\delta_{m,n})a_n = 0 \tag{28}$$

2. for $m = N$, the $(N+1)$-st term contains only the T-matrix element and a_{N+1} is replaced by its asymptotic form :

$$\sum_0^N (T_{N,n} + V_{N,n} - E\delta_{N,n})a_n + T_{N,N+1}(f_{N+1} + t\, g_{N+1}) = 0 \tag{29}$$

3. for $m = N+1$, all V-matrix elements are neglected and also a_{N+2} is replaced by its asymptotic value :

$$T_{N+1,N}a_N + (T_{N+1,N+1} - E)(f_{N+1} + t\, g_{N+1}) + T_{N+1,N+2}(f_{N+2} + t\, g_{N+2}) = 0 \tag{30}$$

4. for $N+2 \leq m \leq \infty$, only the asymptotic forms of the coefficients occur:

$$T_{m,m-1}(f_{m-1} + t\, g_{m-1}) + (T_{m,m} - E)(f_m + t\, g_m) + T_{m,m+1}(f_{m+1} + t\, g_{m+1}) = 0 \tag{31}$$

The equations of subset (31) are identical with Eq. (16) and hence automatically satisfied if N is chosen in the exterior region. The subsets (28), (29) and (30) constitute a set of $N+2$ linear inhomogeneous equations determining the $N+2$ unknown coefficients a_n and t.

The extension of the above arguments to a many-body channel depending on one degree of freedom (such as the intercluster distance) requires the discussion of the correct asymptotic behaviour [8]. The treatment of a multi-channel situation is straightforward : one has to generalize from a description in terms of the phase-shift to one in terms of the scattering matrix $S_{i,j}$ [9].

3 Application

We investigate the transfer reaction $h(t,d)\alpha$ at low energies. This reaction is of great interest to astrophysics and thermonuclear physics. Experimental data in the energy region of interest concern the elastic scattering [10] and the so-called "charge asymmetry" effect [11]. Similar reactions such as $h(h,2p)\alpha$, or $t(t,2n)\alpha$, for which experimental data as well as theoretical calculations are available, are not directly suitable for comparison because they involve states of different isospin quantum number. Thus one has to rely strongly on theoretical analysis for the investigation of this reaction.

^6Li decays into the binary cluster channels (α,d) and (h,t) above the thresholds of approximately $1.5MeV$ and $16MeV$ respectively. The monopole channel opens at about $34MeV$. We shall be interested in the energy region just above the (h,t) threshold. Here we compare two calculations : 1) a two channel model in which only the cluster channels are taken into account and 2) a four channel model in which the cluster modes are coupled to the monopole and quadrupole modes described earlier.

Because the first excited state of the pure collective motion is situated slightly above the (h,t) threshold, we expect that there will be an important effect of the collective motion on the scattering and reaction parameters in the cluster channels. We concentrate here on the cross section of the reaction $h(t,d)\alpha$ at low centre-of-mass energy above threshold, i.e. $0 < E < 1MeV$. We use the so-called Modified Hasegawa-Nagata nucleon-nucleon interaction [12] with Majorana exchange parameter $m = 0.397$ and including Coulomb forces.

In order to evaluate the reliability of the method we have first performed the calculations on the elastic scattering in the (d,α) channel, at low energies above threshold. This is a well-established case for which experimental data are available. In figure 1 we present the phase shift for the single-channel calculation, for the calculation including coupling to the (h,t) channel and also for coupling to both (h,t) and collective channels, as a function of the centre-of-mass energy. The results of all theoretical approximations satisfactorily reproduce the phase shift analyses of experimental data [13]. Thus we are confident that our approach is suitable for this type of calculation.

From figure 1 we see that the effect of including the coupling with the collective modes is to increase the phase shift. The qualitative explanation of this result is obvious. When new channels are introduced, the ^6Li ground

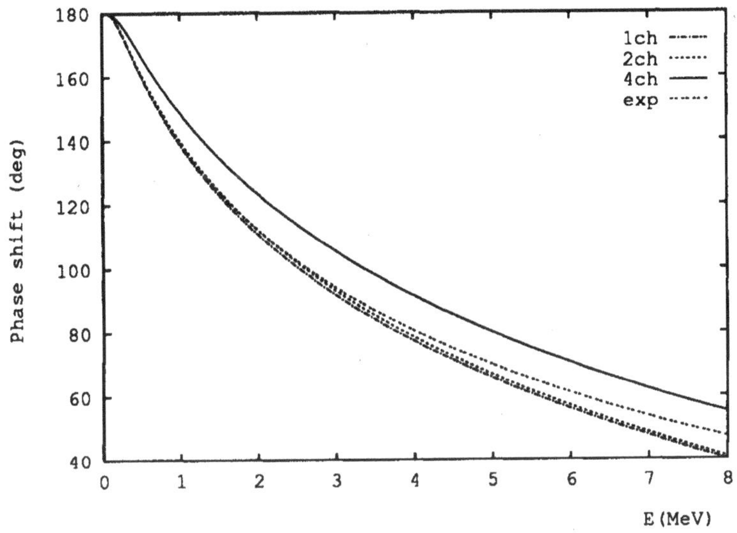

Figure 1: Phase shift δ of elastic (d, α) scattering versus centre-of-mass energy E

state energy is lowered due to the variational principle; thus the scattering length decreases from the large positive value it has when the ground state lies near the threshold. The $^{3}0_1$ phase shift equals π at the energy of the (α, d) threshold, and subsequently decreases. The inclusion of the collective modes is of little consequence here, but this is to be expected as the energy range is well below that of the first excited collective state.

We now turn our attention to the calculations for the energy region above the (h, t) threshold. In this region both the cluster channels are open. We limit ourselves to low energies $(0 < E < 1MeV)$ above the threshold. Here the collective channels are closed. We compare the phase shift for elastic scattering in the (h, t) channel and the cross section for the reaction $h(t, d)\alpha$ in the 2-channel as wel as in the 4-channel approximations. Both quantities result from the solution of the discretized Schrödinger equation in the Algebraic Method as referred to above. In figure 2 we give the phase shift δ. In figure 3 we display the cross section σ.

The cross section increases very fast above threshold and reaches a value of about $30mb$ at $0.4MeV$ above treshold in the 2-channel case. In the 4-channel case there is a dramatic increase to about $280mb$ and the cross section shows a well developed maximum. Also the behaviour of the phase shift reveals the

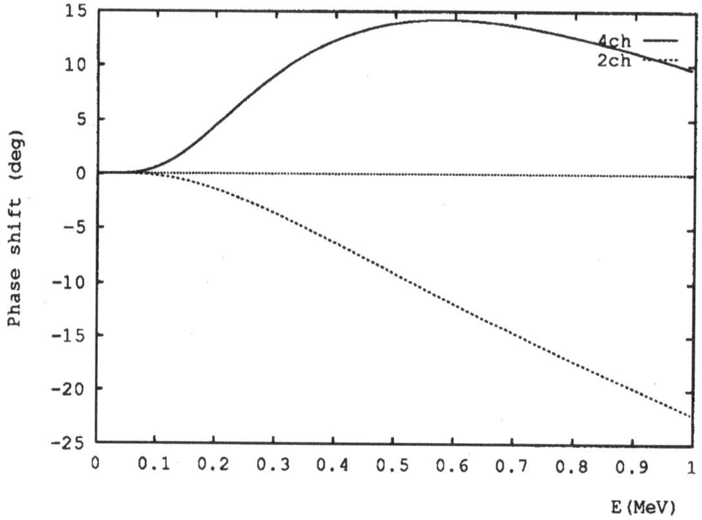

Figure 2: Phase shift δ of elastic (t, h) scattering versus centre-of-mass energy E

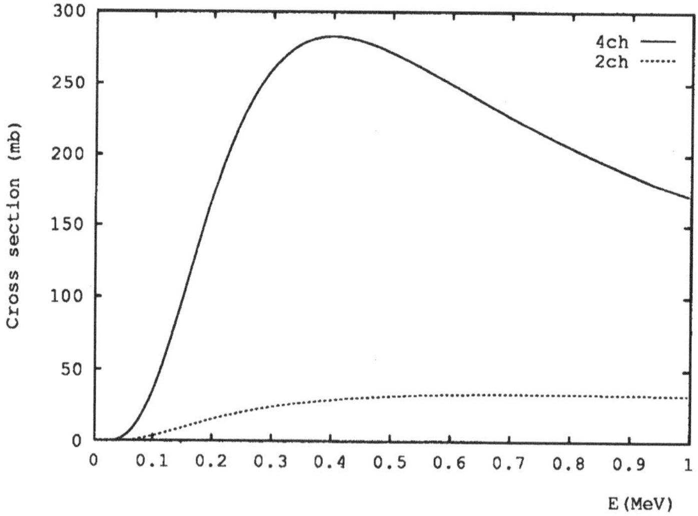

Figure 3: Cross section σ of the reaction $h(t, d)\alpha$ versus centre-of-mass energy E

influence of the collective channels. In the 4-channel calculation it has a broad maximum, rising to about 14 degrees at $0.6MeV$.

The cross-section shows a well pronounced peak. This indicates the existence of a resonance above the (h,t) threshold, the origin of which must be ascribed to the vicinity of a quasi bound state in the collective spectrum. If the quasi-bound states associated with the monopole and quadrupole modes are considered i.e. without coupling to the open channels, it is found that they heve a first excited state at an excitation energy of approximately $19MeV$, which is close to the (h,t) threshold. Therefore we expect the coupling to these modes to the cluster channels to affect the cross section strongly. The same effect is also seen in the elastic scattering where the negative scattering length indicates the presence of a quasi-bound state near the threshold. One can also compare with the experimental data in the paper by Nocken [11] on the differential cross section. By crude averaging over the angular dependence one may infer that our calculation, at peak value, exceeds the experimental value by a factor of about fifteen. However, the point of this paper is to show that certain collective channels can give sizeable effects and to draw attention to this coupling. As always in nuclear physics, when deviations with experiment occur we have the dilemma of ascribing these to the model or to the nuclear force that is used. We have ascertained that we could manipulate the parameters of the interaction so as to fit the experiment much more closely, but instead we choose to focus on bringing out the relevance of the collective subspace.

Another important quantity that also exhibits the collective effects discussed above is the astrophysical factor $S(E)$ related to the reaction cross section by the simple relation $S(E) = E\sigma(E)\exp(2\pi\kappa)$ (where κ is the Coulomb parameter) A polynomial fit to the cross-section in the 2 channel calculation yields (in units $keVb$) for the 2-channel respectively 4-channel calculation

$$S^{(2)}(E) = 949.4 - 1.2E - 0.0004E^2 + ... \quad (32)$$

$$S^{(4)}(E) = 7715.2 + 0.6E - 0.0152E^2 + ... \quad (33)$$

We conclude that inclusion of the collective modes leads to a significant increase. These values can be compared to $200keVb$ for $t(t,2n)\alpha$ and $5500keVb$ for the $h(h,2p)\alpha$ reactions, each at their threshold.

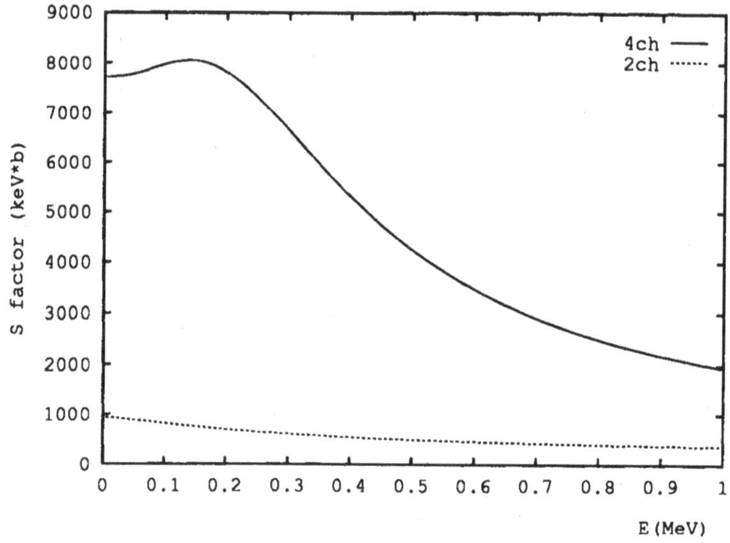

Figure 4: Spectroscopic factor of the reaction $h(t,d)\alpha$ versus centre-of-mass energy E

4 Conclusion

In this paper we have proposed a study of the continuum of light nuclei in terms of a joint description of cluster and collective modes. We have used the Algebraic Method for scattering which treats both kinds of modes in a uniform way. The feasability of this method is demonstrated by an application to an energy region in the continuum of the ^6Li system.

We have shown that, at energies where a channel of a binary process is open and the collective channels are closed, there exists a coupling mechanism whereby the collective degrees of freedom are excited. This excitation reflects back on the binary process. In cases where the collective excitation energy is close to the threshold of the binary reaction the effects can be quite drastic, resulting in a considerable enhancement of the cross section.

References

[1] R. Ceuleneer and P. Vandepeutte, *Phys. Rev.* **C31** (1985) 1528

[2] K. Wildermuth, Y. C. Tang, *A Unified Theory of the Nucleus* (Vieweg, Braunschweig, 1977)

[3] F. Arickx et al., *Nucl. Phys.* **A318** (1979) 269;
Ibid., **A377** (1982) 121

[4] J. Heller and H. A. Yamani, *Phys. Rev.* **A9** (1974) 1201

[5] G. F. Filippov and I. P. Okhrimenko, *Sov. J. Nucl. Phys.* **32** (1981) 480

[6] M. Abramowitz and I. A. Stegun, *Handbook of Mathematical Functions* (National Bureau of standards, 1964)

[7] Y. I. Nechaev and Y. F. Smirnov, *Sov. J. Nucl. Phys.* **35** (1982) 808

[8] G. F. Filippov, *Sov. J. Nucl. Phys.* **33** (1981) 488

[9] V. S. Vasilevsky and Yu. Rybkin,, *Sov. J. Nucl. Phys.* **50** (1989) 411

[10] R. Vlastou et al., *Nucl.Phys.* **A292** (1977) 29;
N. Jarmie,*Phys. Rev.* **C16** (1985) 2151; ibid. **C15** (1985) 1613;
W. Grüebler et al., *Nucl.Phys.* **A242** (1975) 285; ibid. **A334** (1980) 365.

[11] E. E. Gross et al., *Phys. Rev.* **C5** (1972) 602;
U. Nocken et al., *Nucl. Phys.* **A213** (1973) 97;
Schnieder et al., *Nucl. Phys.* **A247** (1975) 203;
G. Rai et al., *Phys. Rev.* **C38** (1988) 2036;
M. Bruno et al., *Phys. Rev.* **C38** (1988) 521; ibid. **C43** (1991) 99

[12] A. Hasegawa and S. Nagata, *Prog. Nucl. Phys.* **45** (1971) 1786

[13] V. M. Krasnopolsky et al., *Phys. Rev.* **C43** (1991) 822

HARTREE-FOCK APPROXIMATION TO NUCLEAR SYSTEMS WITH RELATIVISTIC NON-LINEAR MODELS

P. Bernardos[a], V.N. Fomenko[b], N.Van Giai[c], M.L. Quelle[d], S. Marcos[a], R. Niembro[a] and L.N. Savushkin[e]

[a] Departamento de Física Moderna, [d] Departamento de Física Aplicada
Universidad de Cantabria, 39005 Santander, Spain
[b] Department of Mathematics, Institute for Railway Engineering, St.Petersburg, Russia
[c] Division de Physique Théorique, Institut de Physique Nucléaire, F-91406, Orsay, France
[e] Department of Physics, Institute for Telecommunications, 191065 St. Petersburg, Russia

Abstract

Scalar field self-interaction terms are included in the framework of a relativistic $\sigma + \omega + \pi + \rho$ Hartree-Fock approach to nuclear structure. The model provides a good description of binding energies and charge densities, improving previous relativistic Hartree and Hartree-Fock approximations.

1. Introduction

The most important nuclear ground state properties have been qualitatively reproduced within the linear $\sigma + \omega$ relativistic model suggested by J.D. Walecka[1]. The original version of the model included only Yukawa-type mesonic fields coupled directly to the corresponding nucleon currents. However, in order to obtain a more accurate quantitative description it appeared necessary to extend the Walecka model by introducing into the Lagrangian some additional density dependence terms.

In the non-linear $\sigma + \omega$ model[2,8], this is given by the self-interactions of the scalar field responsible for many-body forces which play an important role in the nuclear matter saturation properties. The form of the non-linear self-interactions is prompted by requirement of a renormalizable theory[3]. One further motivation of a Lagrangian containing non-linear self-couplings may be found in the framework of chiral symmetry[4].

Extended investigations of nuclear matter and finite nuclei properties in the non-linear $\sigma + \omega$ model have been carried out in the Dirac-Hartree approximation[5,6]. In these papers it was shown that for doubly-magic nuclei the model gives a rather good description of nuclear properties. In spite of the success of this model, we must point out some shortcomings. The absence of exchange terms is responsible of a too small value of the symmetry energy if the experimental g_ρ coupling constant value is chosen. Furthermore, the effective spin-spin interaction coming from the nucleon identity through the antisymmetrization of the wave function is missing. This effect is more important in spin-unsaturated systems[7]. Finally, though the spin-orbit force is globally rather well described, the model is unable to reproduce reasonably well the reduction of d-states splitting when going from ^{40}Ca to ^{48}Ca.

On the other hand, in refs.[8,9] the structure of infinite nuclear matter and finite nuclei has been investigated in the Dirac-Hartree-Fock (DHF) approximation, where self-interactions were not considered. A satisfactory description of the single-particle binding energies and densities was obtained for doubly-magic nuclei. The exchange terms were shown in this investigation to play an essential role in determining the symmetry energy and spin-orbit interaction. However, the nuclei were not bound enough, this defect being more important for light nuclei. This fact is connected with the high compression modulus K of nuclear matter inherent to this model. A better agreement with experiment should be expected by introducing additional density-dependent terms in the Lagrangian in order to decrease the value of K.

In this contribution we consider an effective Lagrangian density $\mathcal{L}$ containing scalar field self-interactions (SI) and we develop a Dirac-Hartree-Fock approach to nuclear matter and finite nuclei. Although in ref.[10] an attempt was made to take into account the exchange effects in the framework of the non-linear $\sigma + \omega$ model, it was only applied to calculate the equation of state of dense nuclear and neutron matter.

2. The model

The effective Lagrangian density $\mathcal{L}$ is a sum of a free Lagrangian $\mathcal{L}_0$ and an interaction Lagrangian $\mathcal{L}_I$. $\mathcal{L}$, except for the scalar part in $\mathcal{L}_0$, is identical to the corresponding one in ref.[8]. Now $\mathcal{L}_0$ contains the phenomenological potential-energy functional[5]:

$$U(\sigma) = \frac{1}{2}m_\sigma^2\sigma^2 + \frac{1}{3}b\sigma^3 + \frac{1}{4}c\sigma^4. \tag{1}$$

With the notation of ref.[8], that will be used throughout this paper, $\mathcal{L}_I$ reads:

$$\mathcal{L}_I = -g_\sigma\bar{\psi}\sigma\psi - g_\omega\bar{\psi}\gamma^\mu\omega_\mu\psi - g_\rho\bar{\psi}\gamma^\mu\vec{\rho}_\mu.\vec{\tau}\psi - \frac{f_\rho}{2M}\bar{\psi}\sigma^{\mu\nu}\partial_\mu\vec{\rho}_\nu.\vec{\tau}\psi - \frac{f_\pi}{m_\pi}\bar{\psi}\gamma_5\gamma^\mu\partial_\mu\vec{\pi}.\vec{\tau}\psi. \tag{2}$$

The Euler-Lagrange equations give the Dirac equation for the nucleon field:

$$\left[i\gamma^\mu\partial_\mu - M - g_\sigma\sigma - g_\omega\gamma^\mu\omega_\mu - g_\rho\gamma^\mu\vec{\rho}_\mu.\vec{\tau} - \frac{f_\rho}{2M}\sigma^{\mu\nu}\partial_\mu\vec{\rho}_\nu.\vec{\tau} - \frac{f_\pi}{m_\pi}\gamma_5\gamma^\mu\partial_\mu\vec{\pi}.\vec{\tau}\right]\psi(x) = 0, \tag{3a}$$

whereas for the scalar field we have:

$$(\Box + \hat{m}_\sigma^{*2})\sigma + g_\sigma\bar{\psi}\psi = 0, \tag{3b}$$

where $\hat{m}_\sigma^*$ is given by the relation $\hat{m}_\sigma^{*2} = \frac{1}{\sigma}\frac{\partial U(\sigma)}{\partial\sigma}$.

Eq. (3a) enables us to write down an equivalent Hamiltonian density corresponding to Lagrangian $\mathcal{L}$ as: $\mathcal{H} = \mathcal{H}_0 + \mathcal{H}_1$, where $\mathcal{H}_0$ is linear in the meson fields:

$$\mathcal{H}_0 = \bar{\psi}(-i\vec{\gamma}.\vec{\nabla} + M)\psi + \frac{1}{2}\bar{\psi}(g_\sigma\sigma + g_\omega\gamma^\mu\omega_\mu + ...)\psi, \tag{4}$$

whereas $\mathcal{H}_1 = U - \frac{1}{2}\sigma\frac{\partial U}{\partial\sigma}$ contains only non-linear terms like σ^3 and σ^4. The non-linear character of $\mathcal{H}_1$ makes the calculation of their exchange contribution to the energy rather complicated.

In order to write down the Dirac equation for nucleons in a DHF approach, one could use eq. (3b), and similar equations for other mesons, to eliminate the meson fields in eq. (3a). In practice, this is difficult because $\hat{m}_\sigma^*$ is an operator. In this work we consider the static limit of the meson fields and we make the approximation of replacing the σ field in $\hat{m}_\sigma^*$ and $\mathcal{H}_1$ (but not in $\mathcal{H}_0$) by its ground state expectation value σ_0. Then, eq. (3b) reads:

$$(\Box + m_\sigma^{*2})\sigma + g_\sigma \overline{\psi}\psi = 0, \tag{5}$$

where $m_\sigma^{*2} = m_\sigma^2 + b\sigma_0 + c\sigma_0^2$. The quantity m_σ^* plays the role of the effective mass of the σ meson and becomes the bare σ mass in absence of non-linearity ($b = c = 0$). Similarly,

$$\mathcal{H}_1 = -b\frac{1}{6}\sigma_0^3 - c\frac{1}{4}\sigma_0^4. \tag{6}$$

We have thus linearized the scalar meson field equation with respect to the σ operator. Then, the σ field can be cast in the following form:

$$\sigma = -g_\sigma \int S_\sigma(x,y)\overline{\psi}(y)\psi(y)d^4y, \tag{7}$$

where S_σ is the scalar propagator corresponding to eq. (5).

For the $\omega_\mu, \vec{\pi}$ and $\vec{\rho}_\mu$ fields one can write similar expressions to eq. (7) with the corresponding propagators S_i and source terms. Note that the σ effective mass m_σ^* depends on r and consequently S_σ does not have the simple Yukawa form of S_i ($i = \omega, \pi, \rho$).

Introducing expression (7) for the σ field and the corresponding ones for the $\omega_\mu, \vec{\pi}$ and $\vec{\rho}_\mu$ fields in the Dirac equation and $\mathcal{H}_0$, we can write both of them in a form where only the nucleon fields $\psi(x)$ are present[11]. Then, the Dirac equation can be obtained from the Hamiltonian density $\mathcal{H}_0$. Thus, in our approach, $\mathcal{H}_1$ plays the role of a first order perturbation.

The nucleon field $\psi(x)$ can be expanded on a complete set of stationary single-particle Dirac spinors $\{ f_\alpha(\vec{x})e^{-iE_\alpha t} \}$. Assuming the tree approximation and the static limit for the meson field, we obtain the Dirac equation for the spinor $f_\alpha(\vec{x}_1)$:

$$\left[-i\vec{\alpha}.\vec{\nabla} + \beta M + \beta\Sigma(\vec{x}_1)\right] f_\alpha(\vec{x}_1) = E_\alpha f_\alpha(\vec{x}_1). \tag{8}$$

This equation is formally identical to the corresponding one of ref.[8]; *only the σ-propagator is different.*

In order to calculate the self-energy Σ entering the Dirac equation, we can use the expressions developed in ref.[8] replacing the Yukawa propagator of the σ meson by the new one appearing in eq. (7). In nuclear matter Σ is momentum dependent and can be written as[11]:

$$\Sigma(p) = \Sigma_s(p) + \gamma_0\Sigma_0(p) + \vec{\gamma}.\hat{p}\Sigma_v(p). \tag{9}$$

The total binding energy is obtained by adding to the corresponding part of $\mathcal{H}_0$, the non-linear contribution coming from the energy density $\mathcal{H}_1$. We can write it as:

$$E = \frac{1}{2}\sum_{\alpha}(T_\alpha + E_\alpha - M) + \int \mathcal{H}_1(\sigma_0)d^3r, \tag{10}$$

where T_α is the relativistic kinetic energy of the orbital α.

3. Results

We have carried out calculations of infinite and finite systems considering three models. In the first one (Hartree-Fock with self-interactions, or HFSI) the $\delta(\vec{r})$ parts of the π and ρ induced interactions have been removed from the Hamiltonian as in ref.[8]. The $\delta(\vec{r})$ components have been taken into account, however, in the other two models: the linear case HF(δ), for which $b = c = 0$, and the non-linear one HFSI(δ).

3.1 Nuclear Matter

In our calculations, the bare nucleon mass, the π, ω and ρ meson masses and the f_π and g_ρ coupling constants assume their empirical values: M=939 MeV, m_π=138 MeV, m_ω=783 MeV and m_ρ=770 MeV, $f_\pi^2/4\pi$=0.08 and $g_\rho^2/4\pi$=0.55. The ratio f_ρ/g_ρ=3.7, as given by the vector meson dominance model, was taken for model HFSI (this choice corresponds to set (e) of ref.[8]) while we have taken f_ρ/g_ρ=6.6, as given by πN scattering data, for HF(δ) and HFSI(δ) models.

The values of g_σ, b, c are determined by the saturation condition of symmetric nuclear matter: ρ_0=0.14 fm^{-3} for HFSI model and ρ_0=0.1484 fm^{-3} for HF(δ) and HFSI(δ) models, E/A=-15.75 MeV, and the compression modulus K. Finally, m_σ is adjusted to get the right r.m.s. charge radius of ^{16}O, while g_ω is chosen to have reasonable values of spin-orbit splittings.

In the following, we shall always refer to the results of ref.[8] [case (e)] for comparison. Table 1 summarizes some nuclear matter properties. It shows that, in order to bring about a realistic K value, the models with self-interactions need only a small renormalization in g_σ and g_ω coupling constants, that can justify to treat $\mathcal{H}_1$ in a perturbative form.

Table 1. Adjusted parameters of the model and properties of symmetric nuclear matter. $\bar{b} = b/Mg_\sigma^3$ and $\bar{c} = c/g_\sigma^4$ are dimensionless parameters[5]. The value of the nucleon scalar effective mass M^* is given at the Fermi surface. a_4 is the symmetry energy parameter.

Model	$g_\sigma^2/4\pi$	$g_\omega^2/4\pi$	$m_\sigma(MeV)$	$\bar{b}*10^3$	$\bar{c}*10^3$	$K(MeV)$	M^*/M	$a_4(MeV)$
ref.[8]	4.16	11.18	440	0	0	465	0.56	38.6
HF(δ)	7.19	8.22	525	0	0	399	0.60	35.0
HFSI	4.005	10.4	412	-6.718	-14.61	250	0.61	35.0
HFSI(δ)	6.92	7.5	515	-1.52	-2.62	300	0.63	33.9

The momentum dependence of the self-energy components for HFSI model is shown in Fig. 1. The Σ_s and Σ_0 components depend very little on the momentum and they are very similar in all models. The Σ_v component has a stronger momentum dependence. In HF(δ) and HFSI(δ) models it remains positive for all momentum values[11].

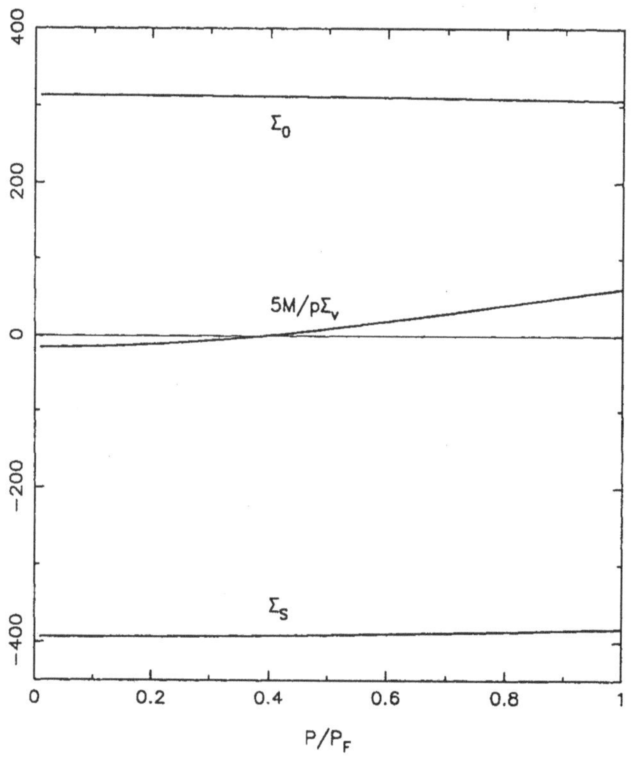

Fig. 1. Components of the self-energy Σ (in MeV) for HFSI model as functions of the nucleon momentum.

3.2 Finite Nuclei

We have performed calculations of the doubly-closed shell nuclei ^{16}O, ^{40}Ca, ^{48}Ca, ^{90}Zr and ^{208}Pb assuming spherical symmetry. The set of DHF equations is solved in coordinate space following the method of ref.[8]. In Table 2 are shown some calculated and experimental finite nuclei properties. We have not calculated the center-of-mass corrections (C.M.) to the total binding energy. As indicative values, the non-relativistic center-of-mass corrections taken from ref.[13] are also shown.

As we have mentioned before, the main problem of ref.[8] is the small total binding energy. We can see that all the present models (HF(δ), HFSI, HFSI(δ)) improve this property and the spin-orbit splittings in ^{16}O and ^{40}Ca too, especially the HFSI model. As for the ^{48}Ca nucleus, all models reproduce rather well the spectacular reduction of the d-state spin-orbit splitting. This is done, essentially, by the isovector meson contributions.

Table 2. Comparison of present calculations (HF(δ), HFSI and HFSI(δ)), including Coulomb, with experimental values and ref.[8] for finite nuclei. The energies are given in MeV and the rms charge radii r_c are in fm.

	^{16}O			^{40}Ca			^{48}Ca			^{90}Zr		^{208}Pb	
	$-E/A$	r_c	Δ_{LS}	$-E/A$	r_c	Δ_{LS}	$-E/A$	r_c	Δ_{LS}	$-E/A$	r_c	$-E/A$	r_c
ref.[8]	5.61	2.73	7.3	6.82	3.47	8.0	7.10	3.47	4.1	7.40	4.26	6.74	5.47
HF(δ)	6.36	2.74	6.17	7.73	3.46	7.33	7.96	3.47	3.55	8.34	4.24	7.74	5.44
HFSI	7.43	2.73	6.4	8.33	3.48	7.05	8.45	3.48	3.27	8.58	4.26	7.78	5.52
HFSI(δ)	6.97	2.73	5.62	8.22	3.44	6.88	8.37	3.45	3.12	8.69	4.21		
EXP	7.98	2.73	6.3	8.55	3.48	7.2	8.67	3.47	4.3	8.71	4.27	7.87	5.5
C.M.	0.61			0.20			0.18			0.08		0.02	

Although the different models give very similar charge radii, there are differences in the details of predicted charge distributions. In Fig. 2 are displayed charge densities calculated with the three models considered in this paper. Experimental charge densities are shown for comparison. The results of the model without self-interactions (HF(δ)) follow more closely the experimental data in the nuclear interior region. The HFSI densities exhibit more oscillations due to pronounced shell effects. This may be related to the lower compression modulus of HFSI compared to that of HF(δ) and HFSI(δ). However, the lack of the zero-range $\delta(\vec{r})$ interaction in HFSI probably contributes also to increase these oscillations. All models reproduce satisfactorily well the surface and tail regions of the charge densities.

On the whole, it is clear that the inclusion of self-interaction effects together with contributions of π and ρ mesons (both Hartree and Fock terms) improves substantially the overall description of nuclear ground state properties.

4. Conclusions

Although relativistic phenomenological models provide a rather good global picture of the nuclear properties, it was not possible until now to obtain at the same time a good description of the spin-orbit interaction and the total binding energy. In this work we have considered, as a starting point, an effective Lagrangian density containing scalar field self-interactions and we have developed a DHF approach in order to investigate the problem within this framework.

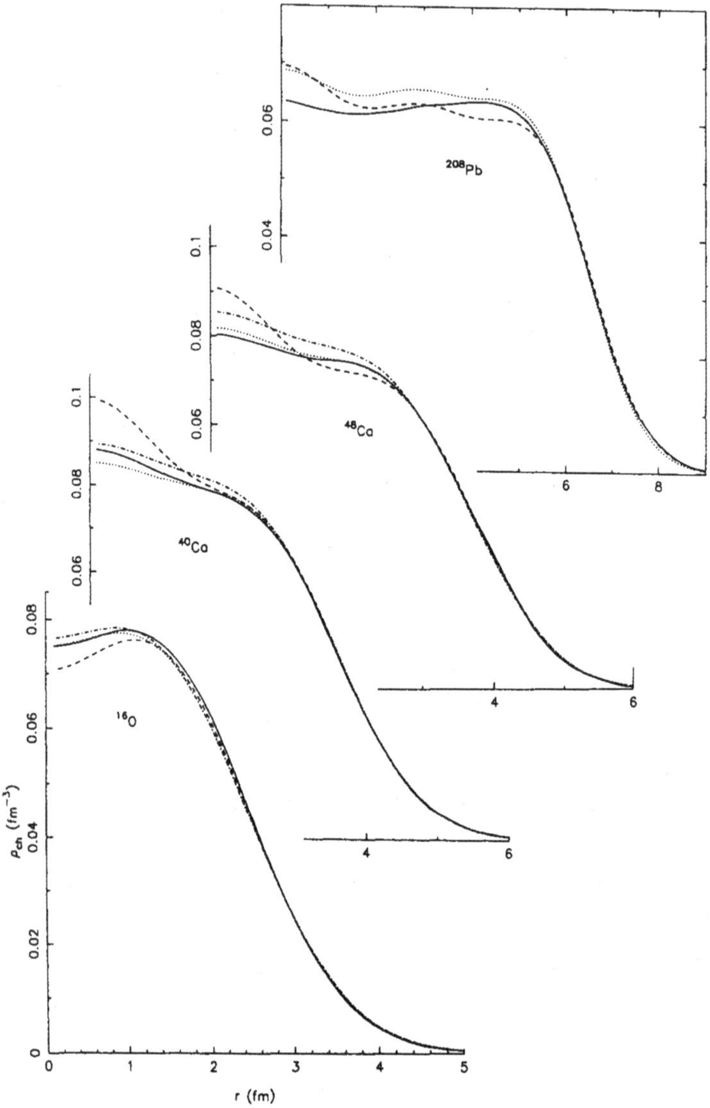

Fig. 2. Calculated and experimental[14] (solid line) charge distributions for the indicated nuclei. The doted, dashed and doted-dashed lines correspond to HF(δ), HFSI and HFSI(δ) models, respectively.

One of the manifestations of the scalar field self-interactions is the dressing of the scalar meson in the nuclear medium. In this case the σ-meson mass is different from the bare mass and becomes density dependent. The self-interactions introduce also a strong density dependence in the energy density that lowers the value of the compression modulus and improve the total binding energy in finite nuclei. Furthermore, the exchange terms and the isovector mesons are very important for a quantitative description of the symmetry energy in nuclear matter and the spin-orbit interaction in nuclei.

Although in the present investigation we have not made a systematic search of the best parameter set by doing an optimun fit to the data we have found a good description of the nuclear ground state properties in finite nuclei which improves the existing non-linear Dirac-Hartree as well as the linear DHF models.

The method proposed here in order to include scalar self-interactions in the framework of a DHF approach can also be applied to chiral models[14] and other relativistic non linear models.

5. Acknowledgments

This work is dedicated to the memory of *Alain Bouyssy*, who was for many years to us a source of inspiration in physics and of friendship in life.

6. References

1. J.D. Walecka, *Ann. Phys.* **83** (1974) 491. B. D. Serot and J.D. Walecka, Adv. *Nucl. Phys.* **16** (1986) 1.
2. J. Boguta and A.R. Bodmer, *Nucl. Phys.* **A292** (1977) 413. J. Boguta and S.A. Moszkowski, *Nucl. Phys.* **A403** (1983) 445.
3. S. A. Chin, *Ann. Phys.* **108** (1977) 301.
4. W. Bentz, L.G. Liu and A. Arima, *Ann. Phys.* **188** (1988) 61.
5. A. Bouyssy, S. Marcos and Pham Van Thieu, *Nucl. Phys.* **A422** (1984) 541.
6. M. Rufa, P.G. Reinhard, J.A. Maruhn, W. Greiner and M.R. Strayer, *Phys. Rev. C* **38** (1988) 390.
7. S. Marcos, R. Niembro, M.L. Quelle and J. Navarro *Phys. Lett.* **271** (1991) 277.
8. A. Bouyssy, J.F. Mathiot, N.V. Giai, and S. Marcos, *Phys. Rev. C* **36** (1987) 380.
9. A. Bouyssy, S. Marcos, J.F. Mathiot, and N.V. Giai, *Phys. Rev. Letts.* **55** (1985) 1731.
10. J. Ramschutz, F. Weber and M.K. Weigel, *J. Phys. G: Nucl. Part. Phys.* **16** (1990) 987; F. Weber and M.K. Weigel, *J. Phys. G: Nucl. Part. Phys.* **15** (1989) 765.
11. P. Bernardos, V. Fomenko, N.V. Giai, M.L. Quelle, S. Marcos, R. Niembro and L.N. Savushkin, to appear in *Phys. Rev. C* .
12. X. Campi and D. W. L. Sprung, *Nucl. Phys.* **A194** (1972) 401.
13. B. Frois, *Proc. of the International Conference on Nuclear Physics*, Florence, Italy, 1983, ed. by P. Blasi and R.A. Ricci (Tipografia Compositori, Bologna, 1983), p. 221.
14. V.N. Fomenko, L.N. Savushkin, S. Marcos, R. Niembro, M.L. Quelle, Preprint UCS/FMN 93-2, Santander (1993).

MODEL INDEPENDENT SINGLE PARTICLE WAVE FUNCTIONS AND INFORMATION THEORY

M. CASAS[1], A. PLASTINO[2] and A. PUENTE[1]

[1]Departamento de Física, Universitat de les Illes Balears, 07071 Palma de Mallorca, Spain

[2]Physics Department, National University La Plata, CC 67, 1900 La Plata, Argentina

Abstract

A method for the construction of single particle wave functions is presented. The formalism is based on the maximization of a quantal pseudoentropy, subject to the constraints posed by the knowledge of some experimental expectation values of each shell. The algorithm is tested with the proton level of some light nuclei.

1. Introduction

The mean field theories provide an effective one particle description of many fermion systems. In the nuclear physics case, several mean field models, running from the simple Woods-Saxon potential to the more sophisticated density-dependent Hartree-Fock (DDHF), provide an accurate description of the ground state (gs) properties of both spherical and deformed nuclei[1,2].

Several zero range interactions[1,3−5] can be used in the DDHF calculations, and all of these give very similar predictions for both the charge density[1] and the momentum distribution of the g.s.[6], suggesting that its gross features are only dependent on very general aspects of the mean field.

Using different Skyrme interactions in a DDHF calculation one may find significant differences concerning the single-particle (sp) properties i.e. the sp radii and energies.

Until the last years, this was not an important problem, because the only experimental informations on single particle states were the quantum numbers, single particle energies and radii of a few states around the Fermi surface[7,8].

Recently, by recourse to the lepton-nucleus scattering in the quasi-elastic channel, using the (e, e' p) reactions one can obtain information about the sp properties like valence orbits and also the deeper lying levels. From the structure function, the momentum distribution and the mean square radius $< r^2 >$ of each proton shell of some spherical nuclei, have been recently available[9]. In many cases these $< r^2 >$ values are adjusted by selecting a different Woods-Saxon well for each orbit[10]. It would be desirable to have a unified theoretical description for all shells and this would be important to decide what is the interaction that better reproduce the experimental results.

The aim of this work is to present a method based on a modified Maximum Entropy Principle (MEP), that allows to construct the single particle wave function (spwf) of a given nucleus from the knowledge of a few sp expectation values. These sp wave functions must be model independent if one employs a model-independent experimental input.

The plan of the article is as follows: in Section 2 we present a summary of the method for inferring wave functions by using a modified Maximum Entropy Principle, in order to describe pure states. In section 3 we present a test of our method using the mean values of different operators obtained in a DDHF calculation with the SIII interaction. In particular we have applied our method to some spherical light nuclei: ^4He, ^{16}O and ^{40}Ca, and we emphasize the fact that with an appropriate choice of the operators one can guarantee the correct asymptotic behavior of the wf. Finally some conclusions are drawn in Section 4.

2. The inference of the single particle wave functions.

It has been shown[11] that the g.s. of a system described by a Hamiltonian $\hat{H}$ can be inferred in the form

$$|g.s.>=\sum_j C_j^{(0)}|j> \tag{1}$$

with

$$|C_j^{(0)}|^2 = exp\{-[\lambda_0+\sum_{i=1}^{N}\lambda_i g_i(j)]\} \tag{2}$$

where $g_i(j) =< j|\hat{g}_i|j >$ are the expectation values of N conmuting operators, diagonal in the basis $|j>$ and λ_i are a set of optimizable parameters.

Taking

$$\lambda_0 = ln\sum_j exp[-\sum_{i=1}^{N}\lambda_i g_i(j)] \tag{3}$$

we ensure the normalization of (1) and the expression (2) maximizes the *quantal entropy*[11–12]

$$S=-\sum_j |C_j^{(0)}|^2 ln|C_j^{(0)}|^2 \tag{4}$$

subject to the constrains

$$g_i =< g.s.|\hat{g}_i|g.s. > \tag{5}$$

The Lagrange multipliers λ_i are to be obtained as the solutions of the system of equations

$$\frac{\partial\lambda_0}{\partial\lambda_i} = -g_i \tag{6}$$

The information entropy (4) differs from the thermodynamic entropy and measures the lack of information concerning the probability distribution of the inferred wave function.

The method can be applied to infer the radial part $u_{nl}(r)/r$ of the sp wave function of the spherical nuclei.

We have taken

$$u_{nl}(r) = r^{n+l+1} exp[-1/2(\lambda_0 + \sum_{i=1}^{N} \lambda_i g_i(r))] \qquad (7)$$

where the principal quantum number starts at $n = 0$ and the $(l+1)$ factor accounts for, according to Bayesian rules[13], our knowledge concerning the behaviour of the radial wf near the origin. Notice that the wf proposed in (7) satisfies a relative maximum-entropy principle[14].

For the single particle states with the same angular momentum l and different quantum number n, the position of the pertinent zeros must be obtained by recourse to the Gramm-Schmidt procedure. So that a set of functions with the factors $r^{l+1}, r^{l+2}, r^{l+3}$, and the $\lambda_i's$ parameters corresponding to the sp level with $n = 0$ is orthogonalized. In this way, approximate values of the nodes of the spwf are obtained[14–16]. By recourse to the expectation values $g_i(r)$ corresponding to the sp states with $n > 0$, we optimize the wave functions of these states by adjusting the Lagrange multipliers λ_i. In this step we loss some orthogonality; nevertheless the iteration of this process up to the selfconsistency gives inferred wave functions of excellent quality, whereas the orthogonality is preserved.

Assuming spherical symmetry and making the Fourier transform the same recipes hold in k space, replacing the r-variable by the momentum k.

Our approach allows to construct the spwf in r or in k space by recourse to a small knowledge of information in these spaces.

3. Applications

In order to test the possibilities of this MEP inference approach, we take as input values g_i of Eq. (7) those obtained in a DDHF calculation, using the SIII interaction[3]. For simplicity, we work with two operators ($N = 2$) taking the most simple functions in r and k-space, $g_1(r) = r$ and $g_2(r) = r^2$, and the same choice in k-space.

We have applied our method in order to infer the spwf of all proton levels of doubly-closed shell nuclei: ^4He, ^{16}O and ^{40}Ca. Using only the information concerning $< r >$ and $< r^2 >$ (or $< k >$ and $< k^2 >$) obtained with a DDHF calculation and the SIII interaction, one can infer the pertinent wf whose overlaps with the HF exact ones are shown in Table I. Notice that all of these overlaps are of the order of a 99%, the worst result corresponds to the $1S_{1/2}$ of ^{40}Ca, this is due to the loss of orthogonality in the process described in Section 2. The iteration of this process up to the selfconsistency improves the quality of the results as it is shown in Fig. 1 where we have plotted the HF, the inferred and the selfconsistent probability densities for the $1S_{1/2}$ proton level of ^{40}Ca in r-space.

The quality of the results in r and k-space is also shown in Figs. 2a and 2b where the HF probability densities of the $0P_{3/2}$ proton level of ^{16}O in r and k space are compared to the corresponding MEP densities. The discrepancies between the

TABLE I. Overlaps between HF and inferred wave functions corresponding to all proton levels of ^4He,^{16}O and ^{40}Ca using $g_1(r) = r$ and $g_2(r) = r^2$.

	Shell	r- space	k-space
^4He	$0S_{1/2}$	0.9991	0.9991
^{16}O	$0S_{1/2}$	0.9998	0.9997
	$0P_{3/2}$	0.9993	0.9989
	$0P_{1/2}$	0.9991	0.9996
^{40}Ca	$0S_{1/2}$	0.9998	0.9986
	$0P_{3/2}$	0.9997	0.9987
	$0P_{1/2}$	0.9999	0.9994
	$0D_{5/2}$	0.9995	0.9981
	$1S_{1/2}$	0.9917	0.9887
	$0D_{3/2}$	0.9993	0.9995

TABLE II. Comparison between HF and inferred values $< r^4 >$ and $< k^4 >$ for each single particle proton level of ^4He,^{16}O and ^{40}Ca using $g_1(r) = r$ and $g_2(r) = r^2$.

	Shell	Quantal $< r^4 >$	Inferred $< r^4 >$	Quantal $< k^4 >$	Inferred $< k^4 >$
^4He	$0S_{1/2}$	21.65	20.82	0.739	0.720
^{16}O	$0S_{1/2}$	33.69	33.90	0.413	0.409
	$0P_{3/2}$	79.66	78.65	0.988	0.986
	$0P_{1/2}$	97.44	95.31	0.870	0.866
^{40}Ca	$0S_{1/2}$	67.30	67.93	0.198	0.197
	$0P_{3/2}$	131.62	132.23	0.556	0.558
	$0P_{1/2}$	126.81	126.84	0.583	0.586
	$0D_{5/2}$	215.70	214.86	1.138	1.140
	$1S_{1/2}$	293.47	284.13	1.245	1.247
	$0D_{3/2}$	235.18	232.52	1.085	1.091

TABLE III. Comparison between HF and inferred values $< r^2 >$, $< r^6 >$ for each single particle proton level of ^{16}O using $g_1(r) = r$ and $g_2(r) = rexp(-\alpha r^2)$ with ($\alpha = 0.04$).

Shell	Quantal $< r^2 >$	Inferred $< r^2 >$	Quantal $< r^4 >$	Inferred $< r^4 >$	Quantal $< r^6 >$	Inferred $< r^6 >$
$0S_{1/2}$	4.553	4.552	33.69	33.84	345.31	349.68
$0P_{3/2}$	7.565	7.561	79.66	78.66	1120.0	1051.6
$0P_{1/2}$	8.164	8.155	97.44	95.37	1640.7	1483.9

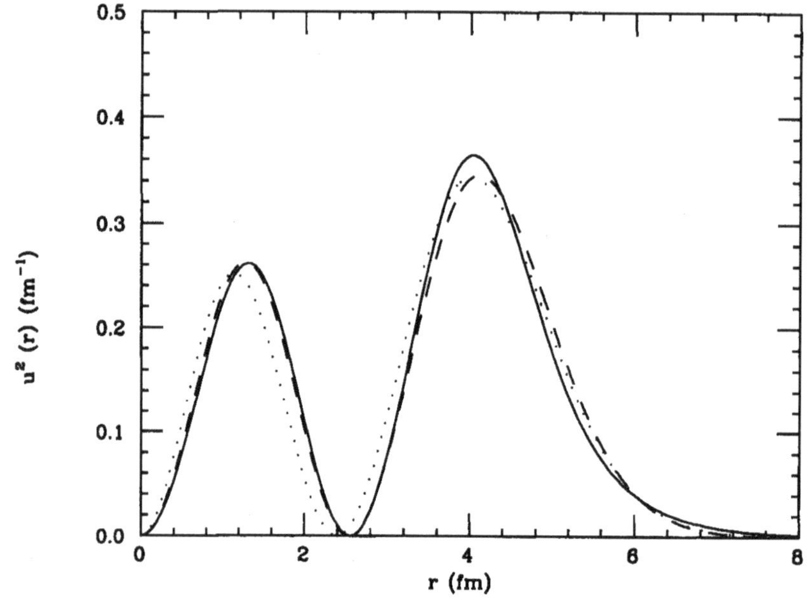

Fig. 1. Comparison between HF (continuous line), inferred (dotted line) and selfconsistent (dashed line) probability densities for the $1S_{1/2}$ proton level of ^{40}Ca ($g_1(r) = r$ and $g_2(r) = r^2$).

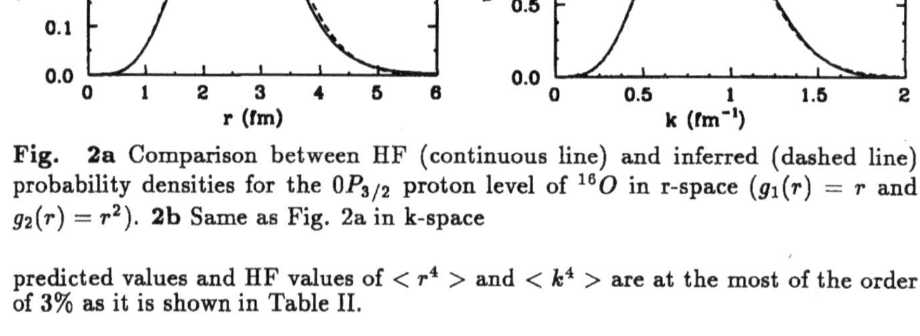

Fig. 2a Comparison between HF (continuous line) and inferred (dashed line) probability densities for the $0P_{3/2}$ proton level of ^{16}O in r-space ($g_1(r) = r$ and $g_2(r) = r^2$). **2b** Same as Fig. 2a in k-space

predicted values and HF values of $< r^4 >$ and $< k^4 >$ are at the most of the order of 3% as it is shown in Table II.

It is clear from (7) that the choice of the functions $g_1(r) = r$ and $g_2(r) = r^2$ does not guarantee the correct asymptotic behavior of the spwf in r-space. One can obtain the correct asymptotic behavior taking $g_2(r) = rexp(-\alpha r^2)$.

For the ^{16}O choosing $\alpha = 0.04$ the discrepancies between HF and MEP results of $< r^2 >$, $< r^4 >$ and $< r^6 >$ are lower than 0,1 %, 2% and 10% respectively (see Table III).

4. Conclusions

Using an information theory algorithm and the knowledge of few expectation values of some operators for each single particle level one can obtain the sp wave functions in r and in k-space. In this method the experimental measurements of some nuclear observables for each shell, using the (e, e' p) reactions become of great significance, because they pave the way to obtain sp wave functions independent of any theoretical mean field interaction, if the experimental input is also model independent. Even if the experimental information of the deepest states is not available, the wf obtained using the experimental results around the Fermi surface can be used to choose the best mean field interaction.

Acknowledgments

This work was performed under the auspices of both the DGICYT-PS90-0212 grant (Spain) and the General Agreement between the Universities of Illes Balears and National de La Plata. A. Plastino acknowledges support from CONICET (Argentina)

References

1. D. Vautherin and D. M. Brink, *Phys. Rev.* C **5** (1972) 626.
2. P. Quentin and H. Flocard, *Ann. Rev. Nucl. Part. Sci.* **28** (1978) 523.
   J. L. Friar and J. W. Negele, *Adv. Nucl. Phys.* **8** (1976) 219.
3. M. Beiner, H. Flocard, Nguyen van Giai and P. Quentin, *Nucl. Phys.* **A 386** (1982) 79.
4. H. S. Kholer, *Nucl. Phys.* **A258** (1976) 301.
5. J. Bartel, P. Quentin, M. Brack, G. Guet and H. B. Hakanson, *Nucl. Phys.* **A 386** (1982) 79.
6. M. Casas, J. Martorell, E. Moya de Guerra and J. Treiner, *Nucl. Phys.* **A 473** (1987) 429.
7. R. A. Broglia, O. Hansen and C. Reidel, *Adv. Nucl. Phys.* **6** (1973) 287.
8. P. W. Woods, R. Chapman, J. N. Mo, P. Skensved and J. A. Kuchner, *Phys. Lett.* **B116** (1982) 320.
9. E. N. M. Quint, PHD. Thesis, University of Amsterdam (1988) and references therein.

10. J. Mougey, M. Bernheim, A. Bussière, A. Gillebert, Phan Xuan Hô, M. Priou, D. Royer, I. Sick and G. J. Wagner, *Nucl. Phys.* **A 262** (1976) 461.

11. N. Canosa, A. Plastino and R. Rossignoli, *Phys. Rev.* **A 40** (1989) 519.

12. M. Canosa, R. Rossignoli and A. Plastino, *Nucl. Phys.* **A 512** (1990) 492; N. Canosa, R. Rossignoli and A. Plastino, *Phys. Rev.* **A 43** (1991) 1445.

13 E. T. Jaynes, in *Maximum-Entropy and Bayesian Methods in Science and Engineering* eds. G. J. Erikson and C. Raysmith (Kluwer, Dordrecht, 1988), Vol. 1, pp. 25-30; M. Tribus, ibid., pp. 31-52; S. F. Gull, ibid., pp. 53-74.

14. N. Canosa, R. Rossignoli and A. Plastino, *Nucl. Phys.* **A 550** (1992) 453.

15. N. Casas, A. Plastino, A. Puente, N. Canosa and R. Rossignoli, *Phys. Rev.* **A 47** (1993) 3530.

16. N. Casas, A. Plastino, A. Puente, *Phys. Rev.* **C 48** August (1993).

SEMICLASSICAL DESCRIPTION OF NUCLEAR SURFACE PROPERTIES IN RELATIVISTIC MODELS*

M. CENTELLES and X. VIÑAS

Departament d'Estructura i Constituents de la Matèria, Facultat de Física, Universitat de Barcelona, Diagonal 647, E-08028 Barcelona, Spain

ABSTRACT

The surface energy, surface thickness and curvature energy of semi-infinite nuclear matter are studied in the framework of the relativistic mean field theory. Systematic calculations are performed for non-linear $\sigma-\omega$ models using a semiclassical relativistic extended Thomas–Fermi method. Comparisons with non-relativistic calculations for Skyrme interactions are also made.

1. Introduction

Under the assumption that a nucleus of mass number A approaches a leptodermous system, with a constant density in the bulk which falls off to zero in a thin surface layer, the nuclear binding energy can be expanded in a power series of $A^{1/3}$.[1-3] For a symmetric nucleus without Coulomb repulsion, the energy can then be expressed through a mass formula as a sum of a volume contribution a_v, a surface term a_s, a curvature term a_c and higher-order corrections[1]:

$$E = a_v A + a_s A^{2/3} + \left(a_c - 2a_s^2/K\right) A^{1/3} + \dots, \quad (1)$$

where $2a_s^2/K$ accounts for the surface compression effects and K is the bulk nuclear incompressibility. The coefficients a_s and a_c can be obtained from semi-infinite nuclear matter . Most of the calculations of a_s and a_c have been carried out within the non-relativistic density functional theory, employing effective density-dependent Skyrme interactions and the semiclassical extended Thomas–Fermi (ETF) method.[4,5]

The value of the curvature energy a_c obtained from theoretical calculations is of the order of 10 MeV.[1,5-8] On the contrary, the empirical coefficient determined from the analysis of experimental data on nuclear masses and fission barriers is compatible with a vanishing value.[2,3] Several explanations have been suggested to solve this so-called curvature energy puzzle. For instance, the consideration of finite-range interactions, Friedel oscillations, ground-state correlations or relativistic effects. The influence of Friedel oscillations as well as the effect of the finite range have been analyzed in Ref. 8 and it has been found that they are not responsible for the discrepancy.

In recent years the relativistic investigation of nuclear systems has attracted much interest.[9,10] The phenomenological $\sigma-\omega$ model of Walecka[9,11] has been widely

*Work supported in part by the DGICYT (Spain) under grant PB89-0332. The Supercomputing Center of Catalonia (CESCA) is gratefully acknowledged for supplying computing time.

used to describe many nuclear properties in the mean field Hartree approximation with considerable success.[12-14] The Lorentz covariance of the mean field theory automatically leads to important density dependence and non-local effects in the nuclear interaction. In addition, the spin-orbit force is taken into account from the outset and the meson exchange gives rise naturally to finite-range interactions. All these effects are very important in the description of the surface of a nuclear system.[15] The surface energy in the $\sigma-\omega$ model has often been estimated by fitting a mass formula to calculations of finite nuclei.[9] Semi-infinite nuclear matter calculations of the surface energy and the surface thickness for the $\sigma-\omega$ model have been performed in the Hartree approximation[11,16,17] or in the Thomas-Fermi approximation.[11,17]

Our purpose here is to systematically study the surface properties (surface energy, surface thickness and curvature energy) of semi-infinite symmetric nuclear matter in relativistic mean field theory. For the calculations we use the relativistic extended Thomas-Fermi (RETF) method of Refs. 18 and 19. It includes gradient corrections of order $\hbar^2$ in the particle density and the scalar field, which provide a better description of the nuclear surface than the simpler Thomas-Fermi model.[19-21] A complete account of this work has been presented elsewhere.[22]

The application of the RETF method to the $\sigma-\omega$ model and the basic formalism to compute the surface and curvature energy coefficients is reviewed in Sec. 2. Sec. 3 is devoted to the discussion of the results. In Sec. 4 a comparison with non-relativistic calculations with Skyrme interactions is made. Finally, our conclusions are drawn in Sec. 5.

2. Theory

2.1. *Relativistic Extended Thomas-Fermi Equations for the $\sigma-\omega$ Model*

Our calculation of semi-infinite nuclear matter is based on the relativistic non-linear $\sigma-\omega$ model which includes scalar self-interactions.[9,11] In the standard relativistic mean field approximation, the meson fields are replaced by their ground-state expectation values and exchange effects and negative energy states are neglected. Imposing time-reversal symmetry, only the time-like component V_0 of the vector field contributes. To second order in $\hbar$, the RETF energy density for symmetric matter is

$$\begin{aligned} \mathcal{E} &= \mathcal{E}_0 + \mathcal{E}_2 - m\rho + \frac{1}{2}\left[(\nabla\phi_o)^2 + m_s^2\phi_o^2\right] + \frac{1}{3}b\phi_o^3 + \frac{1}{4}c\phi_o^4 \\ &\quad + g_v V_0\rho - \frac{1}{2}\left[(\nabla V_0)^2 + m_v^2 V_0^2\right], \end{aligned} \tag{2}$$

where

$$\mathcal{E}_0 = \frac{1}{4\pi^2}\left[k_F\varepsilon_F^3 + k_F^3\varepsilon_F - m^{*4}\ln\frac{k_F+\varepsilon_F}{m^*}\right] \tag{3}$$

is the standard ($\hbar^0$) Thomas-Fermi term and

$$\mathcal{E}_2 = B_1(k_F,m^*)\,(\nabla\rho)^2 + B_2(k_F,m^*)\,(\nabla\rho\cdot\nabla m^*) + B_3(k_F,m^*)\,(\nabla m^*)^2 \tag{4}$$

corresponds to the gradient corrections of order $\hbar^2$ (see Ref. 19). In the above equations ρ is the baryon density, $k_F = (3\pi^2\rho/2)^{1/3}$ is the local Fermi momentum, $\varepsilon_F = \sqrt{k_F^2 + m^{*2}}$, and the effective mass m^* is related to the scalar field by $m^* = m - g_s\phi_o$.

The ground-state density and meson fields are obtained by solving the Euler–Lagrange equations ensuing from the variational principle applied to Eq. (2):

$$\varepsilon_F - m + g_v V_0 + \frac{\delta \mathcal{E}_2}{\delta \rho} - \lambda = 0\,, \tag{5}$$

$$(\Delta - m_v^2)V_0 = -g_v \rho\,, \tag{6}$$

$$(\Delta - m_s^2)\phi_o = -g_s\rho_s + b\phi_o^2 + c\phi_o^3\,, \tag{7}$$

where λ is the chemical potential and ρ_s is the scalar density.[19,22] When the gradient corrections of order $\hbar^2$ are included in ρ_s, we have not always been able to find stable numerical solutions of Eq. (7) for parameter sets which have an effective mass at saturation below $m^*/m = 0.60$ and, at the same time, a scalar mass above $m_s = 500$ MeV (see also Ref. 20). A reliable scheme to overcome this problem has been suggested very recently.[23]

2.2. *Surface and Curvature Energy Coefficients*

The surface energy a_s and the curvature energy a_c (without the compression term) entering the mass formula (1) can be obtained from semi-infinite nuclear matter as follows[1]:

$$a_s = 4\pi r_\infty^2 \int_{-\infty}^{\infty} dz\, \mathcal{E}_s(z) \tag{8}$$

$$a_c = 8\pi r_\infty \int_{-\infty}^{\infty} dz (z - z_0)\, \mathcal{E}_s(z) + 8\pi r_\infty \int_{-\infty}^{\infty} dz \left. \frac{\partial \mathcal{E}_s(z)}{\partial \kappa} \right|_{\kappa=0}\,, \tag{9}$$

where $\mathcal{E}_s(z) = \mathcal{E}(z) - a_v\rho(z)$ is the surface energy density. The quantity κ is the curvature ($\kappa = 2/R$ for a sphere of radius R), $r_\infty = (4\pi\rho_\infty/3)^{-1/3}$ is the nuclear matter radius, and the constant z_0 gives the location of the effective sharp surface. Note that due to self-consistency, a_v equals the chemical potential λ.

The two contributions to the curvature energy in (9) are called geometrical and dynamical, respectively. In general the surface energy density depends on the curvature κ in two different ways. First, $\mathcal{E}(z)$ may have an explicit dependence on the laplacian operator Δ which in the limit of $R \to \infty$ reads $d^2/dz^2 + \kappa d/dz$. Second, the density ρ and the meson fields V_0 and ϕ_o contain an implicit dependence on κ.[22] For the equilibrium density and meson fields, the energy density (2) is stationary with respect to changes in the density and in the fields, and therefore the contribution of the implicit curvature dependence of ρ, V_0 and ϕ_o vanishes.[22] On the other hand, the energy density (2) has no explicit dependence on curvature since the laplacian terms in $\mathcal{E}_2$ have been removed by partial integration.[19] Consequently, if the functional (2) with $\mathcal{E}_2$ given by Eq. (4) is used, a_c is given only by the geometrical contribution.

3. Results

As has been discussed by other authors in the past, the linear $\sigma-\omega$ model is not able to give a satisfactory description of the nuclear surface.[9,10] The deficiencies of this model can be cured by including non-linear self-couplings of the scalar field.[11] In this section the question will be addressed under which conditions it is possible to obtain at the same time an acceptable surface energy a_s and surface thickness t (90% to 10% fall-off distance of the density profile), and the corresponding behaviour of the curvature energy a_c will be analyzed.

We have performed RETF calculations of semi-infinite nuclear matter for a number of parameter sets which differ in the saturation properties. The corresponding results for a_s, t and a_c are given in Figs. 1, 2 and 3, respectively. The energy per particle and the density at saturation are fixed at $a_v = -15.75$ MeV and $\rho_\infty = 0.16$ fm^{-3}. As shown in Ref. 22 changing a_v or ρ_∞, if within the standard values, does not significantly modify the discussion which follows. The range of variation of m^* (effective mass at saturation) and K (nuclear matter incompressibility) covers the commonly accepted values: m^*/m goes between 0.55 and 0.80, and we use $K = 200$, 300 and 400 MeV. Different values of the scalar mass m_s are considered, which range from $m_s = 400$ to 550 MeV. The nucleon mass and the vector meson mass take their experimental values: $m = 939$ MeV and $m_v = 783$ MeV. For the lower values of m^*/m problems of convergence of the numerical solution were encountered with some m_s, and hence these results are not represented in the figures. In Figs. 1 and 2 the horizontal dashed lines approximately indicate the region compatible with the empirical data: the surface energy a_s should be around 16.5–21 MeV and the surface thickness t around 2.2–2.5 fm.

A common trend in Figs. 1–3 is that on increasing the scalar mass, a_s, t and a_c decrease, which is more visible for small m^*. This strong dependence on the scalar mass is not surprising since m_s sets the range of the attractive interaction and, therefore, directly modifies the surface profile. For large m_s the range of the scalar interaction is shortened, resulting in a steeper surface.

For constant m^* and m_s, the surface energy clearly increases with the incompressibility. Figures 2 and 3 show, however, that the behaviour of t and a_c as a function of K is more complicated. The global tendency of the surface thickness is to decrease with increasing K, but t increases with K if $m^*/m \geq 0.7$ and $m_s \geq 500$ MeV. The calculated a_c is brought down by increasing K if $m^*/m \leq 0.7$, whereas the opposite behaviour is found for $m^*/m \geq 0.7$. In any case, it must be noted that the variation of t and a_c as a function of K is weaker than the variation of a_s.

The way that a_s and a_c change as a function of m^* is not simple either. As a general trend, the surface energy goes down with increasing m^*, but this tendency is less apparent as K or, especially, m_s increase. The curvature energy exhibits the same behaviour with m^* as a_s, but less markedly. From Fig. 2 it is seen that the surface thickness always shows a downward tendency with increasing m^* for the encompassed values of K and m_s. In understanding this result it is useful to take into account that the effective mass gives a measure of the non-local effects, whose contribution tends

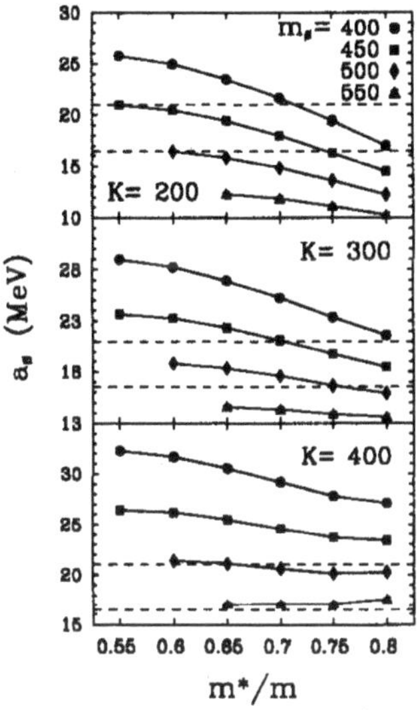

Fig. 1. Surface energy a_s of semi-infinite nuclear matter in the relativistic non-linear $\sigma-\omega$ model as a function of the scalar mass m_s (MeV), and of the incompressibility K (MeV) and the effective mass m^*/m at saturation. Note the different values on the vertical axis.

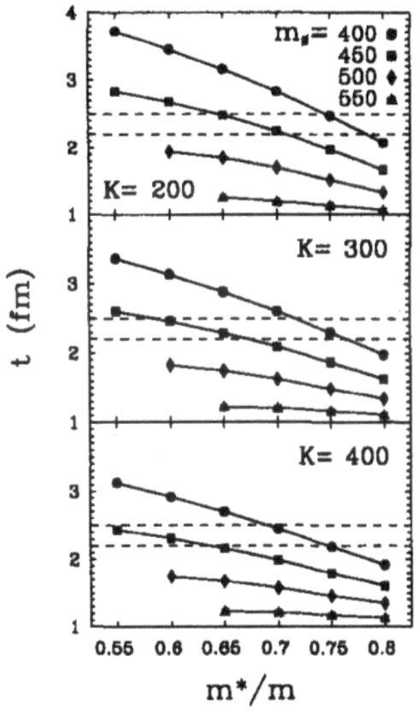

Fig. 2. Same as Fig. 1 for the surface thickness t of the semi-infinite profile.

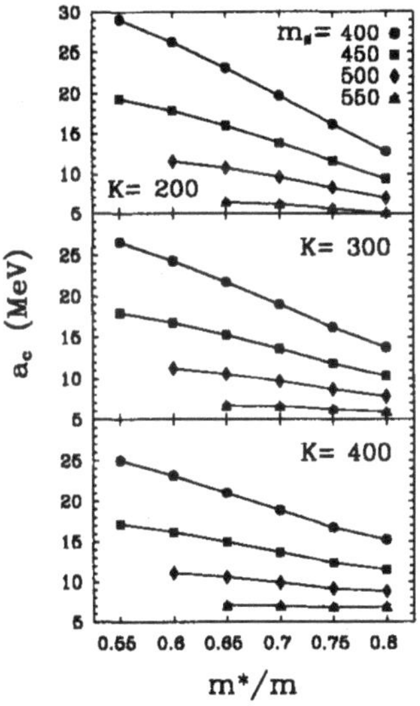

Fig. 3. Same as Fig. 1 for the curvature energy a_c of semi-infinite nuclear matter.

to make the surface more diffuse.[15,24] A larger value of the effective mass is associated with less non-locality, and therefore favours a sharper surface.

If one demands that both a_s and t lie within the empirical region, indicated in rough by the dashed lines in Figs. 1 and 2, the incompressibility is restricted to be around K = 200 MeV and the scalar mass to m_s = 400–500 MeV. The sets with K = 300 MeV or K = 400 MeV are clearly unable to yield an acceptable prediction for a_s and t simultaneously. Indeed, with K = 200 MeV not only it is possible to get the right a_s and t, but it is also possible to obtain approximately the same values of a_s and t (and a_c) with different m^* and m_s. For instance, we have found $a_s \approx 18$ MeV and $t \approx 2.3$ fm with several pairs $(m^*/m,\ m_s)$; e.g., with (0.6, 480 MeV), (0.7, 450 MeV) or (0.8, 380 MeV). Typically, a small value of m^* favours $m_s \sim 500$ MeV, while $m_s \sim 400$ MeV is preferred by the higher m^*.

Concerning the curvature energy, it must be pointed out that in the region of parameter sets where both a_s and t take a realistic value, the calculated a_c is systematically around 13–15 MeV. This result is far from the empirical coefficient, which is compatible with a vanishing value. One can obtain lower values for a_c in agreement with the empirical coefficient by choosing moderate values of K and increasing m^* and m_s, but then a_s and t come out unrealistically small. We thus conclude from this analysis that in the relativistic mean field model there is no mechanism able to lower the curvature energy while keeping the right surface energy and surface thickness.

4. Comparison with non-relativistic calculations

In a sense the relativistic mean field model constitutes the relativistic analogue of the non-relativistic Hartree–Fock method with density-dependent interactions. Both models achieve about the same agreement with experiment.[10,13,14] It is therefore appealing to compare the two models also from the semiclassical perspective of this work. Table 1 presents the surface energy, surface thickness and curvature energy, as well as the corresponding saturation properties, for several Skyrme interactions: SI–SVI[25] and SkM*,[5] ordered according to their incompressibility. The results reported in this table have been calculated within the non-relativistic ETF theory, using the semiclassical energy density functional of order $\hbar^2$ which includes effective mass and spin–orbit contributions[4,5]:

$$\mathcal{E} = \frac{f}{2m}\left[\frac{3}{5}(3\pi^2/2)^{2/3}\rho^{5/3} + \beta\frac{(\nabla\rho)^2}{\rho} + \frac{1}{3}\Delta\rho - \frac{1}{12}\rho\frac{(\nabla f)^2}{f^2} + 2m^2\rho\frac{\boldsymbol{W}^2}{f^2}\right] + V(\rho,\ \nabla\rho) - 2m\rho\frac{\boldsymbol{W}^2}{f}, \qquad (10)$$

where $V(\rho,\ \nabla\rho)$ is the Skyrme potential energy density, $\boldsymbol{W}$ is the spin–orbit potential, $\beta = 1/36$, and $f = m/m^*$, with m^* the non-relativistic effective mass.

From Table 1 it is seen that all the forces displayed (except SI) give a reasonable surface energy, around 16–17.5 MeV, although the surface thickness is rather small in several cases. This is typical of the semiclassical calculations performed with the

Table 1. Nuclear matter properties and surface properties of several Skyrme interactions.

	a_v (MeV)	ρ_∞ (fm^{-3})	K (MeV)	m^*/m	a_s (MeV)	t (fm)	a_c (MeV)
SkM*	−15.77	0.1603	217	0.79	16.00	2.23	10.53
SV	−16.05	0.1551	306	0.38	17.52	1.97	9.34
SIV	−15.96	0.1509	325	0.47	16.95	1.85	8.38
SII	−15.99	0.1484	341	0.58	17.56	1.90	8.45
SIII	−15.85	0.1453	355	0.76	16.47	1.73	7.33
SVI	−15.76	0.1435	364	0.95	15.98	1.64	6.82
SI	−16.00	0.1553	370	0.91	14.85	1.41	5.99

non-relativistic functional of order $\hbar^2$, due to the value $\beta = 1/36$ of the coefficient of the Weizsäcker term in the kinetic energy density.[26] For the considered Skyrme interactions the curvature energy lies between 6 and 10.5 MeV. This value of a_c is a little smaller than the result obtained using the relativistic parameterizations which yield the right a_s and t.

To explore more closely the comparison with non-relativistic calculations, we have fitted a non-linear $\sigma-\omega$ parameter set to exactly reproduce the nuclear matter properties of the SkM* interaction (Table 1). The mass m_s of the scalar meson has been chosen so that the semi-infinite calculation gives approximately the same surface energy as SkM*. We have performed calculations with SkM* for two values of the coefficient β of the Weizsäcker term in the non-relativistic functional (10); namely, for $\beta = 1/36$ which is the standard value,[4,5] and for $\beta = 1/18$ which somehow imitates the effect of $\hbar^4$-order corrections.[5,6] The relativistic set fitted to the saturation properties and the surface energy of SkM* with $\beta = 1/36$ will be called RSk1, and that for $\beta = 1/18$ will be called RSk2. The surface properties corresponding to these parameterizations are displayed in Table 2.

If the saturation properties and a_s are the same, it is seen from Table 2 that the coefficient a_c also comes out very similar in the relativistic and non-relativistic calculations, for $\beta = 1/36$ and for $\beta = 1/18$. However, the surface thickness turns out to be appreciably smaller in the calculation with the relativistic sets, which means a steeper density profile at the surface in the relativistic case. The fact that the relativistic calculation of the nuclear thickness yields a smaller value than SkM* cannot be attributed only to the finite range of the relativistic interactions.[22] Peforming

Table 2. Surface properties for SkM* and for the relativistic parameter sets RSk1 and RSk2.

	SkM* ($\beta = 1/36$)	RSk1	SkM* ($\beta = 1/18$)	RSk2
a_s (MeV)	16.00	15.99	17.92	17.91
t (fm)	2.23	1.81	2.44	2.09
a_c (MeV)	10.53	10.62	13.75	13.13

the same type of comparison in calculations of finite nuclei with Skyrme interactions and relativistic parameter sets,[22] the situation is similar to the case of semi-infinite nuclear matter: the agreement is better for the energies than for the charge radii, which are smaller in the relativistic calculations. This hints at a difference between relativistic and non-relativistic determinations of the density profile.

5. Conclusions

The surface properties of semi-infinite nuclear matter have been investigated with a recently developed semiclassical formulation of the relativistic mean field theory which includes gradient corrections of order $\hbar^2$.[18,19] The dependence of a_s, t and a_c on the scalar mass m_s and on the incompressibility K and the effective mass m^* at saturation, has been found to be stronger than the dependence on the saturation energy a_v and density ρ_∞, when these quantities are changed within usually accepted values.

In contrast with the common belief that few parameterizations of the $\sigma-\omega$ model are consistent with the empirical values of the surface energy and surface thickness, it has been shown that the right a_s and t can be obtained simultaneously with a rather wide region of non-linear parameter sets, if they have $K \sim 200$ MeV and m_s lies in the range 400–500 MeV (for $0.55 \le m^*/m \le 0.80$). It is interesting to note that the favoured value of the incompressibility ($K \sim 200$ MeV) lies in the low part of the range of accepted values by many authors.

In the region of realistic a_s and t the curvature energy always falls roughly in the range 10–15 MeV. It is possible to bring a_c down to zero, but the corresponding a_s and t are then unacceptably small. It seems that the anomaly in the curvature energy cannot be solved within current relativistic mean field models. Even though there is no clear explanation for this discrepancy (see e.g. Refs. 7 and 8), it is plausible that it could be traced back to the mean field approximation, either in the non-relativistic or relativistic case.

For comparison, non-relativistic calculations with Skyrme interactions have been made. A tendency to thinner surfaces has been found in the relativistic model (see also Ref. 10). Such difference between both theories has some consequences on quantities which are very sensitive to the shape of the nuclear surface, as in the case of the level density parameter.[27]

References

1. W. D. Myers and W. J. Swiatecki, *Ann. Phys.* **55** (1969) 395; *Ann. Phys.* **84** (1974) 186.
2. W. D. Myers, *Droplet Model of Atomic Nuclei* (IFI/Plenum, New York, 1977).
3. P. Möller, W. D. Myers, W. J. Swiatecki and J. Treiner, *At. Data Nucl. Data Tables* **39** (1988) 225.
4. B. Grammaticos and A. Voros, *Ann. Phys.* **123** (1979) 359; *Ann. Phys.* **129** (1980) 153.

5. M. Brack, C. Guet and H. -B. Håkansson, *Phys. Rep.* **123** (1985) 275.
6. J. Treiner and H. Krivine, *Ann. Phys.* **170** (1986) 406.
7. W. Stocker, J. Bartel, J. R. Nix and A. J. Sierk, *Nucl. Phys.* **A489** (1988) 252.
8. M. Durand, P. Schuck and X. Viñas, *Z. Phys.* **A346** (1993) 87.
9. B. D. Serot and J. D. Walecka, *Adv. Nucl. Phys.* **16** (1986) 1.
10. P. -G. Reinhard, *Rep. Prog. Phys.* **52** (1989) 439.
11. J. Boguta and A. R. Bodmer, *Nucl. Phys.* **A292** (1977) 413.
12. A. Bouyssy, S. Marcos and Pham Van Thieu, *Nucl. Phys.* **A422** (1984) 541.
13. P. -G. Reinhard, M. Rufa, J. Maruhn, W. Greiner and J. Friedrich, *Z. Phys.* **A323** (1986) 13.
14. Y. K. Gambhir, P. Ring and A. Thimet, *Ann. Phys.* **198** (1990) 132.
15. J. L. Friar and J. W. Negele, *Adv. Nucl. Phys.* **8** (1975) 219.
16. D. Hofer and W. Stocker, *Nucl. Phys.* **A492** (1989) 637.
17. W. Stocker and M. M. Sharma, *Z. Phys.* **A339** (1991) 147.
18. M. Centelles, X. Viñas, M. Barranco and P. Schuck, *Nucl. Phys.* **A519** (1990) 73c.
19. M. Centelles, X. Viñas, M. Barranco and P. Schuck, *Ann. Phys.* **221** (1993) 165.
20. M. Centelles, X. Viñas, M. Barranco, S. Marcos and R. J. Lombard, *Nucl. Phys.* **A537** (1992) 486.
21. D. Von-Eiff and M. K. Weigel, *Phys. Rev.* **C46** (1992) 1797.
22. M. Centelles and X. Viñas, *Nucl. Phys.* **A563** (1993) 173.
23. C. Speicher, E. Engel and R. M. Dreizler, *Nucl. Phys.* **A562** (1993) 569.
24. X. Campi and S. Stringari, *Nucl. Phys.* **A337** (1980) 313.
25. D. Vautherin and D. M. Brink, *Phys. Rev.* **C5** (1972) 626; M. Beiner, H. Flocard, Nguyen Van Giai and P. Quentin, *Nucl. Phys.* **A238** (1975) 29.
26. O. Bohigas, X. Campi, H. Krivine and J. Treiner, *Phys. Lett.* **B64** (1976) 381.
27. M. Centelles, X. Viñas and P. Schuck, *Nucl. Phys.* **A**, in press.

ANALYSIS OF THE NUCLEAR SURFACE AND CURVATURE ENERGIES

M. DURAND, P. SCHUCK
Institut des Sciences Nucléaires, Université Joseph Fourier and IN2P3
F - 38026 Grenoble-Cédex, France

and

X. VIÑAS
Departament d'Estructura i Constituents de la Materia,
Facultat de Fisica, Universitat de Barcelona, E - 08028 Barcelona, Spain

ABSTRACT

The surface and curvature energy coefficients a_s and a_c of the nuclear mass formula are studied. Semiclassical and quantal results are in good agreement and a_c is found to be close to the empirical value for a Fermi gas bounded by an external Woods-Saxon potential, but this is shown to be due to the lack of selfconsistency of the potential. A selfconsistent calculation using finite range forces does not change the previous theoretical situation : a_c is found to be about an order of magnitude above the empirical value.

1. Introduction

The aim of this work is a detailed analysis of the curvature term in the nuclear mass formula. The binding energy per nucleon E_0/A of a nucleus with A nucleons reads[1] :

$$E_0/A = a_v + a_s A^{-1/3} + a_c A^{-2/3} + \dots \qquad (1)$$

where the dots stand for additional terms and for the Coulomb energy which will not be considered here since we shall only deal with the surface a_s and curvature a_c coefficients. Such mass formulas are also developed for metallic and ^3He clusters[2]. The empirical value[1] of a_c is usually simply put to zero but all the past calculations[3,4,5,6] report a curvature energy $a_c \sim 10$ MeV. Friedel oscillations were suggested[3,7] to be responsible for this difference : they will be evaluated first by a comparison between quantal and semiclassical calculations for a Fermi gas in a Woods-Saxon potential and second in a selfconsistent approach generalised to finite range forces.

2. Definition of the curvature energy

The groundstate energy E_0 of a spherical nucleus can be expressed by means of the surface energy density $\varepsilon_s(r)$:

$$E_0 = a_v A + 4\pi R^2 \int_0^\infty dr\, \frac{r^2}{R^2}\, \varepsilon_s(R,r) \quad ; \quad \varepsilon_s(R,r) = H(R,r) - a_v \rho(R,r) \qquad (2)$$

where H is the hamiltonian density of the nucleus, R is the radius of a spherical nucleus with a sharp surface (in the limit $R \to \infty$, $R = r_0 A^{1/3}$) and a constant density, containing the same number A of particles, a_v is the energy per particle of infinite matter and ρ is the matter density. For not too small nuclei, the integrand in Eq.(2) is peaked around the nuclear surface: first, with the new variable $u = r - R$, the geometrical factor r^2/R^2 gives a descending series in R, then ε_s is expanded as a function of the curvature $\kappa = 2/R$. This leads to the asymptotic expansion of E_0 Eq.(1). Thus, in the geometry of half infinite matter with z the coordinate perpendicular to the surface, i.e. the limit of u when $R \to \infty$, the coefficients in Eq.(1) read :

$$a_s = 4\pi r_0^2 \int_{-\infty}^{+\infty} dz\ \varepsilon_s(\kappa, z)\Big|_{\kappa=0} \quad ; \quad a_c = a_c^{geo} + a_c^{dyn} \tag{4a}$$

$$a_c^{geo} = 8\pi r_0 \int_{-\infty}^{+\infty} dz\ z\ \varepsilon_s(\kappa, z+z_0)\Big|_{\kappa=0} \quad ; \quad a_c^{dyn} = 8\pi r_0 \int_{-\infty}^{+\infty} dz\ \frac{\partial \varepsilon_s(\kappa, z)}{\partial \kappa}\Big|_{\kappa=0} \tag{4b}$$

The constant z_0 in a_c^{geo} stands for the effective location of the sharp surface at $r = R$. For technical reasons, it will be convenient to write :

$$a_c^{geo} = \tilde{a}_c^{geo} - 2\frac{z_0}{r_0}\, a_s \ ; \quad \tilde{a}_c^{geo} = 8\pi r_0 \int_{-\infty}^{+\infty} dz\ z\ \varepsilon_s(\kappa, z)\Big|_{\kappa=0} \tag{5}$$

The curvature is also implicitly contained in the Laplacians which, for a spherical system of radius $R \to \infty$, can be written as :

$$\lim_{R\to\infty} \Delta_r = \frac{d^2}{dz^2} + \kappa\frac{d}{dz} \tag{6}$$

The Laplacians stem from the Schrödinger equation and from the the two body forces, but also, eventually, from the kinetic energy density which can be written in two ways:

$$\tau_1(r) = -(\hbar^2/2m)\textstyle\sum_{i=1}^{A}\ \varphi_i^*(r)\ \Delta\varphi_i(r)$$
$$\tau_2(r) = (\hbar^2/2m)\textstyle\sum_{i=1}^{A}\ |\nabla\varphi_i(r)|^2$$

Both expressions for $\tau(r)$ always give the same kinetic energy: for finite nuclei as for half infinite nuclear matter, in quantal calculation as in any semiclassical approximation. In spite of the same implicit dependence of both definitions on κ coming from the wave functions, the two forms of τ will lead to two different expressions and values for a_c^{dyn} Eq.(4b) (because of an explicit Δ-operator in only one of them) and also to different values of a_c^{geo} : however, the sum $a_c^{geo}+a_c^{dyn}$ will always be the same. This is true for any linear combination of τ_1 and τ_2 weighted to unity and thus there exists a great arbitrariness in the partitioning of a_c into a_c^{geo} and a_c^{dyn} : here τ_2 is selected since then a_c^{dyn} is determined only by the curvature dependence of the wavefunctions and not by the explicit Laplacian appearing in τ_1. This implies that a_c^{dyn} is the part of a_c which can not be extracted from a fully quantum mechanical calculation in any straightforward manner.

3. Curvature energy in a model

The model consists of infinite nuclear matter embedded in an external Woods-Saxon potential $V(z) = -V_0/(1+e^{z/a})$ in the z-direction perpendicular to its surface, and constant in the directions parallel to the surface. For comparisons with later approximations, a quantal calculation is first performed.

3.1. Quantal calculation

The wave functions of this problem are analytic and were given in a previous work[8]. They allow to calculate the local density $\rho(r)$ and the kinetic energy density $\tau(r)$, and then the parameters a_s and a_c^{geo}, by means of the surface energy density : $\varepsilon_s = \tau + V.\rho - a_v\rho$. It must be noted that the energy per particle a_v is not equal to the chemical potential λ, as it is the case in a selfconsistent problem, but instead $a_v = 3/5\lambda - 2/5V_0$. This constitutes one of the differences of the model (external potential) with a more realistic situation.

The Friedel oscillations are found to be weak in both ρ and τ as it was also found in more realistic selfconsistent calculations[9], however they are amplified in the integrands (ε_s and $z.\varepsilon_s$) for a_s and a_c^{geo} as a result of both the Friedel oscillations in ρ and τ and their interfering combination in ε_s. Though oscillating, the surface energy density ε_s drops to zero relatively fast, when $z \to -\infty$. This assures a good convergence of the integral for a_s Eq.(4a). Unfortunately, this is no longer the case for the quantity $z\varepsilon_s$ needed in a_c whose oscillations are only very slowly damped. As a function of the lower bound of the integral Eq.(5), $\tilde{a}_c^{geo}$ asymptotically oscillates like a Si-function which allows to extract the value $\tilde{a}_c^{geo} = 18.23$ MeV. With the chosen potential, the sharp surface position is $z_0 = -0.805$ fm, leading to :

$$a_s = 15.30 \text{ MeV} \quad ; \quad a_c^{geo} = 40.09 \text{ MeV} \tag{7}$$

The latter value is comparable with the one calculated for the spherical billiard[10], $a_{c,SB}^{geo} = 42.84$ MeV and to what has been found in selfconsistent quantal calculations[10] for a_c^{geo}. As already mentioned, the dynamical contribution a_c^{dyn} can not be calculated quantally, since the curvature dependence of the wave functions is unknown, but the quantal value of a_c^{geo} will help to measure the accuracy of approximated calculations within the same model.

3.2. Semiclassical Wigner-Kirkwood calculation

3.2a. geometrical contribution to the curvature energy

The contribution of Friedel oscillations to the value of a_c^{geo} Eq.(7) can be given by a comparison with the curvature energy calculated within the semiclassical $\hbar$ expansion of Wigner and Kirkwood[11,12] which is free from these quantal effects. Now, the $\hbar$ expansion of the local density $\rho_{WK}(r)$ and the kinetic energy density $\tau_{WK}(r)$ up to order $\hbar^4$ are needed[12]. These quantities diverge at the classical turning point, nevertheless, for the calculation of expectation values they are very efficient. For the

Woods-Saxon potential, $a_{s,WK}$, $\tilde{a}^{geo}_{c,WK}$ (and $z_{0,WK}$) are obtained in an almost analytic way[13], after some partial integrations performed in order to avoid non integrable singularities[14].

Table 1 gives the values of $a_{s,WK}$, $\tilde{a}^{geo}_{c,WK}$ and $a^{geo}_{c,WK}$ with the detail of the role of the different contributions in $\hbar$ (for the potential parameters $a = 0.6\ fm$, $V_0 = 56\ MeV$ and $k_F = 1.35\ fm^{-1}$).

Table 1: Surface and curvature energy coefficients for the Woods-Saxon potential, in a quantum mechanical calculation (QM), in the Wigner-Kirkwood (WK) and in the density functional theory (DFT) approximations up to different orders in $\hbar$.

	a_s	$\tilde{a}^{geo}_c$	z_0	a^{geo}_c	a^{dyn}_c	a_c
QM	15.30	18.23	−0.805	40.09	-	-
WK-$\hbar^0$	12.42	-21.29	−0.796	-3.76	0	-3.76
WK-$\hbar^2$	15.48	18.23	−0.808	40.34	−42.31	-1.97
WK-$\hbar^4$	15.31	18.76	−0.810	40.77	−42.31	-1.54
DFT-$\hbar^2$	13.77	15.62	−0.818	35.61	−42.31	-6.70
DFT-$\hbar^4$	14.14	17.61	−0.813	38.01	−42.31	-4.38

The WK series for a_s converges : inclusion of $\hbar^4$-correction allows to reach a perfect agreement with the quantal result within an accuracy better than 0.1%. Although the Thomas-Fermi contribution to $a^{geo}_{c,WK}$ has the wrong sign, the $\hbar^2$-term totally corrects for it and again the $\hbar^4$-correction is almost negligible, so that the sum seems converged to :

$$a_{s,WK} = 15.31 \text{ MeV} \quad ; \quad a^{geo}_{c,WK} = 40.77 \text{ MeV} \tag{8}$$

with $z_{0,WK} = -0.810$ fm. In comparison with the quantal result Eq.(7) the accuracy on a^{geo}_c is still good. The difference between the semiclassical results and the quantal ones is due to the Friedel oscillations : their influence on a_s is totally negligible while they give a contribution of about 2% to a^{geo}_c in the present case.

The value of $a^{geo}_{c,WK}$ can be compared, in the limit $a \to 0$ and $V_0 \to \infty$, with the same quantity for the billiard[10], $a^{geo}_{c,SB} = 42.84$ MeV. Only contributions of order $\hbar^2$ survive and the result :

$$\lim_{V_0\to\infty} \lim_{a\to 0} a^{geo}_{c,WK} = 8\pi r_0\, 4\, (\hbar^2/2m)\, k_F^3\, (1/20\pi^2 + 23/900\pi^2) = 44.26 \text{ MeV} \tag{9}$$

is very close to the exact one Eq.(7) : this further adds to the confidence in the validity of the WK approach.

The conclusion is thus that the Friedel oscillations may have only a weak contribution to the geometrical part of the curvature energy.

3.2b. dynamical contribution to the curvature energy

In general the evaluation of $\partial\varepsilon_s/\partial\kappa$ in Eq.(4b) is not trivial but it becomes easy in the framework of the WK approximation since the Laplacians and therefore the curvature κ, due to Eq.(6), explicitly appear in ρ_{WK} and τ_{WK}. One gets:

$$a_{c,WK}^{dyn} = -8\pi r_0 \; (\hbar^2/2m) \; k_F^3 \; (13/45\pi^2) = \; - \; 42.31 \text{ MeV} \tag{10}$$

It is remarkable that $a_{c,WK}^{dyn}$ only depends on the nuclear matter quantities k_F and r_0 and not on the shape of the external potential and that the $\hbar^4$ order vanishes. For the spherical billiard, $a_{c,SB}^{dyn}$ turns out to be identical to the WK result.

Altogether, the total curvature energy up to order $\hbar^4$ in our model is :

$$a_{c,WK} = \; - \; 1.54 \text{ MeV} \tag{11}$$

in good agreement with the value for the billiard[10,15] : $a_{c,BS} \sim 0.5$ MeV.

Thus, with a semiclassical expansion up to order $\hbar^4$, the geometrical curvature energy is in excellent agreement with the quantal result for finite values of the diffuseness of the Woods-Saxon potential; moreover, the semiclassical total curvature energy extrapolates very well to the result of the billiard. Therefore we can expect that the influence of Friedel oscillations on the total curvature energy $a_c = a_c^{geo} + a_c^{dyn}$ is small, and that they can not be responsible for the nuclear curvature problem. Moreover, the above agreements give confidence in the semiclassical approximations for the densities.

The curvature energy puzzle in nuclear physics being inherently connected with the semiclassical density functional theory (DFT), this method is investigated in the following section.

3.3. Density Functional Theory calculation

The following Euler-Lagrange equation has to be solved :

$$\frac{\delta\tau[\rho]}{\delta\rho} + V(z) - \lambda = 0 \tag{12}$$

where $\tau[\rho]$ is the semiclassical functional for the kinetic energy density[4,12]. The solution of this equation allows to calculate the geometrical part of the curvature energy : the results are presented in Table 1. They are very similar to the Wigner-Kirkwood ones though definitely less well converged. Thus the above conclusions on the unimportance of the Friedel oscillations in the surface and geometrical curvature energies remain valid in the case of the density functional theory.

The Laplacians contained in the functional $\tau[\rho]$ contribute, as in the WK case, directly to a_c^{dyn} . On the other hand, the density ρ owns an implicit κ dependence : altogether, $a_{c,DFT}^{dyn}$ is shown to analytically equal $a_{c,WK}^{dyn}$. Thus, the evaluation of the total curvature energy works as well in the DFT case as it does in the WK case : therefore, the DFT can not be responsible for the curvature energy puzzle.

In $a^{dyn}_{c,DFT}$ the implicit dependence of the density on κ due to the difference between λ and a_v contributes for -9.77 MeV : this means that the lack of selfconsistency in the case of an external potential brings an extra contribution of ~ -10 MeV, leading to the low value of a_c Eq.(11). Consequently, selfconsistent calculations must be performed, in order to get a more realistic approach to the curvature energy, with inclusion of finite range effects, which was not done previously.

4. Selfconsistent calculation with finite range forces

The curvature energy of various Skyrme forces has been discussed within DFT at length[3,4,5,16]. We here investigate whether finite range forces can change anything to the fact that with zero range Skyrme forces the curvature energy is too high with respect to the empirical value[1] $a_c \sim 0$.

In the selfconsistent case, the energy per particle in infinite nuclear matter a_v equals the chemical potential at saturation ($a_v = \lambda$) and $\rho(z)$ is obtained in making the surface tension stationary with respect to the changes in the density :

$$\frac{\delta H}{\delta \rho} - \lambda = 0 \tag{13}$$

Contrary to what was the case for the external potential, the implicit dependence on κ of the selfconsistent density does not play any role in the dynamical part since:

$$a_c^{dyn} = 8\pi r_0 \int_{-\infty}^{+\infty} dz \left\{ \frac{\partial}{\partial \kappa}[H(z) - \lambda\rho(z)] + \left(\frac{\delta H}{\delta \rho} - \lambda \right) \frac{\partial \rho}{\partial \kappa} \right\} \Bigg|_{\kappa=0} \tag{14}$$

and due to Eq.(13) the coefficient of $\partial\rho/\partial\kappa$ is zero. For the extraction of the curvature energy coming from the finite range part, the finite range interaction is expanded in a series of distributions :

$$v(r) = \delta(r) \int d^3r'\, v(r') \; + \; \frac{1}{6}\Delta\delta(r) \int d^3r'\, r'^2\, v(r') \; + \ldots \tag{15}$$

then the Hartree contribution to a_s and a_c^{dyn}, with $\rho_0 = \rho(z \to -\infty)$, is :

$$a_{s,H} = a_{s,H}(\kappa = 0) - 4\pi r_0^2 \frac{\kappa}{24} \rho_0^2 \int d^3r'\, r'^2\, v(r') \tag{16}$$

$$a^{dyn}_{c,H} = -8\pi r_0 \frac{1}{24} \rho_0^2 \int d^3r'\, r'^2\, v(r') \tag{17}$$

Table 2 shows the energy per particle a_v, the surface energy a_s, the curvature energy a_c and its contributions coming from the geometrical a_c^{geo}, kinetic dynamical $a^{dyn}_{c,kin}$ and potential dynamical $a^{dyn}_{c,pot}$ parts for all the forces analysed[17,18,19].

Table 2: Volume, surface and curvature energy coefficients in a selfconsistent calculation with different forces.

Force	a_v	a_s	a_c^{geo}	$a_{c,kin}^{dyn}$	$a_{c,pot}^{dyn}$	a_c	K_∞	$\tilde{a}_c$
BKN[17]	-16.134	17.012	21.300	-29.694	15.271	6.877	369	5.308
BKNM[17]	-16.000	15.590	23.831	-31.755	18.061	10.137	201	7.719
Gogny[18]	-16.323	20.178	18.760	-32.610	28.602	14.752	230	11.204
SkVII[19]	-15.776	17.698	23.281	-29.508	14.151	7.924	366	6.212

The effect of the finite range of the force partially cancels the strong negative contribution coming from the kinetic energy and the agreement with the same densities calculated with the Skyrme force is qualitatively good. The values of a_c^{geo} are around 20 MeV, instead of about 40 MeV found with an external potential : this is due to the fact that in this latter case there was no contribution from the potential. As for the Skyrme forces[4,5], all values of a_c range roughly between 7 and 15 MeV.

We also analysed how the surface and curvature energies change when the range and the strength of the finite range term in the BKN force[17] are varied keeping the nuclear matter properties unchanged : when the curvature energy is decreased, then the surface energy is decreased too and a curvature energy around zero corresponds to an unrealistic surface energy around 10 MeV.

As a conclusion, the finite range in the nuclear effective interaction does not decrease the curvature energy value below ~10 MeV and consequently this effect is not responsible for the curvature energy puzzle.

Conclusion

The nuclear curvature energy has been studied for a Fermi gas embedded in an external Woods-Saxon potential and for the selfconsistent case using finite range forces, in order to try to elucidate the fact that the empirical value, $a_c \sim 0$, for the curvature energy coefficient in the nuclear mass formula is not in agreement with theoretical calculations within a semiclassical version of the Skyrme Hartree-Fock approach (the density functional formalism) giving $a_c = 9 - 15$ MeV, depending on the zero range force.

In the case of an external potential, semiclassical calculations (Wigner-Kirkwood expansion and Density Functional Theory) are in a very good agreement with quantal results, even in the limit of the billiard, and they are also compatible with the expected value $a_c \sim 0$. However, the agreement with this empirical value seems due to the lack of selfconsistency in the model ($\lambda \neq a_v$). The conclusions are first that Friedel oscillations have very little influence on the curvature energy and therefore they can not be responsible for the curvature energy puzzle, second that DTF is a very reliable method and it should give correct results for a_c also in the selfconsistent case.

In a second part, the selfconsistent case is generalised to finite range forces : our study indicates that finite range interactions can not modify the general conclusion[3,4,5] that the curvature energy lies in the range $7 - 15$ MeV which is practically an order of magnitude higher than the empirical value.

The conclusion is therefore that the curvature energy puzzle of nuclear physics remains unsolved using conventional effective phenomenological density dependent forces. Relativistic effects were also suggested[5] but a detailed study[20] shows that they do not change the situation either. All these analyses seem to show that the difference between the calculated and empirical values of a_c lies beyond the mean field theory. Consequently, groundstate fluctuations due to the surface modes could be a candidate.

Acknowledgements

Help of J. Dechargé, M. Farine and M. Pi is gratefully acknowledged.

References

1. W. D. Myers, *Droplet Model of Atomic Nuclei* (IFI/Plenum, N.Y.,1977)
   P. Möller, W. D. Myers, W. J. Swiatecki and J. Treiner, *Proc. 7th Int.Conf. nuclear masses and fundamental constants* (AMCO-7), (Darmstadt-Seeheim, Lehrdruckerei Darmstadt,1984) p. 457, ibid. *Atom. and Nucl. Data Tables* **39** (1988).
2. E. Lipparini and A. Vitturi, *Z. Phys.* D**17** (1990) 57.
   M. Seidl, M. E. Spina and M. Brack, *Z. Phys.* D**19** (1991) 101.
   E. Engel and J. Perdew, *Phys. Rev.* B**43** (1991) 1331.
   A. Guirao, M. Pi and M. Barranco, *Z. Phys.* D**21** (1991) 185.
3. W. Stocker, J. Bartel, J. R. Nix and A. J. Sierk, *Nucl. Phys.* A**489** (1988) 252.
4. M. Brack, C. Guet and H. B. Hakansson, *Phys. Rep.* **123** (1985) 275.
5. J. Treiner and H. Krivine, *Ann. Phys.* (N.Y.) **170** (1986) 406.
6. M. Farine, *Z. Phys.* A**320** (1985) 337.
7. M. Brack, *Phys. Rev.* B**39** (1989) 3533 and private communication.
8. A. Ayachi, M. Durand, P. Schuck and V. S. Ramamurthy, *Z. Phys.* A**347** (1987) 141.
9. J. Côté and M. Pearson, *Nucl. Phys.* A**304** (1978) 104.
10. W. Stocker and M. Farine, *Ann. Phys.* (N.Y.) **159** (1985) 255.
11. E.P. Wigner, *Phys. Rev.* **40** (1932) 749, ibid. *Phys. Rev.* **46** (1934) 1002.
    J. G. Kirkwood, *Phys. Rev.* **44** (1933) 31.
    B. K. Jennings, *Ph. D. Thesis*, McMaster University (1976).
    P. Ring and P. Schuck, *Nuclear Many Body Problem* (Springer-Verlag) 1980.
12. B. Grammaticos and A. Voros, *Ann. Phys.* (N.Y.) **123** (1979) 359; ibid. **129** (1980) 153.
13. M. Durand, P. Schuck and X. Viñas, *Z. Phys.*, A**346** (1993) 87.
14. H. Krivine, M. Casas and J. Martorell, *Ann. Phys.* (N.Y.) **200** (1990) 304.

15. R. Balian and C. Bloch, *Ann. Phys.* (N.Y.) **60** (1970) 401.
16. M. Farine, *Z. Phys.* A**331** (1988) 375.
17. P. Bonche, S. Koonin and J. W. Negele, *Phys. Rev.* C**13** (1976) 1226.
D. Dalili, J. Nemeth and C. Ngô, *Z. Phys.* A**321** (1985) 335.
J. Nemeth, M. Barranco, C. Ngô and E. Tomasi, *Z. Phys.* A**323** (1986) 419.
18. J. Dechargé and D. Gogny, *Phys. Rev.* C**21** (1980) 1568.
19. M. J. Giannoni and P. Quentin, *Phys. Rev.* C**21** (1980) 2076.
20. M. Centelles, X. Viñas, M. Barranco and P. Schuck, *Ann. Phys.* **221** (1993) 165.
M. Centelles and X. Viñas, *Nucl. Phys.* A in press.

ANALYTIC DETERMINATION OF COULOMB DYNAMIC POLARIZATION POTENTIALS

J. Gómez-Camacho and M. V. Andrés

Departamento de FAMN. Facultad de Física.
Universidad de Sevilla. Apdo.1065, 41080 Sevilla, Spain.

Abstract

We present a method to determine analytic expressions for local L-independent potentials that describe the dynamical polarization effects due to the coupling to excited states. The method consists in determining the radial dependence of the polarization potential so that its first-order semiclassical scattering amplitudes coincide for all the scattering angles with the second-order amplitudes corresponding to the coupling to the excited state. For the case of Coulomb excitation, ignoring the excitation energy, the second-order amplitudes are analytic functions of the angle. The polarization potential is expanded in powers of $1/r$, so that the first order transition amplitudes become analytic functions of the angle. Comparing both expressions, the coefficients in the expansion of the polarization potential can be obtained. We have obtained the imaginary polarization potential due to the quadrupole coupling of a 0^+ ground state to a 2^+ state. The effect of this polarization potential describes adequately the result of exact coupled channels calculations. Comparisons with other polarization potentials in the literature have been made.

1 Introduction

Coulomb excitation has an important effect in the scattering of heavy ions. The description of this effect could be done, in principle by means of a coupled channels calculation that includes explicitly the coupling to the low lying collective states. However, the long range of the Coulomb form-factors, together with the large number of partial waves that should be considered in the case of heavy ions, makes the standard coupled channels calculation awkward, and the interpretation of the results in terms of simple mechanisms obscure.

If one is mostly interested on the effects of the ground state, it is more convenient to describe the effect of Coulomb excitation by means of a Dynamic Polarization Potential (DPP). Formally, the DPP is a non-local L-dependent operator that depends to second order on the coupling potential to the excited states and on the propagator through the excited states. Of course, the use of the exact expression of the DPP is equivalent to perform the exact coupled channels calculation. However, the DPP can be simplified to give a more manageable operator, that can be included in standard optical model calculations to describe, approximately, the effect of Coulomb excitations.

Many authors have derived approximate expressions for the Coulomb DPP. Love *et al.*[1] proposed a polarization potential for quadrupole Coulomb interaction which is local and L-independent (LTS hereafter). It is purely imaginary. If the Coulomb radius is neglected, it behaves as r^{-5}. Although it reproduces the deviations of the experimental elastic cross-sections from the standard Fresnel shape, it is derived using plane wave propagators, which are not adequate for heavy ion scattering. Also, LTS polarization potential diverges when the distance between the center of the colliding nuclei corresponds to the turning point of a head-on collision.

Baltz *et al.*[2] (BGKP hereafter) improve LTS result avoiding the use of plane waves and using instead Coulomb wave functions. Although BGKP polarization potential allows for a good description of experimental data, it has the inconvenience of been angular momentum dependent. This makes it difficult to use, because most optical model codes just allow for radial dependence. Fröbrich *et al.*[3] obtained an L-independent polarization potential equivalent to BGKP using an inversion procedure. However, the previous knowledge of the scattering matrix is needed, and so one cannot get the polarization potential a priori.

Recently, Derner and Grawert [4] have calculated Coulomb polarization potentials in the semiclassical limit. They obtain an expression for the polarization potential that depends on the scattering angle. Their expressions apply not only to the central polarization potential in a spin-zero ground state, but also to the polarization potentials affecting to transition potentials due to the coupling to intermediate states. They consider approximately the effect of the excitation energy by means of a gaussian approximation to the Coulomb integrals.

In this work we calculate a local, L-independent polarization potential due to quadrupole Coulomb excitation. The radial dependence of the potential is not given just by r^{-5}, as in previous cases, but it is written as a power series expansion. As in the case of Derner and Grawert[4], we make use of the semi-classical approximation. Our expressions have the advantage of being only a function of the radius, and we make use of the exact expressions of the Coulomb integrals. However, we ignore so far the excitation energy of the excited state. It should be noticed that the technique that we present here can be used to derive non-central and transition polarization potentials[5]. The expression for the DPP is derived in section 2. In section 3 we compare our results with the LTS potential and with exact coupled-channel calculations. Section 4 is for the summary and conclusions.

2 Dynamic Polarization Potentials

The aim of this work is to obtain an L-independent local potential that reproduces the second order matrix elements due to the coupling to excited states. To do that, we require that the first order matrix elements of the polarization potential are equal to the second order matrix elements of the coupling potential, summed over all the excited states considered. Furthermore, we will use a semi-classical description, so that each angular momentum is described by a classical trajectory, corresponding to a given scattering angle. Thus, we get:

$$b_{00}^{(2)}(\theta) = \frac{1}{(i\hbar)^2}\sum_n \int_{-\infty}^{\infty} V_{0n}(\mathbf{r}(\theta,t))e^{-i\omega_n t}dt \int_{-\infty}^{t} V_{n0}(\mathbf{r}(\theta,t'))e^{i\omega_n t'}dt' \quad (1)$$

$$b^{pol}(\theta) = \frac{1}{i\hbar}\int_{-\infty}^{\infty} V^{pol}(\mathbf{r}(\theta,t))dt \quad (2)$$

It should be stressed that the coupling potentials, as well as the polarization potential, are local functions, that depend only on the relative co-ordinate $\mathbf{r}$. This depends on the scattering angle and the time through the function $\mathbf{r}(\theta,t)$ that defines the trajectory for a given scattering angle θ. The functional form of $V^{pol}(\mathbf{r})$ is determined so that $b^{pol}(\theta)$ coincides with $b_{00}^{(2)}(\theta)$ for all the scattering angles. It should be noticed that, in general, $b_{00}^{(2)}(\theta)$ will be complex, and thus the polarization potential will have real and imaginary parts. However, if the excitation frequency ω_n can be ignored, and if the coupling potentials are hermitian, as in the case of Coulomb excitation, then $b^{(2)}(\theta)$ is purely real, and is given by

$$b^{(2)}(\theta) = -\frac{1}{2}\sum_n |b_{0n}^{(1)}(\theta)|^2 \quad (3)$$

where $b_{0n}^{(1)}(\theta)$ is the first-order transition amplitude to the state n, given by

$$b_{0n}^{(1)}(\theta) = \frac{1}{i\hbar}\int_{-\infty}^{\infty} V_{0n}(\mathbf{r}(\theta,t))dt \quad (4)$$

Let us consider the case of quadrupole Coulomb excitation of a 0^+ state, to a 2^+ state. There five magnetic substates to be considered, labelled by m. The first order transition amplitudes are:

$$|b_{0m}^{(1)}(\theta)|^2 = K(d_{m0}^2(\pi/2)I_{2m}(\theta,0))^2 \quad (5)$$

where $I_{2m}(\theta,0)$ is the semiclassical Coulomb integral [6], and K is an adimensional constant given by

$$K = \frac{4\pi Z^2e^2}{25\hbar^2v^2a_0^4}B(E2,0^+\rightarrow 2^+) \quad (6)$$

a_0 is half the distance of closest approach, and v the velocity. Let us write the polarization potential as:

$$V^{pol}(r) = -i\frac{K\hbar v}{2r}\sum_{n=0}^{\infty} c_n\frac{a_0^n}{r^n} \quad (7)$$

Then, the first order integral of V^{pol} can be written in terms of Coulomb integrals[6], so that one gets:

$$b^{pol}(\theta) = -\frac{K}{2}\sum_{n=0}^{\infty} c_nI_{n0}(\theta,0) \quad (8)$$

Thus, the expansion coefficients of the polarization potential are defined through the relation

$$\sum_n c_n I_{n0}(\theta,0) = \frac{3}{4} I_{22}(\theta,0)^2 + \frac{1}{4} I_{20}(\theta,0)^2 \qquad (9)$$

Note that all the expressions here are analytic functions, although they get complicated as n increases. The coefficients c_n can be obtained considering that $I_{n0}(\theta,0)$ is proportional to $\tan(\theta/2)^n$ for small values of θ. Thus, expanding both sides of the previous equation in powers of $\tan(\theta/2)$, and equating the coefficients, one can obtain the values of c_n. The second half of eq. 9 is proportional to $\tan(\theta/2)^4$ for small theta. So, the first non-vanishing c_n is the one for n=4. The values that we obtained for c_n, however, increased with n, so that the expansion did converge slowly for values of the radius close to the distance of closest approach, $2a_0$. Thus, we redefined the series expansion of the polarization potential as:

$$V^{pol}(r) = -i\frac{4\pi Z^2 e^2 B(E2, 0^+ \to 2^+)}{50\hbar v r^5} \sum_{m=0}^{\infty} d_m \left(\frac{a_0}{r - a_0}\right)^m \qquad (10)$$

We have seen that, taking just the first five terms in the expansion, the error is less than 1 per thousand at the turning point, and smaller for larger distances. The corresponding coefficients are:

$$d_0 = 1; \quad d_1 = \frac{4}{3}; \quad d_2 = -\frac{11}{12} + \frac{15\pi^2}{64}; \quad d_3 = -\frac{41}{5} + \frac{15\pi^2}{16}; \quad d_4 = -\frac{2611}{120} + \frac{285\pi^2}{128}. \qquad (11)$$

The following coefficients in the expansion are considerably smaller than these, and alternate in sign.

The treatment of the analytic expressions of the Coulomb integrals, as well as the series expansion and the expansion coefficients was done using the symbolic algebra code MATHEMATICA.

3 Results and Comparison with other Models

We have chosen the $^{18}O + {}^{184}W$ system to make the comparison of our results with coupled–channel calculations and with other models. That is because in that case the cross–sections are greatly influenced by Coulomb excitation and because it has become an standard touchstone since the pioneering work of Love *et al*[1]. Reorientation effect is not taken into account, only Coulomb excitation is allowed, the $B(E2, 0 \to 2)$ of ^{184}W is taken as 3.76×10^4 e^2fm^4, while the excitation energy of the 2^+ state is ignored. The nuclear optical potential was taken[1] as a Woods–Saxon with the same geometry for the real and the imaginary part (V = 40 MeV, W = 25 MeV, $r_0 = 1.29$ fm, a = 0.508 fm). Calculations were made by using FRESCO code[7]. We considered two energies: 90 MeV, as in [1], and 75 MeV, just at the Coulomb barrier. In Figure 1 we compare the polarization potential calculated in this work (full line) with the

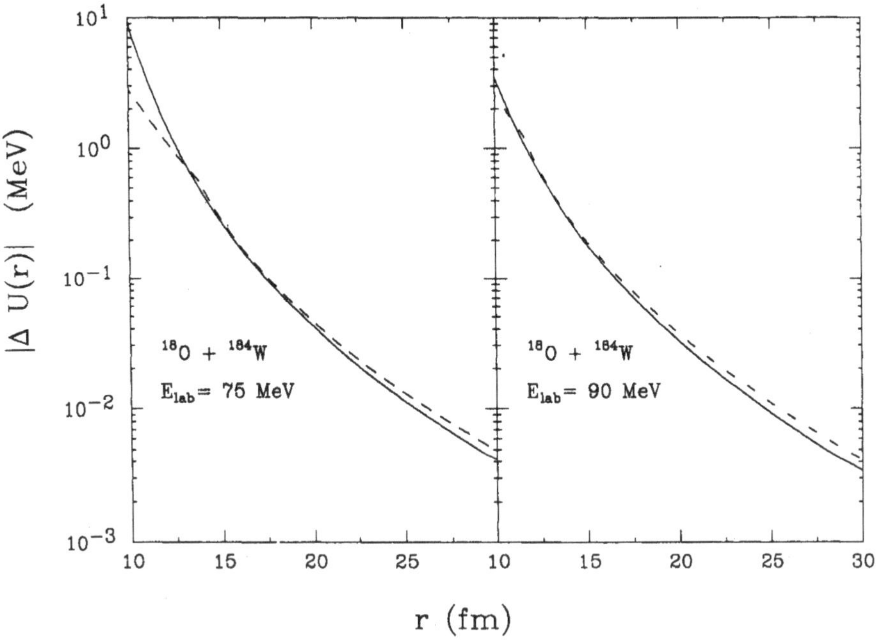

Figure 1: Dynamic polarization potential calculated in this work (full line) and the LTS potential[1] (dashed line), for $^{18}O + ^{184}W$ at 75 MeV and 90 MeV laboratory energy.

LTS polarization potential (dashed line). The agreement of both expressions is fairly good. However, for large distances the LTS potential is larger than ours by a factor 4/3. This overestimation of the LTS potential was already pointed out by Baltz *et al.*[2]. Also, for a distance $2a_0/0.9$, there is a discontinuity in the LTS potential. This is due to the fact that the LTS potential is inversely proportional to the local momentum. Thus, to avoid singularities at the turning point, Love *et al.*[1] introduce ad hoc the prescription of using the local momentum at $r = 2a_0/0.9$ for smaller values of the radius. In Figure 2 we compare coupled channels calculations (crosses) with the optical model calculations using the potential calculated in this work (full line) and the LTS potential (dashed line). At 90 MeV, the agreement of both optical model calculations with the coupled channels result is very good in general. For the small angle region, the LTS potential produces absorption slightly larger than the coupled channels calculation, that agrees remarkably well with the results of our potential. This is consistent with the fact that LTS overestimate the true polarization potential for large distances. At 75 MeV, the situation at forward angles is similar to the case of 90 MeV. However, at backward angles, there is a discrepancy between both optical model calculations and the coupled channels results. A similar situation was found by Franey *et al.*[9] while studying the $^{16}O + ^{152}Sm$ system at a laboratory energy of 72

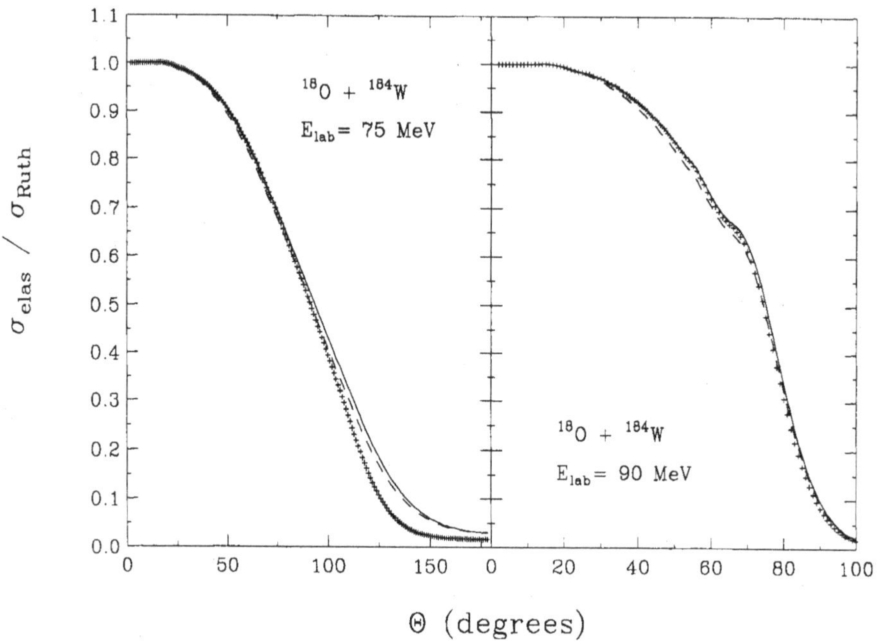

Figure 2: Elastic cross sections for $^{18}O + ^{184}W$ at 75 MeV and 90 MeV laboratory energy, calculated in exact coupled channels including Coulomb excitation to the 2^+ state in ^{184}W (crosses), and in optical model calculation using the potential calculated in this work (full line) and the LTS potential[1] (dashed line)

MeV: neither LTS nor BGKP polarization potential were able of reproducing the CC elastic cross–section at angles bigger than 90 degrees. To reproduce the backward angle region one had to consider the real polarization potential. In fact, if one calculates the polarization potential from the coupled channels calculation averaging the trivially equivalent local potential[10] over all the J-values, with weights given by the elastic wavefunction times the partial reaction cross-section for that J[8], one finds that the imaginary part is in very good agreement with our calculation, but there is a real part confined to distances below $2a_0$. This real potential, that for Coulomb trajectories would not have any effect, has an important effect in the backward cross sections, due to the distortion of the trajectory by the nuclear potential, and to quantum effects, because the scattering wavefunctions have finite values beyond the classical turning point.

4 Summary and Conclusions

We have described a procedure to obtain local L-independent polarization potentials to describe dynamical polarization effects, making use of the semiclassical description of the trajectory. We require that the first-order scattering amplitudes of the polarization potential are equal to the second-order amplitudes obtained coupling to the intermediate state. If the coupling potentials are hermitian, and if the excitation energy can be ignored, the second-order amplitudes are purely real and proportional to the square of the first-order amplitudes. Then, the polarization potential is purely imaginary.

For the case of Coulomb excitation, the first-order amplitudes are analytic functions of the scattering angle. The polarization potential can be expanded in powers of $1/r$, so that its matrix elements can also be written as a series of analytic functions of the angles. Doing a further expansion of these analytic functions for small scattering angles, one can obtain the coefficient of the expansion of the polarization potentials in powers of $1/r$. This expansion, however, converges slowly for values of r close to the turning point, $2a_0$. Thus, we re-write the series expansion in terms of powers of $a_0/(r - a_0)$. The convergence of this series is very fast, so that with 5 terms one gets an accuracy of $1/1000$ at the turning point.

The potential that we obtain is similar to the one proposed by Love *et al*[1]. However, our derivation is more soundly based, because we do not have to use plane wave propagator. Our potential does not have any singularity or discontinuity at the turning point. Also, the behavior of our potential for large distances, overestimated by the potential of Love *et al*[1], is the correct one.

Our potential is similar to the L-dependent potential of Baltz et al[2], or to the angle-dependent potential of Derner and Grawert[4], in the fact that the first-order matrix elements of the polarization potential reproduce second-order matrix elements of the coupling. Nevertheless, our potential has the advantage of being independent on L (or the angle). We consider that, if one is interested in the description of coupling effects on the elastic scattering, and if this effect is a smooth function of the orbital angular momentum, it should always be possible to find an L-independent local dynamic polarization potential that produces this effect. Thus, there is no need to introduce L-dependent polarization potentials.

Our potentials reproduce adequately the coupled channels results, with the exception of backward scattering close to the Coulomb barrier. There, the real part of the dynamic polarization potential plays an important role, even if it is confined to the classically forbidden region. Of course, in a treatment based on Coulomb trajectories, we cannot obtain the potentials in the classically forbidden region. An extension of this work considering the exact expressions of the matrix elements of r^{-n} potentials with could help to describe the origin or the real polarization potential.

This work does not consider the effect of the excitation energy of the intermediate state. This effect could be taken into account as in Love *et al* by the introduction of a global renormalization of the potential, corresponding to the ratio of the total inelastic cross section considering the excitation energy to the one without considering it. Nevertheless, we expect that a finite excitation energy will give rise to a real polarization potential, and will also modify the geometry of the imaginary polarization potential.

Acknowledgements

This work has been partially supported by the spanish DGICyT under contract PB89-0636. The authors wish to acknowledge fruitful discussions with M. A. Nagarajan.

REFERENCES

[1] W.G. Love, T. Terasawa and G.R. Satchler, Nucl. Phys. **A291** (1977) 183.

[2] A.J. Baltz, S.K. Kauffmann, N.K. Glendenning and K. Pruess, Nucl. Phys. **A327** (1979) 221.

[3] P. Fröbrich, R. Lipperheide and H. Fiedeldey, Phys. Rev. Lett. **43** (1979) 1147.

[4] J. Chr. Derner and G. Grawert, Nucl. Phys. **A531** (1991) 404.

[5] M. V. Andrés and J. Gómez-Camacho, to be submitted to Nucl. Phys. **A**

[6] K. Alder and A. Winther, Electromagnetic excitation: Theory of Coulomb excitation with heavy ions (North–Holland, Amsterdam, 1975)

[7] S. Wolfram, MATHEMATICA: A program for doing mathematics by computer, Addison Wesley (1988).

[8] I. J. Thompson, FRESCO, Comp. Phys. Reps. **7** (1988) 167.

[9] M. A. Franey and P.J. Ellis, Phys. Rev. **C23** (1981) 787.

[10] G. R. Satchler, Direct Nuclear Reactions (Oxford University Press, Oxford, 1983) p. 821

FINITE SIZE EFFECTS IN THE ELECTROMAGNETIC QUASI-ELASTIC RESPONSES OF NUCLEI

A.M. LALLENA and J.E. AMARO
Departamento de Física Moderna, Universidad de Granada,
E-18071 Granada, Spain

and

G. CO'
Dipartimento di Fisica, Università di Lecce and INFN, Sezione di Lecce,
I-73100 Lecce, Italy

ABSTRACT

The effects of the finite size of the nucleus in the electromagnetic excitation of the quasi-elastic peak have been studied by comparing the responses calculated for ^{12}C with the Fermi gas and the shell model. The contribution of the Meson Exchange Currents has also been considered. We have investigated the possibility to improve the Fermi gas model with approximate estimations of the finite size effects.

1. Introduction

The study of the electromagnetic quasi-elastic responses of the nucleus has received a lot of attention in the last decade because the technical improvements of the electron accelerators have allowed to perform experiments where the longitudinal and transverse responses have been separated[1]. The understanding of these experimental data is still an open problem. At present there is no nuclear model able to describe consistently both responses. The longitudinal responses are overestimated and the transverse ones are underestimated.

Many possible explanations of the problem have been proposed. They can be roughly grouped in three categories: many-body effects, relativistic effects and modifications of the nucleon properties inside the nucleus[2].

In the traditional non relativistic many-body approach a large body of theoretical work has been describing the nucleus as an infinite system of nucleons.

The pragmatical reason for this choice is that a proper description of the quasi-elastic dynamics should take into account subtle nuclear effects such as final state interactions (FSI), meson exchange currents (MEC), short-range correlations, Δ isobar excitations, etc. The treatment of these effects is simplified in an infinite system where, due to translational invariance, the single particle wave functions are plane waves. This allows to perform analytically part of the calculations and simplifies the numerics.

From the physics point of view there are good reasons to believe that, in the study of the quasi-elastic responses, the nucleus can be reasonably described as an infinite system. In this region the order of magnitude of the distances probed by the virtual photon exchanged by the electron and the nucleus is smaller than 1 fm. Since surface and shell effects are extended over bigger distances their contribution to the quasi-elastic excitation is expected to be small.

Obviously, nuclei are not infinite systems of nucleons, and surface and shell effects are playing an important role in the nuclear dynamics.

In this contribution we present the result of a work we have done in order to estimate the finite size effects in the description of the quasi-elastic excitation. We have carried on this investigation comparing shell model results in ^{12}C with those obtained using various approximations of the infinite system model, all of them based upon a Fermi gas description of the nucleus.

2. The model

In Plane Wave Born Approximation the inclusive electron scattering cross section can be separated in two terms depending respectively from the longitudinal

$$R_L(q,E) = \sum_f \mid \langle f \mid \rho(\mathbf{q}) \mid i \rangle \mid^2 \delta(E_f - E_i - E), \tag{1}$$

and transverse

$$R_T(q,E) = \sum_f \mid \langle f \mid \mathbf{J}_T(\mathbf{q}) \mid i \rangle \mid^2 \delta(E_f - E_i - E) \tag{2}$$

responses, where $|i\rangle$ and $|f\rangle$ indicate the initial and final states of the nuclear system and ρ and $\mathbf{J}_T$ are respectively the nuclear charge and the transverse term of the nuclear current.

In our model we have considered Meson Exchange Currents (MEC) which are acting only on the transverse response. This means that in addition to the ordinary one-body currents (convection and magnetization) we have considered MEC generated by the exchange of one pion: the seagull, or contact, term and the pionic, or pion in flight, term.

The shell model responses have been calculated using a Woods-Saxon well as described in ref.3. We have considered the one nucleon emission term by a proper treatment of the continuum.

In the literature there are quite a few Fermi gas calculations[4-6] of the quasi-elastic response, but comparing them with our shell model results we found some relevant discrepancies. In addition, we found that the Fermi gas results are not consistent among themselves. For example the calculations of refs.5 and 6 produce contributions of the pionic current to the transverse response with different sign. We were interest in clarifying these problems. For this reason we have recalculated the Fermi gas responses obtaining slightly different expressions from the usual ones[7].

MEC are a small perturbation with respect to the main one-body piece of the nuclear current. For this reason the major contribution of the MEC is given by the interference with the one-body term. The contributions given by the pure MEC are extremely small[3,7], and we have neglected them in the Fermi gas calculations.

3. Results

In Fig.1a we compare the Fermi gas (dashed and dotted lines) and the shell model (full lines) transverse responses for three different values of the momentum transfer. The dashed lines have been obtained with the traditional value of the Fermi momentum for nuclear matter, $k_F = 272$ MeV/c. These responses are remarkably different from the shell model ones. This is due to the fact that in finite nuclei the average density is smaller than the nuclear matter density. We took care of this by fixing an effective value of the Fermi momentum given by :

$$\frac{\langle k_F \rangle = (3\pi^2)^{1/3} \int \rho(r)^{4/3} r^2 \, dr}{\int \rho(r) r^2 \, dr,} \tag{3}$$

where $\rho(r)$ is the proton density of the finite nucleus. For ^{12}C we obtained $k_F = 215$ MeV/c and the dotted lines have been calculated with this effective value. The agreement with the shell model has greatly improved. We noticed that this value is very close to that empirically fixed in ref.8 to reproduce the full inclusive cross section.

Having seen the good results obtained using the effective Fermi momentum, we have tried to improve the agreement between Fermi gas and shell model using the Local Density Approximation (LDA). In this approach the finite nucleus response is described at any place by the nuclear matter response calculated at a proper density. This approximation is widely used in calculating both electromagnetic and hadronic responses of the nuclei.

In Fig.1b we compare the LDA (dashed and dotted lines) and the shell model (full lines) transverse responses. The dotted lines have been obtained using the proton density given by the shell model, and in spite of the fact that the LDA produces a tail at high excitation energies, like in the shell model responses, the agreement between the two models is rather poor.

In order to understand this disagreement we have further studied the sensitivity of the LDA response to the variations of the density $\rho(r)$. We have modified the proton density using a Fermi distribution whose parameters have been adjusted with the least squares method to reproduce the shell model response. The results of this calculation is presented by the dashed lines in Fig.1b. The proton density profile producing the adjusted curves is shown in Fig.2 together with the shell model

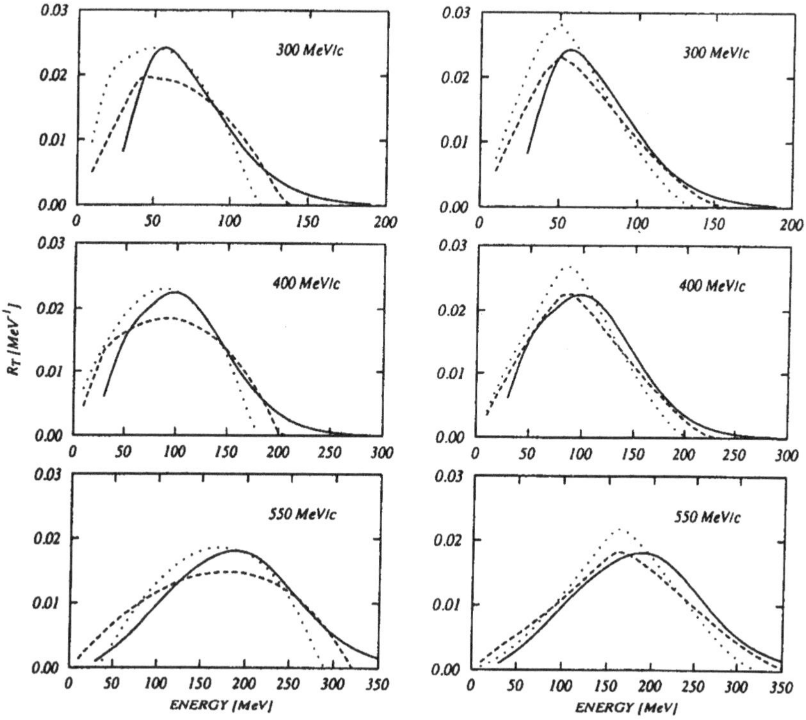

(a) (b)

Figure 1. Transverse response functions of ^{12}C for various momentum transfer. The full lines represent the shell model results. (a) The dotted and dashed lines have been obtained with the Fermi gas model using $k_F = 215$ MeV/c and $k_F = 272$ MeV/c respectively. (b) The dotted and dashed lines have been obtained with the LDA using the shell model density and the adjusted Fermi density respectively.

proton distribution. One can see that the two densities are very different and this generates doubts on the validity of the LDA at least in light nuclei.

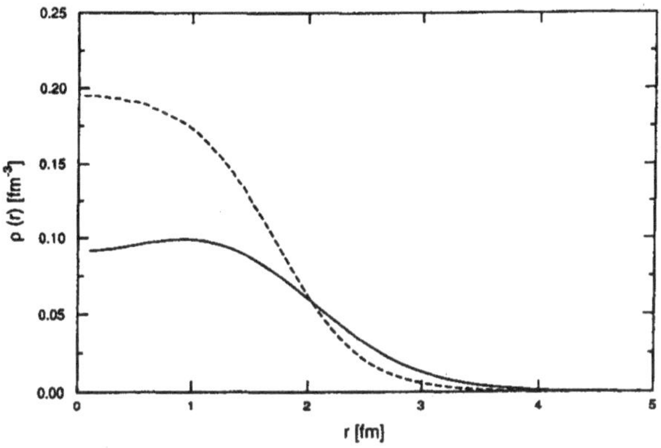

Figure 2. Proton density distributions of ^{12}C used in LDA calculations. The full line is the shell model density and the dashed line the adjusted Fermi density.

For the longitudinal responses we obtain results very similar to those shown in the previous figures[7].

The interference terms between MEC and one-body currents are shown in Fig.3a where the dashed lines are presenting the Fermi gas results (with the effective Fermi momentum k_F=215 MeV/c), and the full lines are showing the shell model responses. All the positive curves are given by the interference with the seagull term, while the negative ones correspond to the pionic interference. This behavior is in agreement with the calculations of ref.5.

Since the agreement between the MEC contributions obtained in Fermi gas and in the shell model is rather satisfactory, we have considered a model which takes care of finite size effects at the one-body level and uses the Fermi gas results for the MEC.

The effective response containing the MEC contributions is calculated considering the total Fermi gas response (one-body plus MEC), subtracting the one-body Fermi gas response and then adding the one-body shell model term. The results obtained with this model are shown in Fig.3b by the dashed lines. One can remark the excellent agreement with the shell model calculation (full lines).

4. Conclusions.

From our investigation we draw the following conclusions.

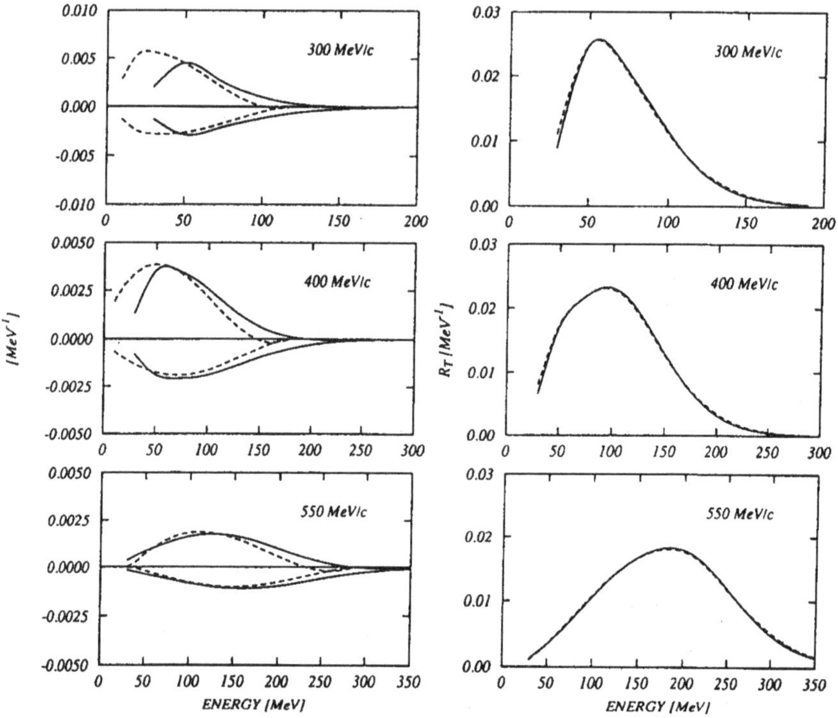

Figure 3. (a) Interference terms between one-body and MEC. The positive curves correspond to the one-body seagull contributions, while the negative lines to the one-body pionic ones. The full lines have been obtained with the shell model and the dashed lines with the Fermi gas model using k_F = 215 MeV/c. (b) Transverse responses obtained with one-body and MEC. The full lines are the shell model results. The dashed lines have been obtained using the effective model described in the text.

1) The results of the Fermi gas model are strongly dependent from the value of the Fermi momentum k_F and we have found that using an effective value of it one can obtain a good agreement with the shell model results.
2) We found that LDA is not a reliable approximation in light nuclei. We achieved a reasonable description of the shell model results only with densities rather different from those produced by the shell model.
3) The effects of MEC in the quasi-elastic peak are rather well described by the Fermi gas calculations. Since their contributions are relatively small with respect to the one-body responses, we proposed a model where finite size effects are considered only at one-body level. The results obtained are in excellent agreement with those of the shell model.

5. Acknowledgments.

This work has been developed in the frame of the CICYT-INFN agreement. The work of J.E.A. and A.M.L. has been partially supported by the DGICYT (Spain) under contract PB92-0927 and by the Junta de Andalucía (Spain).

6. References

1. R.Altemus et al. , Phys. Rev. Lett. 44 (1980), 965.
   P.Barreau et al. , Nucl. Phys. A402 (1983), 515.
   M.Deady et al. , Phys. Rev. C28 (1983), 631.
   Z.E.Meziani et al. , Phys. Rev. Lett. 52 (1984), 2130.
   Z.E.Meziani et al., Phys. Rev. Lett. 54 (1985), 1233.
   M.Deady et al., Phys. Rev. C33 (1986), 1897.
2. P.M. Boucher and J.W. Van Orden, Phys. Rev. C43 (1991) 582, and references therein.
3. J.E.Amaro, G.Co', E.M.V.Fasanelli and A.M.Lallena, Phys. Lett. B277 (1992) 249.
   J.E.Amaro, G.Co' and A.M.Lallena, Ann. Phys. (N.Y.) 221 (1993) 306.
4. W.M.Alberico, P.Czerski, M.Ericson and A.Molinari, Nucl. Phys. A462 (1987) 269.
   M.J.Dekker, P.J.Brussaard and J.A.Tjon, Phys. Lett. B266 (1991) 249.
5. M.Kohno and N.Ohtsuka, Phys. Lett. B98 (1981) 335.
6. W.M.Alberico, T.W.Donnelly and A.Molinari, Nucl. Phys. A512 (1990) 541.
7. J.E.Amaro, Ph. D. thesis, Universidad de Granada (1993)
   J.E. Amaro, G. Co' and A.M. Lallena. Universidad de Granada preprint, UG-DFM-7/93.
8. E.J.Moniz et al., Phys Rev. Lett. 26 (1971) 445.

A NUCLEAR FLUID-DYNAMICAL DESCRIPTION OF LOW-LYING COLLECTIVE MODES AT A FINITE TEMPERATURE

J. da Providência Jr.
Departamento de Física, Universidade de Coimbra
P-3000 Coimbra, Portugal

ABSTRACT

At zero temperature a fluid-dynamical model has been considered some years ago which describes surface collective modes whose energy is directly dependent on the surface tension.

One of the interesting features of this model, which we have investigated again, is that at a finite temperature it provides an additional dispersion relation which leads to a larger fragmentation of the strength. The present model, due to its parametrization, provides a better description of the nuclear dynamics than similar models and predicts an appreciable redistribution of transition strength with temperature, in agreement with recent experimental findings.

1. Model, Lagrangian and Equations of Motion

Our aim is to study the extension to a finite temperature of the model proposed in refs.[1-3]. This model may be solved analitically, allows for a rather large strength distribution and leads to a description of giant resonances and also to a description of low-lying collective modes which depend strongly on the surface tension and include important rotational components. The energy of the surface modes is directly related to the surface tension.

We will follow closely ref.[4] where a simpler version of the model of refs.[1,2] was studied and extended to a finite temperature.

We start from the classical limit of a variational formulation[5] of the Liouville-von Neumann equation of statistical mechanics in order to obtain a mean field approximation appropriate to mixed states. The lagrangian

$$L = i\hbar \; tr(\dot{U} D_0 U^\dagger) - tr(U D_0 U^\dagger H), \tag{1}$$

determines the time-evolution of the density matrix $D = U D_0 U^\dagger$, where D_0 stands for the density matrix describing a state of equilibrium and U is a unitary operator. If we write $U = \exp\left(\frac{i}{\hbar}\hat{\mathcal{G}}\right)$, the lagrangian L reduces, for small amplitude oscillations, to the following harmonic lagrangian

(2)

$$L^{(2)} = tr\left(D_0\left(\frac{i}{2\hbar}[\hat{\mathcal{G}},\dot{\hat{\mathcal{G}}}] - \frac{1}{2\hbar^2}[\hat{\mathcal{G}},[\hat{H},\hat{\mathcal{G}}]]\right)\right). \tag{3}$$

We will consider the classical limit of the lagrangian in Eq. 3, retaining only the leading order terms in a Wigner-Kirkwood expansion in powers of $\hbar$. We will assume density dependent Skyrme forces. In the energy functional characterizing the equilibrium state we do not include terms with derivatives of the density. These terms tend to compress the central density and determine the surface profile. We replace such terms by a temperature dependent surface tension $\sigma(T)$ which decreases with increasing temperature in order to mock up the fact that with increasing temperature the surface profile of the nucleus becomes smoother. We note that a sharp surface profile arises in our calculations because we are neglecting terms with derivatives of the density in the equilibrium energy density. In the equilibrium state the energy is,

$$E^{(0)} = \frac{4\pi}{3}R_0^3(\tau_0 + \sum_\sigma a_\nu \rho_0^\nu) + \sigma 4\pi R_0^2, \tag{4}$$

where R_0 stands for the radius of the nucleus and ρ_0 and τ_0 stand respectively for the equilibrium density and the equilibrium kinetic energy density. The equilibrium distribution function is a Fermi type distribution function

$$f_0 = \frac{\Theta(R_0 - r)}{1 + \exp[(\frac{p^2}{2m} - \varsigma_0)/T]}, \tag{5}$$

where R_0 and ς_0 are determined by the particle number A and by the minimization of the free energy (see ref.[4]) which leads to the equilibrium condition

$$\frac{2}{3}\tau_0 + \sum_\nu a_\nu(\nu - 1)\rho_0^\nu - \frac{2\sigma}{R_0} = 0, \tag{6}$$

where

$$\rho_0 = g\int \frac{d^3p}{(2\pi\hbar)^3} f_0, \tag{7}$$

$$\tau_0 = g\int \frac{d^3p}{(2\pi\hbar)^3} f_0 \frac{p^2}{2m}, \tag{8}$$

and g stands for the spin-isospin multiplicity. We note that $\varsigma_0 = \mu - U_0$, where μ stands for the chemical potential and $U_0 = \sum_\nu a_\nu \nu \rho_0^\nu$ is the equilibrium selfconsistent potential.

In the classical limit we have $\hat{\mathcal{G}} = \sum_{i=1}^A \mathcal{G}(\mathbf{x}_i, \mathbf{p}_i, t)$, where the generator $\mathcal{G}(\mathbf{x},\mathbf{p},t)$ may be decomposed into a time-even term Q and a time-odd term P ($\mathcal{G} = Q + P$). A parametrization by means of a scalar field $\psi(\mathbf{x},t)$ and a tensor field $\phi_{\alpha\beta}(\mathbf{x},t)$ is chosen for the generator Q

$$Q = \psi(\mathbf{x},t) + \frac{1}{2}p_\alpha p_\beta \phi_{\alpha\beta}(\mathbf{x},t). \tag{9}$$

We assume P is the generator of a canonical transformation such that the following equation is approximately satisfied

$$f_0 + \{f_0, P\} + \frac{1}{2}\{\{f_0, P\}, P\} + \dots$$
$$= \frac{\Theta(R_0 + R_1(\theta,\phi,t) - r)}{1 + \exp[(\frac{p^2}{2m} - \varsigma_0 - \varsigma_1(\mathbf{x},t) + \frac{p_\alpha p_\beta}{2m}\chi_{\alpha\beta}(\mathbf{x},t))/T]}. \tag{10}$$

This equation generalizes to a finite temperature the definition of the generator P given in refs.[1,2]. We remark that R_1 describes the surface displacements. It is clear that at a finite temperature no generator P exists such that Eq. 10 holds exactly. This means that, at a finite temperature, R_1, ς_1 and $\chi_{\alpha\beta}$ are not an adequate parametrization since these quantities can not be generated by a canonical transformation. However the errors involved in the deviation from canonicity will be minimized if we require that the fields R_1, ς_1 and $\chi_{\alpha\beta}$ are such that the entropy remains constant.

From the distribution function f,

$$f = f_0 + \{f_0, \mathcal{G}\} + \frac{1}{2}\{\{f_0, \mathcal{G}\}, \mathcal{G}\} + \dots, \tag{11}$$

we derive the physical quantities, such as the density, the current and the energy. Up to first order in the variational fields we have the following expressions for the density

$$\rho = \rho_0 + \mathcal{A}_1\varsigma_1 - \frac{\rho_0}{2}\chi_{\alpha\alpha}, \tag{12}$$

where

$$\mathcal{A}_1 = \frac{\partial\rho_0}{\partial\varsigma_0}, \tag{13}$$

and for the current[6]

$$j_\alpha = \frac{\rho_0}{m}\partial_\alpha\psi + \frac{2\tau_0}{3}\left(\frac{1}{2}\partial_\alpha\phi_{\mu\mu} + \partial_\beta\phi_{\alpha\beta}\right). \tag{14}$$

In order to arrive at the expression 14 we had to impose the following boundary condition[6]

$$x_\alpha\phi_{\alpha\beta}|_{r=R_0} = 0. \tag{15}$$

We ensure that the boundary condition 15 is fulfilled and that the entropy S and the particle number A are conserved on the average by means of the Lagrange multipliers $\underline{\xi}$, T and μ. From the classical limit of the lagrangian in Eq. 3, with the choices for the generators Q and P implied by eqs. 9 and 10 and taking into account the subsidiary conditions which ensure the boundary condition 15, the conservation of the entropy S and of the particle number A, the following effective lagrangian is obtained

$$\mathcal{L}^{(2)} = (L - \mu A + TS)^{(2)} + \oint_{\Sigma_0} d\Sigma\,\xi_\alpha n_\beta\phi_{\alpha\beta}, \tag{16}$$

where $n_\beta = x_\beta / r$. For the chosen parametrization, the lagrangian is given by

$$\mathcal{L}^{(2)} = -\int_{D_0} d^3x \left\{ m\dot{\phi}_{\alpha\beta} \left[\delta_{\alpha\beta} \left(\frac{\tau_0}{3} + \frac{\rho_0}{2}\varsigma_1 \right) - \frac{\tau_0}{3} \left(\chi_{\alpha\beta} + \delta_{\alpha\beta}\frac{1}{2}\chi_{\mu\mu} \right) \right] \right.$$
$$\left. + \dot{\psi} \left(\rho_0 + \mathcal{A}_1\varsigma_1 - \frac{\rho_0}{2}\chi_{\mu\mu} \right) \right\} - \oint_{\Sigma_0} d\Sigma R_1 \left(\rho_0 \dot{\psi} + \frac{m\tau_0}{3}\dot{\phi}_{\mu\mu} \right)$$
$$+ \oint_{\Sigma_0} d\Sigma \xi_\alpha n_\beta \phi_{\alpha\beta} - T^{(2)}[\psi, \phi_{\alpha\beta}] - \mathcal{E}^{(2)}[\varsigma_1, \chi_{\alpha\beta}, R_1], \tag{17}$$

$$T^{(2)}[\psi, \phi_{\alpha\beta}] = \int_{D_0} d^3x \left\{ \frac{\rho_0}{2m} (\nabla\psi)\cdot(\nabla\psi) + \frac{2\tau_0}{3} (\partial_\alpha \psi) \left(\frac{1}{2}\partial_\alpha \phi_{\mu\mu} + \partial_\beta \phi_{\alpha\beta} \right) \right.$$
$$\left. + \frac{2mG}{15} \left[\left(\frac{1}{2}\partial_\alpha \phi_{\mu\mu} + \partial_\beta \phi_{\alpha\beta} \right)^2 + \frac{1}{6} (\partial_\alpha \phi_{\beta\gamma} + \partial_\beta \phi_{\gamma\alpha} + \partial_\gamma \phi_{\alpha\beta})^2 \right] \right\}, \tag{18}$$

(see ref.[6]),

$$G = g \int \frac{d^3p}{(2\pi\hbar)^3} f_0 \left(\frac{p^2}{2m} \right)^2 \tag{19}$$

and

$$\mathcal{E}^{(2)}[\varsigma_1, R_1, \chi_{\alpha\beta}] = \int_{D_0} d^3x \left[\frac{1}{2}\mathcal{A}_1\varsigma_1^2 + \frac{\tau_0}{6} \left(\chi_{\alpha\beta}\chi_{\alpha\beta} + \frac{1}{2}\chi_{\mu\mu}\chi_{\nu\nu} \right) \right.$$
$$\left. - \frac{\rho_0}{2}\chi_{\nu\nu}\varsigma_1 + \frac{\mathcal{C}_3}{2} \left(\mathcal{A}_1\varsigma_1 - \frac{1}{2}\rho_0\chi_{\nu\nu} \right)^2 \right] + \frac{\sigma}{2R_0^2} \oint_{\Sigma_0} d\Sigma R_1^2 [\ell(\ell+1) - 2]. \tag{20}$$

Here $\mathcal{C}_3 = \sum_\nu a_\nu \nu(\nu-1)\rho_0^{\nu-2}$.

We see therefore that our lagrangian, which is given by Eqs. 17-20, contains information referring to the equilibrium distribution function since it is expressed in terms of the moments of f_0, in momentum space, of order zero (ρ_0), of second order (τ_0) and of fourth order (G).

The dots over the variational fields stand for time-derivative. By requiring that the action integral is stationary for arbitrary variations of the fields ψ, $\phi_{\alpha\beta}$, ς_1, $\chi_{\alpha\beta}$ and R_1, the following equations and boundary conditions are obtained

$$\mathcal{A}_1\dot{\varsigma}_1 - \frac{\rho_0}{2}\dot{\chi}_{\mu\mu} + \frac{\rho_0}{m}\Delta\psi + \frac{2\tau_0}{3} \left(\frac{1}{2}\Delta\phi_{\nu\nu} + \partial_\alpha\partial_\beta\phi_{\alpha\beta} \right) = 0, \tag{21}$$

$$\left\{ \rho_0 \dot{R}_1 - n_\alpha \left[\frac{\rho_0}{m}\partial_\alpha\psi + \frac{2\tau_0}{3} \left(\frac{1}{2}\partial_\alpha\phi_{\mu\mu} + \partial_\gamma\phi_{\alpha\gamma} \right) \right] \right\}_{r=R_0} = 0, \tag{22}$$

$$\delta_{\alpha\beta}\frac{m\rho_0}{2}\dot{\varsigma}_1 - \frac{m\tau_0}{3} \left(\delta_{\alpha\beta}\frac{1}{2}\dot{\chi}_{\nu\nu} + \dot{\chi}_{\alpha\beta} \right)$$
$$+ \frac{2\tau_0}{3} \left(\delta_{\alpha\beta}\frac{1}{2}\Delta\psi + \partial_\alpha\partial_\beta\psi \right) + \frac{2mG}{15} [\delta_{\alpha\beta} \left(\frac{1}{2}\Delta\phi_{\mu\mu} + \partial_\mu\partial_\nu\phi_{\nu\mu} \right)$$
$$+ \Delta\phi_{\alpha\beta} + \partial_\beta\partial_\alpha\phi_{\mu\mu} + 2\partial_\alpha\partial_\gamma\phi_{\gamma\beta} + 2\partial_\beta\partial_\gamma\phi_{\gamma\alpha}] = 0, \tag{23}$$

$$\Big\{\delta_{\alpha\beta}\Big[-\frac{m\tau_0}{3}\dot{R}_1+n_\gamma\Big(\frac{\tau_0}{3}\partial_\gamma\dot{\psi}+\frac{2mG}{15}(\frac{1}{2}\partial_\gamma\phi_{\mu\mu}+\partial_\mu\phi_{\mu\gamma})\Big)\Big]$$
$$+n_\alpha\Big[\frac{\tau_0}{3}\partial_\beta\psi+\frac{2mG}{15}\Big(\frac{1}{2}\partial_\beta\phi_{\mu\mu}+\partial_\mu\phi_{\mu\beta}\Big)-\frac{1}{2}\xi_\beta\Big]$$
$$+n_\beta\Big[\frac{\tau_0}{3}\partial_\alpha\psi+\frac{2mG}{15}\Big(\frac{1}{2}\partial_\alpha\phi_{\mu\mu}+\partial_\mu\phi_{\mu\alpha}\Big)-\frac{1}{2}\xi_\alpha\Big]$$
$$+n_\gamma\frac{2mG}{15}(\partial_\alpha\phi_{\beta\gamma}+\partial_\beta\phi_{\gamma\alpha}+\partial_\gamma\phi_{\alpha\beta})\Big\}_{r=R_0}=0, \tag{24}$$

$$\mathcal{A}_1\dot{\psi}+\frac{m\rho_0}{2}\dot{\phi}_{\alpha\alpha}+(1+\mathcal{A}_1\mathcal{C}_3)\Big(\mathcal{A}_1\varsigma_1-\frac{1}{2}\rho_0\chi_{\nu\nu}\Big)=0, \tag{25}$$

$$\delta_{\alpha\beta}\frac{\rho_0}{2}\Big[\dot{\psi}+\varsigma_1+\mathcal{C}_3\Big(\mathcal{A}_1\varsigma_1-\frac{\rho_0}{2}\chi_{\nu\nu}\Big)\Big]+\frac{\tau_0}{3}\Big[\delta_{\alpha\beta}\frac{1}{2}(m\dot{\phi}_{\mu\mu}-\chi_{\mu\mu})+m\dot{\phi}_{\alpha\beta}-\chi_{\alpha\beta}\Big]=0, \tag{26}$$

$$\Big\{\rho_0\dot{\psi}+\frac{m\tau_0}{3}\dot{\phi}_{\mu\mu}+\frac{\sigma R_1}{R_0^2}[\ell(\ell+1)-2]\Big\}_{r=R_0}=0. \tag{27}$$

While at zero temperature this model leads to four sound velocities which were previously obtained in refs.,[1,2] at a finite temperature an additional sound velocity is obtained so that an additional degree of freedom arises. This happens due to the fact that at finite temperature the profile of the Fermi surface in an infinite medium acquires a finite width. As we will see the appearence of a larger number of sound velocities implies also a larger number of eigenmodes.

2. Numerical Results and Conclusions

We have considered a potential energy density of the form previously used in refs.,[6,7]

$$u=a_2\rho^2+a_{2+1/6}\rho^{2+1/6}, \tag{28}$$

where

$$a_2=-\frac{3}{8}\times 3075.8 \quad \text{MeV fm}^3, \tag{29}$$

and

$$a_{2+1/6}=\frac{1}{16}\times 20216.4 \quad \text{MeV fm}^{3+\frac{1}{2}}. \tag{30}$$

We have additionally considered a temperature dependent surface tension. Since the surface of the nucleus becomes more diffuse when the temperature increases, we have considered (see ref.[8]) that the surface tension σ decreases with temperature according to $\sigma(T)=\sigma(0)(1-xT^2)$, where $\sigma(0)=1.0$ MeV fm^{-2} and $x=6\times10^{-3}$ MeV^{-2}.

In the present paper we are mainly interested in extending the results of refs.[1,2] to a finite temperature taking into account the effect of the surface tension. We arbitrarily consider a nucleus with A=208. In table 1, we give for different values of the temperature, the equilibrium values of ρ_0, τ_0, ς_0, μ, $E^{(0)}/A$, the surface energy and the entropy S for the system with A=208.

Table 1. Values of some of the ground state quantities for different values of T: the equilibrium density ρ_0, the kinetic energy density τ_0, $\varsigma_0 = \mu - U_0$, the chemical potential μ, $E^{(0)}/A$, the surface energy and the entropy.

T [MeV]	0	1	2	3	4	5
ρ_0 [MeV fm^{-3}]	0.1479	0.1476	0.1465	0.1447	0.1422	0.1389
τ_0 [MeV fm^{-3}]	3.104	3.102	3.096	3.086	3.068	3.041
ς_0 [MeV]	34.97	34.89	34.66	34.25	33.67	32.90
μ [MeV]	-11.76	-11.85	-12.10	-12.52	-13.11	-13.87
$E^{(0)}/A$ [MeV]	-10.79	-10.74	-10.58	-10.31	-9.926	-9.424
$4\pi R_0^2\sigma$ [MeV]	607.0	604.3	596.2	582.6	563.3	538.1
S	0	29.37	58.88	88.66	118.8	149.6

For the electric modes we consider the operators

$$D(\mathbf{x}) = r^2 \text{ for } \ell = 0, \quad D(\mathbf{x}) = (r - r^3/R_0^2)Y_{10} \text{ for } \ell = 1, \text{ and}$$
$$D(\mathbf{x}) = r^\ell Y_{\ell 0} \text{ for } \ell > 1. \qquad (31)$$

In Table 2 we present the energies and the percentages of the energy wheighted sum rule (m_1) corresponding to several eigenmodes for a nucleus with A=208.

For $\ell = 0$ and T=0 MeV we observe that 95.15% of the m_1 sum is concentrated at 15.87 MeV, while in the simpler model of ref.[4] we obtain 92.69% of the m_1 sum located at 14.63 MeV. At a finite temperature we see that new eigenmodes appear, 0_1^+, 0_4^+, 0_8^+... The new vibrational mode 0_1^+ absorves 23.13% of m_1 at T=5 MeV. The eigenmode 0_3^+ at T=0 MeV exhausts 2.25% of m_1 and at T=5 MeV it exhausts 6.26% of m_1. So it is clear that the strength becomes more distributed for $\ell = 0$ when we increase the temperature.

For $\ell = 2$, we note that most of the m_1 sum is concentrated in the state 2_2^+, which corresponds to the giant quadrupole resonance, and in the low-lying mode 2_1^+.

For $\ell = 3$, the m_1 sum is basically distributed through several modes, but still the two states exhausting a larger fraction of m_1, at T=0 MeV, are 3_1^- which is the low-lying mode, and exhausts 34.10% of m_1 at T=0 MeV, and 3_3^- which exhausts 43.44% of the m_1 sum at T=0 MeV and corresponds to the giant octupole resonance (GOR). It is however very interesting to note that at T=5 MeV this picture has changed considerably. While we still have the low lying state 3_1^-, which at T=5 MeV exhausts 40.95% of the m_1 sum, the strength corresponding to the GOR is now quite spread since the mode 3_3^- exhausts only 27.93% of the m_1 sum at T=5 MeV. This seems to indicate that the giant resonances disappear at high temperatures.

For $\ell = 4$ and $\ell = 5$ we see that the strength is distributed in a larger number of states being the low-lying mode the one which exhausts a larger fraction of the m_1 sum rule. We note that at a finite temperature the new modes 4_8^+ and 5_8^- appear. In particular at T=5 MeV the mode 5_8^- exhausts 20.93% of the m_1 sum.

Table 2. Energies and fractions of the energy wheighted sum rule (considering the excitation operators given by Eq. 31, are given for the electric modes listed in the first column. A nucleus with A=208 was considered.

ℓ^π	T=0 MeV		T=3 MeV		T=5 MeV	
	$\hbar\omega_i$	m_1 [%]	$\hbar\omega_i$	m_1 [%]	$\hbar\omega_i$	m_1 [%]
0_1^+	—	—	13.62	9.752	12.90	23.13
0_2^+	15.87	95.15	15.65	84.05	15.25	68.07
0_3^+	18.95	2.256	18.75	3.566	18.16	6.261
0_4^+	—	—	26.62	4.789×10^{-1}	25.13	6.550×10^{-1}
0_5^+	28.14	2.838×10^{-1}	28.46	4.940×10^{-2}	27.86	5.977×10^{-2}
0_6^+	36.83	3.428×10^{-1}	36.39	2.801×10^{-1}	35.02	2.496×10^{-1}
2_1^+	3.473	30.90	3.641	35.23	3.810	40.73
2_2^+	11.70	64.19	11.99	59.55	12.33	53.00
2_3^+	17.45	2.165	17.28	2.525	16.80	3.680
2_4^+	20.54	1.094	20.53	1.209	20.40	1.213
2_5^+	21.12	1.002	21.36	8.183×10^{-1}	21.70	6.560×10^{-1}
2_6^+	—	—	24.75	9.638×10^{-5}	23.49	6.407×10^{-3}
3_1^-	2.923	34.10	2.960	37.24	2.931	40.95
3_2^-	8.428	2.879×10^{-4}	8.724	6.333×10^{-3}	9.083	2.035×10^{-1}
3_3^-	18.53	43.44	18.72	38.08	18.73	27.93
3_4^-	22.80	10.88	22.68	13.58	22.37	20.00
3_5^-	25.15	3.847	25.43	3.428	25.81	2.916
3_6^-	26.87	5.178	26.99	5.300	27.02	5.185
3_7^-	—	—	29.65	1.958×10^{-1}	28.03	3.874×10^{-1}
4_1^+	4.505	33.85	4.581	37.45	4.557	41.78
4_2^+	12.26	2.046	12.62	1.008	13.07	2.428×10^{-1}
4_3^+	23.36	22.39	23.35	18.16	22.96	11.97
4_4^+	27.64	8.859	27.78	11.82	27.89	16.63
4_5^+	29.67	17.10	29.90	15.22	30.21	11.96
4_6^+	33.45	8.450	33.49	4.588	32.13	2.033×10^{-3}
4_7^+	35.38	4.184	34.43	7.182	34.10	10.27
4_8^+	—	—	36.21	1.434	35.46	3.949
5_1^-	6.208	33.18	6.319	36.88	6.293	41.41
5_2^-	15.82	2.880	16.23	1.667	16.74	6.312×10^{-1}
5_3^-	27.24	13.11	27.13	10.64	26.54	7.286
5_4^-	31.63	2.676	31.93	3.661	32.29	5.602
5_5^-	34.76	22.45	34.92	20.41	34.92	14.16
5_6^-	38.96	3.778×10^{-6}	38.27	7.887×10^{-1}	36.55	4.140
5_7^-	40.75	20.08	40.70	5.370	39.72	1.855×10^{-1}
5_8^-	—	—	41.17	15.01	41.38	20.93

We see that when the temperature increases there is a spreading of the strength corresponding to the giant resonances and that large fractions of the m_1 sum are concentrated at the energies of the low-lying modes 2_1^+, 3_1^-, 4_1^+, 5_1^-.

The states 3_1^-, 4_1^+ and 5_1^- have energies which are approximately proportional to $A^{-1/2}$ as is also the case in the simpler model of ref.,[4] but now have lower energies and strong transverse components in the current. However for $\ell = 2$ we were not able to find an analogous state, since the state 2_1^+, as well as the other states, have an energy which depends on A approximately as $A^{-1/3}$. The exceptions to the "$A^{-1/3}$ rule" are the low-lying states with $\ell \geq 3$.

An interesting feature of this model is that new states appear when we consider a finite temperature. For low temperatures, as the temperature decreases, these states exhaust a smaller fraction of the energy wheighted sum rule. Examples of the new states at finite temperature are 0_1^+, 0_4^+, 0_8^+, 1_6^-, 1_{11}^-, 2_6^+, 3_7^+, 4_8^- and 5_8^+. At low temperatures, the fractions of the m_1 sum corresponding to these modes, are proportional to T^2.

The present model predicts an appreciable redistribution of transition strength with temperature, in agreement with recent experimental findings.[9,10]

It may be shown that the present model fulfills the sum rules m_1, m_{-1} and m_3 for electric modes and also the sum rules m_1 and m_3 for magnetic modes.

We note that a nuclear fluid-dynamical formalism presents itself as a natural way of introducing temperature in the calculation of vibrational modes in an atomic nucleus. Since above 4 MeV the shell effects are washed out and the nucleus is more like a Fermi gas, the use of nuclear fluid dynamical models appears to be also quite appealing at finite temperatures.

References

1. L. Brito and C. da Providência, *Phys. Rev.* **C32** (1985) 2049; L. Brito, *Phys. Lett.* **177B** (1986) 251.
2. J. da Providência, L. Brito and C. Providência, *Nuovo Cimento* **87** (1985) 248.
3. H. Pascoal and L. Brito, *Phys. Rev.* **C39** (1989) 158.
4. J. da Providência Jr., *Nucl. Phys.* **A523** (1991) 247.
5. J. da Providência and C. Fiolhais, *Nucl. Phys.* **A435** (1985) 190.
6. J. P. da Providência and G. Holzwarth, *Nucl. Phys.* **A398** (1983) 59; J.P. da Providência, *J. Phys.* **G13** (1987) 783.
7. J.P. da Providência, *Journal de Physique* (1984) C6-333; J.P. da Providência and G. Holzwarth , *Nucl. Phys.* **A439** (1985) 477.
8. C. Guet, Strumberger and M. Brack, *Phys. Lett.* **B205** (1988) 427.
9. J. J. Gaardhøje et al., *Phys. Rev. Lett.* **56** (1986) 1783.
10. D. R. Chakrabarty et al., *Phys. Rev.* **C36** (1987) 1886.

SPECTROSCOPIC FACTORS FROM A RELATIVISTIC ANALYSIS OF (e,e'p)

J.M. Udías, P. Sarriguren, E. Moya de Guerra, E. Garrido and J.A. Caballero

Instituto de Estructura de la Materia,
Consejo Superior de Investigaciones Científicas,
Serrano 119, E-28006 Madrid, Spain

Abstract

We derive spectroscopic factors for the outermost shells in ^{40}Ca and ^{208}Pb from a fully relativistic analysis of $(e, e'p)$ data. We discuss various effects related to Coulomb distortion, the choice of the nucleon current operators, and the relativistic treatment of the outgoing-distorted and bound nucleon wave functions. We find spectroscopic factors of about 70%, consistent with theoretical predictions, and compare them with previous results.

1. Introduction

It is well known that the quasielastic $(e, e'p)$ reaction constitutes a very well suited tool to extract experimental information on absolute spectroscopic factors. Modern facilities allow to study this reaction in detail and high precision measurements of cross sections are available [1, 2]. Although the advantages of the $(e, e'p)$ process over other possible reactions to study spectroscopic factors are widely recognized, the extraction of these factors from experiment is still not free of ambiguities. A reliable determination of spectroscopic factors requires an accurate knowledge of the mechanism of the reaction and, in this context, the exact treatment of the Coulomb distortion of the electrons is important, especially in heavy nuclei such as lead. In this work we treat the Coulomb distortion of the electron in an exact way and use a relativistic formalism for both the leptonic and the nuclear vertices. Differences and similarities with previous calculations are discussed.

In what follows, we summarize the formalism used to describe the coincidence $(e, e'p)$ reaction in impulse approximation (IA). We denote by k_i^μ (k_f^μ) the four-momentum of the incoming (outgoing) electron and by q^μ the four-momentum of the exchanged photon. The ejected proton four-momentum is denoted by P_F^μ. We use the notation and conventions of Ref. [3] as well as $\hbar = c = 1$. The electromagnetic transition is treated in Born approximation, although we take into account the effect of the nuclear Coulomb potential to all orders by using electron distorted wave functions.

The differential cross-section for this process is then written as [4]:

$$\frac{d\sigma}{d\epsilon_f d\Omega_f dE_F d\Omega_F} = \frac{\delta(\epsilon_i + E_A - \epsilon_f - E_F - E_{A-1})}{(2\pi)^5} 4\alpha^2 \epsilon_f^2 E_F |\vec{P}_F| \overline{\sum} |W_{if}|^2 , \quad (1)$$

where $\overline{\Sigma}$ indicates sum (average) over final (initial) polarizations and

$$W_{if} = \int d\vec{x} \int d\vec{y} \int \frac{d\vec{q}}{(2\pi)^2} j_\mu^e(\vec{x}) e^{-i\vec{q}(\vec{x}-\vec{y})} \frac{(-1)}{q_\mu^2} J_N^\mu(\vec{y}) . \quad (2)$$

In this expression j_μ^e and J_N^μ stand for the electron and nuclear currents, respectively. The electron current is given by the well known point-like Dirac particle expression:

$$j_e^\mu(\vec{r}) = \bar{\psi}_f^e(\vec{r}) \gamma^\mu \psi_i^e(\vec{r}) , \quad (3)$$

where ψ_i^e, ψ_f^e stand for initial and final electron wave functions, which are solutions of the Dirac equation with the Coulmob potential. In IA and within an independent particle model picture, the nuclear current can be written in terms of the nucleon current operator $\hat{J}_N^\mu$

$$J_N^\mu(\vec{r}) = \bar{\psi}_F^N(\vec{r}) \hat{J}_N^\mu \psi_B^N(\vec{r}) , \quad (4)$$

with ψ_B^N, ψ_F^N the wave functions for the initial bound nucleon and final nucleon, respectively, and $\hat{J}_N^\mu$ a nucleon current operator to be specified later. The bound state wave functions for the proton ψ_B^N are spinors with well defined angular momentum quantum numbers. They have been computed within the framework of the relativistic independent particle shell model. The mean field in the Dirac equation is determined through a Hartree procedure from a phenomenological relativistic lagrangian with scalar and vector S-V terms. We use the parameters of Ref. [5], which are fitted to reproduce nuclear matter properties and the charge radius in ^{40}Ca, and the TIMORA code [6]. The wave function of the detected proton ψ_F^N is a scattering solution of a Dirac-like equation, which includes S-V global optical potentials, obtained by fitting elastic proton scattering data [7].

For the current operator we consider the two choices *cc*1 and *cc*2 introduced by de Forest [8] in momentum space

$$\hat{J}_N^\mu(cc1) = (F_1 + \bar{\kappa} F_2)\gamma^\mu - \frac{\bar{\kappa} F_2}{2M}(P_F + \bar{P}_I)^\mu , \quad (5)$$

$$\hat{J}_N^\mu(cc2) = F_1 \gamma^\mu + i \frac{\bar{\kappa} F_2}{2M} \sigma^{\mu\nu} q_\nu , \quad (6)$$

where F_1 and F_2 are the nucleon form factors and $\bar{P}_I$ is the four-momentum of the initial nucleon assuming on-shell kinematics [8]. As it is well known [3], Eqs. (5) and (6) are equivalent when the initial and final nucleons are on-shell. For off-shell nucleons, as is our case, both expressions lead to different results and do not satisfy current conservation.

The operator $cc2$ was used in previous fully relativistic calculations [9, 10]. In this work we use both $cc1$ and $cc2$ operators. Comparisons between half-off-shell electron-proton cross-sections obtained with both operators ($cc1$ and $cc2$) have previously been made using plane waves for initial and final nucleons [8, 11]. In Refs. [8] and [11] small differences (a few percent) were found at the kinematical conditions usually attained in the experiments. In the next section we present similar comparisons when realistic spinors are used to obtain the cross-section.

Different notations have been introduced in the literature by different authors to distinguish various degrees of approximations to the reaction mechanism. To avoid confusion we shall specify separately the approximation taken at each vertex, i.e., at the electronic and at the nuclear vertices. To distinguish the case when we use plane waves from the case when we use distorted waves in the electron current (Eq. (3)), we use the notations PWBA and DWBA, respectively. On the other hand, to distinguish the cases when we use plane waves and distorted waves for the ejected proton, we use the notation PWIA and DWIA, respectively. Unless otherwise specified, the results presented are obtained with proton distorted waves (DWIA).

2. Results

We present results for the $3s_{1/2}$ and $2d_{3/2}$ shells in ^{208}Pb and for the $2s_{1/2}$ and $1d_{3/2}$ shells in ^{40}Ca and compare with data in parallel kinematics ($\vec{q} \parallel \vec{P}_F$). We have chosen to study these shells for several reasons: *i)* they correspond to experimentally well separated peaks, *ii)* for these doubly magic nuclei the theoretical description is simpler and there are available optical potentials, *iii)* this choice allows us to study different mass regions where the Coulomb distortion of the electron waves is expected to play a different role, and *iv)* for these shells we can compare our results with experimental data as well as with other relativistic and/or nonrelativistic theoretical results. All the calculations have been done for a fixed value of the kinetic energy of the outgoing proton ($T_F = 100$ MeV). We also take a fixed value of the incoming electron energy, $\epsilon_i = 412$ MeV (375 MeV) for ^{208}Pb (^{40}Ca).

We first discuss results on the $3s_{1/2}$ shell in more detail because this shell is the most extensively studied in $(e, e'p)$, both experimentally and theoretically. To start with, it is worth clarifying the situation with regards to the disagreement between the PWBA results in Refs. [9] and [10]. As pointed out in Ref. [10], both calculations being the same in PWBA should lead to the same results. We found that the reason for the disagreement is due to the fact that different wave functions were used for the bound proton. In our calculations we used the code by Horowitz and Serot (TIMORA) with the standard set of initial conditions and obtained a result (see dashed line in Fig. 1a) in agreement with that in Ref. [10]. With the same code but changing the initial conditions to those used by McDermott [9, 12], we find a result (see dotted line in Fig. 1a) that agrees with the result in Ref. [9]. In what follows we use always for the bound proton the wave function obtained with the standard set of initial conditions. This wave function corresponds to a binding energy of 5.7 MeV.

In Fig. 1a we also show our result (solid line) in PWBA corresponding to the choice *cc*1 for the current operator.

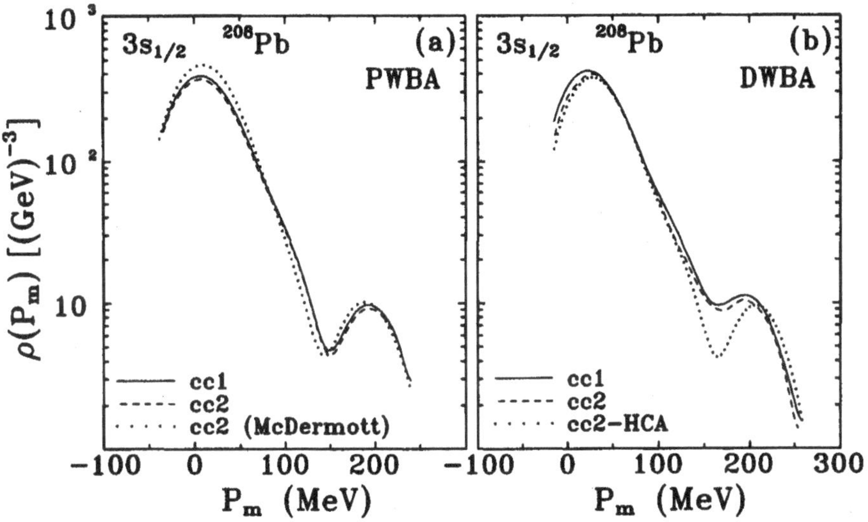

FIG.1 : Reduced cross-sections for the shell $3s_{1/2}$ of ^{208}Pb in DWIA. Results in PWBA (a) and in DWBA (b) are shown.

In Fig. 1b we show our DWBA results obtained with the *cc*1 operator (solid line) together with the DWBA results of Ref. [10] (dashed line) that was obtained with the *cc*2 operator. Also shown in this figure is the result obtained with the code of McDermott (HCA), used in Ref. [9] (dotted line). In this figure, the three curves are obtained using the same relativistic optical potential [7] and also the same wave function for the bound proton. One can see that our results (solid line) and those from Ref. [10] (dashed line) are similar and show up the same qualitative changes with respect to the corresponding PWBA results shown in Fig. 1a. These changes are: *i)* a shift in P_m, *ii)* an increase at the maxima, and *iii)* a filling of the minimum at $P_m \sim 150$ MeV. Note that the result obtained within HCA (dotted line in Fig. 1b) is not able to reproduce the two last mentioned effects (focussing effects) but it reproduces adequately the shift in P_m.

In Fig. 2 we show in more detail the comparison between our PWBA and DWBA results shown by the dotted and solid lines, respectively. Also shown in Fig. 2 is the

PWBA result with an effective q-value (dashed line) q_{eff}. With q_{eff} the shift in P_m caused by electron Coulomb distortion is well accounted for.

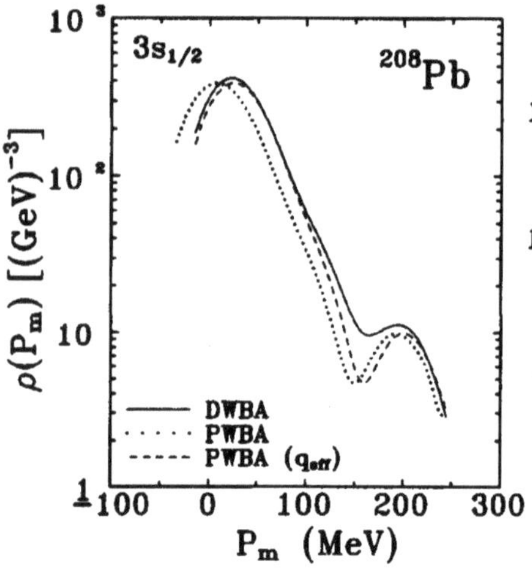

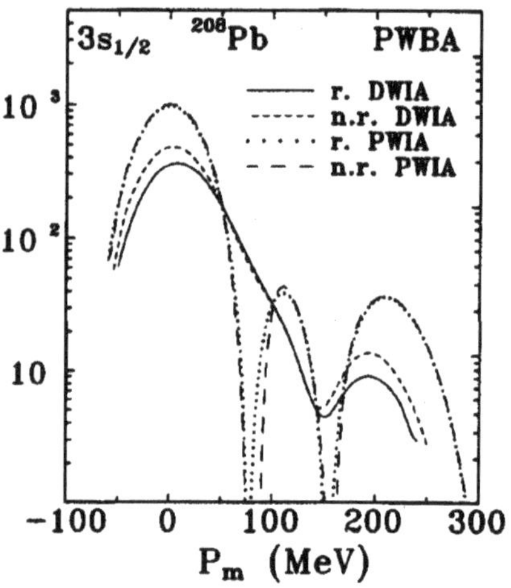

FIG.2 : DWIA results in parallel kinematics. FIG.3 : PWBA results from relativistic and nonrelativistic calculations.

The focussing effects found in our calculations agree with those of Ref. [10], which also treat Coulomb distortion in an exact way. On the other hand, the nonrelativistic results based on the code DWEEPY [13], which treats the Coulomb distortion in an approximate way making use of an expansion in powers of $(Z\alpha/k)$, overestimates the focussing effects. The spectroscopic factors are very sensitive to focussing effects because these factors are derived by scaling the theoretical reduced cross-section to the experimental one. Therefore, DWBA calculations are necessary to deduce reliable spectroscopic factors in heavy nuclei such as lead. In comparing relativistic and nonrelativistic calculations, it is also important to know how different approximations for the proton wave functions affect the reduced cross-section. To this end we compare in Fig. 3 the results of relativistic and nonrelativistic calculations in PWBA because in this case there is no effect from electron Coulomb distortion and the differences come only from the various approximations at the nuclear vertex.

In Fig. 3 the dotted and dashed lines show the PWIA results of the relativistic and nonrelativistic calculations, respectively. The solid and short-dashed lines show the results in DWIA of relativistic and nonrelativistic calculations. For the relativistic calculations we use the nuclear current operator J_{cc2} and the standard TIMORA solution [6] for the relativistic bound proton wave function. For the nonrelativistic bound proton wave function we use the upper component of the relativistic one, properly normalized, in the standard nonrelativistic form of the nucleon current, based on *cc*2. The relativistic optical potential, as previously mentioned, has been taken from Ref. [7]. The nonrelativistic optical potential has been taken from Ref. [1] and fits proton elastic scattering data from this nucleus at 98 MeV. As seen in Fig. 3, both relativistic and nonrelativistic calculations give practically the same result in PWIA. The differences between relativistic and nonrelativistic calculations are much more noticeable in DWIA. As seen in Fig. 3, the relativistic potential leads to stronger absorption than the nonrelativistic one (about a 20% difference at the peaks). This is at first sight surprising since both potentials reproduce well the elastic scattering of protons from this nucleus at an energy of 98 MeV. Therefore, the difference seen between the DWIA results in Fig. 3 must be attributed to details of the optical potential to which the elastic proton scattering is not sensitive. As it is well known [5] the relativistic S-V potential gives rise to non-local terms in the nonrelativistic reduction. The effect of these terms in the $(e, e'p)$ reaction was investigated by Boffi and collaborators [14] concluding that these terms affect only the inelastic processes, where the nuclear interior is important, producing about a 15% increase of the absorption due to final state interactions in the $(e, e'p)$ reaction. This conclusion is in agreement with the increased absorption observed in Fig. 3 in going from the nonrelativistic DWIA to the relativistic DWIA.

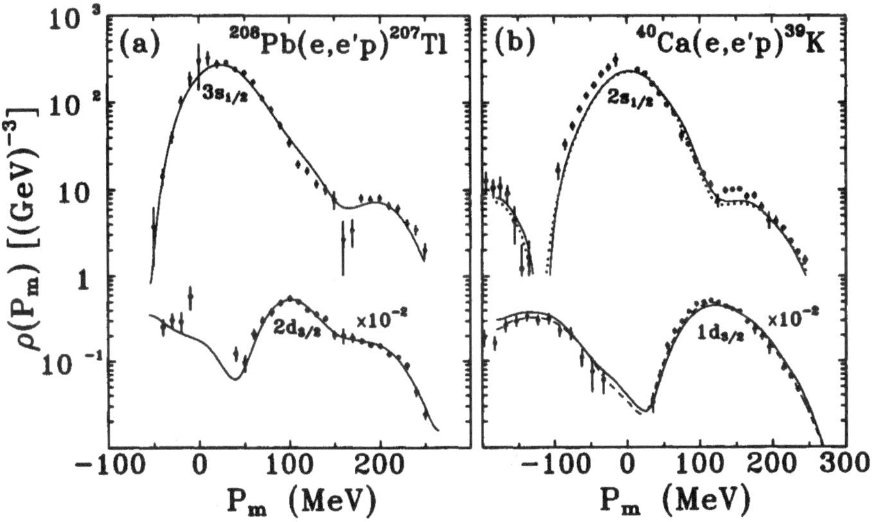

FIG.4: Theoretical and experimental reduced cross-sections. See text.

The main results of this work are presented in Table I and in Fig. 4, where we compare our relativistic DWBA results to the experimental data, in parallel kinematics. Our theoretical results are given by the solid lines in Fig. 4 and have been scaled by the spectroscopic factors presented in Table I.

Our spectroscopic factors are given in Table I, corresponding to calculations with *cc*1 and *cc*2 operators. They have been obtained by scaling our theoretical results on $\rho(P_m)$ to the experimental data from Refs. [1, 2] shown in Fig. 4. For each shell the overall scale factor (spectroscopic factor) has been obtained by means of an error weighted least squares procedure. Also given for comparison in Table I are the spectroscopic factors obtained from other calculations. The spectroscopic factors reported by Jin *et al.* [10] have been obtained with a fully relativistic DWBA formalism similar to the one used in this work, using the *cc*2 current operator. As discussed in the previous section, in the case of the $3s_{1/2}$ shell in ^{208}Pb the authors of Ref. [10] use the same bound nucleon wave function used here. For ^{40}Ca the bound nucleon wave functions used in Ref. [10] are different [15] from the standard TIMORA solutions used here. We also quote the spectroscopic factors obtained by McDermott [9] using also the *cc*2 operator. The last column of Table I contains the spectroscopic factors obtained with the nonrelativistic analyses reported in Refs. [1, 2].

It should be stressed here that in our analysis the determination of the spectroscopic factors has no free parameters. All of the parameters entering in the relativistic potentials were obtained from independent considerations [7, 5]. Taking this into account it is remarkable the good quality of the fits ($\chi^2/dof < 3$) to the experimental data seen in Fig. 4a for the two shells in ^{208}Pb. Our results in Fig. 4a correspond to the calculations with the *cc*1 operator, using the spectroscopic factors in column three of Table I. Similar results are obtained using the *cc*2 operator and the spectroscopic factors of column four in Table I.

In the case of ^{40}Ca the focussing effect is negligible and the only noticeable effect of electron Coulomb distortion is a shift in P_m. This is illustrated in Fig. 4b for the $2s_{1/2}$ orbital where we plot the DWBA result (solid line) and the result in PWBA with q_{eff} (dotted line), both using the *cc*1 operator and the same spectroscopic factor ($S = 0.44$, see Table I). Although not shown in the figure, the same is true with regards to DWBA and PWBA calculations for the $1d_{3/2}$ orbital in ^{40}Ca.

As seen in Fig. 4b the agreement between our theoretical results for the $2s_{1/2}$ and for the $1d_{3/2}$ shells in ^{40}Ca with the *cc*1 current operator is not as good as that

Table I: Comparison of our spectroscopic factors obtained with *cc*1 and *cc*2 operators and with other relativistic and nonrelativistic analyses.

		cc1	cc2	Jin *et al.* [10]	McDermott[9]	nonrel.
^{208}Pb	$3s_{1/2}$	0.65(4)	0.70(4)	0.71	0.65	0.50(5) [1]
	$2d_{3/2}$	0.66(4)	0.73(4)	–	–	0.53(4) [1]
^{40}Ca	$2s_{1/2}$	0.44(3)	0.51(3)	0.75	–	0.50(6) [2]
	$1d_{3/2}$	0.60(3)	0.76(4)	0.80	0.66	0.65(5) [2]

obtained for the shells in ^{208}Pb. If we use instead the *cc*2 operator, the agreement is not improved for the $2s_{1/2}$ shell, but it improves considerably for the $1d_{3/2}$ shell. This is seen in Fig. 4b where we show our results for the $1d_{3/2}$ shell with the *cc*1 and the *cc*2 operators by solid and dashed lines respectively. For the case of the $2s_{1/2}$ shell χ^2/dof are larger than 7 in all cases, showing the poor quality of the fit. For the $1d_{3/2}$ shell χ^2/dof is almost 5 in the case of the *cc*1 operator and is less than 2 in the case of the *cc*2 operator.

It is important to realize that all the cases where our spectroscopic factors in columns three and four of Table I are low correspond to situations in which the shape of the theoretical curve does not match well the trend in the experimental data and χ^2/dof are large, i.e., the fits are poor. This is particularly the case for the $2s_{1/2}$ shell in ^{40}Ca, where our spectroscopic factor is as low as that obtained in the nonrelativistic analysis [2]. For the shell $1d_{3/2}$ in ^{40}Ca, where the fit with the *cc*2 operator is very good, we get a spectroscopic factor $S = 0.76$, similar to that found in ^{208}Pb, larger than the one obtained in the nonrelativistic analysis [2], and in better agreement with theoretical calculations of short range effects.

References

[1] E.N.M. Quint, Ph. D. Thesis, University of Amsterdam (1988).

[2] G.J. Kramer, Phys. Lett. **B227** 199 (1989); Ph.D. Thesis, University of Amsterdam (1990).

[3] J.D. Björken and S.D. Drell, *Relativistic Quantum Mechanics* (Mc Graw-Hill, N.Y. 1964).

[4] J.M. Udías, Ph.D. Thesis, Universidad Autónoma de Madrid (1993).

[5] C.J. Horowitz and B.D. Serot, Nucl. Phys. **A368**, 503 (1981); Phys. Lett. **B86**, 146 (1979).

[6] C.J. Horowitz, D.P. Murdock and B.D. Serot, *Computational Nuclear Physics*, (Springer-Verlag, Berlin, 1991).

[7] S. Hama *et al.*, Phys. Rev. C**41**, 2737 (1990);

[8] T. de Forest, Jr., Nucl. Phys. **A392**, 232 (1983).

[9] J.P. McDermott, Phys. Rev. Lett. **65**, 1991 (1990).

[10] Y. Jin, D.S. Onley and L.E. Wright, Phys. Rev. C**45**, 1311 (1992).

[11] J.A. Caballero, T.W. Donnelly and G.I. Poulis, Nucl. Phys. **A555**, 701 (1993).

[12] J.P. McDermott, *private communication.*

[13] C. Giusti and F. Pacati, Nucl. Phys. **A473**, 717 (1987); Nucl. Phys. **A485**, 461 (1988); M. Traini, Phys. Lett. **B213**,1 (1988).

[14] S. Boffi, C. Giusti, F.D. Pacati and F. Cannata, Nuovo Cimento **98**, 291 (1987).

[15] Y. Jin, Ph.D. Thesis, Ohio University (1991).

SUMMARY

P. SCHUCK

Institut des Sciences Nucléaires, 53 Avenue des Martyrs
38026 Grenoble Cédex, France

This meeting has been the tenth in a series of workshops on nuclear dynamics with phase space concepts. As at previous occasions the meeting has been largely open to adjacent fields. More specifically this time a whole session was devoted to the physics of finite clusters, indeed a subject bearing strong affinities with the physics of the nucleus.

Inspite of the broad variety of talks clearly four main topics have emerged :

1) Multifragmentation and fluctuations

On the theoretical side X. Campi and S. Leray reported on their latest results of the percolation approach partly related also to the problem of intermittency. J. Randrup told us how to include fluctuations into the BUU approach. A formalism of van Kampen developed for classical gases was extended to the Fermi gas, an aspect which has previously been developed by the Nantes group. In a numerical simulation in 2 dimensions using a Boltzmann Langevin equation, it has been found that the corresponding results agree with the analytical analysis. An important problem for unstable nuclear matter is the question of which wavelengths are involved in the instability modes. We heard two contradictory talks on this subject : P. Chomaz was claiming that only one wave vector was involved, whereas A. Rapisarda showed that essentially all wave numbers are participating. I think it is one of the crucial points to be cleared up in the future. The contribution of Rapisarda contained furthermore the very interesting analysis of the spinoidal instability in terms of chaotic trajectories. New ideas on surface instabilities or stabilities were advanced by J. Nemeth and L. Moretto. The former explained the possibility of "surface fingers" while the latter treated sheat instabilities of nuclear matter and stabilisation of nuclear bubbles due to Coulomb and surface energies. C. Dorso reported on his latest ideas on fragment formation during time evolution and M. Colonna studied how break up in pure BUU dynamics depends on the number of pseudoparticles. Fermionic molecular dynamics is taylored to include fragment formation and even shell effects. Unfortunately it is limited so far to the low mass region. H. Feldmeier exposed his latest results for relatively heavy clusters ($A \cong 12$). For binary break-up Y. Abé has developed a theory for "fission plus dissipation".

As a personal remark, I would like to stress a point of view which has come out in discussion with P. Danielewicz. The fluctuations on top of BUU have so far mostly been considered from the stochastic, two body collisional point of view. However for fragmentation, as was stressed long time ago by J. Knoll, the fluctuations of the mean field may be equally, if not more, important. One therefore should develop a transport theory with a collision term where the density fluctuations collide with the particles :

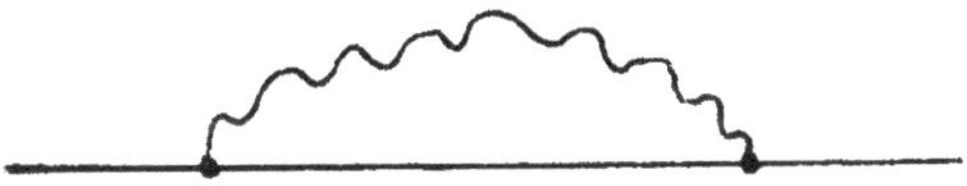

An attempt in this direction has recently been given by L. Münchow (Z. Physik) and one may also learn from the Lennard Balescu approach in plasma physics.

On the experimental side we heard two very nice status reports, one by D. Durand on the gradual evolution of evaporation to multifragmentation and another by J. Pochodzolla on "Correlations and Fluctuations in Nuclear Multifragmentation" where he reported on the results form the ALADIN experiment. His analysis favors statistical break-up.

2) Mesons, Deltas, γ's in excited nuclei; particle production

U. Mosel gave us an extensive overview on the physics involved at SIS energies with mesons in matter, Δ-excitation, γ-production presenting the latest results of the Giessen-group. W. Nörenberg and myself reported on the very recent subject of the rho-meson in hot and dense matter and its potential detectability from dilepton production. The theoretical models are not conclusive as yet : while the GSI group predicts only a broadening and no shift of the rho-meson we conclude that there is broadening and upward shift. On the other hand the controversial density dependence of the bare rho mass (QCD sum rules) can still completely overthrow the situation. In addition to the rho meson the η-meson (W. Nörenberg) and the "sigma-meson" were studied with surprising effects. A very detailed study on the shift of the delta peak in nuclear reactions induced by protons or light ions has been presented by E. Oset. He convincingly disentangled all contributions (coherent and incoherent) and concluded that the coherent pion production plays a crucial role for the explanation of the shift of the delta peak. J. Knoll made a progress report on his investigation of coherence effects in γ production. Among other things he presented an exactly solvable hard sphere model where he is able to bear out his ideas. It turned out in the discussion that the object he is studying is in fact the nuclear structure function S (q, ω) and it would be interesting to confront his results with those known from linear response theory. At SIS energies of $\cong$ 1 GeV/n we can expect that a large fraction of nucleons are transformed into Δ's. W. Ehehalt presented the results of the relativistic transport model developed in Giessen. In the central region of a Au+Au reaction the density of Δ's reaches $\cong$ 0.15 fm^{-3} so that one can talk about delta matter. Two other talks which were more remotely linked to the present subject were given by Y. Ivanov (two fluid descriptions of relativistic heavy ion collisions) and by A. Ramos (Λ's in nuclei).

On the experimental side we also had several contributions. First V. Kuehn gave a very comprehensive review talk on the present state of the art. The first measurements of dilepton spectra at the Bevalac have been reanalyzed and C. Naudet presented much clearer spectra than the ones first published. This is especially helpful for better comparison with theoretical calculations which increasingly are being performed for in matter rho meson decay and other sources of dileptons. J.

Diaz reported on hard photons from the Kr + Au reactions and P. Sapienza showed what we can learn from probing the nucleus with γ's and light particles. Correlation functions of $\pi^+ \pi^-$, partly from old Diogène data, were analysed by J. Pluta and unexpected positive correlations at low invariant mass were found. It turns out, however, that these correlations are mainly due to shadowing effects from the heavy ion.

3) Transverse nuclear flow

Nuclear flow is an ongoing active field. There is new indications that we here can indeed get substantial insight into the nuclear equation of state. P. Danielewicz explained how one can deduce from the flow of low mass clusters, in employing a BUU equation for each species, collectivity in the flow. Such information had always escaped us so far. F. Sébille showed results for the flow with the non local Gogny force for the Nb+Nb reaction in the 200 to 400 MeV/n range. "Gogny" forces corresponding to various compressibilities of nuclear matter were used and the analysis seems to indicate that the incompressibility modulus K_∞ should be around 200 MeV. This conclusion corroborates an earlier one by P. Danielewicz. It seems therefore that the flow in this energy range can give a clear fingerprint on the nuclear EOS, a very important result from heavy ion physics. C. Fuchs presented a relativistic model where out of equilibrium effects of the interpenetrating nuclei can be taken into account.

We had two talks from experimentalists : J.P. Alard reported on the data from Saturne and GSI for Au+Au in the 150-1000 MeV/n energy range. G. Rai presented the latest and unfortunately last data from the Bevalac, also for the Au+Au reactions. Inclusive spectra, collective flow and azimuthal correlations in relation with the nuclear EOS were analysed.

4) Clusters

The talks on clusters once again demonstrated how closely related these finite Fermi systems are with the physics of the nucleus. C. Brechignac gave an excellent overview on fission of metallic clusters. F. Garcias presented a Thomas-Fermi calculation for the fission of sodium clusters, an approach very well known from nuclear physics. J.P. Martin talked about the latest developments in pinning down the existence of supershells in clusters, a feature which we are missing in nuclei since they are too small. E. Lipparini reported on his recent work on non local effects in the plasmon dispersion of alkali clusters. He has solved the two electron Bethe-Goldstone equation and used the non local G-matrix as an input to the RPA equation. This improved very much the agreement with experiment over previous investigations using the bare Coulomb force. Two subjects were treated by S. Hernandez. In the first part she developed a density functional approach to ^3He drops which includes finite range effects and in the second part the very interesting idea of triplet pairing in finite ^3He drops was presented and a schematic one shell model solved.

5) Conclusion

Many other topics were present at the conference. One of particular actuality are the hot giant resonances and their width. A talk on this subject was given by M. di Toro. Other aspects of linear response theory were treated by J. da Providencia, E. Moya de Guerra and A. Lallena.

On the whole, I think, that the conference was very fruitful; for instance the interaction of nuclear physics with other related fields is always very rewarding. Open problems turned out to persist in multifragmentation. Mesons in nuclei is a subject which has gotten a very strong impetus recently whereas the information from transverse nuclear flow seems finally to give a hint to the nuclear equation of state.

I think it is acknowledged by all participants that this was a very stimulating and pleasant meeting. Many thanks to the organisers !

www.ingramcontent.com/pod-product-compliance
Lightning Source LLC
LaVergne TN
LVHW082253241026
850861LV00008B/38
* 9 7 8 9 8 1 0 2 1 5 5 6 9 *